T0189822

Communications in Computer and Information Science   **1465**

More information about this series at https://link.springer.com/bookseries/7899

David Gerber · Evangelos Pantazis ·
Biayna Bogosian · Alicia Nahmad ·
Constantinos Miltiadis (Eds.)

Computer-Aided Architectural Design

Design Imperatives: The Future is Now

19th International Conference, CAAD Futures 2021
Los Angeles, CA, USA, July 16–18, 2021
Selected Papers

 Springer

Editors
David Gerber
University of Southern California
Los Angeles, CA, USA

Evangelos Pantazis
University of Southern California
Los Angeles, CA, USA

Biayna Bogosian
Florida International University
Miami, FL, USA

Alicia Nahmad
University of Calgary
Calgary, Canada

Constantinos Miltiadis
Aalto University
Espoo, Finland

ISSN 1865-0929   ISSN 1865-0937 (electronic)
Communications in Computer and Information Science
ISBN 978-981-19-1279-5   ISBN 978-981-19-1280-1 (eBook)
https://doi.org/10.1007/978-981-19-1280-1

This Springer imprint is published by the registered company Springer Nature Singapore Pte Ltd.
The registered company address is: 152 Beach Road, #21-01/04 Gateway East, Singapore 189721, Singapore

Preface

The CAAD Futures 2021 Conference, with the theme "Design Imperatives: The Future is Now", was hosted by the University of Southern California (USC) Viterbi School of Engineering and took place virtually during July 16–18, 2021. The conference call, formulated during the beginning of the COVID-19 global pandemic, focused on highlighting how computational design research could address the immediate and imminent issues influencing the built environment. This call attracted 97 paper submissions that went through a rigorous double-blind peer-review process resulting in 33 accepted and presented papers through the conference, which attracted 90 registered attendees. The selected papers were presented within seven tracks and are included in the present publication.

In addition to the paper presentations, 28 guest speakers and co-panelists contributed to 13 sessions on a wide range of topics, which complemented and elaborated on the conference theme. Furthermore, 10 selected workshops were conducted involving 26 tutors and 88 participants. The following editorial offers more information and contextualizes the conference.

2021 Conference History and Organization

CAAD Futures 2021 was initiated by David Gerber (of USC) on behalf of the CAAD Futures Foundation, who established the general theme "Design Imperatives: The Future is Now." By mid-2020, the organizing committee was assembled to include Biayna Bogosian (of USC and later Florida International University), Evan Pantazis (of USC and IBI Group), Alicia Nahmad (of the Architectural Association and later the University of Calgary), and Constantinos Miltiadis (of Aalto University).

The team commenced conference organization during the height of the COVID-19 pandemic, carrying out weekly remote meetings from July 2020 until July 2021, when the conference took place. Undoubtedly, the ongoing pandemic was an immense challenge for the conference and its organization, which necessitated the team to plan for the contingencies of the circumstances.

The first uncertainty was whether the conference could be hosted in person at USC or if we had to resort to a remote, teleconferencing format. Our initial planning suggested a hybrid format, in which the participants could join remotely or in-person. However, by March 2021, following health and safety directives, it was decided that the conference had to be conducted in a remote virtual format.

Another uncertainty was a concern regarding the impact of the pandemic on research. The regulations imposed on universities across the globe for restricting the spread of the virus had significant repercussions on how the research was carried out. On the one hand, access to labs, infrastructure, and physical spaces for conducting research and experiments was limited. On the other hand, day-to-day activities that inform research, such as teaching, interpersonal communication, and peer-to-peer feedback, also had to adapt to the challenging remote conditions.

In light of the above, the commitment of the organizing committee was to actualize the conference despite all the uncertainties and attempt to organize a significant event for the research community. The drive for the conference organization was based on the fact that the CAAD Futures conferences have been essential discussion hubs for the computational design community and the pandemic, more than anything, highlighted the importance of community building. Also, the distributed nature of the global pandemic has brought to the forefront several previously discussed matters, such as the notion of virtual space and the digitization of our profession. Thus, it was an opportunity to look at these critically and in the light of the conference theme.

On the Open Call and CAAD Futures Conference

The open call was drafted in August 2020 and was developed upon the founding mission of CAAD Futures, established in 1985. The CAAD Futures organization was formed to promote "the advancement of Computer Aided Architectural Design in the service of those concerned with the quality of the built environment," to play an active role in promoting research interactions and collaborations, and to facilitate the dissemination of research.

Our principal consideration, 35 years later, was how to reframe promoting the advancement of CAAD when computation and related tools permeate all facets of architecture—practice, education, and research—while computation itself has diversified immensely to account for sub-genres, disciplines, and methodologies. Our intention with the call (shared below) was to be inclusive and affirmative to all facets related to computation, to account for the phenomenon at large, and for the conference to serve as a convergence for the different contemporary and relevant streams of research, and their respective researchers.

The suggested thematic areas included relatively recent strands such as machine learning and already established ones such as fabrication and design automation, as well as inter-and cross-disciplinary domains topical to computational practices like shared economies, matters of ethics, policy, equity, and sustainability. The call also provisioned for matters of research methodology, education, and literacy, in addition to open-source tools and initiatives. Finally, diametrically opposed to the conference theme, the track 'Past Futures' was intended to account for computation as an evolving historical phenomenon, its heritage, and critique.

The call was shared in September 2020 and remained open until March 2021, collecting 97 full papers submitted for review. Next to the call for papers, a call for workshops (including technical tutorials, seminars, and hackathons), collected 26 proposals. Of the 10 workshops selected, nine were conducted remotely with one conducted in person in Los Angeles, USA.

Lastly, the organizing committee planned a series of panel discussions with a number of invited field experts from both academia and the industry. These took place between conference paper presentations, reflecting and elaborating on different aspects of the open call.

We provide the open call for submissions and brief descriptions of the discussion panel topics in the following subsections.

Keynotes and Panel Discussions

Day 1: July 16, 2021

The conference began with an opening keynote presentation by Chris Luebkeman, who gave his interpretation of the conference theme discussing contemporary necessities such as the democratization of tools, the popularization of techniques, and the ubiquity of data, proposing regeneration and circularity principles additional to sustainability.

The first panel discussion of the day, titled "Digital Regionalism & Other Computational Futures", featured Theodora Vardouli, Rodrigo Orchigame, and Daniel Cardoso Llach. Vardouli opened the panel by discussing the archeology of graph theory in architecture and its impact on topology and geometry. Orchigame followed with a presentation of past paradigms in computation and logics developed in Brazil, Cuba, and India, elaborating on inconsistency, uncertainty, and partiality, respectively. Cardoso Llach discussed the pitfalls in the history of computation, stating that "after 50 years of computer-aided and computational design, we are still not drawing things together" and proposing a more sincere and transparent view on design media.

The second panel discussion of the day, titled "AI, the Built Environment & Human Wellbeing", with participants Chanuki Illushka Seresinhe, John Law, and Nono Martinez Alonso discussed how different deep learning techniques could be used to understand how we perceive our cities and how they affect us. Illushka Seresinhe argued that machine learning could help us to consider the built environment in terms of human subjectivities and social wellbeing. She further discussed the use of big datasets and deep learning to understand how the aesthetics of the environment influence human wellbeing. Law presented AI techniques for quantifying the perception of urban typologies by their users and their effects on design and human settlements, questioning how we might design future cities conducive to our wellbeing. Martinez Alonso discussed different techniques to engage with machine learning processes from a design perspective.

The third panel discussion of the day, titled "Computation and the State of Practice", consisted of Jason King, Pablo Zamorano, Silvan Oesterle, and Maryanne Wachter. They attempted to draw a picture of the contemporary situation in computational design teams in larger firms and smaller design practices. Following brief presentations, the discussion addressed opportunities within large organizations to apply computational techniques and methods and how these can leverage small practices. It also highlighted transparency in developing processes, workflows, and longitudinal integration. It advocated for the long-term development of tools that can be used across scales and projects—an alternative to frequent practices of developing ad hoc and project-based solutions.

The closing keynote of the day, titled "An Intro to Native IFC and Open Source AEC", was presented by Dion Moult, initiator of Blender BIM and contributor to open-source projects such as the IfcOpenShell and Ladybug tools. Moult discussed the imperative and inevitability of developing open standards, workflows, and open-source tools with and within the design community, emphasizing their social and political implications. His presentation concluded with a demonstration of Open IFC in practice and an invitation to participate in OS Arch projects.

Day 2: July 17, 2021

The second day started with a keynote by Cristiano Ceccato, who presented the design and construction process of several large-scale airport terminals built recently by Zaha Hadid Architects. His talk discussed the importance of advanced design tools in their practice and drew parallels between aircraft manufacturing and modern-day construction in terms of complexity. He concluded the talk by addressing how new technologies can help overcome challenges and outlined several parameters for shaping what he called more "hopeful futures."

The first panel discussion of the day, titled "Virtual Worlds", included Konstantinos Dimopoulos, Jose Sanchez, and Kate Parsons. Dimopoulos elaborated on a definition of video game urbanism and its contribution to worldbuilding stemming from his practice. Next, Sanchez discussed his upcoming video game 'Common'hood' intended to promote practices of commoning and social engagement through collaboration in the digital domain. Lastly, Parsons presented her methodology for designing immersive environments and her interest in NFT art production and dissemination. The discussion revolved around how social engagement in the virtual domain contributes to the 'real' physical world.

The second panel discussion of the day, titled "Digital Literacy", was a discussion with Vera Bühlmann and Ludger Hovestadt hosted by Constantinos Miltiadis and Evan Pantazis on 'claiming architectonic questions again.' It started with a brief introduction of the Chair for CAAD (now Digital Architectonics) and its legacy and continued with examples from the guests' current pedagogical project. Hovestadt emphasized that technology today is too easy. Contemporary architectonics requires rhetoric and philology to engender 'digital renaissance'. Bühlmann highlighted the need to construct instruments to find stability in the contemporary world, embracing as ethics the art of abstract gnomonics.

David Benjamin conducted the closing keynote of the day. He presented his research methodology, which unifies his teachings at Columbia GSAPP and his practice The Living, to rethink the concept of innovative and sustainable design processing. The work presented touched upon a lot of topics which included synthetic biology, sustainable material science, AI-driven computational design workflows, and environmental sensing. Most importantly, Benjamin discussed how we should be thinking about designing for change over time and how computational methods and data can help manage complexity by actively correlating visible and invisible parameters.

Day 3: July 18, 2021

The third day started with a keynote by Martha Tsigkari who introduced recent technologies and tools in the areas of form-finding, machine learning, AR/VR, robotics, optimization, and analysis used by the Applied Research and Development (ARD) group within Foster + Partners. She outlined how such tools extend the designers' intuition and enhance the performance-driven design and remote design collaboration. Tsingari also discussed the importance of data for the design process and applications that enable designers' real-time feedback. She concluded by presenting a vision of technology and design tools as collaborators and enhancers of human intuition and tools that allow problem-solving.

The first panel discussion of the day, titled "Designing Materials and Interactions", consisted of Behnaz Farahi, Felicia Davis, and Maddy Maxey. Through her work and the critical lens of feminism, emotion, bodily perception, and social interaction, Farahi focused on the relationship between the human body and its embodied space. Davis presented her work on computational textiles and how responsiveness and aesthetics could embed socio-spatial and political dynamics. Finally, Maxey presented research on wearables and e-textile prototypes that combine interactive design with haptics to communicate complex human and machine relationships.

The "Data and Policy" panel included Wendy W. Fok, Matthias Standfest, Maider Llaguno-Munixta, and Yannis Orfanos. Starting the discussion, Standfest presented his research on architectural and urban performance modeling and the role of machine learning in advancing the field. Next, Llaguno-Munixta presented her research on studying air pollution and the importance of data resolution in developing novel design practices that improve environmental health. Orfanos followed up by presenting his research on using data and information visualization to expand the way we think about spatial design and analysis in various scales. Finally, Fok highlighted that in conversations about data and design, it is essential to understand the role of digital property, authorship, and ownership.

The conference's closing keynote was given by Marcos Novak, who elaborated on the conference call and reflected on the past 30 years of computation in architecture and urban design. His presentation, spanning from classical to contemporary times, traced characteristics of each era's avant-garde, speculating on which those could be for the 21st century. He concluded by outlining *THEMAS* (Technologies, Humanities, Engineering, Mathematics, Arts and Sciences), a holistic and integrative model for research practice and pedagogy.

Acknowledgements

The organizing committee would like to express its gratitude to all contributors and attendees who made this conference possible. We would like to thank the authors for their scientific contributions to the field and conference theme; the tireless group of peers who reviewed conference submissions and accredited their quality through its rigorous double-blind process; workshop tutors and workshop participants who allowed us to explore the conference theme through computational thinking and design; all the guest speakers who shared their research and expanded on the conference theme; and the audience that followed the three intense days of the conference. We would like to thank the Viterbi School of Engineering at USC for allowing us to host, and we would like to thank our sponsors Autodesk Inc and IBI Group for their generous sponsorships, which facilitated the conference and this publication.

September 2021

David Gerber
Alicia Nahmad
Biayna Bogosian
Constantinos Miltiadis
Evangelos Pantazis

CAAD Futures 2021 Open Call for Submissions

The theme "Design Imperatives: The Future is Now" is an instigation and reflection onto imminent matters that we are confronted with as designers, architects, planners, engineers, innovators, and policymakers of the living environment. More than ever, our discipline is facing challenging imperatives including rapid and pervasive digitization and automation, the overwhelming rate of data availability, the question of continuous growth as well as the diminished resources, and the impending environmental crisis. These are conditions that generate even greater degrees of uncertainty in conceiving design strategies.

Against this backdrop, the augmentation of new affordances and sensibilities into design practice is deemed necessary for the definition of new paradigms of architectural relevance.

Establishing a design practice capable of addressing both issues of the social and material world mandates the reformulation of computational design thinking, the renewal of our methodologies, and the reevaluation of frameworks for interdisciplinary synergy and social participation. Central to this new relevance is a critical, unambiguous, and active positioning against the face of pressing imperatives. As key considerations, we identified our global ecological health; our duty of care towards helping ourselves and future generations to better manage the planetary boundaries; and the betterment of the human condition at large as well as the support of resilient and sustainable architectural futures.

Under these circumstances, CAAD Futures 2021 aims to broaden its inquiry into the socio-economic, political, and environmental imperatives as they pertain to space, and the capacity of computational design to intervene. Extending this call, the conference intends to raise these issues within the design-computing discourse and foster synergistic relationships for their investigation through the lens of design. To instigate discussions we pose the following questions:

- What is the heritage of design computing and what futures can we imagine?
- Can the reconsideration of computational design thinking serve towards the formulation of a plan of action to address planetary crises underway?
- How does design maintain its disciplinary unity against the situatedness of its knowledge and the multi-modality of its practices?
- What is the future of design-computing practice within the architectural profession?
- How can we advocate for novel funding models and policies that foster scalable and responsible research in line with a resilient and ethical practice?

Organization

Organizing Committee

David Gerber (Chair)	University of Southern California, USA
Alicia Nahmad (Co-chair)	University of Calgary and R-ex, Canada
Biayna Bogosian (Co-chair)	Florida International University, USA
Constantinos Miltiadis (Co-chair)	Aalto University, Finland
Evangelos Pantazis (Co-chair)	IBI Group and Topotheque, USA
Thanasis Farangas (Assistant)	National Technical University of Athens, Greece

Keynote and Panel Speakers

Benjamin, David	Autodesk Research and The Living, USA
Buhlmann, Vera	TU Vienna, Austria
Ceccato, Cristiano	Zaha Hadid Architects, UK
Davis, Felecia	Pennsylvania State University, USA
Dimopoulos, Konstantinos	SAE Institute, The Netherlands
Farahi, Behnaz	California State University, USA
Hovestadt, Ludger	ETH Zurich, Switzerland
King, Jason	IBI Group, USA
Llach, Daniel Cardoso	Carnegie Mellon University, USA
Llaguno-Munixta, Maider	Princeton University, USA
Law, Stephen	Alan Turing Institute, UK
Luebkeman, Chris	ETH Zurich, Switzerland
Miltiadis, Constantinos	Aalto University, Finland
Maxey, Maddy	E-textiles, USA
Martinez Alonso, Nono	Autodesk Inc., USA
Moult, Dion	LendLease, Australia
Novak, Marcos	University of California, Santa Barbara, USA
Ochigame, Rodrigo	Leiden University, The Netherlands
Oesterle, Silvan	ROK, Switzerland
Parsons, Kate	Pepperdine University, USA
Sanchez, Jose	University of Michigan, USA
Standfest, Matthias	Archilyse AG, Switzerland
Sereshine, Shanuki	Alan Turing Institute, UK
Tsigkari, Martha	Foster + Partners, UK
Vardouli, Theodora	McGill University, Canada
Fok, Wendy	WE-DESIGNS and Pratt Institute, USA

Wachter, Maryanne   Independent Researcher, Switzerland
Orfanos, Yannis   WeWork, USA
Zamorano, Pablo   Heatherwick Studios, UK

Scientific Committee

The conference chairs would like to thank the following individuals for their contribution
to the scientific process.

Achten, Henri	Czech Technical University in Prague, Czech Republic
Adel, Arash	University of Michigan, USA
Ampanavos, Spyros	Harvard Graduate School of Design, USA
Anzalone, Phillip	City University of New York, USA
Aparicio, German	Cal Poly Pomona, USA
Artopoulos, Georgios	The Cyprus Institute, Cyprus
Attia, Shady	Université de Liège, Belgium
Aviv, Dorit	University of Pennsylvania, USA
Baharlou, Ehsan	University of Virginia, USA
Bao, Dingwen	RMIT University, Australia
Bernal, Marcelo	Universidad Técnica Federico Santa María, Chile
Bogosian, Biayna	Florida International University, USA
Braumann, Johannes	Association for Robots in Architecture, Austria
Bruscia, Nick	University at Buffalo, SUNY, USA
Castellón González, Juan José	Rice University, USA
Celani, Gabriela	Unicamp, Brazil
Dan Zhang, Catty	UNC Charlotte, USA
De Azambuja Varela, Pedro	CEAU/FAUP, Portugal
Donohue, Mark	California College of the Arts, USA
Doyle, Shelby	Iowa State University, USA
Dubor, Alexandre	Institute for Advanced Architecture of Catalonia, Spain
Economou, Athanassios	Georgia Institute of Technology, USA
Erdine, Elif	Architectural Association, UK
Erkal, Emre	Erkal Architects, Turkey
Feghali, Yara	University of California, Los Angeles, USA
Fox, Michael	Cal Poly Pomona, USA
Fricker, Pia	Aalto University, Finland
Fukuda, Tomohiro	Osaka University, Japan
Gannon, Madeline	Carnegie Mellon University, USA
Gardner, Guy	University of Calgary, Canada
Gerber, David	University of Southern California, USA
Gerber, Andri	ZHAW, Switzerland
Gonzalez, Paloma	Massachusetts Institute of Technology, USA

Gonzalez de Canales, Francisco	Architectural Association, UK
Groenewolt, Abel	KU Leuven, Belgium
Han, Yoon J.	Aalto University, Finland
Herr, Christiane M.	Xi'an Jiaotong-Liverpool University, China
Herrera, Pablo C.	Universidad Peruana de Ciencias Aplicadas, Peru
Herruzo, Ana	Arizona State University, USA
Hosmer, Tyson	University College London and Zaha Hadid Architects, UK
Jones, Nathaniel	Arup, USA
Kalantar, Negar	California College of the Arts, USA
Kieferle, Joachim	Hochschule RheinMain, Germany
Kilian, Axel	Massachusetts Institute of Technology, USA
Kladeftira, Marianthi E.	ETH Zurich, Switzerland
Koehler, Daniel	University of Texas at Austin, USA
Koerner, Andreas	Universität Innsbruck, Austria
Kontovourkis, Odysseas	University of Cyprus, Cyprus
Körner, Axel	University of Stuttgart, Germany
Kostourou, Fani	University College London, UK
Kotnik, Toni	Aalto University, Finland
Koutsolampros, Petros	University College London, UK
Kretzer, Manuel	Anhalt University of Applied Sciences, Germany
Krishnamurti, Ramesh	Carnegie Mellon University, USA
Llaguno-Munitxa, Maider	Princeton University, USA
Maierhofer, Mathias	Universität Stuttgart, Germany
Manninger, Sandra	SPAN, USA
Marcus, Adam	California College of the Arts, USA
Marengo, Mathilde	Institute for Advanced Architecture of Catalonia, Spain
Marincic, Nikola	ETH Zurich, Switzerland
Martinez Alonso, Nono	Harvard University, USA
Mavasaityte, Rasa	University of Texas at Austin, USA
Mayer, Matan	IE University, Spain
Miltiadis, Constantinos	Aalto University, Finland
Mohite, Ashish	Aalto University, Finland
Morel, Philippe	EZCT Architecture & Design Research, France
Mostafavi, Sina	Delft University of Technology, The Netherlands
Nahmad, Alicia	University of Calgary, Canada
Orbey, Betul	Dogus University, Turkey
Orozco, Jorge	ETH Zurich, Switzerland
Ozkar, Mine	Istanbul Technical University, Turkey
Pantazis, Evangelos	Topotheque, Greece
Papanikolaou, Dimitris	UNC Charlotte, USA

Park, Hyoung-June	University of Hawaii at Manoa, USA
Peterson, Eric	Florida International University, USA
Petzold, Frank	Technical University of Munich, Germany
Pinochet, Diego	Massachusetts Institute of Technology, USA
Piskorec, Luka	Aalto University, Finland
Popescu, Mariana	ETH Zurich, Switzerland
Poustinchi, Ebrahim	Kent State University, USA
Puckett, Nick	AltN Research + Design and OCAD University, Canada
Ramirez, Jorge	Interela, Mexico
Reffat, Rabee	Assiut University, Egypt
Rockcastle, Siobhan	University of Oregon, USA
Roman, Miro	ETH Zurich, Switzerland
Romano, Christopher	University at Buffalo, USA
Santos, Luis	Kent State University, USA
Schneider, Sven	Bauhaus-Universität Weimar, Germany
Schwartz, Mathew	New Jersey Institute of Technology, USA
Scroggin, Jason	Scroggin Studio, USA
Senske, Nick	Iowa State University, USA
Slocum, Brian	Universidad Iberoamericana, Mexico
Smith, Ian	Ecole Polytechnique Fédérale de Lausanne, Switzerland
Soderberg, Catherine	Texas Tech University, USA
Standfest, Matthias	Archilyse, Switzerland
Stouffs, Rudi	National University of Singapore, Singapore
Sugihara, Satoru	ATLV, USA
Summers, Martin	University of Kentucky and PLUS-SUM Studio, USA
Symeonidou, Ioanna	University of Thessaly, Greece
Tucker, Shanta	Atelier Ten, USA
Vardouli, Theodora	McGill University, Canada
Verebes, Thomas	New York Institute of Technology, USA
Vermillion, Joshua	University of Nevada. Las Vegas, USA
Walmsley, Kean	Autodesk Inc., USA
Wiltsche, Albert	Graz University of Technology, Austria
Wit, Andrew	Temple University, USA
Xydis, Achilleas	ETH Zurich, Switzerland
Zacharias, Maroula	Harvard Graduate School of Design, USA

Sponsors

We would like to thank our industry sponsors Autodesk Inc. and IBI Group for their generous contributions to the conference.

Contents

Architectural Automations and Augmentations: Design

Architectural Automations and Augmentations: Fabrication

Architectural Automations and Augmentations: Environment

Architectural Automations and Augmentations: Spatial Computing

Past Futures and Present Futures:
Research and Pedagogy

Digital Hot Air: Procedural Workflow of Designing Active Spatial Systems in Architecture Studio

Catty Dan Zhang[✉]

University of North Carolina at Charlotte, 9201 University City Blvd, Charlotte, NC 28223, USA
cattydanzhang@uncc.edu

Abstract. Providing inhabitants the protected relationship with the matter of air has been the fundamental drive in almost all aspects of architectural productions, while in some cases clear boundaries are implemented for controlled environment, in other situations they are more porous and ephemeral. This paper presents an attempt of challenging volume and boundary relationships in architecture by designing forms of airflow using state of art 3D visual effect technologies. A computational workflow was established for a fourth-year undergraduate architecture topics studio using 3D VFX software. The workflow integrates smoke simulation, motion operation, collision behavior, and raster data processing, outlined in two simulation exercises. The pedagogical investigation aims to inspire innovative typologies of spatial connectivity and separation by designing fixed forms and amorphous mediums simultaneously. It offers an experimental protocol for using airflow as an architectural material in intuitive design processes. The outcomes of the simulation experiments led to discussions of the visual, tactile, olfactory opportunities of the invisible medium in architectural speculations.

Keywords: Active spatial systems · Airflow form-making · CFD

1 Introduction

Fabricating environment that activates the behavior of amorphous and invisible mediums has been explored theoretically as well as in innovative practices in the past decades, in which case architecture is no longer considered as stable objects passively revealing its environmental conditions [1]. Rather, it is understood as relations defined by adaptive boundaries that track movements of flow-fields within and around [2], and by active systems that modulate the transformation and the evolvement of such flow. Vapor, light, and heat– as a few examples– have become materials for construction, relating human body and the atmosphere in physiological aspects such as thermal comfort or multisensorial perceptions. Such investigations create spatial volumes and boundaries dynamically by altering homogenous mediums, as Phillipe Rahm coined in meteorological architecture approach where climate of the space is considered as a new architectural language "slipping from the solid to the void, from the visible to the invisible, from metric composition to thermal composition" where "limits fade away and solids evaporate" [3]. Furthermore,

© Springer Nature Singapore Pte Ltd. 2022
D. Gerber et al. (Eds.): CAAD Futures 2021, CCIS 1465, pp. 3–14, 2022.
https://doi.org/10.1007/978-981-19-1280-1_1

with the public perception of void spaces being heavily impacted by health and safety measures in light of the COVID-19 pandemic, the tendency of designing active mediums and adaptive boundaries urges new sets of rules and constrains to be established, which are related to both space making strategies as well as methods of visualizing and perceiving invisible activities within the void.

In her article "Capturing the Dynamics of Air in Design", Jane Burry emphasizes the critical role of "a panoply of tools and approaches spanning digital and analogue simulations, real time or otherwise intuitively engaging analytical feedback and the opportunity", as they are used to "make design changes and rapidly witness their influence on the complex atmospheric and atmosphere-borne phenomena" [4]. Today, a diverse set of computational tools for energy assessments, ventilation strategies, and sensorial awareness has transformed design methods around invisible mediums. Multiphysics software (CFD) for studying flow behaviors, as an example, provides designers platforms to virtually test out configurations of fixed forms at various scales in relationship to turbulence, temperature, humidity, and velocity of air, where the form and the function of the design are articulated following the climate [5]. Sensor network and advanced display systems such as mixed reality, on the other hand, allow multisensorial aspects of the invisible microclimate in physical spaces to be made perceptible and comprehensible to inhabitants [6]. Nevertheless, the complexity and limitations of integrating CFD in architectural-related computational design frameworks with most of mainstream CFD software packages and plugins [7] demand very clearly defined design problems that are mostly building performance related such as optimizing enclosures and building profiles, or populating spaces based on the distribution of fresh air, wind flow, and thermal comfort. Or, they are limited to the design of pre-imagined effects, such as in *Cloudscapes* by TRANSSOLAR and Tetsuo Kondo Architects. In that case, simulation modeling was conducted to study the wind and solar impacts for constructing a semi-external cloud, where surprises of unintended air behavior were experienced in the built installation rather than being explored as design potentials in the simulation stage [8]. Moreover, even though the development of plugins such as RhinoCFD offer better integration with modeling platforms throughout the design workflow, due to the need of external post-processing step and their data visualization characteristics, these tools usually require designers to have sophisticated knowledge to interpret quantified datasets not only about flow behavior but also on performance driven design techniques.

The goal of this paper is to demonstrate a methodology of designing active spatial systems that considers forms of airflow as the agency. It avoids the production of performative architectural outcome in favor of exploratory design by diving into the basics of form-making logics. This leads to experimentations on alternative toolsets which use airflow as a fundamental architectural material to generate innovative forms as well as artistic and perceptual effects grounded in scientific accuracy of the material behavior. The work is outlined through a pedagogical investigation in an architectural design studio in fall 2020. The objective of the research is to establish and test a computational workflow to (1) expand design opportunities that foreground the materiality and form potentials of airflow; (2) be integrated in intuitive design process during conceptual stage that is not necessarily driven by building performance criteria; and (3) provide a core framework for specific design tools to build upon when adapting to various architectural context and agenda.

2 Background

2.1 VFX as Architecture Design Tool

In his article "A Disaggregated Manifesto: Thoughts on the Architectural Medium and Its Realm of Instrumentality", Nader Terhani points out that "computational realm has offered code and rule-based functions" that "produce systemic variations that can proliferate options while absorbing a great deal of complexities", and "unlike ever before, we do have ways of connecting phenomena across scales, to see them side by side and to imagine consequentiality across disciplines" [9]. The advances of visual effects software have offered a new set of protocols in architecture which expanded the designer's ability in understanding and producing aesthetic applications, complex natural systems, performance-driven iterative design processes and so on, where the interplay between natural science and built forms has become closer than ever the gaming of virtual geometry and material attributes. Emerging trends in design with VFX simulation have been greatly benefited from ever-evolving solvers with optimized algorithms. These tools, capable of producing realistic effects, approximate the behavior of objects, mediums, and materials in real world with desirable accuracy and fast computing time [10]. Other than their capacity of visualizing special effects as final architectural representation, diverse design practices favor the immediacy of such tools to pursue innovative forms through the process of simulation. *PULSUS* by INVIVIA, as an example, explores the translation of material properties into form effects as a way to discusses aliveness in responsive installations. Maya simulation of fabric draping onto rigid forms leads to the fabrication of a series of four fabric-like human-form concrete surfaces, enhancing public interactions by establishing intimacy through the play of form and sensorial augmentations using responsive technology [11]. Such investigations avoid reproduction of known systems, rather, they celebrate poesies over accuracy, looking for opportunities to "multiply and make tangible futures that break away from the established momenta of thinking and doing" [12].

With their verified efficiency and precision in CFD simulations [10], the capacity and speed of producing hyperrealistic visualization of effects, as well as the experimental design opportunities, visual effect technologies and their architectural potential are explored in this research through a series of simulation-based design exercises.

2.2 Identifying Key Relationships for Airflow Form-Making

Creating forms with air– a medium with complex physics mechanics– does not follow singular or linear processes. Although building science related approaches have greatly benefited from computational tools, as a material favored by artists and designers due to its mythological and ephemeral qualities, its form characteristics in artistic applications has not been systematically explored with computational processes. The production of physical installations or artifacts usually relies on combining certain machineries to

establish physics models. Fog vortex ring or tornado– typologies often found in art installations– are created based on well-known physics principles. Animated and texturized
forms in Little Wonder's *10 Kinds of Fog*, as another set of examples, are produced by
moving fog through various modulators, filters, or textured surfaces with an experimental process interacting air, water, and machines [13]. Establishing a computational setup
for airflow form-making is to translate sophisticated machinery into a set of dynamic
systems and relations in a procedural workflow to broaden the typological potentials.

Previous research conducted by the author has showcased effective methods to chore-
ograph airflow through composing heterogeneous contexts such as temperature fluctuations, dynamic external force fields, as well as interactions between rigid and fluid
materials. A series of robotic thermal devices was designed to produce a visual cat-
alog of forms of hot air constructed via scripted heat and motion patterns (Fig. 1.a.).
Employing physical computing techniques, these instruments consist of (1) points of
heat source generating convective flows, (2) motion systems, and (3) surfaces which air
currents collide with. The investigation proofed the potential of air for constructing geometric primitives in three-dimensional space. The research outcome– visualized using
Schlieren photography– demonstrated the versatility of the material as well as intricate
mechanisms and control methods needed [14].

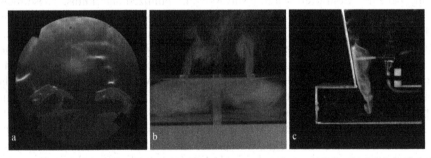

Fig. 1. a: Forms of hot air sculpted by scripted heat and motion patterns using robotic thermal
devices, visualized using Schlieren photography. b and c: Physical fog instruments designed and
assembled by students in author's fall 2019 studio demonstrating effects of fog colliding with
various types of surfaces. Work by Emily Dobbs and Eliot Ball (b), and Andrew Shook (c).

This investigation also builds upon a set of analogue experiments in the author's
fall 2019 undergraduate topics studio titled "Respirators", which examined the moving
air as a spatial agency to populate territories and habitats shared among human, nature,
and machine. Conceptual instruments were constructed by students to study forms of
dynamic voids enclosed by visible particles of air (Fig. 1.b and 1.c.). The class developed
procedures of reassembling respirators abstractly with kits of parts to create integrated
systems that produced spatial effects. Currents of fog colliding at various angles onto
articulated surfaces that were layered, perforated, or texturized, resulted in visual and
tactile volumes and boundaries modulated by rigid forms while being constantly altered
by external forces.

These analogue machines revealed the qualitative interplay among the medium, the
form, and the environment through hands-on experiments and directly observed effects,

which informed logics of the computational setup. Identified key systems and their relationships are:

- Hot air dynamics: the source and the behavior of hot air controlled by parameters such as source geometry, particle size, density, temperature, gravity, and environmental forces.
- Rigid body systems: rigid objects with kinematics, motion, and transformation potential.
- Collision relationships: dynamic interactions between air currents and rigid objects.

3 Computational Setup

3.1 Architectural Context

Employing VFX software as tools for architectural speculation, this paper documents the investigation in a 4th year undergraduate architecture studio in fall 2020 titled "Filters Applied: Medium Hybrids and Architectural Voids", taught remotely due to the COVID-19 pandemic. The studio intended to respond to urgencies of global public health crises by envisioning possible futures of meat processing facilities– hot spots during the pandemic due to the transmission of invisible particles within the public void and across species along production and packaging lines. In this regard, we seek new concepts of spatial planning and design to establish adaptable and active systems that modulate the behavior of amorphous and invisible mediums to produce architectural, technological, and social effects.

This context led to considerations of designing the distribution of airborne particulate matter, motion paths of their sources, kinematics of objects and machineries.

3.2 Workflow

Prior to the design of the final architectural intervention with specific site and program requirements, the studio set up two exercises of constructing experimental simulation models with particulate air, rigid objects, and void spaces in SideFX Houdini [15], both of which adopted the previously identified key relationships. The first exercise focuses on forms created with airflow at an object level, exploring sources of hot air and its collision relationship with surfaces in motion; and the second exercise looks at image compositing methods, through which 2D raster data are translated three-dimensionally into collective spatial relations. This simulation workflow centers around Houdini Pyro solver which tackles smoke and flame effects based on attributes of points describing the physical properties such as density, temperature, velocity, and so on. With the procedural node-based architecture of the software, the workflow establishes a set of translations among geometry, physics, kinematics system, and raster data (Fig. 2), leading towards various spatial logics interrogating the notion of shared atmosphere.

Exercise 1: Surface/Air. The Surface/Air exercise explores: (1) the creation, the movement, and the transformation of hot air in geometry nodes (SOPs); (2) the form and materiality of solid surfaces functioning as collision objects using solver combinations

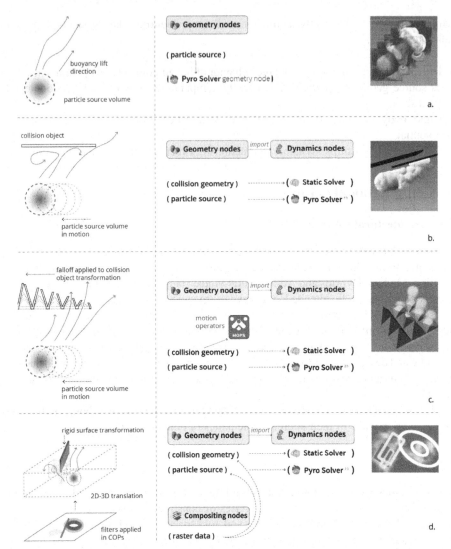

Fig. 2. Procedural workflow established in four simulation steps exploring translations among geometries, physics, kinematics systems, and raster data.

in dynamic node (DOP) networks; and (3) the coupling of the source motion and the surface transformation using motion operators (MOPs)– a free, open source suite of Houdini Digital Assets (HDAs) built initially in 2018 to simplify and automate common motion graphics processes [16]. By playing with the curvature, porosity, motion trajectory, and geometric transformation of collision objects, forms are to emerge interactively as currents of hot air being split, diffused, and redirected (Fig. 3).

Exercise 2: Image/Air. The Image/Air exercise practices spatial and environmental planning process through image-to-form and/or image-to-air translations with Houdini

Fig. 3. Smoke emitted from a spherical source colliding onto an irregular belt moving counterclockwise controlled using MOPs modifier nodes. Work by Nick Sturm and Dante Gil Rivas.

compositing networks which contain compositing nodes (COPs) for manipulating 2D pixel data. By setting up "lines through ellipses" scenarios based on abstractions of movements, circulation, proximity, shared public zones and enclosed individual spaces, this exercise intends to rethink the typical plan-based design process through image-making techniques. Animated raster outputs from the interplay between linear and circular 2D shapes are then loaded into dynamic networks for Pyro simulations (Fig. 4). Expanding the discoveries from exercise 1, singular and localized air collision scenarios are to be developed into collective forms, systems, and spaces, where dynamic volumes and barriers emerge autonomously from image manipulations.

Fig. 4. Animated 2D shapes generated by COPs being translated into form and air. Work by Kyle Brodfuehrer and Michael Allen.

4 Outcomes and Reflection

4.1 Typologies of Spatial Connectivity and Separation

The exercises resulted in diverse design outcomes revealing the roles of rigid objects and amorphous mediums played in the search for typologies of spatial connectivity and separation. By categorizing medium hybrids as volumes and boundaries, effective results could be interpreted as conditions illustrated in Fig. 5. Successful simulation models captured multiple of these conditions sequentially with curated sets of motion and collision events throughout time, presenting the capacity of generating both subtle and drastic variations by altering a simple set of parameters.

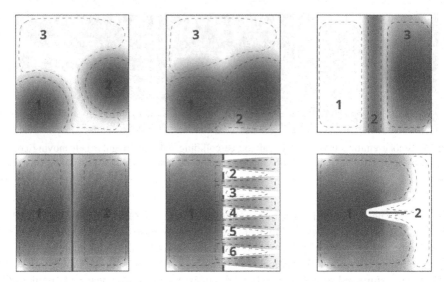

Fig. 5. Subdividing voids with connected/separated air volumes, and rigid/soft boundaries.

Figure 6 and Fig. 7 showcase examples from exercise 1, where form sequences derive from rigid surfaces splitting, venting, and colliding with hot air in a particular order. In the first example (Fig. 6), the initial state of the collision object consists of two flat surfaces intersecting at 90°. By applying a series of operations including pivot rotation and tapered extrusion, the vertical portion of the surface moves downwards while the horizontal piece turning into a 4 × 4 grid of 3D modules each with an opening at its center. This transformation follows a gradual pattern starting from the far end and moving towards the origin of the smoke. The directionality of flow, the timing and duration of the surface movements, are among parameters being tested. The horizontal surface subdivides the volume of smoke into thinner plumes when panels open up, resulting in turbulences and vortices due to the disturbance generated in the velocity field. These plumes colliding onto the vertical surface afterwards, become puffs merging back into a whole. The second example (Fig. 7) practices a similar set of relationships in a different configuration. Smoke emitted from a spherical source being sculpted and redirected by a series of moving arcs positioned between the source and a horizontal perforated surface, results in streams of smoke emerging above with fluctuating intensity. These soft thresholds or thick barriers define ephemeral "rooms" above the perforated panel with dynamic borders.

Another major trajectory focuses on position relationships of particle sources. Figure 8 showcases an investigation based on composing images with circular and rectangular shapes in motion using a series of Houdini compositing nodes to create dynamic inversion, blur, and gradient color effects. The color attributes of the resultant images are imported into geometry nodes, parametrically controlling the spatial distributions of smoke sources and collision objects using pixel sorting techniques. Figure 9 presents multiple particle emitters within contained geometries. Rectangular and round corner conditions, proportions of horizontal and vertical linear obstacles, and proximities

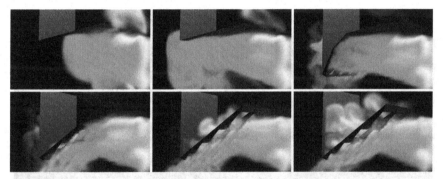

Fig. 6. Sequential volumes created by smoke colliding onto kinetic surfaces. Work by Michael Allen and Kyle Brodfuehrer.

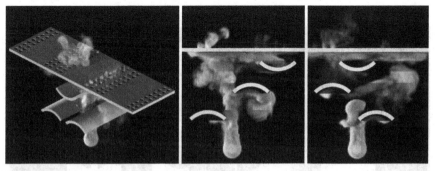

Fig. 7. Dynamic smoke barriers formed above perforated panel. Work by Hunter Thurlo and Lea Mitchem.

between the sources of the emittance are a few parameters for generating a taxonomy of animated distribution patterns. As the direction of flow being altered dramatically while colliding onto the enclosure of each rigid volume from within, redirected currents of air interact with existing particle volumes, resulting in diverse volumetric subdivisions within the homogeneous medium.

4.2 Speculative Applications

Based on computational experiments conducted in both exercises, students then translated simulation logics into architectural strategies when designing their final projects. These design proposals adopt previously defined air/surface/image relationships in various program scenarios, building components, and spatial organizations. Roof system filtering exhausted odors, façade penetration allowing thermal byproducts from meat processing to serve as visual barriers, and circulation planning considering proximity for controlled social interactions, are a few examples of final projects tested in part in Houdini as proof of concepts (Fig. 10). The discussion of these design proposals is beyond the scope of this paper, however, the investigation at architectural scale demonstrates promising opportunities of direct and indirect translations between forms of air

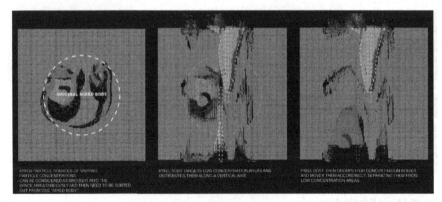

Fig. 8. Applying pixel sorting operation on composed images which were then turned into smoke source and collision objects in Pyro simulation. Work by Michael Allen and Kyle Brodfuehrer.

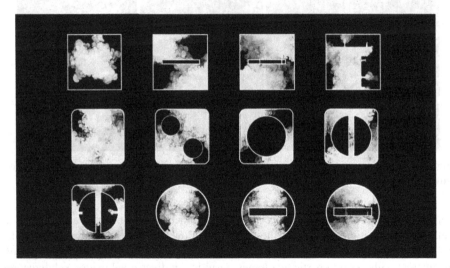

Fig. 9. Smoke distribution studies in partial circulation spaces with various types of corner conditions. Work by Amelia Gates and Carson Edwards.

and architectural spaces. Following a procedural workflow and with the immediacy of effects generated, the work produced in this studio breaks away from conventions and preconceptions of the solid and the void, the interior and the skeleton, the fixed and the ephemeral, projecting a new set of relations between architecture and atmosphere.

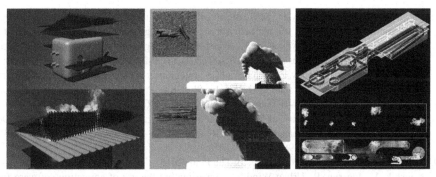

Fig. 10. Speculative design applications tested in Houdini: (left) roof filtration system, work by Hunter Thurlo and Lea Mitchem; (middle) steam penetrating through façade dynamically based on sorted items in the building, work by Michael Allen and Kyle Brodfuehrer; and (right) aggregated circulation spaces considering human emittance within, work by Amelia Gates and Carson Edwards.

5 Conclusion and Future Work

Borrowing VFX tools, this paper presented a computational workflow of designing digital hot air to rethink mediums, visible forms, and integrated systems through intuitive and artistic process. This methodology translates principles of physics in airflow form-making into key relationships in the computational setup, demonstrating effective visual outcomes that are beneficial in exploratory design processes. While the research mainly focuses on using air simulations as experimental design tools to expand new forms of architectural knowledge, the outcomes also suggest the potential of future scientific developments with more articulated parameters to examine specific conditions and to generate pragmatic applications. However, without built-in functions of comprehensive quantitative data analysis in VFX software, additional steps would need to be carefully tested for such purpose in future work. Due to limited access to machines with higher computing powers during a semester of remote teaching and learning, those simulations were preliminary. Nevertheless, the procedural workflow established in the process allows easy refinements and adaptions when tackling larger scale and higher complexity situations in the future.

References

1. Gissen, D.: Subnature: Architecture's Other Environments. Princeton Architectural Press, New York (2009)
2. Moe, K.: Nonlinear perspective. Log **47**, 119–130 (2019)
3. Filipendin, M.U., Bradić, H.: The meteorological architecture of Philippe Rahm. Int. J. Contemp. Arch. The New ARCH. **1**, (2014)
4. Burry, J.: Capturing the dynamics of air in design. In: De Rycke, K., et al. (eds.) Humanizing Digital Reality, pp. 103–110. Springer, Singapore (2018). https://doi.org/10.1007/978-981-10-6611-5_10
5. Rahm, P.: Thermodynamic practice. In: de Rycke, L., et al. (eds.) DMSB 2017, Humanizing Digital Reality. pp. 74–83. Springer, Singapore (2018)

6. Latifi, M., Burry, J., Prohasky, D.: Make the invisible microclimate visible: mixed reality (MR) applications for architecture and built environment. In: Gengnagel, C., Baverel, O., Burry, J., Ramsgaard-Thomsen, M., Weinzierl, S. (eds.) Impact: Design with All Senses. pp. 325–335. Springer, Berlin (2020). https://doi.org/10.1007/978-3-030-29829-6

7. Chronis, A., Dubor, A, Cabay, E., Roudsari, M.S.: Integration of CFD in computational design: an evaluation of the current state of the art. In: ShoCK! - Proceedings of the 35th eCAADe Conference – vol. 1. pp. 601–610, Sapienza University of Rome, Rome, Italy (2017)

8. Abdessemed, N.: On the nature of thermodynamic models in architecture and climate. In: De Rycke, K., et al. (eds.) Humanizing Digital Reality, pp. 85–91. Springer, Singapore (2018). https://doi.org/10.1007/978-981-10-6611-5_8

9. Tehrani, N.: A Disaggregated Manifesto. The Plan (2016)

10. Kaushik, V., Janssen, P.: Urban Windflow: investigating the use of animation software for simulating windflow around buildings. In: Real Time - Proceedings of the 33rd eCAADe Conference, pp. 225–234, Vienna University of Technology, Vienna, Austria (2015)

11. Song, H., Luo, O., Sayegh, A.: Attributes of aliveness: a case study of two interactive public art installations. Presented at the DeSForM 2019 November 15 (2019). https://doi.org/10.21428/5395bc37.eaef01da

12. Jahn, G., Morgan, T., Roudavski, S.: Mesh agency. In: Proceedings of the 34th Annual Conference of the Association for Computer Aided Design in Architecture (ACADIA), pp. 135–144. Los Angeles (2014)

13. Sadar, J.S., Chyon, G.: 10 Short films about fog. JAE **73**(1), 115–119 (2019)

14. Zhang, C.D., Sayegh, A.: Multi-dimensional medium-printing: prototyping robotic thermal devices for sculpting airflow. In: Computing for A Better Tomorrow - Proceedings of the 36th eCAADe Conference, vol. 1. pp. 841–850, Lodz University of Technology, Lodz, Poland (2018)

15. Houdini. SideFX

16. Foster, H., Schwind, M.: MOPs: motion operators for Houdini

Comprehending Algorithmic Design

Renata Castelo-Branco$^{(\boxtimes)}$ ⓘ and António Leitão ⓘ

INESC-ID/Instituto Superior Técnico, University of Lisbon, Lisbon, Portugal
{renata.castelo.branco,antonio.menezes.leitao}@tecnico.ulisboa.pt

Abstract. Algorithmic Design (AD) allows for the creation of form through algorithms. Its inherent flexibility encourages the exploration of a wider design space, the automation of design tasks and design optimization, considerably reducing project costs and environmental impact. Nevertheless, current AD uses representation methods that radically differ from those used in architectural practice, creating a mismatch that is further exacerbated by the inadequacy of current programming environments. This creates a barrier to the adoption of AD, demotivating architects from its use.

We propose to address this problem by coupling AD with adequate representation methods for designing complex architectural projects. To this end, we explore three essential concepts: storytelling, interactive evaluation, and reactivity. These concepts can be both complementary and mutually exclusive, which means compromises must be made to accommodate them all. We outline a strategy for their integration with the AD workflow, highlighting the advantages and disadvantages of each one, and pinpointing their intersection. Finally, we evaluate the proposed strategy using computational notebooks as programming environments.

Keywords: Algorithmic design · Program comprehension · Documentation · Storytelling · Liveliness · Interactive evaluation · Reactivity · Computational notebooks

1 Introduction

Algorithmic Design (AD) defines the creation of designs through algorithmic descriptions, i.e., computer programs with instructions for the machine to perform [8]. Despite its numerous advantages, there is a comprehension barrier demotivating many architects from its use. The goal of this research is to make AD a more accessible design process, allowing the industry to benefit from AD's potential to combine design creativity and design optimization. In order to do so, we will dive into program comprehension research, identifying strategies that aid the construction and comprehension of AD programs.

1.1 Algorithmic Design

AD allows the designer to delegate repetitive tasks to the computer, accelerating the production process and reducing human errors [8]. It also supports rapid change with little effort, providing considerable cost savings to the industry [51].

© Springer Nature Singapore Pte Ltd. 2022
D. Gerber et al. (Eds.): CAAD Futures 2021, CCIS 1465, pp. 15–35, 2022.
https://doi.org/10.1007/978-981-19-1280-1_2

Another important advantage is AD's influence on performance-based design [20], which is gaining ever more emphasis with the growth of climate change awareness and cost-reduction needs. While the incorporation of analysis data in early design stages alone already helps architects achieve better performing solutions, the connection between AD and simulation tools opens the door to much faster optimization processes [34].

Despite the numerous advantages AD presents to the process of architectural creation, it relies on algorithms, written using programming languages. This representation method radically differs from the ones traditionally used in design and is therefore difficult to use by professionals of creative fields [32].

Visual programming languages present a friendlier approach to AD that simplifies the learning process and offers more adequate graphical features for the task at hand. However, they also lack scalability [7], meaning that as programs grow in complexity they become hard to understand and navigate [9,11], hindering their use in large-scale projects [28].

Additionally, and in either case, AD programs (visual or textual) frequently turn up to be the unstructured product of successive *copy&paste* of program elements, which result from the experimentation process that characterizes design thinking [51]. Hence, programming architects find it difficult to understand AD programs that represent complex designs, particularly those developed by others [30] or by themselves in the past. This research targets the development of AD programs that describe complex 3D models, hence we focus on text-based AD.

1.2   Integrated Development Environments

Much of the programming struggle mentioned above can be alleviated with an Integrated Development Environment (IDE): a computer application that aims at facilitating the programming task. Sadly, existing IDEs are tailored for traditional software development processes. As such, architects find it hard to use them as design tools.

We propose to make AD more akin to the traditional architectural practice by integrating more adequate methods of developing and maintaining algorithmic representations of complex architectural projects in current IDEs. To that end, we explore two ideas that facilitate program comprehension: (1) documentation - the task of explaining a project to facilitate its comprehension; and (2) liveliness - the ability to live test the program as it is being developed or modified to facilitate comprehension of the impact of changes.

Documentation and liveliness are two very broad concepts, which we further subdivide to better suit architectural design. More specifically, we explore: (i) storytelling, (ii) interactive evaluation, and (iii) reactivity. These three concepts contribute to the comprehension of AD programs in very different ways and they can be both complementary and mutually exclusive. In this article, we outline a strategy for their integration with the AD workflow, which we evaluate using computational notebooks as the base IDE for the experiments.

2   Related Work

The difficult task of understanding computer programs is far from being limited to the architectural community. Research in the field of program comprehension has long aimed at modeling a cognitive theory on how programs are understood [48]. The following sections present some of the theories developed over time.

2.1   Program Comprehension

This section addresses two main branches of program comprehension: the understanding of computer programs through explanatory human-readable text and the use of graphical representations to illustrate programs' structure and behavior.

Program Documentation. In 1984, Donald Knuth proposed Literate Programming [25], a system that encourages users to build programs as a structured web of ideas, that is, by creating program parts and stating the relationships between them as they go, in whatever order they find best for the comprehension of the work. The problem with this solution is the perceived lack of efficiency: having to explain, in advance, the intended program, delays the programming task. Documentation is generally agreed to be one of the most useful, yet dreadfully tiresome parts of the job, and consequently one of the most avoided [5].

Naturally, given its importance for software maintenance, automatic documentation tools have also been developed [16]. Machine learning techniques, such as neural networks, in particular, are currently achieving considerable success in the automation of documentation [21,36]. However, for the specific case of architecture, traditional program documentation fails to ensure program comprehension, as it is geared to the production of hyperlinked texts. In AD programs, documenting complex geometry requires not textual explanations, but rather the equivalent sketch translation.

Bret Victor [50] argued that, when designing, artists think visually. However, when writing code, they must think linguistically. Not only is this translation process hard on design ideas, but most often it is also impossible to convey the entirety of artistic meaning with words. He also states that just as we created writing to make thoughts visible (a user interface for reason), or mathematical notation to make mathematical structures visible (a user interface for algebra), we must also develop IDEs that can make our designs visible [49] (Fig. 1).

Program Visualization. This specific branch of program comprehension research focuses on the use of graphical systems to facilitate the comprehension of programs. It began to have some expression in the '60s to help programmers deal with the complexity of (what was then considered) modern software [2]. The proposals included the use of various diagrammatic techniques to explain programs [18,22,33] (examples in Fig. 2).

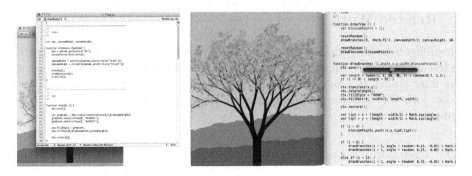

Fig. 1. Bret Victor's Inventing on Principle: from a simple code editor (left) to an enhanced IDE (right) providing an "immediate connection" to what is being created (adapted from [49]).

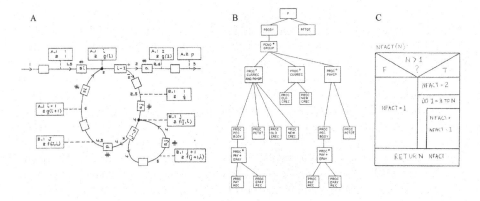

Fig. 2. A - Goldstine and von Neumann's flow-diagram [18]; B - Jackson diagram for a process payment program [22]; C - Nassi-Shneiderman diagram for the factorial function [33].

As technology evolved, so did the aspirations of the scientific community [12] and the availability of better displays, colors, and, later on, 3D representations, motivated a growing emphasis on animations over static representations of programs. Early on, Brown developed several systems that offered programmers dynamic displays of the program's fundamental operations [6].

Naturally, different programming paradigms motivate different types of visualization. For instance, declarative programming visualization requires presenting what the program does, while imperative programming also requires visualizing how it does it. Following this line of thought, different authors focused on different aspects of program visualization. Myers [32] classified program visualization systems according to what they illustrate (data, code, or algorithm) and how (statically or dynamically). Price et al. [40] further subdivided the field into a more hierarchical taxonomy, and many others followed [4, 43].

Diehl [12] stressed that it is not only important to comprehend what a program does and how it does it, but also how we got to that point in the development process. The current widespread use of version control systems for both individual and collaborative work stands as proof to this, although their reliance on purely textual mechanisms hardly places them within program visualization.

A program visualization feature that fully exploits the visual nature of AD is traceability, that is, the identification of which parts of the model correspond to which parts of the program, and vice-versa [29]. This connection established between the program and respective result is of utter importance for the comprehension of AD programs [49], and several AD tools offer it already, e.g., Dynamo and Rosetta [29].

2.2 Liveliness

The dynamic component of program visualization introduced terms such as liveliness, interactivity, or reactivity, which became the main agenda for IDE developers invested in program comprehension.

Live Coding. The introduction of live coding blurred the lines between programming and explaining, advocating a real-time connection between program and result [42]. Live coding is frequently described as a creativity technique centered upon the writing of interactive programs on the fly [42]. In many cases, the liveliness requirement addresses the timing needs of live performances, such as musical ones [47]. However, its application to the programming task also helps users relate the changes in the program to their respective impact on results, thus aiding program comprehension.

In the case of AD, live coding requires the model to be quickly recomputed for every change applied to the program, which can be difficult to ensure when the program generates a complex model. For simple cases, however, there are already competent solutions integrated with AD, such as Grasshopper and Luna Moth [1].

Interactive Evaluation. Another perspective on liveliness that also aids comprehension without implying a scalability issue is interactive evaluation. By opposition to batch-evaluation, where the entire program must be loaded prior to execution, interactive evaluation motivates users to test small program fragments, debugging the program as it is being constructed [23]. It is frequently used for exploratory programming [42], a workflow typically employed when the requirements of the program are not fully defined. Hence the programming task becomes more of an exploratory experiment, a common scenario in AD.

Despite the advantages, there is a catch, particularly for beginners. Interactive evaluation promotes *ad-hoc* program construction that can introduce confusing bugs for developers unfamiliar with the obstacles of program state [15]. Experienced programmers avoid this by using a batch-oriented style in interactive systems, resetting the program after any major change, an approach that

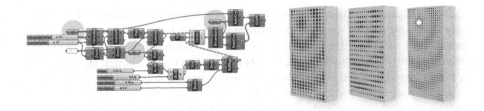

Fig. 3. Grasshopper diagram from [45].

is enforced by some pedagogical IDEs [14,15]. Once more, the programming paradigm used is also relevant, as declarative programming languages minimize the negative effects of program state when compared to imperative ones.

Reactivity. Reactivity is another take on liveliness that also advocates exploratory programming. However, the focus lies on tracking the dependencies in the program so that any change updates the entire state, as it happens in spreadsheets, for instance. The central idea is to have the system automatically manage dependencies to free the user from such burden [3]. With reactivity, since any change to the program triggers the re-evaluation of all dependencies, the program state is always consistent. As such, this paradigm also provides an alternative solution for the state problem introduced by interactive evaluation.

The data flow paradigm [27] is a good example of reactive programming. Here, programs are described by graph structures with nodes representing the functions and the wires that connect them representing the data that flows between components whenever something changes upstream. Most visual programming languages used in architecture, such as Grasshopper, are literal interpretations of this paradigm (Fig. 3). Note, however, that in what regards program comprehension, the data flow paradigm is flawed, since the node structure hides program complexity instead of explaining it, while also promoting program fragment repetition.

3   The Program Comprehension Dyad

To promote the use of AD, we propose a new design medium that allows for the creation of algorithmic descriptions in a live and documented way, making the task of understanding and changing AD programs easier for both the creator and others. To this end, we explore a program comprehension dyad: (1) documentation and (2) liveliness. These are two very broad concepts, which we fine-tune for an ideal IDE for the development of AD projects. Documentation is narrowed down to the concept of (i) storytelling, and liveliness is subdivided into (ii) interactive evaluation and (iii) reactivity. The following paragraphs summarize the definitions we consider for each of these concepts.

(1) **Documentation**, in the programming context, refers to the task of explaining the program, to facilitate later comprehension to the authors and other readers. The concept encompasses most of the research made under the program comprehension and visualization umbrellas, with emphasis on static visualizations. Documentation can be done prior, during, and after the program development, and it can also be internal and external to the program.

(i) Our proposal for the integration of documentation in the context of AD narrows this concept down to **storytelling**: the creation of a tale for the program's evolution. Storytelling thus contemplates internal documentation done while programming and aims at creating a narrative of the program development history, with program and respective documentation intertwined.

(2) **Liveliness** is the ability to live-test programs. Liveliness accommodates the dynamic or animated part of program visualization and subsequent branches. Liveliness can manifest itself in many ways, including live coding, interactive evaluation, and reactivity. Given that live coding tends to suffer from scalability issues, we focus on the two latter concepts.

(ii) **Interactive evaluation** allows a two-way flow of information between the computer and the user. With the user controlling just how lively a system is, the scalability problem becomes less relevant, although it increases the risk of program state inconsistencies.

(iii) In the context of AD, we consider **reactivity** as an automatic response that maintains state consistency by reevaluating all program parts that depend, directly or indirectly, from a part that was changed. This is more efficient than reevaluating the entire program, as is typically done in live coding, thus delaying but not eliminating the scalability problems that tend to affect complex AD programs.

Storytelling, interactive evaluation, and reactivity intersect in many ways. They can both complement and hinder each other. In the following sections, we explore this relationship and point out mechanisms to elude possible conflicts. Two case study project adaptations will be used to illustrate the proposed concepts: BIG Architects' Business Innovation Hub for the Isenberg School of Management in Amherst, Massachusetts (Fig. 4, left); and Santiago Calatrava's Liège-Guillemins railway station in Liège, Belgium (Fig. 4, right).

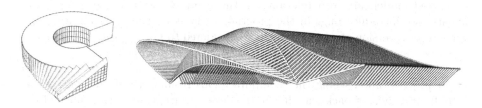

Fig. 4. Case study projects: Isenberg on the left and Liège-Guillemins on the right.

3.1 Storytelling

In an ideal IDE for AD, designers should be allowed to keep the artifacts produced along the design process in an organized fashion, explaining and documenting the resulting program. Storytelling intends to transform the program into a creative journal of the design's development that includes not only textual documentation but also sketches the architect made during a creative sprout [13]. Drawings are *"essential to both public discourse about architecture and the development of the architect's thinking"* [46, pp. 58]. Following this reasoning, we argue that the drawings produced when idealizing the design and the way it translates into a program (Fig. 5) can help document the program's evolution and expected outcome.

Fig. 5. Sketches made while developing the Isenberg school program.

Static images of the program's output, such as snapshots or renders of the generated model, are also relevant for the comprehension process. Very frequently, we introduce bugs in the program without noticing. Having a correct version (or a version of what the author considered to be the proper behavior of the program) available for comparison, may prove vital in debugging.

Withal, it is important to note that architectural design is by no means a linear process: upon seeing the (computational) results of their designs, architects frequently step back and change the design concept, rendering much of the documentation out of date. Storytelling embraces this issue by applying Diehl's concept of software evolution [12] to the context of architectural design: new documentation artifacts should be added whenever the program changes its intended behavior, thus keeping the history of design changes in the AD program.

3.2   Interactive Evaluation

Designers should feel motivated to construct and test their programs interactively, that is, executing programs parts immediately after (re)writing them. Interactive evaluation transforms what could otherwise be a very abstract process into a tangible and relatable one.

Interactive evaluation of AD programs typically produces visual results, which can become documentation artifacts. Given the proposed storytelling approach, we adapt our take on liveliness to accommodate the narrative style as well. Figure 6 presents the interactive development process of the slabs in the Isenberg project: an initial iteration of the slab function creates the intended contour; modifications are applied to contemplate the ground-floor exception; and later iterations convert the contour into 3D objects distributed along the height of the building. Each change is followed by a test that produces these visual results and the entire process is kept and documented in the AD program.

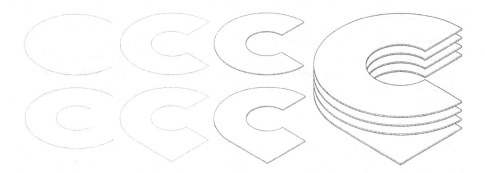

Fig. 6. Interactive test results from the Isenberg slab function evolution process.

Nevertheless, there are two major setbacks to this intersection of interactive evaluation and storytelling: (1) the resulting verbosity, and (2) the aggravated state problems.

(1) The step-by-step development process often results in an accumulation of localized tests and repeated or scattered program fragments. Refactoring and outlining techniques can help architects rearrange some of the scruffiness to obtain a cleaner and more intelligible program in the end.

(i) *Refactoring* is commonly defined as the process of improving the structure of existing programs without changing their semantics or external behavior [17]. There are several semi-automatic refactoring tools [31] that can be adapted to an AD context. Of particular interest to the case are those that help the programmer join scattered code in summarized functions.

(ii) *Outlining* techniques, i.e., the structuring and identification of what each part of the program is or does, may also serve to hide parts of the program for particular audiences and/or purposes. For instance, an outlining mechanism that identifies the tests the user wishes to activate or deactivate in any run considerably improves the program's performance and allows the same program to serve multiple presentation purposes. This goes farther than the outlining system typical of visual programming languages, such as Grasshopper (which allows readers to deactivate individual nodes on demand), by supporting a structured system of goal-oriented active/deactivated groups of program fragments.

Naturally, by using these mechanisms to reorganize the program, we are partially relinquishing the development history. However, we believe this is a necessary trade-off, since the order in which we create a story may not necessarily correspond to the order in which we wish to tell it to others.

(2) State inconsistency is a direct consequence of the use of interactive evaluation, which gets further exacerbated by the permanence of repeated definitions in the program. Outlining may ease part of the problem by deactivating outdated definitions. To avoid it entirely, we propose reactivity, which is discussed in the following section.

3.3   Reactivity

AD programs are systems of hierarchical relations between parts of the design. The functions that compose the separate parts of the building are interdependent so that the entire ensemble can function as a whole. Consequently, all building elements should morph appropriately when design parameters are changed (Fig. 7).

Fig. 7. Liège-Guillemins project: three possible variations of the model with different lengths, widths, and height for the central hangar.

This workflow essentially means that changes to the program frequently influence more parts of that program, other than the one we are modifying. As such, interdependencies must be considered when applying changes. However, in large programs, keeping a mental track of this system of relations is close to impossible. Hence, the program should make these relations visible for designers.

This can be achieved, for instance, by having the entire set of active tests in the program reacting to the changes that are being applied. If we can immediately visualize the impact of the changes all over the program, we are more likely to succeed in maintaining the necessary interdependencies in the AD project. Doing so in an interactive manner, using user-friendly mechanisms, such as buttons or sliders, to explore the influence of design parameters in the project's design space, further adds to the comprehension layer.

Reactivity offers this capacity by keeping track of dependencies in the program. Any change to the program triggers a reevaluation of the dependent parts, ensuring state consistency. Naturally, this means there can be no outdated definitions in the program. Once more, users must resort to outlining and refactoring to deactivate old versions of existing definitions or relinquish storytelling all together for reactivity to work. Furthermore, while reactivity does not necessarily imply the regeneration of the entire model, it does imply the constant re-processing of the dependency graph, which can become computationally intensive in large programs. Liveliness always was and will continue to be a double-edge knife. Hence, it should be used in accordance with the compromises it offers.

4   Exploratory Application

In this section we elaborate on a practical implementation of the proposed concepts in an AD workflow - storytelling, interactive evaluation, and reactivity - using computational notebooks as the base IDE for the experiment.

4.1   Computational Notebooks

Many of the problems described above are shared by the scientific community. In science, reproducibility is critical but the increasing specialization of the different areas is making it hard to reproduce published scientific results. To address this issue, the scientific community is embracing new methods of experimentation that rely more and more in computational simulation and analysis of different phenomena.

Computational notebooks have emerged in this context, promoting method transparency and data availability in the form of "executable papers" [26]. By supporting the description of computational experiments and the analysis of simulated or experimental results in an explanatory and reproducible way [44], they became a critical tool in science [24,37], not only to understand the scientific breakthroughs being presented, but also to reenact them. The following paragraphs describe computational notebooks' interpretation of the dyad concepts: (1) documentation and (2) liveliness.

(1) Computational notebooks were designed to support computational narratives, allowing users to simultaneously execute, document, and communicate their experiments through the intertwining of code and textual and visual documentation. The same notebook can serve multiple purposes, such as tutorial, interactive manual, presentation, or even scientific publication [26,38].

(2) These tools also promote interactive evaluation, a workflow that greatly benefits experimentation with data [38]. This is typically achieved with the input and output cell system: users write a fragment of code in a cell and run it immediately after to observe the result. The cell layout provides a half-solution to the heavy computation agenda of liveliness, since the code lodged in each cell is only run at the user's request.

However, interactive evaluation in notebooks motivates exploratory programming and, in this paradigm, the hunger for immediate results frequently overpowers the interest in organizing the program. Developers can add cells in a non-linear order, definitions can be intertwined with tests in the same code cells, and repeated code bits can be spread along the document. This disarray creates a complex dependency net with hidden program state, whose results frequently confuse the developer. This fact has been identified as one of the major pain points in the use of computational notebooks [10,19] and reactive notebook solutions have also been put forward to respond to the criticism [35,39].

Given the above-mentioned benefits and burdens of the use of computational notebooks as comprehensive IDEs, we propose to explore their use in the context of AD to evaluate storytelling, interactive evaluation, and reactivity. To that end, the two case studies presented above were developed in two different notebooks: the Isenberg School model in Jupyter [41], and the Liège-Guillemins station model in Pluto [39], both using the Julia programing language and the Khepri AD tool [45]. Both projects relied on notebooks' natural tendency for documentation and liveliness, yet the implementation of the three sub-concepts had to be adapted to each notebook, as they operate very differently.

Jupyter [41] is an open-source and web-based notebook. Although it was originally developed to support the programming languages Julia, Python, and R [38], nowadays, Jupyter not only offers a wide range of programming languages but it also allows users to mix them in the same notebook. Jupyter is based on the input/output cell approach and it supports repeated definitions, which allows us to easily apply the storytelling strategy.

Pluto [39] is a computational notebook conceived specifically for the Julia programing language, with one outstanding difference from typical notebooks: reactivity. Pluto was inspired by Observable [35], a reactive computational notebook for JavaScript. Both recognize dependencies between cells, so that when one of them is changed, all dependent ones are automatically updated. This means there can be no repeated definitions, which renders storytelling difficult without outlining mechanisms.

4.2   Storytelling

Storytelling defends program documentation as a way to embrace the tale of the program's evolution. When organized chronologically, the artifacts we produce to document our programs can tell the narrative of the design development process for a better comprehension of the final design solution. Figure 8 presents the modeling history of the Liege-Germmins' project.

To tell the tale of design development, instead of building on existing definitions, storytelling states that developers should consider keeping the definition history intact. In the Jupyter notebook, used to develop the Isenberg project, consecutive versions of the same functions were kept in the document in chronological order. For instance, the project is composed of C-shaped slabs, as shown in Fig. 6. All versions of the above-mentioned `slab` function were kept in the notebook, properly documented, textually and visually.

Fig. 8. Liege-Germmins' project development history: modeling sequence.

In Pluto, the same effect can only be obtained if we give each version of the function a new name, or if we deactivate old versions (by commenting them or using other outlining techniques). Being a reactive notebook, Pluto needs to maintain state consistency. Hence, it does not allow for repeated definitions.

Regarding the integration of artifacts in the program, computational notebooks have several mechanisms available. For textual documentation, we can use markdown, which can help both explain and structure the program (Fig. 9 left), and simple code comments to explain things inside function definitions (Fig. 9 right). Markdown also supports mathematical notation, which is one of the best ways to describe parametric shapes. Figure 9 presents, on the left, part of the Pluto notebook, where the `sinusoidal` function parameters are explained. This function defines the shape of the station's roof and arches.

Visual documentation can be added via HTML or through IPython's display package for Jupyter and PlutoUI for Pluto. This type of documentation can consist of drawings created during design development that reveal the author's intention towards the program's expected behavior and results (Fig. 9 right) or images generated for the sole purpose of explaining parts of the program. The latter case includes snapshots and rendered images of the generated model. These images are useful devices to make sure the program is producing the results it should. Figure 10 presents some of the snapshots saved in the Jupyter notebook after running test cells on the Isenberg program that generate isolated elements or the building as a whole.

4.3   Interactive Evaluation

Computational notebooks natively promote interactive evaluation as a form of liveliness controlled by the user. Immediate visual results can be obtained by running cells that produce geometry whenever the user wishes to test a particular code snippet. Ideally, this occurs after every new definition. Figure 11 presents an application of the interactive evaluation workflow to the the train station in Pluto: (1) the programmer (re)defines the `canopie_bars` functions, (2) implements a test case, (3) runs the test, and (4) saves the visual result as

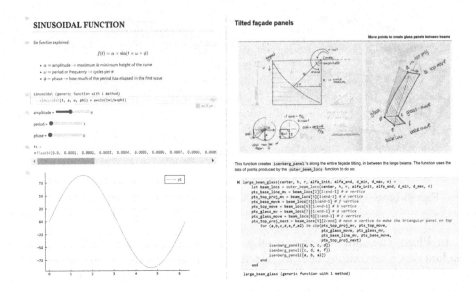

Fig. 9. Pluto, on the left, exploring and explaining the sinusoidal function. Jupyter, on the right, illustrating Isenberg's tilted façade panels' function.

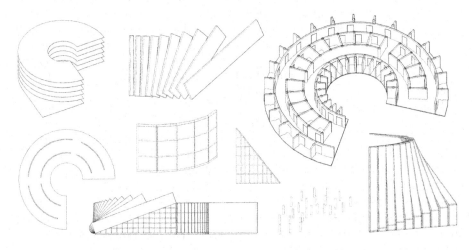

Fig. 10. Snapshots of function tests run from the Jupyter notebook, saved as documentation in the Isenberg program.

documentation in the program. Steps 1 and 3 may be retaken several times until a satisfactory result is achieved. Only then, should the user move on to step 4.

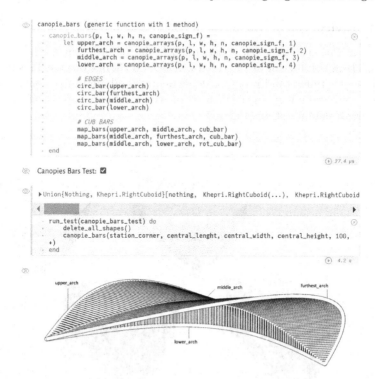

```
canopie_bars (generic function with 1 method)
  canopie_bars(p, l, w, h, n, canopie_sign_f) =
    let upper_arch = canopie_arrays(p, l, w, h, n, canopie_sign_f, 1)
        furthest_arch = canopie_arrays(p, l, w, h, n, canopie_sign_f, 2)
        middle_arch = canopie_arrays(p, l, w, h, n, canopie_sign_f, 3)
        lower_arch = canopie_arrays(p, l, w, h, n, canopie_sign_f, 4)

        # EDGES
        circ_bar(upper_arch)
        circ_bar(furthest_arch)
        circ_bar(middle_arch)
        circ_bar(lower_arch)

        # CUB BARS
        map_bars(upper_arch, middle_arch, cub_bar)
        map_bars(middle_arch, furthest_arch, cub_bar)
        map_bars(middle_arch, lower_arch, rot_cub_bar)
    end
```
27.4 μs

Canopies Bars Test: ☑

```
▶ Union{Nothing, Khepri.RightCuboid}[nothing, Khepri.RightCuboid(...), Khepri.RightCuboid
◀ ███████                                                                              ▶
  run_test(canopie_bars_test) do
    delete_all_shapes()
    canopie_bars(station_corner, central_lenght, central_width, central_height, 100,
  +)
  end
```
4.2 s

Fig. 11. Pluto notebook: interactive evaluation applied to the canopie bars' function in the Liege-Germmins' project.

In this case, the user improved the resulting snapshot by identifying each bar in the image with the corresponding name in the function definition. The drawing thus became a more relevant piece of documentation, not only because it shows what the expected result of the test is, but also because it visually explains the body of the function it illustrates.

Both Jupyter and Pluto provide sets of commands to move cells up and down, and merge and split cells. These can be used for refactoring purposes. As for outlining, we developed two different mechanisms of identifying test cells in the notebooks. In Jupyter, since cells only run at user request, we simply wrapped all test examples with an outlining mechanisms that bound them to a global variable. The state of this variable defines what happens when a test cell is run: it either produces results, or the test is ignored. For Pluto, separate variables manage the state of each test, else the dependency graph would have them all running simultaneously. These variables are presented using PlutoUI's checkbox widget (Fig. 11). Checking a test box will prompt a test cell to run. While the box is checked, the test cell will re-run if any change in the program affects the code within it.

4.4 Reactivity

Reactivity means keeping track of dependencies in the program, so that immediate feedback on results can be provided whenever a change occurs. This is essential for users to understand program dependencies and guarantee consistency in exploratory programming. In the context of AD, reactivity is particularly useful for parametric manipulation (Fig. 12).

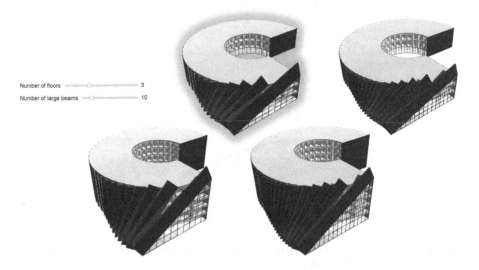

Fig. 12. Isenberg model test with sliders binding the building's parameters. Four variations: 3 to 5 floors variation vertically, and 10 to 15 tilted beams variation horizontally.

Our implementation relies on two extensions of the chosen computational notebooks: Interact for Jupyter and PlutoUI for Pluto, which allow the use of sliders, toggles, and other widgets to visually manipulate data. However, the core difference in the nature of the two notebooks means reactivity is used very differently between them.

In an innately reactive environment like Pluto, widgets bound to global variables provoke immediate updates on dependent cells. This means that in the example shown in Fig. 9 (left), when the sliders bound to the sinusoid parameters are changed, the graph shown beneath is immediately updated, as will be other cells in the notebook that are dependent upon these variables. If they are not specifically signaled as ignorable, they will render new results at any change in the sliders. When one dependent cell causes side-effects, such as the creation of architectural objects that is typical of AD, the cell's reevaluation needs to undo the previous side-effects to avoid accumulating them.

To best adapt reactivity to AD, we modified the Pluto notebook to track down not just cell dependencies but also their side-effects. As such, the notebook's dependency graph knows which geometry was affected by each change, so that it can be selectively deleted and regenerated, maintaining state consistency with good interactive performance.

In Jupyter, we cannot hope to achieve reactivity at the level of the notebook. Since cells are only evaluated on user demand, we may strive for partial reactivity. Interactive features, in this case, are best used for particular tests inside one cell only. Figure 12 presents an example: the widgets are bound to local variables in a test cell used to generate the Isenberg model. Any change to the sliders will prompt the regeneration of the model in that test cell.

5   Conclusion

In this paper, we explored the program comprehension dyad - documentation and liveliness - as a means to improve the comprehension of AD programs. We proposed the integration of more adequate methods of developing and maintaining algorithmic representations of complex architectural projects by fine-tuning the two concepts to better suit the architectural scenario. Three ideas emerged: (1) storytelling - explaining the program through human-readable text and meaningful imagery, intertwined with the code itself to tell the narrative of design development; (2) interactive evaluation - a scalable way of incorporating liveliness in programming, providing feedback on program results upon user demand; and (3) reactivity - a systematic approach to liveliness, which offers immediate feedback on program changes and reveals program dependencies. We proposed to incorporate the three concepts in an AD workflow and we evaluated their application using computational notebooks as the base programming environment.

The three ideas intersect in many ways, starting with interactive evaluation and reactivity being two different ways of applying liveliness to the programming task. Interactive evaluation offers a more scalable option, albeit with program state issues, whereas reactivity resolves any state problems and promotes a more interactive workflow, however, failing to scale to large projects without proper outlining mechanisms. A trade-off is needed between efficiency and consistency, but reactivity in itself will always suffer from scalability issues.

Both interactive evaluation and reactivity promote exploratory programming, which can cause program disarray. Storytelling, on the other hand, defends an organized approach to the programming endeavor, which means refactoring mechanisms must be used for the three approaches to co-exist.

Storytelling also appeals to the preservation of the development history as part of the narrative. This directly conflicts with reactivity's requirements for a consistent program state. The two can only operate simultaneously if outdated definitions are disabled. Interactive evaluation, in turn, greatly contributes to storytelling by producing artifacts that can be used as documentation, namely tests results in the form of imagery.

In sum, the three ideas provide different sets of advantages and trade-offs to the process of developing AD projects. Their integration in the AD workflow requires support from the programming environment according to the architect's own needs at any stage of the development process, as well as considering the ultimate goal for the program. AD programs may be a means of attaining a single constructive goal, they may represent a set of ideas to be re-used in the future, they may serve as presentation mechanisms, etc. Taking this in mind, some of the concepts may or may not apply to each case.

As future work we plan on investigating solutions for the ever-lasting compromise between liveliness and scale. Given that architects do not need to visualize the entirety of the project all the time, nor the entirety of the project's detail, we expect Levels Of Detail (LODs) and/or selective generation based on project areas or view cones to yield good results. This may allow architects to keep using reactive mechanisms later along the project's development. We also plan on incorporating traceability research to the proposed methodology, studying mechanisms of integrating it with the remaining concepts in a functional workflow.

Acknowledgments. This work was supported by national funds through *Fundação para a Ciência e a Tecnologia* (FCT) (references UIDB/50021/2020, PTDC/ART-DAQ/31061/2017) and a PhD grants under contract of FCT (DFA/BD/4682/2020).

References

1. Alfaiate, P., Caetano, I., Leitão, A.: Luna moth: supporting creativity in the cloud. In: 37th ACADIA Conference, pp. 72–81. Cambridge, Massachusetts, USA (2017)
2. Baecker, R., Price, B.: The early history of software visualization. In: Stasko, J., Domingue, J., Brown, M.H., Price, B.A. (eds.) Software Visualization: Programming as a Multimedia Experience, chap. 2, pp. 29–34. MIT Press (1998)
3. Bainomugisha, E., Carreton, A.L., Van Cutsem, T., Mostinckx, S., De Meuter, W.: A survey on reactive programming. ACM Comput. Surv. **45**(4), 34 (2013). https://doi.org/10.1145/2501654.2501666
4. Ball, T., Eick, S.G.: Software visualization in the large. Computer **29**(4), 33–43 (1996). https://doi.org/10.1109/2.488299
5. Bass, L., Kazman, R., Clements, P.: Software Architecture in Practice. Pearson Education, New Jersey, United States (2012)
6. Brown, M.H.: Zeus: A system for algorithm animation and multi-view editing, SRC reports, vol. 75. Digital, Systems Research Center (SRC), Palo Alto, California (1992)
7. Burnett, M.M.: Visual programming. In: Webster, J.G. (ed.) Wiley Encyclopedia of Electrical and Electronics Engineering, pp. 275–283. John Wiley & Sons, Inc. (1999). https://doi.org/10.1002/047134608X.W1707
8. Burry, M.: Scripting Cultures: Architectural Design and Programming. John Wiley & Sons, Inc., Architectural Design Primer (2011)
9. Celani, G., Vaz, C.E.V.: CAD scripting and visual programming languages for implementing computational design concepts: a comparison from a pedagogical point of view. Int. J. Architectural Comput. **10**(1), 121–137 (2012). https://doi.org/10.1260/1478-0771.10.1.121
10. Chattopadhyay, S., Prasad, I., Henley, A.Z., Sarma, A., Barik, T.: What's wrong with computational notebooks? Pain points, needs, and design opportunities. In: CHI Conference on Human Factors in Computing Systems, pp. 1–12. ACM, Honolulu, HI, USA (2020). https://doi.org/10.1145/3313831.3376729

11. Davis, D., Burry, J., Burry, M.: Understanding visual scripts: improving collaboration through modular programming. Int. J. Architect. Comput. **9**(4), 361–375 (2011). https://doi.org/10.1260/1478-0771.9.4.361

12. Diehl, S.: Software Visualization: Visualizing the Structure, Behaviour, and Evolution of Software. Springer, Berlin Heidelberg (2007). https://doi.org/10.1007/978-3-540-46505-8

13. Do, E.Y.L., Gross, M.D.: Thinking with diagrams in architectural design. Artif. Intell. Rev. **15**(1–2), 135–149 (2001). https://doi.org/10.1023/A:1006661524497

14. Felleisen, M., Findler, R.B., Flatt, M., Krishnamurthi, S.: How to Design Programs: An Introduction to Programming and Computing. The MIT Press, Cambridge (2001). https://doi.org/10.4230/LIPIcs.SNAPL.2015.113

15. Findler, R.B., Flanagan, C., Flatt, M., Krishnamurthi, S., Felleisen, M.: DrScheme: a pedagogic programming environment for scheme. In: Glaser, H., Hartel, P., Kuchen, H. (eds.) PLILP 1997. LNCS, vol. 1292, pp. 369–388. Springer, Heidelberg (1997). https://doi.org/10.1007/BFb0033856

16. Forward, A., Lethbridge, T.C.: The relevance of software documentation, tools and technologies: a survey. In: Symposium on Document Engineering, pp. 26–33. ACM, McLean, Virginia, USA (2002). https://doi.org/10.1145/585058.585065

17. Fowler, M.: Refactoring: Improving the Design of Existing Code. Object Technology Series. Addison-Wesley, Boston (1999)

18. Goldstine, H.H., Von Neumann, J.: Planning and coding of problems for an electronic computing instrument: Report on the Mathematical and Logical aspects of an Electronic Computing Instrument. Institute for Advanced Study Princeton, New Jersey (1947)

19. Grus, J.: I don't like notebooks (2018). https://www.youtube.com/watch?v=7jiPeIFXb6U&ab_channel=O%27Reilly. Accessed 21 Jan 2021

20. Hensel, M.: Performance-Oriented Architecture: Rethinking Architectural Design and the Built Environment. John Wiley & Sons, Inc., Architectural Design Primer (2013)

21. Iyer, S., Konstas, I., Cheung, A., Zettlemoyer, L.: Summarizing source code using a neural attention model. In: 54th Annual Meeting of the Association for Computational Linguistics, vol. 1, pp. 2073–2083. ACL, Berlin, Germany (2016). https://doi.org/10.18653/v1/P16-1195

22. Jackson, M.A.: Principles of Program Design. Academic Press Inc, USA (1975)

23. Kery, M.B., Myers, B.: Exploring exploratory programming. In: Symposium on Visual Languages and Human-Centric Computing (VL/HCC), pp. 25–29 (2017). https://doi.org/10.1109/VLHCC.2017.8103446

24. Kery, M.B., Radensky, M., Arya, M., John, B.E., Myers, B.A.: The story in the notebook: exploratory data science using a literate programming tool. In: CHI Conference on Human Factors in Computing Systems, pp. 1–11. ACM, Montreal QC, Canada (2018). https://doi.org/10.1145/3173574.3173748

25. Knuth, D.E.: Literate programming. Comput. J. **27**(2), 97–111 (1984). https://doi.org/10.1093/comjnl/27.2.97

26. Lasser, J.: Creating an executable paper is a journey through open science. Commun. Phys. **3**(143), 1–5 (2020). https://doi.org/10.1038/s42005-020-00403-4

27. Lee, E.A., Messerschmitt, D.G.: Synchronous data flow. Proceedings of the IEEE **75**(9), 1235–1245 (1987). https://doi.org/10.1109/PROC.1987.13876

28. Leitão, A., Lopes, J., Santos, L.: Programming languages for generative design: a comparative study. Int. J. Architect. Comput. **10**(1), 139–162 (2012). https://doi.org/10.1260/1478-0771.10.1.139

29. Leitão, A., Lopes, J., Santos, L.: Illustrated programming. In: 34th ACADIA Conference, pp. 291–300. Los Angeles, California, USA (2014)

30. Loukissas, Y.: Keepers of the geometry. In: Turkle, S. (ed.) Simulation and Its Discontents, pp. 153–170. MIT Press, Cambridge (2009). https://doi.org/10.7551/mitpress/8200.003.0014

31. Mens, T., Tourwe, T.: A survey of software refactoring. IEEE Trans. Softw. Eng. (TSE) **30**(2), 126–139 (2004). IEEE. https://doi.org/10.1109/TSE.2004.1265817

32. Myers, B.: Taxonomies of visual programming and program visualization. J. Vis. Lang. Comput. **1**(1), 97–123 (1990). https://doi.org/10.1016/S1045-926X(05)80036-9

33. Nassi, I., Shneiderman, B.: Flowchart techniques for structured programming. SIGPLAN Not. **8**(8), 12–26 (1973). https://doi.org/10.1145/953349.953350

34. Nguyen, A.T., Reiter, S., Rigo, P.: A review on simulation-based optimization methods applied to building performance analysis. Appl. Energy **113**, 1043–1058 (2014). https://doi.org/10.1016/j.apenergy.2013.08.061

35. Observable Inc: Observable: Make sense of the world with data, together (2021). https://observablehq.com. Accessed 21 Jan 2021

36. Oda, Y., et al.: Learning to generate pseudo-code from source code using statistical machine translation. In: 30th International Conference on Automated Software Engineering (ASE), pp. 574–584. IEEE/ACM, Lincoln, NE, US (2015). https://doi.org/10.1109/ASE.2015.36

37. Perez, F., Granger, B.E.: IPython: a system for interactive scientific computing. Comput. Sci. Eng. **9**(3), 21–29 (2007). https://doi.org/10.1109/MCSE.2007.53

38. Perkel, J.M.: Why Jupyter is data scientists' computational notebook of choice. Nature **563**(7729), 145–146 (2018). https://doi.org/10.1038/d41586-018-07196-1

39. van der Plas, F., Bochenski, M.: Pluto.jl (2021). https://github.com/fonsp/Pluto.jl. Accessed 21 Jan 2021

40. Price, B., Baecker, R., Small, I.: An introduction to software visualization. In: Stasko, J., Domingue, J., Brown, M.H., Price, B.A. (eds.) Software Visualization: Programming as a Multimedia Experience, chap. 1, pp. 3–28. MIT Press (1998)

41. Project Jupyter: Jupyter (2021). https://jupyter.org. Accessed 21 Jan 2021

42. Rein, P., Ramson, S., Lincke, J., Hirschfeld, R., Pape, T.: Exploratory and live, programming and coding: a literature study comparing perspectives on liveness. Programm. J. **3**(1), 1:1–1:33 (2018). https://doi.org/10.22152/programming-journal.org/2019/3/1

43. Roman, G.C., Cox, K.C.: A taxonomy of program visualization systems. Computer **26**(12), 11–24 (1993). https://doi.org/10.1109/2.247643

44. Rule, A., Tabard, A., Hollan, J.D.: Exploration and explanation in computational notebooks. In: CHI Conference on Human Factors in Computing Systems, pp. 1–12. ACM (2018). https://doi.org/10.1145/3173574.3173606

45. Sammer, M.J., Leitão, A., Caetano, I.: From visual input to visual output in textual programming. In: 24th CAADRIA Conference, vol. 1, pp. 645–654. Wellington, New Zealand (2019)

46. Scheer, D.R.: The Death of Drawing: Architecture in the Age of Simulation. Taylor & Francis, Milton Park (2014). https://doi.org/10.4324/9781315813950

47. Sorensen, A., Gardner, H.: Programming with time: cyber-physical programming with impromptu. SIGPLAN Not. **45**(10), 822–834 (2010). https://doi.org/10.1145/1932682.1869526

48. Storey, M.A.: Theories, methods and tools in program comprehension: past, present and future. In: 13th International Workshop on Program Comprehension (IWPC), pp. 181–191. IEEE Computer Society, Washington, DC, USA (2005). https://doi.org/10.1109/WPC.2005.38
49. Victor, B.: Inventing on principle (2012). https://vimeo.com/38272912. Accessed 21 Jan 2021
50. Victor, B.: Stop drawing dead fish (2012). https://vimeo.com/64895205. Accessed 21 Jan 2021
51. Woodbury, R.: Elements of Parametric Design. Routledge, Milton Park (2010)

Embedded Intelligent Empathy: A Systematic Review Towards a Conceptual Framework

Yahya Lavaf-Pour[✉] , Merate Barakat , and Anna Chatzimichali

The University of the West of England, Bristol BS16 1QY, UK
{yahya.lavaf,merate.barakat,anna.chatzimichali}@uwe.ac.uk

Abstract. This paper is part of a feasibility project aiming to expand computational design processes to include design empathy. The project is in response to recent valid criticism of computational design overlooking the empathy of the designer. Computational design has a heavier emphasis on the optimization process, inhibiting designers' rational and empathic input. This preliminary phase of the study aims to provoke debates through a systematic literature review (SLR) and hypothesize that empathy could be systematically integrated into computational design rather than disjointed processes. The SLR identifies gaps in knowledge in this transdisciplinary domain. Found current research suggests that technology can abstract and quantify ephemeral design qualities such as soundscape design to generate rich, intelligent designs. To achieve this, we will establish a list of indices/indicators found in literature, as a data set embedded into an algorithm that derives a computational tool.

Keywords: Designer's empathy · Computational design · Soundscape

1 Introduction

Design is a complex problem-solving process that involves high contextual interdependence requirements. In the past two decades, the advancement in generative decision-aiding tools has facilitated the design process to solve complex design problems. However, one of the critical challenges of this methodology is designing spaces and products that trigger emotional connections resulting from the designers' capacity to empathize with the users' potential needs.

This paper aims to realize the central hypothesis of a feasibility project that considers the development of computational algorithms that can create empathetic links with humans and generate designs that trigger emotional connections is possible. By investigating the current state of knowledge and revealing research gaps through a Systematic Literature Review (SLR) to explore if it is possible to embed empathy in computational design systematically? To that end, the objectives of this paper are to 1) identify the extent to which the answer to the research question can be found in the current literature, 2) fine-tune and adjust our hypothesis based on the data avail-able relevant to the topic, and 3) collate relevant evidence from a pre-specified database.

D. Gerber et al. (Eds.): CAAD Futures 2021, CCIS 1465, pp. 36–48, 2022.
https://doi.org/10.1007/978-981-19-1280-1_3

The ultimate goal is to develop a computational tool to embed ephemeral design qualities into computational methods built on a developed theory that can potentially impact practice.

1.1 Design and Science

During the 20th century, design considerations moved from a craft-oriented phenomenon to an emphasized '*scientized*' design [1], a design process based on objectivity and ratio-nality. In the 1960s, Buckminster Fuller proposed "*radical thinking about the future,*" where he claimed that comprehensive design-science innovations are vital for a utopian future [2]. By the mid-20th century, objectivity had already become an inextricable part of the design, regarded as a field of enquiry with its terms, and is independent yet inter-linked with science. The "*design methods movement*" aimed to strengthen the basis of the design process on objectivity and rationality. This movement supported developing a robust and scientifically ground-ed design methodology that can establish the design process as discreet steps with a specific goal.

During the 1980s, the Design Research Society conference proposed that it was time for design to stop learning vicariously from science and perhaps vice versa [3]. There remains confusion regarding the de-sign/science relationship and divided opinions on whether it is a scientific or a non-scientific domain. However, there is a consensus that design could be the subject of scientific investigation, making the process of design a scientific activity (i.e., systematic, reliable investigation).

1.2 Empathy in Design and Interrelationship Between Subjectivity and Objectivity

Design is a discipline with its own rigorous culture distinguishing it from sciences and the arts and humanities due to its empathic values. Science values are often subjective and rational, and art values are subjective and imaginative, where the field of design is concerned with the importance of appropriating empathy in practice. Devecchi and Guerrini [4] characterize Design Empathy as a qualitative human relationship model (i.e., intersubjective model) that is needed to establish an empathic conditions experience. Although many scholars have written about empathy in design, it remains challenging due to its complex and subjective nature. Design practitioners and researchers have extensively explored the links between design solutions and empathy [5], which occurred in tandem with the development and prefiltration of Computational design practices. It can be argued that these two design paradigms have grown apart.

"*A computational approach enables specific data to be realized out of initial abstrac-tion – in the form of codes which encapsulate values and actions*" [6] to solve com-plex problems that would have been arguably impossible using conventional design methods. The design community has long criticized such generative design processes due to algorithmic thinking dominance over the designer's empathy. A large body of research focuses on incorporating design parameters that can be easily measured and processed through computer optimization and simulations. There is a gap in research focusing on more subjective and less tangible values of design. Although the analogy of swarm behavior and the study of social logics [7] was a tipping point into computational design research and human-centric architecture, empathic, emotional and experiential values have not been tested within the computational framework potentially due to their perceived unquantifiable characteristics.

Current research aims to quantify ephemeral soundscape qualities that create rich designs and incorporate them into an intelligent system [8]. Accordingly, this project identifies Soundscape design as a suitable testbed of ephemeral design qualities. The ISO [9] defines Soundscape as *"sound at the receiver from all sound sources as modified by the environment [namely acoustic environment] as perceived or experienced and/or understood by a person or people, in context."*

This project argues that empathy could be more systematically integrated into computational design than a disjointed process. To that end, this paper presents an SLR to develop a robust theoretical framework on the role of empathy in computational design.

2   Literature Review (SLR)

This Systematic Literature Review (SLR) explores epistemological paths of implementing empathy in design. SLR was initially introduced in the medical field to encourage evidence-based knowledge development [10] and has been adopted by other fields, e.g., the management sciences [11] and information systems research [12]. This paper adopts the Preferred Reporting Items (PRISMA) flow to generate a systematic review plan as part of the protocol and establish inclusion criteria. Here, the SLR data provides a systematic flow towards answering the research question that is narrowed to Soundscape design, a testbed or rather a case study to consider the empathy-oriented design. Accordingly, the identified inclusion criteria of this SLR is grounded in three distinct yet interrelated topics, i.e. Computation, Empathy, and Soundscape.

To ensure that only the highest quality academic literature is part of this study, multiple online databases, including Scopus, IEEE Xplore, google scholars, UWE library's database, and connectedpapers.com. Only peer-reviewed journal papers, reviews and books/book chapters in English were filtered in all searches. The search fields for the advanced research strategy were set to "Article title, Abstract, Keywords" unless otherwise stated, which means that selected research terms should be placed within the article's title, abstract or keywords to qualify for the screening. The initial identification phase using all three design fields (i.e., Empathy, Computation and Soundscape) could not yield a single study. Accordingly, each two research terms were considered independent of the third term (i.e., Empathy AND Computational design; Computational design AND Soundscape; Empathy AND Soundscape). The three search returns shown in Table 1 are referred to as Search 1, 2 and 3. Search 1 investigates literature relevant to Empathy and Computational design. Search 2 focuses on computational design and Soundscape, and Search 3 explores Empathy and Soundscape.

A variation of these terms was tested using the selected databases' advanced search option. Table 1 lists the search terms and interventions used in this study, namely 1) Empathy, Emotion, feeling, perception; 2) Computational design, generative design, Parametric Design, Design Algorithms; 3) Sonic, Aural, subjective, or Ephemeral.

The flow chart in Fig. 1 illustrates the returns from 9 different searches with 17 search terms, resulting in 112 documents in the first round of screening. Papers that

Table 1. Search terms and interventions

Search categories	Search terms and interventions
Search 1	Empath* OR Emotion* OR feeling OR perception AND "Computational design" OR "generative design" OR "Parametric Design" "Design Algorithms"
Search 2	"Computational design" OR "generative design" OR "Parametric Design" "Design Algorithms" AND Soundscape OR sound* OR ephemeral OR Aura* OR sonic*
Search 3	Empath* OR Emotion* OR feeling OR perception AND Soundscape OR sound* OR ephemeral OR Aura* OR sonic*

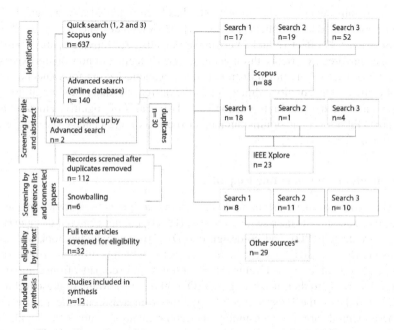

Fig. 1. Flow chart of the systematic review (adopted from PRISMA)

did not fit the conceptual framework and were outside this study's scope were excluded through screening by title and abstract. Further backwards and forwards snowballing using reference lists and citations[1] identified additional six papers. The full-text screening selection was decided based on 1) reviewing the abstract and conclusion and 2) screening conceptual frameworks and methodological approaches. Papers that did not fit into the conceptual framework were excluded, and 12 papers were included in the final synthesized set.

---

[1] By systematically looking at where papers have been published and what they have referenced. Also, where they have been cited in other papers. (Connectedpapers.com was proven to be the most effective for this stage of SLR).

2.1 Empathy in Design

Empathy was first introduced in the field of psychology and philosophy before its introduction to design practice. Within the design discipline, the implementation of empathy had not been a systematic operation. Theodor Lipps' [13] concept of empathy was rooted in the correlation between a cognitive subject and external objects' internal structure. A universally agreed definition and framework of *"Empathy in design"* could not be found in the literature. The discord is about the definition of empathy and the method of implementation in the design process. The scholars' point of agreement is that empathy is a quality of design and supports the process where the inseparable issues of rational and practicality issues are interwoven with personal experiences and private context. To design with empathy, a designer needs to "step into the other's shoes," which seems to be an analogy prevalent in literature.

Since its introduction in design, empathy has been regarded as a design skill that should be considered by designers [14]. Davis [15] views a designer's empathy through two dimensions: 1) affective and 2) cognitive empathy. Affective empathy can be an instinctive, mirrored experience through the designer's feeling of how others experience it [16]. It is a reactive emotional distress and sympathetic feeling for someone at their sight of distress. The cognitive dimension is the designer's understanding of how others experience the designed intervention [17]. Two scales of cognitive empathy being Perspective-Taking (i.e., assuming another's experience) and Fantasy (i.e., experience as a fictional character) [15].

2.2 Methods of Implementing Empathy

There is a variety of methods and tools for helping designers to approach empathy in design practices, such as user-centered design (UCD), human-centered design (HCD), participatory design (PD), and co-design (Co-D) [4]. Found studies look at Socially Responsible Design (SRD) and the inclusivity factor, which is different from empathy [14]. A large body of research develops HCD toolkits to gain empathy from communities to design according to their needs (e.g., IDEO toolkit 2009). Many of these techniques and toolkits rely on the Perspective-Taking skills from a designer's cognitive empathy. These methods are time-consuming and require many resources [5], highlighting the importance of qualitative research to inspire designers to create *'more useful and enjoyable'* products for potential users the de-signer might never meet.

Devecchi and Guerrini [4] determined empathy as the skill to design with another and accepting and acquainting their otherness. The intersubjective relationships are the tools and skills required to develop an empathic experience. The authors indicate that empathy values are intersubjective and sociable dialogue, suggesting a shift from *"design with empathy"* to *"design for empathic experience"*. It can be assumed that to design 'with' empathy, there is a need to devise tools to enable empathic experience conditions to occur [4]. Manzini [18] indicates that these missing tools can be seen as part of the design culture capable of catching a profound sense of sociality.

McDonagh-Philp and Denton [19] coined the term *"empathic horizon"*, which refers to the limitation of a designer's ability to empathies beyond specific characteristics outside their group boundaries such as age, nationality, culture, education and experience

[20]. Literature acknowledges that individuals have different *"emotional intelligent quotient"* at various levels. Indeed, Baron-Cohen and Wheelwright [21] refer to empathy as the measurable Emotional Quotient (EQ) factor that can be changed or improved through training and experience if individuals are willing to engage with empathic values. The ability (EQ) and the willing-ness both play an essential role in *"design empathy"* [22]. The designer's willingness, commitment, or/and claiming responsibility for the project can be based on the empathic horizon and connection with the potential users [20].

The tools and techniques developed in the literature support designers to *"step into the other's shoes"* and *"walk the user's walk"* to design products and spaces that fit the user's life. Various tools have successfully engaged communities through participatory sessions, and design toolkits have also made significant progress in HCD methods. These methods follow purely qualitative processes and heavily rely on individuals' perception.

2.3 Designing Human-Centric (Empathic) Spaces Through Computation

The introducing of computation design provoked new theoretical frameworks on the systemic design processes. August Schmarsaw's [13] concept of *"kinetic perception"* refers to movement through space as essential to gather sensorial experience. With the introduction of design algorithms in the 1960s, many tried to systematize architectural design, where algorithms automatically generated geometric patterns and form. Paul Coates and John Frazer's of the AA School of Architecture searched for space autonomy and developed self-organizing systems through algorithmic thinking. Christopher Alexander's [24] mathematical framework at Cambridge proposed objective representations of topological space, a theory of the process of design. He claims that a form is adapted to the context of human needs and demands the structure of the problem itself, which correspond to the adaptive process's subsystems [24]. Later, Frieder Nake [25] used a Markov chain matrix to generate emergent spatial aesthetics (e.g., a walking algorithm see Fig. 2). Schmarsow's theory, Alexander's mathematical model, and Frieder's algorithmic thinking are all proven conceptual frameworks. Derix and Izaki [26] claim that these frameworks were missing the fundamental definitions of systemic design, developed through computational systems, spatial cognition and spatial analysis.

During the 70s-90s, the interaction between the space and user extended beyond the machine-like closed system, arguing the environment is equally intelligent as the user. Towards the end of the 20th century, swarm intelligence and social animals models became an integral analogy for embedded intelligent design. During this period, space syntax theory was established by Bill Hillier and Julienne Hanson [27] at UCL, encompassing techniques for the analysis of spatial configuration to develop a systemic relation between society and space.

Rudolf Arnheim's [28] Perception of Environmental Form through Gestalt Theories followed a tandem strand of computation design rooted in spatial cognition's psychological aspects. Through physical interaction, Jean Piaget's spatial dimensions, Schmarsow's kinetic perception and Gibson ecological perception all rooted in the psychological theory of 'enaction'. Network analysis and graph theory have been adapted and applied to model spatial phenomena, such as Kevin Lynch's mental map simulation [26]. Juhani Pallasmaa [29], the Architectural theorist, in his essay "Empathic Imagination: Formal and Experiential Projections", refers to the embodied emotional experience as the true

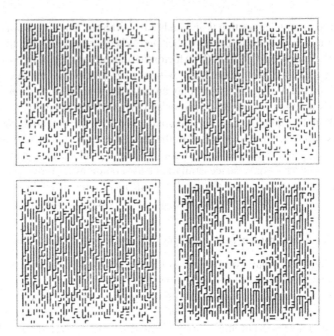

Fig. 2. Walk-Through-Raster, series 2.1, four realizations, 1966, plotter drawings [25].

quality of architectural space. The association between spatiality and experience can generate phenomenological descriptors that define feelings towards space. These preceptive attributes from spatial phenomena lead to design methodology, which, according to Derix and Izaki [26], is vital for evaluating human-centric computational design.

2.4 Design-Machine

Cross [30] claimed that designing with computers could have an adverse effect, but the apparent benefit was the speed at which a decision is made. He [31] continued to question the use of computers in design and conducted a reverse experiment of the 'Turning Test' to search for understanding limitations and requirements (at the time) for future computer-aided systems, entitled *"Can a Machine Design"*. The experiment used human participants (designers/architects) to simulate the way computers are used to design. A team of architects and engineers attempted to answer design questions posed by other participant designers in a separate room who were given a small brief to produce a sketch concept. The participants could ask questions using cards and closed-circuit TV cameras and receive answers. Ten similar experiments were carried out in search of potential emergent systemic behavioral patterns. According to the brief, the designer participants and the helping team's messages were recorded and classified into themes and topics. The data gathering method helped establish the designer's pattern of activity.

One of the surprising conclusions made was that the human-machine interaction produced the least desirable result compared to un-aided human or fully automated designs. Another argument that emerged was the question, *"can a machine make an aesthetic judgement"*. A set of implicit rules of aesthetic judgment that establish a 'bad' design instead of constructing aesthetically 'good' design rules was devised. The rules to evaluate the design and define the 'bad' elements were collected from experts' comments. The conclusion was that a set of simple rules embedded in a system was an effective way to help designers create better designs. Perhaps a more surprising observation was that human experts were inconsistent in applying their 'own' rules. A machine could do/or could help with aspects that are regarded as uniquely human attributes (i.e., making aesthetic judgment) more consistently than the human experts.

While scholars critically questioned the use of the machine, the works of MIT's Architecture Machine Group Cedrick Price (Nicholas Negroponte, Cedric Price, and Christopher Alexander) imposed two questions: 1) whether the designer is a consultant or the author; and 2) if the computer is a tool or the designer. The authorless design was being explored as algorithmic thinking emerged from self-organizing systems of forms. As Coates argued in 1966 *"architects [...] to be systems designers [and] think algorithmically to be able to propose algorithms to a computer in order to develop their thoughts by observing the outcome"* [32]. The argument is that the designer oversees and observes the outcomes. Therefore, the designer's autonomy is intact, but the outcome is not under the designer's authority, expanding the authorship question. It can be argued that to protect the design's autonomy in controlled conditions, the solutions to the same design problem must be isomorphic. An algorithm can control the condition by embedding the rules into a system design rooted in sensorial experiences and structural isomorphism. Thus, the solutions to the same design problem are isomorphic, but the design outputs are not necessarily identical.

No universally agreed answer to the question of the machine's or the designer's role can be found in the literature. It appears that the lack of clarity on these answers has hampered the process of empathic design through computational methods.

2.5 Soundscape as a Testbed

Soundscape design is the field that considers the human response to the sonic environment that is among the major contributing factors of people's perceptual experience of places. Since this research aims to examine data-driven empathy, Soundscape is a good candidate to be a testbed to investigate the concept of intelligent empathy in design.

The Handbook for Acoustic Ecology defines 'soundscape' as *"an environment of sound or sonic environment with an emphasis on how it is perceived and understood by the individual, or by a society"* [33]. In the 2010s, soundscape design evolved as an interdisciplinary field, where the majority of research agrees on the emphasis on perception and interpretation of the society or individuals [34, 35]. Several scholars have attempted to model soundscape perception to identify the sonic environment ephemeral dimensional attributes [8, 36, 37]. Earlier studies tested a limited number of soundscapes and perceptual attributes, finding that preference and pleasantness were the Soundscape's primary characterizations. Some of the later studies added activity and variability as secondary dimensions [38, 39, 40].

Axelsson, Nilsson and Berglund [36] tested a comprehensive set of soundscape excerpts to derive an empirical model of Soundscape based on a large set of descriptive perceptual attributes (e.g., pleasant, calm, eventful, annoying). The study investigated the relationship between perceptual dimensions and physical soundscape properties. Fifty soundscape recordings of ten different locations (urban courtyards, urban motorways, schoolyards, suburban parks, and residential areas), and technological sounds (car alarms, airplane). Three members of the research team independently listened to all fifty excerpts to assess the dominance of either natural, technological and human sound in each of these recordings. Listeners were asked to score these binaural recordings of urban outdoor Soundscape against the 116 characteristics that deemed appropriate soundscape perception attributes. These attributes were selected from a more extensive list of adjectives (n = 189) obtained from an earlier study [41], indicating primary attributes determinants of aesthetic appeals of photographs. Axelsson, Nilsson and Berglund [36] found that the urban outdoor soundscapes are represented by two principal components (un)pleasantness and (un)eventfulness. Their result corresponded with Russell's circumplex model of effect devised based on emotions and environmental psychology [42, 43, 44].

Soundscapes dominated by human sounds (like children playing) were more eventful. Natural sounds where more pleasant, and soundscapes dominated by technological sounds were found to be generally unpleasant - acknowledging that unpleasant natural sounds or pleasant technological sounds probably exist. Cain, Jennings and Poxon [37] used five semantic descriptors from a list of emotional soundscape dimensions. The list of different soundscape dimensions came from three different sources obtained by earlier studies. 1) A lab-based experiments extracting 25 listeners' emotional responses to 6 soundscape recordings. 2) The data source came from sound-walk transcripts from 5 different locations, emphasizing the urban Soundscape description concerning the location and context. 3) The source came from many responses to the question *"what is your favorite sound and why"*.

As a result of a multidisciplinary conversation, five emotional dimensions and their relative semantic descriptors were shortlisted and used in two experiments. In the first experiment, participants listened to 8 different binaural soundscapes recordings and were asked to use semantic descriptors from the five identified dimensions to describe their feelings towards the soundscapes. For each response, the participants would score the recording with the associated descriptor using an SD rating scale of 1–9. The second experiment was set up similar to experiment 1 with different recordings but representative of similar urban settings. Half of the jurors were presented with images of the context while listening to the soundscape recordings (experiment 2A). The other half listened to the audio-only (2B).

After conducting a Principal Component Analysis (Varimax rotation) on rating data, two factors were underlying the five identified semantic descriptors. The two principal dimensions explaining 80% of the variability amongst the original five dimensions were Calmness and Vibrancy. Although the two principal components were referred to as calm and vibrancy, the full semantic descriptors should be used to describe emotions accurately. Therefore dimension 1 (Calmness) also include adjectives such as Relaxation, Comfort and reassurance and intrusiveness. Moreover, Dimension 2 (vibrancy) can refer to arousal. The study concluded that most of the emotional soundscape dimensions could be plotted in a 2-D perceptual space. It is worth noting that different emotional responses from different demographic groups can be a significant factor in defining soundscape perceptions which were not mainly cover in Cain et al. [14]. The study claimed that although Soundscape's emotional component analysis is a qualitative evaluation of the Soundscape, there is a need for quantitative measurements of the sound to be included in the analyses.

Another point that is particularly useful for the study of intelligent empathy was the impact of Soundscape's emotional dimensions and the significant impact on people perceiving a space. An earlier study by Axelsson [41] looked at measuring another aspect of human perception. That is to analyze the aesthetic appeals of photographs. He used MDS through two interlinked experiments. One experiment was a large group of participants assorting Photograph into groups of similar aesthetic appeal. The participant then scaled each photograph based on a scale from zero to 10, indicating the photographs' aesthetic appeal. Attributes with scale were obtained from an experiment I based on a subset of photographs and the MDS method. The resulting attributes with scales were used in Experiment II to explore the reason for the similarity in photographs' aesthetic appeal by analyzing the relationship between the attribute's scales and MDS dimensions [36].

Experiment I included three phases of data collection, sorting scaling an interview; each phase was conducted individually for each participant. There were no time limitations, generally taking two to three hours. 189 attributes were obtained from 564 photographs. The 50 photographs were analyzed through MDS and used as stimuli for EXP II. Exp 2 used these 50 photographs with scales to determine dimensions underlying similarity in aesthetic appeal. Experiment II included 100 participants. Ten different participants assessed each scored five photographs out of the 50 batches, meaning each photograph. The scaling was based on 168 attributes (141 improved sets of attributes + 20% repeated for validation [141 + 27 = 168]). As a result, six principal components were identified to explain all the attribute scales. These components included Hedonic Tone, Expressiveness, playfulness, Amusingness, Eroticism and the six components was not possible to identify at first (after oblique rotation component 6 was identified as familiarity). Hedonic tone and familiarity were the two strongest predictors of the first MDS dimension.

The combined outcome of two interlinked experiments predicted two MDS dimensions (EXP I) and the photographs' average appeal value (Exp I). Three MDS dimensions underlying similarities in photographs' aesthetic appeal were Hedonic tone – Familiarity, absence of color, and expressiveness-Dynamics. Axelsson, Nilsson and Berglund's [36, 37, 41] work is a valuable framework for identifying the necessary attributes that affect soundscape perception. The Soundscape model that emerged from these studies can be used to measure soundscape design for the current project.

Many studies that investigated subjective qualities in search of measuring human perceptual dimension used MDS method. Kerrick et al. [45], Gabrielsson and Sjögren [46], and Bjork [47] used semantic differential scaling to scale perceptual ephemera such as sounds. Cain et al. [37] deployed semantic differential (SD) rating scales (developed by Osgood [48]) to establish emotional dimensions of a soundscape using principal component analysis. Other studies used both techniques (MDS and SD) to identify human psychological dimensions, such as visual perception. However, there is no direct translation from these ephemeral values methods into computational and algorithmic thinking.

3   Conclusions

Designing objects and spaces with the awareness of user behavior and spatial organization has been an essential aspect of design over the past few decades. In recent years computational platforms have enabled designers and architects to create complex forms and architectural spaces. The vast pool of data-enabled design techniques through computational plug-ins means that architects and designers can systematize the design solutions based on the objectives embedded in plug-ins. Computational tools assist designers with structural integrity, spatial configurations, environmental simulations, and more. However, there is no such mechanism to internalize users' experience or create an emotional understanding of the designers.

The interplay between psychology, art theory and computational design is arguably a promising crossroad in creating a future generation of empathic, intelligent spaces. The potential for implementing empathy in computational design requires a reasonable degree of understanding perceptual frameworks within a psychological context. To improve an empathic design process, Kouprie and Visser [20] suggested a psychological framework. They reviewed the definition of empathy within the psychology discipline in the search for further support for the empathic design process.

The SLR presented in this paper reveals specific points that limit conventional methods of designing with empathy. This paper is the initial phased of a project built on the argument that design limitations can be resolved through systematic, algorithmic thinking (i.e., computational design) by quantifying qualitative design values. The project asserts that empathic values can be facilitated or better structured using computational platforms, similar to experience and training. Such a platform can converge a logical nexus into a unique generative process that stimulates designers' ability and willingness to design according to a user's emotions but improving the EQ.

Computational design literature shows the fast development and pervasiveness of the field during the past 20 years. However, the SLR presented in this paper identified a scarcity of research in systematically embedding the user's empathy in the design methodology. In searching for the interconnectedness of the topics and a link (if any), the paper only identified early studies of computational design that are contemporary to when empathy in design was introduced. The SLR indicates that it seems that both methodologies bifurcated early during the end of the 20th century. It also seems that there is a very recent renewed interest in returning computational design to include the ephemeral aspects of design.

This paper aims to open up the discussion amongst peers and canvas more connections that may have yet been published through dissemination and discourse.

Acknowledgements. This study is funded by Connected Everything, an EPSRC-funded research network for the Digital Manufacturing community. We would like to thank Professor Rebecca Cain of Loughborough University for her mentorship, and support of the project.

References

1. Cross, N.: Designerly ways of knowing: design discipline versus design science. Des. Issues **17**(3), 49–55 (2001)
2. Fuller, B.: Utopia or Oblivion: The Prospects for Humanity. Lars Müller Publishers (1969)
3. Cross, N., Naughton, J., Walker, D.: Design method and scientific method. Design Studies, pp. 195–201. Elsevier, 2(4) (1981)
4. Devecchi, A., Guerrini, L.: Empathy and Design. A new perspective. Design J. **20**(sup1) S4357–S4364, Taylor and Francis Ltd (2017)
5. Mattelmäki, T., Vaajakallio, K., Koskinen, I.: What Happened to Empathic Design? (2014)
6. Ahlquest, S., Menges, A.: Computation Design Thinking, pp. 10–29. Wiley (2011)
7. Bonabeau, E., Dorigo, M., Theraulaz, G.: Swarm Intelligence: From Natural to Artificial Systems. Oxford University Press, New York (1999)
8. Barakat, M. A.: Sonic Urban Morphologies: Towards Modelling Aural Spatial Patterns for Urban Space Designers. Vol. Ph.D., London: AA School (2016)
9. International Organization for Standardization: ISO 12913–1:2014(en) Acoustics — Soundscape — Part 1: Definition and conceptual framework (2014)
10. Tranfield, D., Denyer, D., Smart, P.: Towards a Methodology for Developing Evidence-Informed Management Knowledge by Means of Systematic Review, pp. 207–222. British Journal of Management (2003)
11. Denyer, D., Tranfield, D.: Producing a Systematic Review. In: Buchanan; D. and Bryman, The Sage Handbook of Organizational Research Method. London (2009)
12. Okoli, C., Schabram, K.: A Guide to Conducting a Systematic Literature Review of Information Systems Research, SSRN Electronic Journal. Elsevier BV (2012)
13. Lipps, T.: Ästhetik [Aesthetic]. Leopold Voss, Leipzig (1923)
14. Cipolla, C., Bartholo, R.: Empathy or inclusion: a dialogical approach to socially responsible design. Int. J. Des. **8**(2),87–100 (2014)
15. Davis, M.H.: Measuring individual differences in empathy: evidence for a multidimensional approach. Journal of Personality and Social Psychology (1983)
16. Spencer, E.: The Principles of Psychology. Williams and Norgate, London (1881)
17. New, S., Kimbell, L.: Chimps, Designers, Consultants and Empathy: A "Theory of Mind" for Service Design (2013)
18. Manzini, E.: Design, When Everybody Designs: An Introduction to Design for Social Innovation. MIT Press, Cambridge (2015)
19. McDonagh-Philp, D., Denton, H.: Using focus groups to support the designer in the evaluation of existing products: a case study. Des. J. **2**(2), 20–31 (1999)
20. Kouprie, M.,Visser, S.: A framework for empathy in design: stepping into and out of the user's life (2009)
21. Baron-Cohen, S., Wheelwright, S.: The empathy quotient: an investigation of adults with asperger syndrome or high functioning autism, and normal sex differences. J. Autism Developmental Disorders **34**(2), 163–175 (2004)
22. Battarbee, K.: Co-experience: Understanding user Experience in Social Interaction. University of Art and Design, Helsinki (2004)
23. Schmarsow, A.: The Essence of Architectural Creation. Leipzig (1983)
24. Alexander, C.: Nontes on the Sythesis of form. Harvard University Press, Cambridge, Massachusetts (1964)
25. Nake, F.: Computer Grafik. Edition Hansjörg Mayer (1966)
26. Derix, C., Izaki, Å.: Empathic Space: The Computation of Human-Centric Architecture. John Wiley & Sons (2014)
27. Hillier, B., Hanson, J.: The Social Logic of Space. Cambridge University Press (1984)

28. Arnheim, R.: Art and Visual Perception: A Psychology of the Creative Eye The New Version. University of California press, Berkeley (1954)
29. Pallasmaa, J.L.: Empathic Imagination: Formal and Experiential Projection, pp. 80–85. Architectural Design, Conde Nast Publications, Inc., 84(5) (2014)
30. Cross, N.: Impact of computers on the architectural design process. The Architects' J., 623–628, 22 March (1972)
31. Cross, N.: Can a machine design? J. Item 17(4), 44–50 (2001)
32. Coates, P., Derix, C.: The Deep Structure of the Picturesque. Architectural Des. 84(5), 32–37 (2014). Conde Nast Publications, Inc.
33. Truax, B.: Handbook for Acoustic Ecology. Vancouver, Canada (1999)
34. Truax, B.: Acoustic tradition and the communicational approach, 2nd edn. Ablex, Westport (2001)
35. Thompson, E.: Introduction: Sound, modernity, and history: The Soundscape of Modernity. The MIT Press. Cambridge, MA (2002)
36. Axelsson, Ö., Nilsson, M. E., Berglund, B.: A principal components model of soundscape perception. J. Acoustical Soc. Am. Acoustical Soc. Am. (ASA) 128(5), 2836–2846 (2010)
37. Cain, R., Jennings, P., Poxon, J.: The development and application of the emotional dimensions of a soundscape. Appl. Acoustics 74(2), 232–239 (2013). Elsevier Ltd
38. Berglund, B., Nilsson, M.E.: On a tool for measuring sound-scape quality in urban residential areas. J. Acustica 92, 938–944 (2006)
39. Cain, R., Jennings, P.: How learning from automotive sound quality can inform urban soundscape design. Des. Principles Practices 3(6), 197–208 (2009)
40. Kawai, K., Kojima, T., Hirate, K., Yasuoka, M.: Personal evaluation structure of environmental sounds: experiments of subjective evaluation using subjects' own terms. J. Sound Vib. 277, 523–533 (2004)
41. Axelsson, Ö.: Towards a psychology of photography: dimensions underlying aesthetic appeal of photographs. Perceptual Motor Skills 105(2), 411–434 (2007)
42. Russell, J.A.: A circumplex model of affect. J. Personality Soc. Psychol. 39(6), 1161–1178 (1980)
43. Russell, J.A., Snodgrass, J.: Emotion and the environment. In: Handbook of Environmental Psychology, pp. 245–280. Edited by D. Stokols and I. Altman, Wiley, New York (1987)
44. Russell, J.A., Ward, L.M., Pratt, G.: Affective quality attributed to environments: a factor analytic study. Environ. Behav. 13(3), 259–288 (1981)
45. Kerrick, J.S., Nagel, D.C., Bennet, R.L.: Multiple ratings of sound stimuli. 1014–1017. J. Acustica 45(4), 1014–1017 (1969)
46. Gabrielsson, A., Sjögren, H.: Perceived sound quality of sound- reproducing systems. J. Acustica 65, 1019–1033 (1979)
47. Bjork, E. A.: The perceived quality of natural sounds. J. Acustica 57, 185–188 (1985)
48. Osgood, Ch., Suci, G. J., Tannenbaum, P. H.: The measurement of meaning, pp. 185–6, University of Illinois Press, Urbana, Chicago and London, 36–38 (1957)

Screen Techniques:

Oscilloscopes and the Embodied Instrumentality of Early Graphic Displays

Eliza Pertigkiozoglou$^{(\boxtimes)}$

Peter Guo-hua Fu School of Architecture, McGill University, Montreal, QC H3A 0C2, Canada
eliza.pertigkiozoglou@mail.mcgill.ca

Abstract. As computer interfaces, electronic screens monopolize architects' interaction with computational media. However, despite their instrumental role in architectural practice, hardly anyone would think of screens as instruments, although they originated from a scientific instrument: the oscilloscope. The oscilloscope was the first device for the graphic display of electrical signals. It soon surpassed its initial scientific visualization purposes to evolve into early computer graphics screens up until the 1970s. This paper looks at the oscilloscope's transition from an electrical instrument to an electronic display to explore how the instrumentality of displays was altered, and why it became less obvious. Two historical uses of oscilloscopes are examined. The first is the work of American artist Ben F. Laposky, who employed early oscilloscopes to perform "light drawings" in the 1950s. The second is a 1964 proposal by computer scientists Ivan and Bert Sutherland at the Massachusetts Institute of Technology for an oscilloscope display that would "perform drawings." The two cases are analyzed using the interpretive concept of *body techniques,* borrowed from sociology. Body techniques allow for an empirical analysis of screen's instrumentality that does not distinguish between representations and operations, but rather includes representations to an irreducible series of embodied actions. This paper argues that examining the embodiment of graphic techniques can lead to an innovative understanding of design-computing techniques as co-shaped by both the affordances of novel technologies and the visual culture of existing design practices.

Keywords: Oscilloscope · Computer graphics · Computer art · Design techniques · Computational design history and theory

1 Introduction

Electronic screens prevail in today's visual culture. As the interfaces of computers and electronic devices, they monopolize interactions with new media. Not only are screens ubiquitous in everyday life, but also notably within the field of architecture. If, as architectural theorist John May argued, architecture today is immersed in the production of computational images [1], then electronic screens mediate architectural practices. They are the surfaces over which architects labor [1: 33]. However, despite being so instrumental in daily life and architectural practice, hardly anyone today would think of

© Springer Nature Singapore Pte Ltd. 2022
D. Gerber et al. (Eds.): CAAD Futures 2021, CCIS 1465, pp. 49–61, 2022.
https://doi.org/10.1007/978-981-19-1280-1_4

electronic screens as instruments. This contradiction is even more apparent considering that screens originated from a scientific instrument: the oscilloscope. This paper delineates the instrumentality of electronic screens by looking at the transformation of early oscilloscopes into the first displays for computer graphics. What happened in this transition that made the instrumentality of displays less obvious? How did screens become invisible actors in architects' interaction with computational media?

1.1 Oscilloscopes as Early Graphic Displays

The oscilloscope was originally invented as a scientific instrument for displaying electrical signals in waveforms. However, as the first device for electrical "seeing,"[1] it soon surpassed its scientific visualization purposes to evolve into the first displays for television, radar, and the computer. Oscilloscopes with tweaked or added parts (such as custom circuits) proliferated as the early screens of electronic art and computer graphics up until the 1970s [2–4]. This research suggests that examining the adaptation of a primarily scientific instrument in visual arts and computer graphics can illuminate how its instrumentality was altered to adjust to different modes of production, use, control and bodily interaction. The paper looks closely at two stories of appropriation of early oscilloscopes to create graphics.

Fig. 1. Illustrations included in Ben F. Laposky's article "Oscillons: Electronic Abstractions" (1969) [6]: Laposky's oscilloscope with photographic setup (left), "Oscillon 19, 1952" (right)

In 1947, the artist and self-taught mathematician Ben F. Laposky from Cherokee, Iowa, came across an article in Popular Science magazine that proposed to use oscilloscopes (back then, known as electrical testing equipment for television service work) to create abstract geometric patterns [5]. The geometry of these patterns consisted of waveforms similar to the curve traces of mechanically controlled pendulums that Laposky was already using in his artwork. Intrigued by the article's recommendation, Laposky

---

[1] The word oscilloscope derives from the Latin *oscillare*, which means to swing, and the affix *scope*, which generally indicates instruments *'for enabling the eye to view or examine or make observations'* (Oxford English Dictionary Online).

employed early oscilloscopes to develop a technique of creating "light drawings" ("light-paintings" as he called them [6, 7]). These "drawings" were actually dynamic wave pulses, which he photographed using a custom setup with a camera facing the oscilloscope screen (Fig. 1) [6]. Laposky was among the first[2] to use the oscilloscope as a visual arts display. Within few years, he had photographed thousands of light forms, which he called "Oscillons" [5].

A decade later, in the 1960s, oscilloscopes were already adapted as the screens of advanced radar systems, and their potential as graphic displays was explored in pioneering research institutions [3]. Within this context, in a memo of Lincoln Laboratory at Massachusetts Institute of Technology (MIT) dated 1964, computer graphics pioneers Ivan and Bert Sutherland proposed the fabrication of a new vectorscope display (a type of oscilloscope) that would "perform" drawings [8]. Brothers' Sutherland proposal built on computer graphics and display research conducted at MIT at the time [9: 49–50]. Many of the ideas included in the proposal originated in Ivan's PhD research on a novel program for computer-aided-design (CAD) called "Sketchpad" [8]. In fact, the display proposal was written just a year after Ivan completed his famous thesis.

These two stories are not the only early oscilloscope uses in art and design. The paper does not aim to present a comprehensive history of computer screen development. Rather, by putting these two specific events in dialogue, it aims to highlight common patterns during the adaptation of new graphic instruments. The two stories have been selected as the focus of this analysis because they manifest oscilloscope's transition from an electrical scientific instrument to an electronic graphic display. Both Laposky's and Sutherlands' displays were based on the same principle of cathode-ray tube (CRT) technology (an electron gun within a vacuum tube pointing arbitrarily at a phosphor surface producing a light trace). However, during the decade that separated the two stories, storage was added to oscilloscopes, transforming their instrumentality. While Laposky's oscilloscope could just display (analyze) electrical signals the moment they occurred, Sutherlands' storage display could also post-process the signals.[3] As this paper will discuss, storage altered the instrument's relation to time and the capability of post-processing signals enabled new bodily actions.

1.2   Approach: Instrumentality as Embodied Techniques

How can one approach the instrumentality of screens? Existing literature on digital media, visual culture, and electronic displays has focused on the relation between the screen's surface and its inner technological workings. The emphasis has been on uncovering the technical reality of electronic displays—a reality hidden behind seemingly familiar representations [11–13]. For example, recent studies discussed how digital photographs still look like chemical photographs and how computer drawings resemble hand drawings but they function differently, as they are all signals [13, 14]. By illuminating the distinctions between virtual representations and their electronic inscriptions (the zeros and ones of the machine language), these studies exposed how digital media brought

---

[2] Others were the American artist Mary Ellen Bute and the Canadian animator Norman McLaren.

[3] For more details about the difference between electrical and electronic instruments, see Tympas (1996) [10].

latent shifts in design practices. However, they focused solely on ruptures and they did not explain why certain representations and modes of practice persist.

Recently, some architecture scholars turned towards techniques as a lens to examine how habitual (routine) activities associated with technical instruments affect ways of knowing and practicing architecture [15]. This paper builds on this scholarship, while contributing a particular focus on embodiment and on the emergence of new techniques. Instead of looking at the screen's instrumentality as its primary function to present and represent, it brings into the equation how bodies adapted and used early electronic displays. It focuses on how new techniques were developed with (and because of) new instruments such as the oscilloscope. The goal is to address how existing graphic practices were translated into novel displays.

To approach the embodied design practices that surrounded screens, this research employs the concept of *body techniques,* articulated by sociologist Marcel Mauss [16]. According to Mauss, body techniques (techniques such as walking, swimming, reading) are transmitted through education (they require apprenticeship) and they are specific to cultural contexts (rather than universal). Although body techniques might involve instruments, their basic technical means is the human body. Recently, sociologists recognized the importance of this concept as a method for empirical analysis of embodied, culturally situated forms of knowing [17]. This paper suggests that examining techniques in relation to the body shifts the focus beyond the instrument's primary representational logic and toward how bodies made sense of new instruments [18, 19].

Another key concept for this analysis is that of *operational sequences*, which was introduced by anthropologist André Leroi-Gourhan (student of Mauss) and built on body techniques [20]. Leroi-Gourhan analyzed techniques as sequences of material actions—he sort of reverse-engineered techniques from material artifacts. For example, he looked at paleontological stone tools to speculate on the gestural actions that produced them. What makes operational sequences particularly suitable to the discussion about screens and oscilloscopes is that, in operational sequences, gestural behavior (instrument operation) and language (signs and representation) are intertwined. According to Leroi-Gourhan, operational sequences are recorded in both language and gesture. When they are interrupted (for example when old tools are not available), language symbols (representations) repair the disruption by creating a new sequence [20: 230–231]. This interdependence between language and gesture is key for understanding how bodies adapted new instruments.

In this paper, body techniques and operational sequences provide the interpretive framework for analyzing the two stories of oscilloscope's adaptation to create graphics. The focus on body techniques is construed here as the focus on how bodies used instruments. Operational sequences allow the analysis of this use as a combination of gesture (operation) and language (representation). The paper proceeds with a close observation of the techniques developed in each case, and finally, it comments on their relation.

2 A Novel Graphic Technique

When Ben F. Laposky employed the oscilloscope to make art in the 1950s, the instrument was not intended for such use. He had to develop his own technique for creating

Fig. 2. Photographs of Ben F. Laposky switching the controls of his oscilloscope. Simple oscilloscope setup (1952), courtesy of Stanford Museum, Cherokee, Iowa (left), advance setup with color filter wheel, courtesy of Skooby Laposky (right)

abstract geometric patterns with the trace of the oscilloscope electron beam. In doing so, he demonstrated the expressive potential of an instrument which was until then used exclusively in engineering research and electronic service work. The fact that he repurposed an instrument not originally used for graphics to make graphics indicates that there must be something about his technique that lies beyond the manual operation of an instrument.

In his article "Oscillons: Electronic Abstractions" [6],[4] Laposky elaborated on his technique. He explained that the CRT oscilloscope worked by focusing an electrical input signal into an electron beam, producing a light spot or trace on a phosphorescent screen. The artist could control the position, brightness, and sharpness of the light trace by manipulating the intensity and focus controls (a set of rotary switches). By turning the switches, he was regulating signal amplifiers (Fig. 2). Repetitive electrical signals (like pulses) produced wave-like curves. Laposky was combining three basic types of waveforms to create complex results (Fig. 3). In his article, a mathematical description of these basic waveforms was going hand-in-hand with a technical description of how to set up the oscilloscope controls to produce them. As if the waves were the vocabulary of his expressive language, Laposky combined them using the rotary switches and custom-made circuits (Fig. 4) to synthesize intricate light forms (Fig. 1, 3).

How was the artist able to develop a novel graphic technique when his instrument required such intricate and unintuitive operations? Laposky used the metaphor of other drawing instruments to conceptualize his process and introduce his technique to non-specialists, who were not versed in electrical circuits or mathematical waveforms. *'The*

---

[4] Although the article was published in 1969, Laposky had created the included artwork in the 1950s. In particular, his debut in the one-man-show *Electronic Abstractions* in Stanford Museum in Cherokee, Iowa, was in 1952.

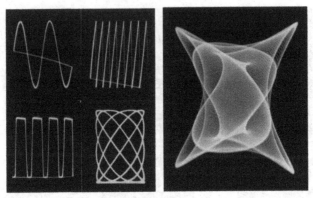

Fig. 3. Explanatory illustrations included in Ben F. Laposky's article "Oscillons: Electronic Abstractions" (1969) [6]. Basic oscilloscope waveforms (left) and "Oscillon 281 (1960)" (right)

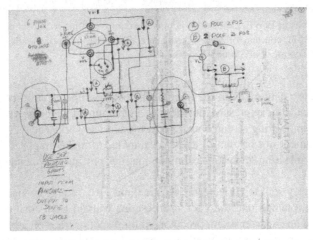

Fig. 4. Ben F. Laposky's sketch of a custom circuit, courtesy of Skooby Laposky

electron beam within the cathode-ray tube is actually the pencil or brush by which these traces are formed,' he explained in his article [6: 435]. This metaphor was illuminating because everyone was familiar with tracing using a brush or pencil. Clearly, one could not actually touch an electron beam and manipulate it as a pencil or brush. However, the word "trace" and some attached images of Laposky's oscilloscope artwork (Fig. 3) offered the necessary clues to the audience. In the images, the readers of the article could recognize lines and shadows that formed surfaces and imagine how a pencil could trace something similar. Thus, they could extrapolate from their own bodily experience with pencils, brushes, and traces to understand this metaphor. The readers' own bodily experience included both seeing and operating the instrument (pencil) inseparably. This suggests that seeing—perceiving the trace—is included in the body technique of drawing. Laposky used the familiar visual language of light traces to make sense of his new instrument and explain his technique to others.

The mastery of the novel technique required continuous experimentation—a process of trial and error. User guides from trade magazines about radio and television electronics gave the artist some insight on circuitry [5]. Practice allowed him to gradually understand of the cause-effect relationship between electrostatic fields and waveforms: how the tuning of the controls affected the appearance of these waves in the oscilloscope screen. To borrow Leroi-Gourhan's concept of operational sequences, initially, the familiar graphic language of pencil traces helped Laposky develop a new sequence using the novel instrument. Through multiple experiments, Laposky established a new mechanical behavior that combined both the operation of the instrument and the visual language of light traces. By learning to "draw" with the oscilloscope, not only he mastered the operation of the instrument, but he acquired a new way of seeing that related light traces with signals, their intensity, and duration.

The takeaways of this analysis are twofold. First, Laposky relied on existing representational references to develop a new graphic technique. Second, this technique was embodied (a new body technique) as it became habitual for the artist through repetition.

3 Instrument Affordances: Oscilloscope's Temporalities

Laposky's technique—the way he was manipulating electrical signals to control the trace of the electron beam—was unfolding in a certain way in time; it had a specific temporality afforded by his instrument. Oscilloscopes, initially, could only display electrical signals that endured in time and were repetitive or periodic like waves. Signals that were too fast or too slow for the eye to see or for the camera to capture were not perceptible. This limitation on the types of signals that could be visualized in early oscilloscopes had certain implications on how Laposky was working with the instrument and on what kind of forms he could produce. The artist had to be in constant dialogue with the oscilloscope, controlling the amplifiers and switches. The forms were kinematic and pulsating. Each experiment, *'each particular figure or kinetic cycle,'* was not repeatable [6: 351]. To be repeated, Laposky would have to record all his control settings very precisely and also regulate the voltage very accurately, which was almost impossible, as he admitted. One could argue that each of his artwork was a performance, as it came into existence by the very act of doing it and ceased to exist right after. It required what Laposky referred to as *'spontaneous creativity'* [6: 351]—a continuous dialogue between the artist and the instrument.

In the 1960s, storage was added in commercial oscilloscopes, fundamentally altering the instrument's relation to time. Although there were many attempts to include storage in CRTs since the 1940s (with the Williams tube the most successful one) [2], storage tubes did not enter the market massively until the 1960s, when an engineer in oscilloscope manufacturing company Tektronix created an inexpensive alternative that could be mass-produced [21, 22].

Tektronix' newsletter article advertised: *'[S]torage may not revolutionize either oscillography or communications. But it will surely provide them both with an exciting new dimension'* [21: 4]. The word "dimension" was used figuratively to emphasize storage's potential. However, read retrospectively, storage did alter an instrument's dimension quite literally: the dimension of time. In the storage oscilloscope the visualization of an electrical signal continued even after the signal ceased to exist. Two additional electron guns within the tube omitted stray electrons which were attracted by the positively charged light trace of the main electron gun, causing the trace to endure. The storage tube technology allowed for the once-fading trace of the electron beam to become more permanent, like writing. This writing capability immediately suggested a possibility for reading. *'The most intriguing storage-related area is electrical readout – transmission of a stored scope signal,'* emphasized the author of the Tektronix article and went on elaborating the possibilities of such readout, including signal magnification, editing, and display on other viewing devices [21: 8].

With the storage oscilloscope, electrical events did not only happen momentarily, but they could also be recorded, analyzed, and post-processed. This capacity for signal post-processing implied that the oscilloscope could become part of a larger technical assemblage that included a computer processor. It is no coincidence that after the introduction of the storage tube, Tektronix created the *'first non-oscilloscope division,'* expanding to the market of *'interactive computer displays'* [22]. Storage turned the oscilloscope from a stand-alone electrical instrument to an intermediary—an interface—between an electronic system and the user. Ivan Sutherland's diagram of the *'typical program structure for computer displays'* [23], for example, illustrates exactly that: how the CRT storage display became part of complex processing systems (Fig. 5). Storage might not have revolutionized oscillography, yet it did alter oscilloscope's instrumentality by altering the instrument's relationship with time. In the following section, the paper discusses how this new temporality paved the way for new body techniques—a new way of seeing and operating graphic displays.

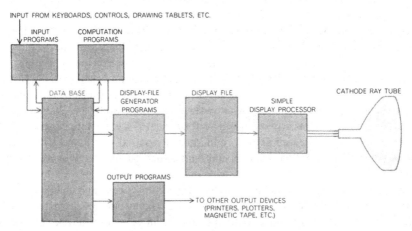

Fig. 5. Diagram of typical program structure for computer displays included in Ivan Sutherland's article "Computer Displays" in Scientific American (1970) [23]. Sutherland explained that the display file is generated by display programs (routines) that *'perform geometric operations.'*

4   Display Routines

The so-called display routines took over the signal processing enabled by the introduction of storage. Display routines were computer programs for early displays. Before the

1970s, screens had not yet crystallized into a single display technology, and they were not commercially available. Instead, multiple custom-made variations based on the basic CRT principles were developed and used in few laboratories and research universities for experimental computer graphics research. When several types of CRT display coexisted, there was a need for customized routines to translate data (signals) into visuals. For example, early video games had their own oscilloscope display routines for drawing the various game objects in alternating frames [4]. Similarly, programs for computer-aided design (CAD), needed different display routines for drawing and interacting with shapes. Display routines automated the generation of computer graphics, mediating between the technical features of each custom display type and the specific needs for interactive graphic visuals. By doing so, they encapsulated part of the creative labor of electronic "drawing."

In a 1964 memo of the Massachusetts Institute of Technology (MIT) Lincoln Laboratory, Ivan and Bert Sutherland issued a proposal for a new display—a vectorscope. Their proposed display would work in tandem with some original, hierarchically structured display routines that they had come up with, aiming to advance Ivan's research in computer graphics. The innovation of this display, according to brothers Sutherland, was that it was going to *'"perform" the display file rather than merely receiving data from it'* [8: 2–3]. Display routines and subroutines would coordinate this performance. Put differently, these routines would automate a sequence of actions, orchestrating the performance of a "light drawing" in the vectorscope's screen.

To convince the lab to invest in the fabrication of their custom display when there was already a "spot-by-spot" display available, brothers Sutherland argued that routines and subroutines for computer drawing would be more powerful in a vectorscope [8: 4]. This is because the vectorscope defined shapes in relation to a starting point and a mathematical function, in contrast to the "spot-by-spot" display, which defined shapes as a pattern of on-and-off spots. In the case of the vectorscope, routines and subroutines would be programs containing the mathematical function of the shape, as well as formulas for moving, rotating, and scaling it. The coordinates of the starting point could be moved or copied *'at several arbitrary positions'* [8: 5]. Thanks to the subroutines, all shapes and functions could be described *'relatively'* to each other and form complex and interactive constellations. Ivan and Bert Sutherland submitted the proposal in January 1964 and by October of the same year the new display—named "Scope 62"—was "on-line" and available for use at MIT's TX-2 computer [8].

As routines and subroutines assumed the performance of complex shapes on the screen, the users of MIT's vectorscope developed new techniques. The display routines inscribed a particular way of "drawing" in which shapes were entities described by a mathematical function and a reference point, relatively associated with each other. What users could now control was shape entities, as opposed to waveforms. Specialized input devices such as the light pen[5] and a set of console buttons corresponding to draw commands of specific curves (lines, circles, polygons, etc.) were the controls at users' disposal instead of the rotary switches and custom circuits of Laposky's oscilloscope (Fig. 6). Through practice and experimentation, the users could develop an understanding

---

[5] *'Light pen: an obsolete pen-like input device that was used with cathode-ray tube display to point at items on the screen or to draw new items or modify existing ones.'* [24].

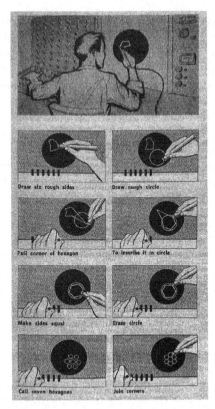

Fig. 6. Trade magazine clipping showing some possible drawing operations using MIT's Display Console, reproduced from a copy found in: Cardoso Llach, 2015 [9: 70]

of the cause-effect relationship between the various control buttons, the light pen's movement, and its effect on shapes. While the vectorscope graphics looked like the ones in engineering hand drawings, the operational sequences that created them were not the same. Like in Laposky's story, the familiar representational language of geometric shapes in engineering drawings served as the link that allowed the establishment of a new operational sequence. It repaired the disruption caused by the interruption of familiar drawing operations, to use Leroi-Gourhan's formulation [20: 230–231]. In the new sequences, the users associated commands to geometric shapes, describing entities relatively to each other. "Drawing" with Sutherland's vectorscope and its display routines was a new technique of seeing and operating the display.

Seen in parallel, the Laposky's and Sutherlands' stories illustrate three things. First, Laposky's oscilloscope artwork performance was, in a way, automated by the display routines. Paraphrasing Leroi-Gourhan, Laposky's custom tools (circuitry) and gestures (rotary operation of switches) were embedded in the machine through these programs.[6]

---

[6] Leroi-Gourhan's quote being translating in the context of this research: *'Both tool and gesture are now embodied in the machine, operational memory in automatic devices, and programing in electronic equipment'* [20: 238].

Second, this process of automating and encapsulating existing operational sequences into routines still created new operational sequences, and eventually, new body techniques. The users of the MIT vectorscope learned to control shapes through button-controlled commands such as "draw shape," "inscribe," "erase" (Fig. 6). They acquired a new way of "drawing" (seeing and operating) that was not less embodied; it was a new body technique that became habitual through repetition. Last, in both cases, existing representational cultures enabled the innovators to establish novel techniques. The graphic languages of pencil and pendulum-like traces in Laposky's case, and the geometric shapes in Sutherlands' case served as points of reference in the translation of existing graphic practices to new electrical and electronic instruments. This interpretation does not mean to undermine Laposky's and Sutherlands' innovative contributions, but rather add to the stories of novelty a narrative of continuities in design cultures.

5   Conclusion

The paper proposed the interpretive framework of body techniques and operational sequences as a novel way to approach the instrumentality of displays. The motivation was to explore why the instrumentality of screens is not obvious, although they originated from a scientific instrument: the oscilloscope. A parallel research objective was also to provide insights into why novel graphic techniques (techniques developed with and because of novel instruments) often perpetuate existing graphic languages. So far, scholarship on design techniques and electronic displays has pinpointed to how digital culture creates ruptures in design cultures, but it has not accounted for continuities.[7] By exploring how bodies made sense and adapted novel instruments in design, this research attempted to address the role that previous design traditions and visual culture play in stories of innovation.

   To address the research questions, the paper focused on two stories of oscilloscope's adaptation as a graphic display, following its transition from a scientific instrument to the first graphic displays. An analysis of the development of novel graphic techniques was presented for each case and discussed comparatively to each other. Putting the stories in dialogue highlighted both common patterns and fundamental shifts. In both cases, existing graphic languages (pencil traces, pendulum waves, geometric shapes) ensured the continuity of meaning during the adaptation of new technologies. At the same time, the technical affordances of the instruments (such as the temporality of storage/non-storage oscilloscopes) implied certain modes of interaction. Oscilloscope's instrumentality was transformed when storage introduced the capacity of signal processing and computation. The oscilloscopes transformed from a stand-alone instrument to an intermediary—an interface. Part of Laposky's performance and engagement with the oscilloscope was embedded into automated programs—the display routines.

   One could speculate that, today, part of this "drawing performance" is embedded in design software and standardized graphic cards. The interpretive framework of body

---

[7] Architecture historians have recently pointed out that histories of digital architecture "gravitate towards breaks, shifts and turn of various kinds," whereas histories of continuities that explain how the common ground was shaped are missing from existing scholarship [25].

techniques could extend to the analysis of other graphic technologies, such as the development of historical or contemporary computer graphics software. Such framework could provide insight into how existing practices were translated into novel technologies by simultaneously illuminating both the transformative role of technologies and the legacy of design cultures. One possible limitation of this method is that embodied knowledge might be distorted when one tries to describe it explicitly [17: 87]. However, design knowledge always involves a tacit and embodied part. Analysis of embodiment is therefore indispensable to understanding design knowledge. The insights gained by the empirical observation of embodied techniques and instrumentation are necessary for understanding the role of computing technologies in design practices.

Acknowledgements. The author wishes to thank professors Jonathan Sterne and Emily I. Dolan, who motivated this research through their graduate seminar on "Instruments and Instrumentalities." Special thanks to Skooby Laposky for generously sharing information and archival material related to the work of his great-uncle Ben F. Laposky. Thanks also to Professor Theodora Vardouli and to graduate students Max Leblanc, Ravi Krishnaswami and Alexander Hardan, who provided valuable and constructive feedback to earlier versions of this work.

References

1. May, J.: Signal. Image. Architecture. Columbia Books on Architecture and the City, New York (2019)
2. Gaboury, J.: The random-access image: memory and the history of the computer screen. Grey Room **70**, 24–53 (2018)
3. Geoghegan, B.D.: An ecology of operations: vigilance, radar, and the birth of the computer screen. Representations **147**, 59–95 (2019)
4. Montfort, N., Bogost, I.: Random and raster: display technologies and the development of videogames. IEEE Ann. Hist. Comput. **31**, 34–43 (2009)
5. Laposky, S.: The Modulated Path: Ben F. Laposky's Pioneering Electronic Abstractions. Vector Hack Festival: Vector Hack 2020, Rijeka (2020)
6. Laposky, B.F.: Oscillons: electronic abstractions. Leonardo **2**, 345–354 (1969)
7. Laposky, B.F.: Oscillographic design. Perspective **2**, 264–275 (1960)
8. Sutherland, I.E., Sutherland, W.R.: A new display proposal. Memo addressed to J. I. Raffel, Lincoln Laboratory, Massachusetts Institute of Technology (1964)
9. Llach, D.C.: Builders of the Vision: Software and the Imagination of Design. Routledge, New York (2015)
10. Tympas, A.: From digital to analog and back: the ideology of intelligent machines in the history of the electrical analyzer, 1870s–1960s. IEEE Ann. Hist. Comput. **18**, 42–48 (1996)
11. Chun, W.H.K.: On software, or the persistence of visual knowledge. Grey Room **18**, 26–51 (2005)
12. Galloway, A.R.: The Interface Effect. Polity, Cambridge (2012)
13. Manovich, L.: The Language of New Media. MIT Press, Cambridge (2001)
14. May, J.: Afterword: architecture in real time. In: Alexander, Z.Ç., May, J. (eds.) Design Technics: Archaeologies of architectural practice, pp. 219–244. University of Minnesota Press, Minneapolis, London (2019)
15. Alexander, Z.Ç., May, J. (eds.): Design technics: archaeologies of architectural practice. University of Minnesota Press, London (2019)

16. Mauss, M.: Body Techniques. In: Sociology and Psychology: Essays, trans. Ben Brewster, pp. 95–123. Routledge and Kegan Paul, Boston (1979)
17. Crossley, N.: Researching embodiment by way of 'body techniques.' Sociol. Rev. **55**, 80–94 (2007)
18. Sterne, J.: Bourdieu, technique and technology. Cult. Stud. **17**, 367–389 (2003)
19. Vardouli, T: Review of Zeynep Çelik Alexander and John May: design technics: archaeologies of architectural practice. J. Arch. Educ. (2020). https://www.jaeonline.org/articles/review/design-technics/, Accessed 28 Jan 2021
20. Leroi-Gourhan, A.: Gesture and Speech. MIT Press, Cambridge (1993)
21. The Storage Story. In: TekTalk, Spring 1966 Issue, Tektronix, Beaverton, pp. 4–22 (1966)
22. Brown, D.: DVST Storage Tube. https://vintagetek.org/dvst-storage-tube/, Accessed 28 Jan 2021
23. Sutherland, I.E.: Computer displays. Sci. Am. **222**(6), 56–81 (1970)
24. Daintith, J., Wright, E.: A Dictionary of Computing. Oxford University Press, Oxford (2008)
25. Touloumi, O., Vardouli, T.: Introduction: toward a polyglot space. In: Vardouli, T., Touloumi, O. (eds.) Computer Architectures: Constructing the Common Ground, pp. 1–11. Routledge, Abingdon (2019)

Rethinking Computer-Aided Architectural Design (CAAD) – From Generative Algorithms and Architectural Intelligence to Environmental Design and Ambient Intelligence

Todor Stojanovski[1]([⊠]) [iD], Hui Zhang[1], Emma Frid[1] [iD], Kiran Chhatre[1] [iD], Christopher Peters[1] [iD], Ivor Samuels[2], Paul Sanders[3] [iD], Jenni Partanen[4], and Deborah Lefosse[5]

[1] KTH Royal Institute of Technology, Stockholm, Sweden
`todor@kth.se`
[2] Urban Morphology Research Group, University of Birmingham, Birmingham, UK
[3] Deakin University, Melbourne, Australia
[4] Tallinn University of Technology, Tallinn, Estonia
[5] Sapienza, Rome, Italy

Abstract. Computer-Aided Architectural Design (CAAD) finds its historical precedents in technological enthusiasm for generative algorithms and architectural intelligence. Current developments in Artificial Intelligence (AI) and paradigms in Machine Learning (ML) bring new opportunities for creating innovative digital architectural tools, but in practice this is not happening. CAAD enthusiasts revisit generative algorithms, while professional architects and urban designers remain reluctant to use software that automatically generates architecture and cities. This paper looks at the history of CAAD and digital tools for Computer Aided Design (CAD), Building Information Modeling (BIM) and Geographic Information Systems (GIS) in order to reflect on the role of AI in future digital tools and professional practices. Architects and urban designers have diagrammatic knowledge and work with design problems on symbolic level. The digital tools gradually evolved from CAD to BIM software with symbolical architectural elements. The BIM software works like CAAD (CAD systems for Architects) or digital board for drawing and delivers plans, sections and elevations, but without AI. AI has the capability to process data and interact with designers. The AI in future digital tools for CAAD and Computer-Aided Urban Design (CAUD) can link to big data and develop ambient intelligence. Architects and urban designers can harness the benefits of analytical ambient intelligent AIs in creating environmental designs, not only for shaping buildings in isolated virtual cubicles. However there is a need to prepare frameworks for communication between AIs and professional designers. If the cities of the future integrate spatially analytical AI, are to be made smart or even ambient intelligent, AI should be applied to improving the lives of inhabitants and help with their daily living and sustainability.

Keywords: Artificial Intelligence (AI) · Computer-Aided Architectural Design (CAAD) · Architectural intelligence · Generative algorithms · Environmental design · Ambient intelligence

© Springer Nature Singapore Pte Ltd. 2022
D. Gerber et al. (Eds.): CAAD Futures 2021, CCIS 1465, pp. 62–83, 2022.
https://doi.org/10.1007/978-981-19-1280-1_5

1   Introduction

The new developments in Information and Communication Technologies (ICT) and Artificial Intelligence (AI) bring revelations of emerging smart cities. AI and robotics, bits and bricks, are becoming integral parts of architecture lexicons [1] and paradigm for smart cities (Michael Batty discusses the role of AI in smart cities and informing urban planning and design [2]). Artificial Intelligence (AI) can be defined as the capability of machines to work intelligently with complex tasks. Intelligently is typically used when the level of tasks reaches some complexity threshold, especially in being able to somehow adapt to unforeseen circumstances. This paper looks at how architects and urban designers can use new computational paradigms in AI and reflects on the role of AI in future digital tools and professional architectural and urban design practice. Computer-Aided Architectural Design (CAAD) can be defined as the application of computational science and technology in the field of architectural design. There are no discussions about Computer-Aided Urban Design (CAUD) even though there are new digital tools for urban design and planning e.g., City Information Modeling (CIM) [3].

CAAD research is characterized by pursuit of architectural intelligence and the development of generative algorithms for buildings and cities. Molly Wright Steenson [4] narrates the story of developing architectural intelligence in the mid-1960s and through the 1970s [5–7] when architectural machines and automated architects became of interest. Inspired by this technological enthusiasm, William J. Mitchell provided a theoretical framework in the book *Computer-Aided Architectural Design* [8, 9], juxtaposing computer systems and architectural practices into algorithms for generative architectural designs and design problem solving. The generative algorithms progressed from parametric models to proceduralism (even applied in CIM for the generation of cities and digital urban planning and design [3, 10]). Commercial city procedural modeling software such as CityEngine (developed by Paskal Müller [11, 12]) and building and city procedural models at the academy (see SkylineEngine [13]) are available.

Architects do not use software that automatically generates buildings and cities, despite the advancements in generative algorithms and proceduralism. The digital tools for architects gradually evolved from Computer Aided Design (CAD) to Building Information Modeling (BIM). CAD is a generic term for programs used for designing, from design of images, logos and other graphics products to the design of machines, buildings and cities. BIM denotes software used by architects and engineers to design, construct, operate and maintain buildings and infrastructures. Even though advertised as BIM, Graphisoft ArchiCAD and Autodesk Revit are used to draw and visualize architectural projects as CAAD (CAD for Architects). This paper aims to broaden the perspective on CAAD (in relation to BIM apps and CAUD/CIM conceptualization) as software for architects that should closely resemble their practices in designing buildings and interiors, neighborhoods and cities. Architects and urban designers have a diagrammatic knowledge and work with design problems on a symbolic level. The digital tools gradually evolved from CAD to BIM software with symbolical architectural elements. The Machine Learning (ML) techniques that characterize development of AI must understand the symbolical architectural and urban elements as well as deliveries from interior design, floor plans, sections and elevations to master plans for neighborhoods. AI has the capability to process data with ML algorithms referred to as Neural Networks (NN) or

Artificial Neural Networks (ANN) and it can interact with designers. Training the various NNs and ANNs with symbolical representations and data from professional practices can create a generation of analytical and interactive AI that can aid design processes. These NNs or ANNs can link symbolical architectural representations with information flows, big data and virtual reconstruct environments. Architects would not only focus on shaping a building in a closed 3D virtual space, but they can harness the benefits of analytical AIs in creating environments.

The morphogenesis of CAAD tools follows advancements in digital technology. The computer and AI reemerge as leitmotifs in architecture and planning every 20 to 30 years with new innovations. The embryonic CAD phase started on mainframe computers in the 1960s with the first program Sketchpad. AI, architectural machines and automated architects were buzzwords in the mid-1960s and through 1970s [5–7]. The developments in Information Technologies (IT) and the widespread of personal computers in the 1980s shaped the CAD, BIM and GIS systems of today. William Gibson published the book *Neuromancer* in 1984. Cyberspace is defined as electronic, invisible space that allows the computer to substitute for urban space and urban experience [14]. The concepts of cyberspace and informational cities [15] became increasingly important. In the last decade, with the widespread use of mobile phones as Information and Communication Technologies (ITCs) there is globalization of cyberspace (as the virtual domain of artificial worlds or as codespace [16]) and AI gets a more prominent role (with learning from big data). The architectural intelligence becomes ambient intelligence that positions architecture globally and in cities. There is a need to prepare frameworks for communication between AIs and professional designers. The following two sections present the historical development of CAAD and digital tools for architects in three morphogenetic periods. The fourth section reviews new development of AI discussing possible application. The fifth section discusses environmental design and ambient intelligence, urban/environmental morphology and design theory. The six section summarizes the history of CAAD and digital tools and discusses future developments. The final section summarizes and concludes the paper.

2   Computer-Aided Design (CAD) and Generative Computer-Aided Architectural Design (CAAD)

Computer-Aided Design (CAD) is a generic term for programs used for designing (from the design of images, logos and other graphics products to designing machines, buildings and cities) or the application of computational science and technology in the field of design. In a context of CAD for Architects there are two histories. This section focuses on "automated design" and generative algorithms for designing buildings and urban environments. The following section describes transition to Building Information Modeling (BIM) as CAD for Architects.

The first CAD program Sketchpad was developed in 1963 by Ivan E. Sutherland [17] at Massachusetts Institute of Technology (MIT). The computational models and computer graphics conceptualizations from Sketchpad for representing points, lines, curves and surfaces remain until today. Sketchpad was developed as a human-computer communication system using the TX-2 computer at MIT Lincoln Laboratory. The human designer communicated with the computer with light pen on the screen that acted as electronic drawing board. Timothy E. Johnson [18] presented Sketchpad III in the same year as a CAD system that was capable of creating three-dimensional designs. Sketchpad was presented at the conference Architecture and the Computer organized in 1964 in Boston where it inspired a debate on automated design. The conference brought together architects like Christopher Alexander and Nicolas Negroponte together with engineers like Marvin Minsky, the cofounder of the MIT's AI lab, and Steven J. Coons who led the MIT's CAD initiative. Walter Gropius (whose assistant Ernst Neufert wrote the influential architecture standardization handbook Architects' Data) opened the conference and the discussion centered on the computer and AI in a context of architectural intelligence. The engineers discussed how computers will change architectural practices and automate designing. Marvin Minsky predicted that computer graphics systems that would be able to sketch, render and generate plans within 10 years. He envisioned that architectural offices would be able to use computer graphics and projected that within 30 years (cited in [4]):

"Computers may be as intelligent, or more intelligent, than people. The machine may be able to handle not only the planning, but the complete mechanical assembly... Eventually computers will have hands, visions and the programs that will make them able to assemble, buildings, make things at a very high rate of speed, economically. Contractors will have to face automation in construction just as the architects will have to face automation of design. Eventually, I believe computers will evolve formidable creative capacity"

Steven A. Coons was professor of mechanical engineering and a researcher in interactive computer graphics. He was involved in advising Ivan E. Sutherland and supervised Timothy E. Johnson who developed Sketchpad. Coons saw CAD programs as digital design tools for engineers focusing on human-computer interfaces. Coons [19] writes:

"By "design" I mean the creative engineering process, including the analytical techniques of testing, evaluation and decision-making and then the experimental verification and eventual realization of the result in tangible form. In science and engineering (and perhaps in art as well) the creative process is a process of experimentation with ideas. Concepts form, dissolve and reappear in different contexts; associations occur, are examined and tested for validity on a conscious but qualitative level, and are either accepted tentatively or rejected. Eventually, however, the concepts and conjectures must be put to the precise test of mathematical analysis. When these analytical procedures are established ones the work to be done is entirely mechanical. It can be formulated and set down in algorithms: rituals of procedure that can be described in minute detail and can be performed by a computer."

The research on CAAD can be tracked through two traditions. There is a critical CAAD tradition that emphasizes design theory and interactions between human designers and computer. Steven A. Coons prioritized computer graphics over automation. He worked on CAD systems that will augment the engineers with new modes of interaction with computers. He was furthermore skeptical of the notion of "automated design" where creativity is transferred from the designer to the creator of the program. He writes [19]:

"There is much talk of "automated design" nowadays, but usually automated design is only part of the design process, an optimization of a concept already qualitatively formed. There are, for example, computer programs that produce complete descriptions of electrical transformers, wiring diagrams or printed-circuit boards. There are programs that design bridges in the sense that they work out the stresses on each structural member and in effect write its specifications. Such programs are powerful new engineering tools, but they do not depend on an internal capability of creativity; the creativity has already been exercised in generating them."

Nigel Cross [20, 21] shows a similar skepticism about automated design, pointing out that studies had suggested that using computers in design might have adverse effects, such as inducing stress, on designers. The only positive effect of CAD was to speed up the design process. In his doctoral thesis from 1974 [21, cited in 20) he concludes:

"The computer should be asking questions of the designer, seeking from him those decisions which it is not competent to handle itself. The computer could be doing all the drawing work, with the designer instructing amendments ... We should be moving towards giving the machine a sufficient degree of intelligent behavior, and a corresponding increase in participation in the design process, to liberate the designer from routine procedures and to enhance his decision-making role."

Nigel Cross summarized his doctoral thesis in the book *Automated Architect*. The book concludes with a CAAD system checklist emphasizing human and machine factors, and their specific roles in the design process. In the same CAAD tradition, Thomas W. Maver founded the research group ABACUS (Architecture and Building Aids Computer Unit, Strathclyde) at the Faculty of Architecture of the University of Strathclyde, Glasgow. He set out a plan to develop CAAD emphasizing the relationship between the computer and the design activity of architects, and he has continuously referred to the "deadly sins" of CAAD [22]. In the 1970s, Tom and Nigel compiled the Bulletin of Computer-Aided Architectural Design (BoCAAD). BoCAAD consisted of a few sheets of Xerox-copied news reports, and they mailed it out free to people working in or interested in CAAD. Tom and Nigel also created TV programs on use of digital tools by architects. Nigel Cross concludes the preface of *Automated Architect* with the words:

"This book is dedicated not to the machines, but to the humans"

The second CAAD tradition enthusiastically embraced "automated design" and developed generative algorithms for designing buildings and cities. William J. Mitchell led a CAD course at University of California Los Angeles (UCLA), and he developed the theoretical framework behind CAAD [8, 9]. He described computer systems and their relationship with architectural practices. He presented algorithms for generative architectural designs and generated floor plans of buildings automatically based on rule-sets for archetypical buildings [23]. He continued writing about computers and the logic of architecture developing a unique design theory that moves from computer graphics to architectural symbolic thinking (as design elements and typologies) [24]. William J. Mitchell will establish the Smart Cities Group at MIT in 2000s expanding the scope from architecture to cities, sadly stopped by his early passing away in 2010. The term CAAD since then has been linked with computer and programming enthusiasm and generative algorithms that characterizes William J. Mitchell research. In the same tradition Philip Steadman [25–27] developed archetypal buildings and "morphospace". Morphospace defines the architectural elements and morphological transformations of an archetypical building.

Between the two traditions stands a group of avant-garde designers starting with Greg Lynn who experimented with digital architecture by using new 3D software [28–32]. Greg Lynn discussed folds and blobs to describe the results of new LOFT tools in 3D modeling software. This tradition furthermore links to the CAD/CAM integration that combines CAD with automated factories and Computer-aided manufacturing (CAM). The 3D printers and robotic factors that emerged in the 1980s helped to create new avant-garde furniture and architecture from the 1990s. Parametrism as a term for this architectural style was coined by Patrick Schumacher, studio partner to Zaha Hadid. The parametricism created stararchitect status e.g., for Frank Gehry and Zaha Hadid. Parametric modeling is simultaneously used for generative algorithms where human designers or computers modify parameters. It established 3D modeling parametric software such as Rhino (supported by Grashopper) as default in architectural education since the 2000s (ArchiCAD had an addon Profiler that acted as a LOFT tool). However, there are differences in the application of the 3D modeling parametric software. Frank Gehry and Zaha Hadid as many other architects used 3D tools to create a unique architectural design. Greg Lynn argued for using parametricism in designing variation (e.g., instead of producing one hundred copies of a same chair, small changes in parameters of furniture design would create one hundred similar, but original chairs). Greg Lynn envisioned "unique generativeness" by manipulating parameters, instead of optimizing design. Lynn's approach has never become the mainstream of parametricism and it is a worthy direction for future thinking and developing new CAAD tools. Sean Keller [32] writes.

"Greg Lynn, for instance, has said that his design was motivated by the desire to use computers in a way that was unpredictable, but not completely arbitrary; and has described the computer as a "pet" which is partially domesticated and partially wild."

Even though CAAD has been used to describe critical approaches, the term predominantly links with computer and programming enthusiasm and generative algorithms that characterizes William J. Mitchell's research or the parametricism of stararchitects like Frank Gehry and Zaha Hadid in practices. A younger generation of researchers continued the parametric modeling and generative algorithm tradition particularly in a context of City Information Modeling (CIM). One research direction of CIM [3] links to generative city algorithms inspired by the shape grammars of George Stiny ([33]; e.g. [34–41]). Independently from CAAD historical developments, the pattern language of Christopher Alexander [42–44] and shape grammars of George Stiny [33] inspired computer scientists to develop procedural models for generating buildings and cities. The proceduralism links to morphological theories. Urban morphology dissects urban elements and factors and composes them in a hierarchical generic structure of streets, lots and buildings in a context of morphologically informed urban design [45–55]. The procedural models use the same generic morphological structure to automatically generate from building façades to entire neighborhoods [11–13, 56–59]. They use hierarchies of urban design and architectural elements and sets of rules to create buildings and urban environments in 3D at various Level of Details (LoDs) [60–62]. The procedural models generate 3D from Geographic Information Systems (GIS) data [13]. AI techniques such as Generative Adversarial Networks (GANs) were used to add geometric and texture details on procedurally generated buildings based on morphological elements [63]. Even though by Christopher Alexander (see the critique [64] and George Stiny moved away from CAAD and did not create computational models from their design theories, the pattern language and shape grammars have had a profound influence on programmers and computer scientists and architects who developed generative algorithms (even more than the works of William J. Mitchell and Philip Steadman that started and continued in the tradition).

3   From Computer-Aided Design (CAD) to Building Information Modeling (BIM) Apps for Architects

The first CAD system came in 1960s. Sketchpad and URBAN5 worked with a light pen as input. URBAN5 included AI as a conversional assistant that helped architects in the design process. The widespread diffusion of personal desktop computers with a computer mouse at the end of the 1970s and 1980s (e.g., Apple II, Apple III and Apple Lisa and x86 series processors by Intel) rendered both the light pen and AI obsolete. Autodesk AutoCAD was released in 1982 and it made possible to draw architectural projects with lines, arcs and dimensions. It recreated the drawing board of architects with the T-square in a digital form with a computer mouse. AutoCAD dominated architectural design practices until the emergence and spread of Building Information Modeling (BIM) software, namely ArchiCAD (initially developed for Apple Lisa in 1984 and transited to Windows in the mid-1990s) and Autodesk Revit (in the 2000s).

The history of BIM started as two parallel developments. Bojár Gábor founded Graphisoft, a programming company in Hungary in the 1980s. Graphisoft launched ArchiCAD, a 3D modeling software that aimed to create photorealistic visualizations of architecture. The difference between ArchiCAD and AutoCAD was that ArchiCAD used building elements as walls, slabs, doors, windows and so on as 2D symbols on a plan and created various 3D representations (including sections, elevations, architectural details, axonometries and perspectives). ArchiCAD digitized the famous handbook *Architects' Data* (often called Neufert, by its author Ernst Neufert) in its building elements that do not show only the 2D symbol, but also the spaces needed to operate (e.g., furniture elements in kitchens or bathrooms). ArchiCAD like AutoCAD worked as digital drawing boards, but the difference was that ArchiCAD created sections, elevations, architectural details, axonometrics and perspectives automatically from the 2D symbols on the plan.

In AutoCAD, architects draw sections, elevations and architectural details manually and they used elevations to create 3D visualizations.

Jonathan Ingram [65] narrates the second BIM history. An engineer by training, he programmed the software packages Sonata and Reflex in the 1980s and 1990s to make a building model. He sold the software to a company that went on to develop Revit. Sonata, Reflex and Revit, like ArchiCAD use architectural elements (walls, slabs, doors, windows and so on), together with generic CAD (lines, arcs, dimensioning, and so on). Autodesk purchased Revit in 2002 to create a competitor on the BIM market that brought success to Graphisoft and ArchiCAD making Revit currently the most popular BIM software. Revit inspired a BIM revolution in the 2000s. The vision of Jonathan Ingram was to create a software that will be used both for drawing architecture and managing construction works. Reflex was programmed to support project management, in contrast to ArchiCAD that could create a list of building elements with price tags, but it posed many difficulties in implementing them for project management. The Industry Foundation Classes (IFC) initiative was intended to create international inventories of building elements together with the construction industry, but the various contexts internationally and even locally never created databases or libraries with products and costs. The complex building models have disadvantages for managing construction projects. Contractors and engineers prioritize completing constructions and reality does not always correspond to architectural drawings and BIM representations. The construction sites are not as tidy as factory floors. Buildings do not work as machines in factories where every building element must be controlled. When constructed, some building elements e.g., walls might not change with centuries. In practice, the BIM model is more a virtual approximation than a digital twin for building management.

While there is evolution from CAD to BIM, the planners and urban designers keep to traditional design skills such as creating scale models, sketching, using notations and drafting over printed two-dimensional cadastral maps that derive from GIS. There is no Computer-Aided Urban Design (CAUD) as expansion of CAAD or BIM, but there is a second stream within the City Information Modeling (CIM) advocacy that aims to create digital drawing boards for urban designers [66–68]. Table 1 shows differences between CAD, BIM and GIS software that is used by architects while designing buildings and cities. The BIM software works with architectural design elements (walls, windows, doors, etc.) instead of geometric elements (points, lines, polygons and solids) as in typical CAD software. GIS like CAD uses geometric elements, but predominantly in two dimensions.

Table 1 shows that generative algorithms, procedural models and AI techniques are not used in the typical packages, even though tools as LOFT or various parametric design tools are integrated in BIM and 3D modeling software (they are available from the 1990s). The two traditions in CAAD, the critical and generative design can be summarized in a context of architectural intelligence as: the automated architect (researched as generative design by Nigel Cross [7]) and architect-machine symbiosis, as analytical and conversational AI who aids the architect (advocated by Nicolas Negroponte [5, 6]). The architect-machine symbiosis links to the critical tradition in CAAD, but it emphasizes AI. Nicolas Negroponte formed the MIT's Architectural Machine Group in 1967, aiming to create (soft) architectural machines that would work together with architects

Table 1. Architecture and urban design software and its linkages with architectural and urbanism practices and AI.

	CAD software	BIM apps	GIS software
Software packages considered	Autodesk AutoCAD, 3DS Catia, 3DS SolidWorks	Autodesk Revit, ArchiCAD	ESRI ArcGIS, QGIS, ESRI City Engine
Human-computer interface	Screen and computer mouse	Screen and computer mouse	Screen and computer mouse (with scripting box)
Programming interface	Not typical for CAD software. Dynamo and Refinery used for visual programming in AutoCAD	ArchiCAD uses GDL as programming interface for designing objects (not common)	Using scrips is common for GIS software e.g., VBscript and Python strips in ArcGIS- QGIS uses Python for scripting-
Design environment	2D (top, down, left, right) and 3D (perspective, axonometry)	2D (plans, sections, elevations, details) and 3D (perspective, axonometry)	2D in GIS. 3D in City Engine (axonometries and perspectives) and 2D-3D in CityGML[a]
Design toolbox	Generic 2D and 3D elements (lines, arcs, solids, nurbs, etc.). AutoCAD uses BLOCKS as generic design elements	Design elements (walls, columns, slabs, windows, doors, etc.) placed on plan	GIS uses generic 2D elements (points, polylines and polygons). City Engine makes hierarchy of streets, lots and buildings
Link to architectural practices	Generic 2D/3D software. AutoCAD uses BLOCKS as generic object	Digital drawing board, use of design (building) elements, visualization architecture	Not used by architects
Link to planning and urban design	Used for drawing master plans	Sometimes used for drawing master plans	GIS is used for drawing master plans and executing spatial and network analyses. CityEngine is sometimes used for visualizations
Procedural generation	Not included	Profiler in ArchiCAD can generate 3D from 2D shapes	CityEngine is a procedural generator of cities. GIS has no procedural models
Artificial intelligence	Not included	Not included	Not included. Some AI techniques are used

[a] There is no software for CityGML CityGML is a data model for storage and exchange of virtual cities aiming to integrate BIM and GIS by combining GIS data models (e.g. shapefiles) with Industry Foundation Classes (IFC).

and learn about architectural practices [5, 6]. In this context, the term "soft" refers not only to architectural intelligence of the computer, but also intelligent environment as evolving organism. Negroponte's architectural machine URBAN2 and URBAN5 were CAD systems who integrated a dialogue with an AI (the conversational program ELIZA), light pen and touchscreen as human computer interface. URBAN5 was an interactive graphic system that engaged in a dialogue with the human designer about the design process. The system was meant to adapt itself to the human designer thus becoming a design partner. However, the AI was not sufficiently sophisticated, and the experiment failed [4, 69]. The unsuccessful experiments with conversational AI discouraged new experiments, despite new developments in conversational and generative AIs.

Even though there was a half century of advocacy for integrating architectural intelligence, architects like Nicolas Negroponte created early CAD systems that interacted directly with a light pen and with the screen and had a dialogue with an AI, but they vanished in the 1980s with new CAD software that used computer mouse and monitors (these became larger and larger and replaced the drawing board). Today the computer pens and touchscreens are reemerging, but that has not influenced CAD, BIM or GIS software. At the same time, there are new developments with AI that can be applied to architecture and urban design.

4 New Developments in Artificial Intelligence (AI) and Possible Applications in Architecture and Urban Design

Artificial Intelligence (AI) can be defined as the capability of machines to work intelligently with complex tasks. The machine mimics humans in setting goals, making plans, considering hypotheses, recognizing analogies and carrying out various other intellectual activities [70]. Artificial neural networks (ANNs) or neural networks (NNs) are computational analogies of the human brain. The ANNs use concepts of logic and analogies to theorize and solve problems, learn from data and improve automatically. AI process data with Machine Learning (ML) algorithms. Training NNs and ANNs includes the input of data. Figure 1 illustrates the evolution from traditional approaches and Machine Learning (ML) to Deep Learning (DL). ML and DL algorithms can be used to achieve classification, regression, clustering and prediction by learning and analyzing large amounts of data. These results can, in turn, be used to help make engineers and designers (architects etc.) to make decisions. In general, ML and DL algorithms may be further classified into three machine learning paradigms, i.e., Supervised learning (S), UnSupervised learning (US) and Reinforcement Learning (RL). Supervised learning makes use of labeled input-output pairs as training data and derives a computational model between input and output data. In contrast to supervised learning, unsupervised learning makes use of training data. Unsupervised methods are appropriate in contexts in which it is difficult to obtain labeled data. More precisely, unsupervised learning can be used to cluster data into groups based on similarity of features, and to identify relationships between such identified clusters. RL is a different way where no training data is provided, and it does not have an explicit training phase. The algorithm builds and updates a DL model based on an agent's interaction with its own environment and develops a strategy to maximize a predefined reward. Figure 1 illustrates AI approaches and applications. Various ANNs or NNs, ML and DL techniques are often used across the S, US and RL spectrum.

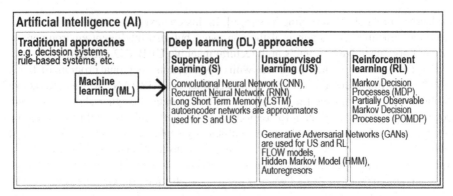

Fig. 1. Evolution of traditional Artificial Intelligence (AI) and Machine Learning (ML) approaches to Deep Learning (DL)

The smart city inspired research on big data and AI in various applications from the scale of the home [71] to the city [72, 73]. DL algorithms have been significantly advancing smart city applications. However, they have not been integrated in the practices of designing buildings or cities. Their application is mostly analytical. Compared to traditional AI approaches; DLs provide improved capacity to deal with big data and to detect and extract more patterns and features. Convolutional Neural Networks (CNNs) have been used for computer vision-based perception and recognition of design elements [74]; procedural modeling and urban simulations [75]; and Deep Graph CNNs for analyzing 3D topological graphs of buildings [76]. ML methods have been applied for predicting land uses [77]. Generative Adversarial Networks (GANs) has been widely used for creating new virtual images and 3D shapes. GANs were firstly introduced by Ian Goodfellow [78] and applied in generating 2D images, videos and 3D content, and image to image translation. Deep Conditional Generative Adversarial Networks (DCGAN) [79] are trained on an objective function that is conditioned on some class labels. GANs are currently being used in digital cities [63], for style transfer of architecture, which extracts a 3D architecture model from the ML generated 2D image. GANs are also used for depth estimation of architecture [80], for generating synthetic building mass models [81] and for generating street scape images [82]. Despite these experiments, the new AI approaches have not been integrated in the CAD or BIM software (or CAAD) that architects work with in their daily practices.

5   Environmental Design and Ambient Intelligence

The computer and AI reemerge as a leitmotif in architecture and planning every 20 to 30 years with new innovations. We are in a new wave of enthusiasm for digital tools. AI gets a more prominent role (with learning from big data) and it can analyze environments (computer vision and image recognition) and text (for example in search engines). Architects can harness the analytical capabilities to deliver not only floor plans and sections of buildings, but environmental designs. AI can improve the lives of inhabitants and help with their daily living and sustainability, by identifying problems at design stages and informing architects and urban designers. However, there is a need to prepare a framework for communication between analytical AIs of environments and professional designers. Architects and urban designers have diagrammatic knowledge and work with design problems on a symbolic level and the data input for ML algorithms and the structure of NN/ANN must follow. Christopher Alexander criticized the application of computers (and AI) as an end [64] developing an architectural and urban design theory emphasizing design problems [42–44, 83]. Architects work with design problems on a symbolic level between context and form (Fig. 2). ML algorithms and the structure of NN/ANN work at a scale of geometry, metrics and attributes (F2/C2). They also work at a level of symbolic representations and patterns, elements and rules that are not always measurable (F3/C3).

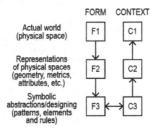

Fig. 2. Understanding the design process/abstractions of the city [83]. Urban designers work with symbolic abstractions.

The concept of design worlds describes the process of designing. The design worlds are environments inhabited by designers when designing. Design worlds act as holding environments for diagrammatic design knowledge [84, 85] where urban design and morphology are entangled [86–89]. The current BIM apps create design toolboxes with geometric elements (points, lines, polygons and solids) and architectural design elements (walls, windows, doors, etc.) that are used in designing and the architectural design elements in BIM represent a symbolic level. However, the symbolic abstractions that are used by architects for urban design are not common in CAD or GIS. There is no Computer-Aided Urban Design (CAUD) that expands the digital toolbox of BIM. GIS has very limited design capabilities and it is in 2D. The symbolic abstractions (referred to as patterns by Christopher Alexander) are complex and involve spatial practices (knowledge/behavior). Ambient intelligence is defined as the capability of humans, computer and robotic systems to understand and represent the environment and the spatial knowledge/behavior of humans. The spatial/spatial behavior includes perceptional layers of nested environments (Fig. 3). Urban morphologists have worked on a morphological structure of cities to inform urban designers [45–53]. Figure 3 illustrates one example of morphological structure (Fig. 3C) supported by three representations (Fig. 3E). The diagrammatic knowledge and expressions of urban designers often implies transforming 3D spaces into 2D symbolic representations. Urban designers in practice commonly combine theories and representations to create design toolboxes.

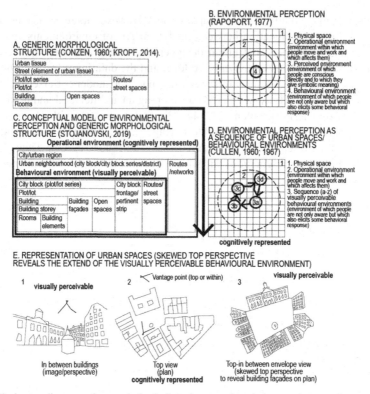

Fig. 3. Understanding generic morphological structure and environmental perception as morphological information for ambient intelligence and environmental design [48, 52, 66, 90–92].

In the generative tradition of CAAD, William J. Mitchell and Philip Steadman started to develop computational and morphological methods and programs for algorithms to generate buildings and cities. Figure 4A illustrates the typical generative algorithm [see 8–9].

A generative algorithm can create millions of variations (in a context of Greg Lynn originality parametric design advocacy), but this produces the "problem of 10 000 bowls of oatmeal" that conflicts the parametricism. Kate Compton, a computer scientist and expert on procedural generation writes on her blog:

> *"I can easily generate 10,000 bowls of plain oatmeal, with each oat being in a different position and different orientation, and mathematically speaking they will all be completely unique. But the user will likely just see a lot of oatmeal. Perceptual uniqueness is the real metric, and it's darn tough."*

The second aspect of generative CAAD is the automatic design process within a computational model as design environment. Table 1 shows the typical design environment for CAD, BIM or GIS that do not go from geometric elements (points, lines, polygons and solids) and architectural design elements (walls, windows, doors, etc.) to a hierarchy of environmental elements. There are various hierarchies and elements. There lies

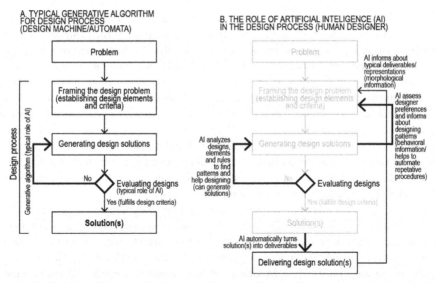

Fig. 4. Artificial Intelligence (AI) in generative algorithms and proposed new roles of AI in informing environmental design, morphological consideration and sustainability [66]. The AI does not deliver solutions, but informs the human designer in a symbiotic relationship as advocated by Nicolas Negroponte [5, 6], but unsuccessfully tested with older conversational AIs [4, 69].

a great potential to use AI not as design machines, but as analysts and programmers of these design environments. While todays CAD, BIM and GIS software acts as IT (closed system), the ICT revolution in the last decades allows to network computers and create AI that can aid designers across the globe. The buildings are nested in layers of environments from the morphology of the plot and adjacent street spaces, location in the neighborhood, city and region, to the global scale from which they draw their resources. While this can be overwhelming to a human designer, AI does not need to deliver the right "oatmeal", but bring analytical information to cook it. AI can link isolated CAD and BIM drawing boards and with analytical algorithms in GIS, and big data analyses of information flows that circulate around smart cities. This can bring environmental layers in architectural design. The AI that works with elements of buildings, cities and environments as human designer will develop ambient intelligence as spatial knowledge. AI mimics humans in setting goals, making plans, considering hypotheses, recognizing analogies and carrying out various other intellectual activities. ANNs/NNs are computational analogies of the human brain that use concepts of logic and analogies to theorize and solve problems, learn from data and automatically improve [70].

6 Discussions

This paper presents two traditions in CAAD. Steven A. Coons, Thomas W. Maver and Nigel Cross are critical to "automated design" and discuss CAD/CAAD that emphasize the human-machine interfaces and the relationship between the computer and the design activity of architects. William J. Mitchell and Philip Steadman started a CAAD tradition to develop morphological methods and programs for algorithms to generate buildings and cities. Between the two traditions stands a group of avant-garde designers starting with Greg Lynn who experimented with digital architecture by creatively using digital tools. This tradition furthermore links to the CAD/CAM integration that combines CAD with automated factories and Computer-aided manufacturing (CAM). CAAD in the future must escape the historical precedents as generative algorithms and create software that enables environmental design and integrates various design and morphological theories and AIs capable of ambient intelligence on various resolutions, from the room as architectural space to the street aligned to the building and its wider local, regional and global context. As a human designer it is impossible to process the large amounts of data and information flows that are currently circulating, but AI has new capability to both analyze data and communicate with human designers.

Even though there was a half century of advocacy for integrating architectural intelligence or automating architectural practices, many architects and urban designers remain reluctant to use software that automatically generates architecture and urban designs. Architects instead use digital tools that gradually evolved from Computer Aided Design (CAD) to Building Information Modeling (BIM) software. A good start for developing computational models are the BIM apps which have integrated building elements and rules on a symbolic level. Architects and urban designers have a diagrammatic knowledge and work with design problems on a symbolic level. The digital tools gradually evolved from CAD to BIM software with symbolical architectural elements. The Machine Learning (ML) techniques that characterize development of AI must understand the symbolical architectural and urban elements as well as deliveries from interior design, floor plans, sections and elevations to master plans for neighborhoods. AI has the capability to process data with ML algorithms that can grasp the symbolics of patterns, elements and rules, if the data input includes hierarchical morphological structures (from a scale of a building with its elements, the city with urban elements, to the environment that includes landscape and transportation flows).

BIM denotes software used by architects and engineers to design, construct, operate and maintain buildings and infrastructures. The BIM representation ideally also creates a digital twin that closely resembles the physical building. However, the BIM application does not correspond closely to the practices of construction engineers. Construction engineers can program and tend to code their own structural models. When leading and managing construction projects they rely on their expertise and professional networks. The buildings do not work as machines and the construction sites are not tidy as factory floors. Contractors and engineers can experience more problems with complex BIM representations for managing construction projects, than architects who need to deliver renderings and technical drawings. In practice, despite the broad BIM applications, the software (as ArchiCAD or Revit) is used like CAAD (CAD for Architects), but without use of generative algorithms or AI techniques that should be integrated in the future and not only to analyze building elements, but also elements of the environment.

This paper describes architecture from a perspective of environmental design (its structure and elements). Ambient intelligence is defined as the capability of humans, computer and robotic systems (that will be embedded in cities) to understand and represent the environment and the spatial knowledge/spatial behavior of humans. The spatial knowledge/spatial behavior includes environmental perception layers of nested environments (Fig. 3). The BIM apps create a hierarchy or structure of building elements, but CAD, CAAD (generative algorithms) or GIS software does not support it. Furthermore, these structures can be very complex, but they are patterns and AI can help in creating a hierarchy or structure if the data is labelled. The CAAD and CAUD must expand to integrate a larger morphological structure (Fig. 3). The current CAD and BIM apps create a building information model where buildings are designed in isolation. The architectural scope of an isolated building excludes the broader environmental design context (a CAAD sin of unsustainability [22]) or insights in the urban/environmental morphology such as relations to adjacent streets and surrounding buildings. Today architects work on buildings in the isolated 2D or 3D virtual environment, while planners and urban designers either manually execute spatial and network analyses in GIS or they sketch over printed GIS backgrounds. Procedural models and AI techniques are not used and they can help with integration of the data. However, there is a need to integrate and label the input data for ML.

To grasp the complexity and symbolics of architecture and urban design the ANNs/NNs must be supervised, meaning it will start with data with predetermined concepts and definitions. Though experience and interaction with humans, machines can be supervised in understanding human concepts and definitions. These machines acquire their data through processing images or text (or code) and there is a need to frame not the problems, but the structure of patterns, elements and rules (that create the diagrammatic knowledge and design symbologies). The following section discusses architects and urban designers in future AI interactions and reflects on the role of AI in future digital tools and professional architectural and urban design practice.

The application of Artificial Intelligence (AI) spans from the scale of the human body (including medicine and bionics) to humanoid robots and artificial humans and control systems for environments including surveillance. Architects and urban designers have also looked at designer AIs not purely from a technician's perspective, but as a competitor (artificial human) that designs buildings and cities. The AI is not the competitor, but programmers who create NNs (that enable generative designs). Architects like Nicolas Negroponte created early CAD systems with conversational AI that interacted directly with a light pen with the screen and had a dialogue with an AI, but they vanished in the 1980s with new CAD software that used computer mouse and monitors (that became larger and replaced the drawing board). Today the computer pens and touchscreens are reemerging, but that has not influenced CAD, BIM or GIS software. The conversational AI are also not preferred and probably might be considered as nuisance. Design studies [20] had suggested that using computers in design might have adverse effects, such as inducing stress, on designers. The only positive effect of CAD was to speed up the design process. The future digital tool should consider the role of AI in human-computer and programming interfaces, design environments and toolboxes (Table 1), but not use design theory as a procedure to generate designs, but to help and interactively automate boring parts of architectural and urban design process.

CAAD as digital tool for architects (even BIM that goes into symbolic representations and inventories of building elements) does not consider the worlds of human designers (theorized by Donald Schön [84] and discussed as computer and human designer factors by Nigel Cross). Architects and urban designers envision and illustrate building and cities artistically. Through practice, they develop unique design toolboxes and worlds. The programmers of BIM and CAD software do not focus on unique design toolboxes and worlds. They tend to be rational problem solvers before eccentric stylists. The digital tools, ICTs and AIs have not addressed the uniqueness and artistry as well as the trademarks of architects and urban designers as artists. Greg Lynn's parametricism in designing variations and use of folds and bulbs is an example of a unique application of the available design toolbox, but it is very difficult to create a unique element (e.g., a series of façades from Gordon Cullen's [91, 92] townscape analyses). Greg Lynn envisioned "unique generativeness" by manipulating parameters, instead of optimizing design. Lynn's approach has never become a mainstream of parametricism and it is a worthy direction for future thinking and developing new digital tools (or creatively using the existing). Lynn worked with tools (LOFT) that were programed before. There is sometimes a need of a typical design toolbox, tailor made, that has not been addressed by programmers But future conversational AI can help not only with analyzing, but also programming design elements and toolboxes.

In the end, the digitization challenges require ethical consideration. Every technology has the potential for both positive and negative impacts on society and this is especially true of inherently disruptive technologies such as AI, pervasive computing and ambient intelligence. When a new technology is applied without due consideration, unintended negative impacts may emerge despite the best intentions of its designers. It is therefore important that the application of such technologies to future cities be considered carefully and comprehensively from ethical perspectives involving a diverse range of actors. Potential misuses or unintended implications of the application of technologies should be considered in addition to their anticipated benefits. There are many challenges for the responsible practical deployment of AI supported tools and infrastructures. These include *transparency,* or how to make artificial systems that enable a better understanding of how they reach decisions, how and why they may fail and what the implications of those failures will be. *Bias*, in which the decisions made by an AI are biased due to limitations in its training data or the assumptions encoded in its decision-making processes. *Privacy*, the need to ensure that individuals from whom data is collected have consented to it and to limit unintended uses of the data. If the cities of the future are to be made *smart* or even somehow *ambient intelligent*, in which they are not just a passive backdrop to inhabitants, but actively shape their lives, then a key question is to what degree should that influence be allowed to extend, how apparent it should be and how to ensure that it will be applied to improving the lives of inhabitants, rather than becoming a means with which to control or exploit them.

7 Conclusions

This paper presents a historical record of Computer-Aided Architectural Design (CAAD) and juxtaposes current digital tools for CAD, BIM and GIS. The term CAAD is closely linked with generative algorithms and architectural intelligence while many architects and urban designers remain reluctant in using software that automatically generates architecture. A generative algorithm can create millions of variations (in a context of

Greg Lynn's original parametric design advocacy), but this produces Kate Compton's "problem of 10 000 bowls of oatmeal" that conflicts with the parametricism. Architects instead use digital tools that gradually evolved from CAD to BIM software, that represent symbolical architectural elements. Architects work with design problems on a symbolic level including unique patterns, elements and rules that are seldom available in CAD. BIM has a large inventory of building elements, but it cannot be used in designing cities. Urban designers tend to sketch over 2D maps printed from GIS. The design elements and symbolic abstractions are not common in CAD or GIS. BIM corresponds to the needs of most architects who design buildings, but this is done in an isolated 3D cubicle without considering the wider environmental context (the morphological structure including adjacent street, the neighborhood, the transportation flows and the city and the global reach). The architects who design buildings do not see the wider environmental layers (as CAUD or environmental design/the cities have global reach today with factories in China and wastelands in the Pacific). The future digital tool should consider the role of AI in human-computer and programming interfaces, design environments and toolboxes (Table 1), but not use design theory as procedure to generate designs, but to help and interactively automate boring parts of architectural and urban design process (e.g., searching for a map in GIS and printing it as a background).

In the future of CAAD must move away from the historical precedents such as architectural intelligence and generative algorithms and perhaps embrace BIM apps as good example of digitizing the design environments and toolboxes for architects who create building plans. But BIM is limited to isolated buildings that are modelled in detail to be further managed. CAAD and CAUD should bring in urban and environmental elements to enable architects in producing environmental designs. There is a need to prepare frameworks for communication between professional designers and AI. This paper presents urban morphological conceptualizations that are start to conceptualizing input data for AI that can learn about environmental design and ambient intelligence. The environment structure includes a layer of nested environments. Ambient intelligence is defined as capability of humans, computer and robotic systems (that will be embedded in cities) to understand and represent the environment and the spatial knowledge/spatial behavior of humans. The morphological/environmental structure of elements should be conceived as data input to AI with ambient intelligence/spatial knowledge across layers of nested environments. The software should integrate various design and morphological theories and AI capabilities to link to big data and develop ambient intelligence at various resolutions, from the room as architectural space to the street aligned to the building and its wider local, regional and global context. As a human designer it is impossible to process large amounts of data and information flows that are currently circulating, but AI has a new capability to both analyze data and communicate with human designers.

In the end, the digitization challenges require ethical consideration. Every technology has the potential for both positive and negative impacts on society and this is especially true of inherently disruptive technologies such as AI, pervasive computing and ambient intelligence. If the cities of the future are to be made *smart* or even somehow *ambient intelligent*, AI should be applied to improve the lives of inhabitants and help with their daily lives and sustainability.

Acknowledgements. The authors of this paper gratefully acknowledge the grants P44455-1 and P44455-2 from the Swedish Energy Agency, Energimyndigheten.

References

1. Ratti, C., Duarte, F.: Data (Are) matter: data, technology, and urban design. In: Melendez, F., Diniz, N., Del Signore, M. (eds.) Data, Matter, Design: Strategies in Computational Design. Routledge, New York (2021)
2. Batty, M.: Inventing Future Cities. MIT Press, Cambridge Mass (2018)
3. Gil, J.: City Information Modelling: a conceptual framework for research and practice in digital urban planning. Built Environ. **46**(4), 501–527 (2020)
4. Steenson, M.W.: Architectural Intelligence: How Designers and Architects Created the Digital Landscape. MIT Press, Cambridge, Mass (2017)
5. Negroponte, N.: The Architecture Machine. MIT Press, Cambridge Mass (1970)
6. Negroponte, N.: Soft Architecture Machines. MIT Press, Cambridge Mass (1975)
7. Cross, N.: The Automated Architect. Pion, London (1977)
8. Mitchell, W.J.: The theoretical foundation of computer-aided architectural design. Environ. Plann. B. Plann. Des. **2**(2), 127–150 (1975)
9. Mitchell, W.J.: Computer-Aided Architectural Design. Van Nostrand Reinhold, New York (1977)
10. Koenig, R., Bielik, M., Dennemark, M., Fink, T., Schneider, S., Siegmund, N.: Levels of automation in urban design through artificial intelligence. Built Environ. **46**(4), 599–619 (2020)
11. Parish, Y.I., Müller, P.: Procedural modeling of cities. In: The Proceedings of ACM SIGGRAPH, pp. 301–308 (2001)
12. Müller, P., Wonka, P., Haegler, S., Ulmer, A., Van Gool, L.: Procedural modeling of buildings. In: The Proceedings of ACM SIGGRAPH, pp. 614–623 (2006)
13. SkylineEngine: http://ggg.udg.edu/skylineEngine/. Accessed 7 July 2021
14. Castells, M.: The Informational City: Information Technology, Economic Restructuring, and the Urban-Regional Process. Basil Blackwell, Oxford (1989)
15. Boyer, M.C.: CyberCities: Visual Perception in the Age of Electronic Communication. Arch. Press, Princeton (1996)
16. Kitchin, R., Dodge, M.: Code/Space: Software and Everyday Life. MIT Press, Cambridge Mass (2011)
17. Sutherland, I.E.: Sketchpad a man-machine graphical communication system. Doctoral Thesis, Massachusetts Institute of Technology (1963)
18. Johnson, T.E.: Sketchpad III, three dimensional graphical communication with a digital computer. Master of Science Thesis, Massachusetts Institute of Technology (1963)
19. Coons, S.A.: The uses of computers in technology. Sci. Am. **215**(3), 176–191 (1966)
20. Cross, N.: Developing design as a discipline. J. Eng. Des. **29**(12), 691–708 (2018)
21. Cross, N.: Human and Machine Roles in Computer Aided Design. Doctoral thesis, University of Manchester (1974)
22. Maver, T.W.: CAAD's seven deadly sins. In: Sixth International Conference on Computer-Aided Architectural Design Futures, pp. 21–22 (1995)
23. Mitchell, W.J., Steadman, J.P., Liggett, R.S.: Synthesis and optimization of small rectangular floor plans. Environ. Plann. B. Plann. Des. **3**(1), 37–70 (1976)
24. Mitchell, W.J.: The Logic of Architecture. MIT Press, Cambridge Mass (1990)
25. Steadman, P., Rooney, J. (eds.): Principles of Computer-Aided Design. Pitman, London (1987)

26. Steadman, P.: Sketch for an archetypal building. Environ. Plann. B. Plann. Des. **25**(7), 92–105 (1998)
27. Steadman, P., Mitchell, L.J.: Architectural morphospace: mapping worlds of built forms. Environ. Plann. B. Plann. Des. **37**(2), 197–220 (2010)
28. Lynn, G.: Folds, bodies and blobs. La Letrre Volée, Bruxelles (1998)
29. Lynn, G.: Animate Form. Princeton Architectural Press, New York (1999)
30. Lynn, G. (ed.): Archaeology of the Digital. Sternberg Press, Berlin (2013)
31. Goodhouse, A.: When is the Digital in Architecture? Sternberg Press, Berlin (2017)
32. Keller, S.: Automatic Architecture: Motivating Form After Modernism. University of Chicago Press, Chicago (2018)
33. Stiny, G.: Introduction to shape and shape grammars. Environ. Plann. B. Plann. Des. **7**(3), 343–351 (1980)
34. Beirão, J.N. Duarte, J.P.: Urban grammars: towards flexible urban design. In: Proceedings of the 23rd Conference on Education in Computer Aided Architectural Design in Europe (sCAADe), pp. 491–500 (2005)
35. Beirão, J.N., Duarte, J.P. Stouffs, R.: Structuring a generative model for urban design: linking GIS to shape grammars. In: Proceedings of the 26th Conference on Education in Computer Aided Architectural Design in Europe (eCAADe), pp. 929–938 (2008)
36. Beirao, J.N., Mendes, G., Duarte, J., Stouffs, R.: Implementing a generative urban design model: grammar-based design patterns for urban design. In: Proceedings of the 28th Conference on Education in Computer Aided Architectural Design in Europe (eCAADe), pp. 265–274 (2010).
37. Beirão, J., Duarte, J., Stouffs, R., Bekkering, H.: Designing with urban induction patterns: a methodological approach. Environ. Plann. B. Plann. Des. **39**(4), 665–682 (2012)
38. Duarte, J.P., Beirão, J.N., Montenegro, N., Gil, J.: City induction: a model for formulating, generating, and evaluating urban designs. In: Arisona, S.M., Aschwanden, G., Halatsch, J., Wonka, P. (ed.) Digital Urban Modeling and Simulation, pp. 73–98, (2012)
39. Gil, J., Montenegro, N., Duarte, J.: Assessing computational tools for urban design - towards a "city information model". In: Proceedings of the 28th Conference on Education in Computer Aided Architectural Design in Europe (eCAADe), pp. 361–369 (2010)
40. Gil, J.A., Almeida, J., Duarte, J.P.: The backbone of a City Information Model (CIM): implementing a spatial data model for urban design. In: Proceedings of the 29th Conference on Education in Computer Aided Architectural Design in Europe (eCAADe), pp. 143–151 (2011)
41. Gil, J., Beirão, J.N., Montenegro, N., Duarte, J.P.: On the discovery of urban typologies: data mining the many dimensions of urban form. Urban Morphol. **16**(1), 27–34 (2012)
42. Alexander, C.: The Timeless Way of Building. Oxford University Press, New York (1979)
43. Alexander, C., Ishikawa, S., Silverstein, M.: A Pattern Language: Towns, Buildings, Construction. Oxford University Press, New York (1977)
44. Alexander, C., Neis, H., Anninou, A., King, I.: A New Theory of Urban Design. Oxford University Press, New York (1987)
45. Conzen, M.R.G.: Alnwick, Northumberland: a study in town-plan analysis. Trans. Papers (Inst. Br. Geogr.) **27**, iii–122 (1960)
46. Moudon, A.V.: Urban morphology as an emerging interdisciplinary field. Urban Morphol. **1**(1), 3–10 (1997)
47. Kropf, K.: Morphological investigations: cutting into the substance of urban form. Built Environ. **37**(4), 393–408 (2011)
48. Kropf, K.: Ambiguity in the definition of built form. Urban Morphol. **18**(1), 41–57 (2014)
49. Scheer, B.C.: The Evolution of Urban Form: Typology for Planners and Architects. Am. Plann. Assoc., Chicago (2010)
50. Scheer, B.C.: The epistemology of urban morphology. Urban Morphol. **20**(1), 5–17 (2016)

51. Stojanovski, T., Östen, A.: Typo-morphology and environmental perception of urban space. In: Proceedings of the XXVth International Seminar for Urban Form, pp. 816–821 (2018)
52. Stojanovski, T.: Urban Form and Mobility-Analysis and Information to Catalyse Sustainable Development Doctoral dissertation, KTH Royal Institute of Technology (2019)
53. Oliveira, V., Monteiro, C., Partanen, J.: A comparative study of urban form. Urban Morphol. **19**(1), 72–92 (2015)
54. Sanders, P.S., Woodward, S.A.: Morphogenetic analysis of architectural elements within the townscape. Urban Morphol. **19**(1), 5–24 (2015)
55. Sanders, P., Baker, D.: Applying urban morphology theory to design practice. J. Urban Des. **21**(2), 213–233 (2016)
56. Roglà, O., Pelechano, Patow, G.N.: Procedural semantic cities. In: CEIG 2017: XXVII Spanish Computer Graphics Conference: European Association for Computer Graphics (Eurographics), pp. 113–120 (2017)
57. Martin, I., Patow, G.: Ruleset-rewriting for procedural modeling of buildings. Comput. Graph. **84**, 93–102 (2019)
58. Vanegas, C.A., Aliaga, D.G., Wonka, P., Müller, P., Waddell, P., Watson, B.: Modelling the appearance and behaviour of urban spaces. Comput. Graph. Forum **29**(1), 25–42 (2010)
59. Vanegas, C.A., Kelly, T., Weber, B., Halatsch, J., Aliaga, D.G., Müller, P.: Procedural generation of parcels in urban modeling. Comput. Graph. Forum **31**(2), 681–690 (2012)
60. Besuievsky, G., Patow, G.: Customizable LOD for procedural architecture. Comput. Graph. Forum **32**(8), 26–34 (2013)
61. Biljecki, F., Ledoux, H., Stoter, J.: Generating 3D city models without elevation data. Comput. Environ. Urban Syst. **64**, 1–18 (2017)
62. Biljecki, F., Ledoux, H., Stoter, J.: An improved LOD specification for 3D building models. Comput. Environ. Urban Syst. **59**, 25–37 (2016)
63. Kelly T., Guerrero, P., Steed, A., Wonka, P., Mitra. N.J.: FrankenGAN: guided detail synthesis for building mass models using style-synchonized GANs. ACM Trans. Graph. **37**(6), 216 (2018)
64. Alexander, C.A.: Much asked question about computers and design. In: Proceedings of the conference Architecture and the Computer, pp. 52–54 (1964)
65. Ingram, J.: Understanding BIM: The Past. Present and Future. Routledge, London (2020)
66. Stojanovski, T., Partanen, J., Samuels, I., Sanders, P., Peters, C.: City information modelling (CIM) and digitizing urban design practices. Built Environ. **46**(4), 637–646 (2020)
67. Stojanovski, T.: City information modeling (CIM) and urbanism: Blocks, connections, territories, people and situations. In: Proceedings of the 4th Symposium on Simulation for Architecture and Urban Design, pp. 86–93 (2013)
68. Stojanovski, T.: City Information Modelling (CIM) and urban design: morphological structure, design elements and programming classes in CIM. In Proceedings of the 36th Conference on Education in Computer Aided Architectural Design in Europe (eCAADe), pp. 507–529 (2018)
69. Negroponte, N., Grossier, L.: Urban 5: A machine that discusses urban design. In: Goore, G.T. (eds.) Emerging Methods in Environmental Design and Planning. MIT Press, Cambridge, Mass., pp. 105–114 (1970)
70. Minsky, M.: The uses of computers in technology. Sci. Am. **215**(3), 246–263 (1966)
71. Guo, X., Shen, Z., Zhang, Y., Wu, T.: Review on the application of artificial intelligence in smart homes. Smart Cities **2**(3), 402–420 (2019)
72. Allam, Z., Dhunny, Z.A.: On big data, artificial intelligence and smart cities. Cities **89**, 80–91 (2019)
73. Ullah, Z., Al-Turjman, F., Mostarda, L., Gagliardi, R.: Applications of artificial intelligence and machine learning in smart cities. Comput. Commun. **154**, 313–323 (2020)

74. Kim, J., Song, J., Lee J-K.: Approach to auto-recognition of design elements for the intelligent management of interior pictures. In: Proceedings of the CAADRIA Conference (2019)
75. Lin, B., Jabi, W., Diao, R.: Urban space simulation based on wave function collapse and convolutional neural network. In: Proceedings of the SimAUD Conference, pp. 145–52 (2020)
76. Jabi, W., Alymani, A.: Graph machine learning using 3D topological models. In: Proceedings of the SimAUD Conference, pp. 427–34 (2020)
77. Xia, X., Tong, Z.A.: Machine learning-based method for predicting urban land use. In: Proceedings of the CAADRIA Conference (2020)
78. Goodfellow, I.J., Pouget-Abadie, J., Mirza, M.: Generative adversarial networks. arXiv preprint arXiv:1406.2661 (2014)
79. Isola, P., Zhu, J.Y., Zhou, T., Efros, A.A.: Image-to-image translation with conditional adversarial networks. IEEE Conference on Computer Vision and Pattern Recognition, pp. 1125–1134 (2017)
80. Ren, Y., Zheng, H.: The Spire of AI - Voxel-based 3D neural style transfer. In: Proceedings of the CAADRIA Conference (2020)
81. Kinugawa, H., Takizawa, A.: Deep learning model for predicting preference of space by estimating the depth information of space using omnidirectional images. In: Proceedings of the ECAADE SIGRADI Conference (2019)
82. Noyman, A., Larson, K.: A deep image of the city: generative urban-design visualization. In: Proceedings of the SimAUD Conference (2020)
83. Alexander, C.: Notes on the Synthesis of Form. Harvard University Press, Cambridge, Mass. (1973 [1964])
84. Schön, D.A.: Designing: rules, types and words. Des. Stud. 9(3), 181–190 (1988)
85. Dovey, K., Pafka, E.: The science of urban design? Urban Des. Int. 21(1), 1–10 (2016)
86. Marshall, S., Çalişkan, O.: A joint framework for urban morphology and design. Built Environ. 37(4), 409–426 (2011)
87. Samuels, I.: A typomorphological approach to design: the plan for St Gervais. Urban Des. Int. 4(3–4), 129–141 (1999)
88. Samuels, I.: Typomorphology and urban design practice. J. Urban Morphol. 12(1), 58–62 (2008)
89. Marshall, S.: Science, pseudo-science and urban design. Urban Des. Int. 17(4), 257–271 (2012)
90. Cullen, G.: The Concise Townscape. Architectural Press, London (1961)
91. Cullen, G.: Notations 1–4. The Architects' J. (supplements) (1967)
92. Rapoport, A.: Human Aspects of Urban Form: Towards A Man—Environment Approach to Urban Form and Design. Pergamon Press, London (1977)

Past Futures and Present Futures: Aesthetics and Ethics of Space

Early-Phase Performance-Driven Design Using Generative Models

Spyridon Ampanavos[1,2(✉)] ⓘ and Ali Malkawi[1,2]

[1] Harvard Graduate School of Design, Cambridge, MA 02138, USA
{sampanavos,amalkawi}@gsd.harvard.edu
[2] Harvard Center for Green Buildings and Cities, Cambridge, MA 02138, USA

Abstract. Current performance-driven building design methods are not widely adopted outside the research field for several reasons that make them difficult to integrate into a typical design process. In the early design phase, in particular, the time intensity and the cognitive load associated with optimization and form parametrization are incompatible with design exploration, which requires quick iteration. This research introduces a novel method for performance-driven geometry generation that can afford interaction directly in the 3d modeling environment, eliminating the need for explicit parametrization, and is multiple orders faster than the equivalent form optimization. The method uses Machine Learning techniques to train a generative model offline. The generative model learns a distribution of optimal performing geometries and their simulation contexts based on a dataset that addresses the performance(s) of interest. By navigating the generative model's latent space, geometries with the desired characteristics can be quickly generated. A case study is presented, demonstrating the generation of a synthetic dataset and the use of a Variational Autoencoder (VAE) as a generative model for geometries with optimal solar gain. The results show that the VAE-generated geometries perform on average at least as well as the optimized ones, suggesting that the introduced method shows a feasible path towards more intuitive and interactive early-phase performance-driven design assistance.

Keywords: Performance driven design · Machine learning · Generative model

1 Introduction

During the design process, an architect strives to reconcile several qualitative and quantitative objectives. Performance-driven design aims to assist in meeting the quantifiable objectives related to a building's performance, most commonly through the use of optimization. To maximize its impact, the performance-driven design methodology needs to be applied from the early design phase[1]. Contrary to its original purpose as a precise problem-solving tool, optimization is increasingly gaining traction as an exploratory

---

[1] Paulson and MacLeamy have both elaborated on the impact of changes along the different phases of design [1, 2]. Similarly, Morbitzer argues that simulation should be used throughout the design process [3].

© Springer Nature Singapore Pte Ltd. 2022
D. Gerber et al. (Eds.): CAAD Futures 2021, CCIS 1465, pp. 87–106, 2022.
https://doi.org/10.1007/978-981-19-1280-1_6

tool in the early design phase [4–10]. However, outside of the research field, the use of optimization in the early design phase has been limited for reasons that relate to: i) time intensity, ii) interpretability, iii) inherent limitations of the required parametric models, and iv) the elusive nature of performance goals in architectural design.

Time Intensity. One major limitation for applying optimization in architecture is the time intensity of the processes involved [6, 11–14]. Environmental or structural simulations can be computationally expensive. Combined with an optimization process that employs a stochastic search method, such as evolutionary algorithms, the calculation time increases by multiple orders of magnitude. In the early design phase, where it is essential to consider multiple design alternatives quickly, the slow speed of optimization disrupts the exploratory process.

Interpretability. When it comes to interpreting multi-objective optimization results, architects can have difficulties in understanding the solution space [15, 16]. Optimization returns a set of high-performing solutions with their corresponding performances; however, the connection between design parameters and performance tradeoffs is not always apparent [16], offering little intuition to the designer.

Limitations of Parametric Models. Parametric models are widely adopted in architecture; however, their applicability in the early design phase has been questioned [4, 17–19]. Davis offers an extensive analysis of how parametric models have certain limits on the changes they can afford before breaking [20]. To accommodate a major change, such as those that often happen during the conceptual phase, the parametric model would need to be replaced by a new one [9, 21, 22]. However, optimization operates on a pre-determined parametric model, and as a result, it conflicts with the nature of the conceptual design stage.

Nature of Performance Goals in Design. Carrara et al. describes the design process as consisting of three operations [23]: i) definition of the desired set of performance criteria, ii) production of design solutions, and iii) evaluation of expected performance. However, they stress that these operations relate in a non-linear way and coevolve during the design process. Others have also discussed the co-evolution of the problem definition and solution in the design process [11, 12, 16, 18, 24, 25]. Therefore, it is expected that the performance goals, parameters, and constraints will be redefined multiple times during the design process.

Consequently, as long as optimization requires from the designer a high investment in terms of time and cognitive effort to create the parametric abstractions and interpret the results, it cannot seamlessly integrate into the early design phase.

This research suggests an alternative method of providing early-phase performance-driven design assistance for optimally performing geometries in real-time and without the need for parametrization. The method makes use of Machine Learning (ML) generative models. It relies on the navigation of a latent space where the results of a series of optimization processes have been encoded in advance.

This paper describes the suggested method and presents a case study where a Variational Auto-Encoder (VAE) is introduced for the generation of geometries with optimal solar gain properties and pre-determined size. The results show that the VAE was able to generate geometries with optimal or close-to-optimal performance for most simulation contexts from the test set, in a fraction of the corresponding optimization time.

This work makes the following contributions to the area of computational design. A novel method for real-time early-phase performance-driven design assistance is introduced, which does not require parametric models. The use of generative models is suggested for the novel task of generating geometries with optimal performance properties. Empirical evidence is provided, suggesting that a VAE can be used as a generative model for optimally performing geometries.

2 Related Work

2.1 Performance-Driven Design

Simulation. Simulations form the basis of performance-driven design. However, a single or a limited number of simulations is not enough to guide design improvements. Systematic simulations [26] attempted to address this subject, but the complex relationship between performance and parameters related to form, together with the time intensity of the calculations, make this an impractical solution. Some template-based tools attempted to give a solution by enabling quick evaluation of alternatives [3, 27, 28], however, they imposed severe restrictions to the range of supported forms and thus were not adopted by the architectural community. Finally, real-time simulations were found to be helpful during the performance-driven form-finding process [29], however, in cases with large design spaces and multiple performance criteria, further guidance is necessary [30].

Sensitivity Analysis. Sensitivity analysis methods can be used to guide exploration based on a single parametric model [31, 32] or to evaluate multiple alternative parametric models [33–35]. However, it has been argued that they do not provide adequate information to lead to optimally performing solutions [13, 36].

Optimization. Optimization processes identify the parameters of a model that result in optimally performing solutions. They have successfully solved engineering or building science problems [5, 6, 37]. However, despite extensive research on optimization for performance-driven design, such methods have not been widely adopted in the architecture practice [5, 6].

Form Exploration. In order to reconcile the engineering nature of optimization with the more exploratory role that designers tend to give to it [4], some research has suggested interactive optimization [8, 9] for integrating performance with designer preferences. Other work has focused on simulation speed, interactivity, and results visualization [16] through the use of surrogate modeling. Last, some recent work has suggested eliminating the parametric modeling overhead by deploying automatic parametrization and data analysis [17].

2.2 Generative Models

Definition. A generative model is a type of ML model that can learn an estimate of a distribution by observing a set of examples from that distribution, i.e., a training set [38]. Once fully trained, sampling a generative model approximates sampling from the original data distribution. For example, a generative model trained on a dataset of faces will generate new faces when sampled[2].

Latent Space. Some generative models work by learning a mapping of the original data to a lower-dimensional space, called the latent space. For example, the Variational Auto-Encoder [41] (VAE) explicitly learns an encoder and a decoder function that maps the original data to and from a latent space. Naturally, similar data points will be located close in the latent space. This characteristic allows for smooth interpolation of data samples by traversing the latent space or even for the composition of new data with specified properties through latent space vector arithmetic, as demonstrated by Wu et al. in the domain of three-dimensional objects [42].

Applications. In the field of architecture, several attempts have been made to use generative models in the creative phase [43–46]. Most such works used Generative Adversarial Networks [47] (GANs), motivated by some impressive results in the field of computer vision [39, 47–49]. However, the subject of performance has not been previously addressed directly in research related to generative models.

3   Approach

Current practices and previous research reveal a lack of support for performance-driven design in the early form-finding process. Almost all related research approaches performance-driven design through the scope of parametric modeling, which imposes severe restrictions of time-intensity, cognitive load, and premature commitment to specific graph topologies [19, 22] to the creative process.

This research suggests that optimal form-finding can be achieved by navigating the latent space of a generative model. A generative model that addresses a specific set of performance metrics can be trained on a dataset where each data point represents both the problem definition and an optimal solution to the problem. When the trained model is sampled, it will generate a new problem definition and an optimal solution following the learned data distribution. In order to generate an optimal solution to a specific problem definition, a sample can be retrieved from the model, constrained by the problem definition of interest. In practice, this can be achieved through search or navigation of the model's latent space. In addition to generating optimal geometric forms from scratch, the same generative model can also be used to suggest optimally performing alternatives that are as close as possible to user-generated forms. For this task, the user-generated geometry becomes part of the constraints that drive the latent space search.

---

[2] See for example the Progressive GAN model [39] trained on the CelebA dataset [40].

The proposed method addresses the current limitation of time intensity associated with performance-driven design. In addition, the ability of ML methods to deal with high dimensional data is used to work directly with geometries from the modeling environment, eliminating the need for geometry parametrization. Last, the method opens up the potential for intuitive, real-time interactivity in a user-guided search for optimal geometries.

The dataset required to train such a model needs to include a diverse set of optimally performing geometries for a wide range of problem definitions. Such datasets do not exist at the moment and are impossible to collect from the real world, so synthetic datasets with the desired characteristics should be created using existing optimization methods. Since a specific model only addresses a pre-determined set of performance metrics, the term "problem definition" refers to the simulation context that drives the optimization process.

Next, a case study is presented, where the performance of interest is related to the solar gain and the size of the building. The case study allows a detailed development and evaluation of the suggested techniques.

4 Case Study

4.1 Problem Scope

In a typical scenario for the design of a new building, the architect would have information including the location, the plot shape and size, the surrounding buildings, and the program of the building. In performance-driven design, maximizing the performance of interest is of primary concern. Then, in the early design phase, where the focus is on form finding, the problem would be expressed as finding a geometry for the building that maximizes the desired performance, given the simulation context (Fig. 1).

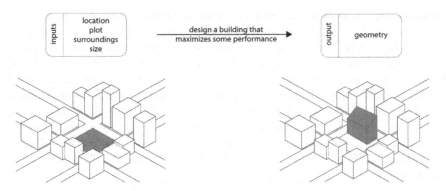

Fig. 1. Optimization: context and expectation for the design of a new building.

In this case study the goal was to generate a geometric form for a building that maximizes the average solar radiation gain during wintertime while keeping the size as close as possible to a predefined target. In more detail, the solar gain objective was defined as the average of the received radiation per area unit of all the mesh faces of a generated geometry. The size was represented by the geometry volume. The location of the building was Boston. Wintertime for the environmental simulation purposes of this case study was defined as any time when the temperature is below 12 C. The plot shape was a square with a side of 10m. In addition, a maximum height of 10m for any building was set.

In this problem, the term boundary condition refers to the configuration of the surrounding buildings, as this was the only part of the solar simulation's boundary condition that varied. The range of the boundary condition was up to one obstructing building on each of the east, south, and west sides of the plot, and up to three total obstructing buildings (Fig. 2). With all obstructions having the same width and height, a total of 342 unique boundary conditions were used.

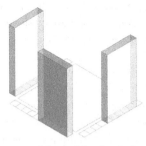

Fig. 2. Range of variable boundary conditions of the simulation. Each of the three parallelepipeds can move on the outlined locations or be omitted.

4.2 Data Generation

Geometry Representation. An optimization algorithm was used during the data generation phase together with a parametric model that generated the geometry. In architectural design, it is common for a parametric model to be created using high-level concepts, such as box or tower, and their transformations, such as scale or twist angle. However, a more neutral and low-level geometry representation is more suitable when no conceptual decisions are assumed. Therefore, a heightmap was used as the geometry generation model. Each parameter of the model controls the height of a point on a two-dimensional grid. This representation provides a simple and intuitive way to describe geometries, with a fair amount of flexibility. One limitation is that it cannot describe certain three-dimensional forms. For example, a height map cannot encode information about cantilevers.

Optimization. *Objective.* As described in the problem definition, the performance goal was to maximize the solar gain during the winter months. At the same time, the total volume was constrained to remain as close as possible to a predefined target ($v = 100\,\mathrm{m}^3$). The volume constraint was used as a proxy for the architectural program, which would prescribe the total surface area in a real-life scenario. In practice, the volume constraint was transformed to a minimization objective, calculated as the squared difference of the target volume from the current geometry volume.

When optimizing a problem with multiple objectives, there are two major approaches. The first is to use scalarization with a single objective algorithm; the second is to use a multi-objective algorithm. This case study used scalarization, combining the two objectives into a single, minimization objective as described in Eq. (1), where a geometry is a mesh instance, J is the minimization objective, AvgRadiation(geometry) evaluates the average of the wintertime incident radiation on all the faces of the geometry mesh, Vol_target is the desired volume, and Vol(geometry) calculates the volume of a geometry.

$$J(geometry) \; = \; - \, AvgRadiation(geometry) \, + \, (Vol\_target - Vol(geometry))^{2*}10^{-3}$$

$$(1)$$

Optimal Solutions Selection. When solving a problem with multiple objectives, the Pareto front, i.e., the set of non-dominated solutions, has been commonly used to identify the best-performing solutions [15, 50–55]. Therefore, the individual objectives on each step of the optimization were recorded, and after the optimization was complete, the Pareto front was calculated, as suggested in relevant work [56]. For each optimization problem, i.e., for each of the 342 boundary conditions, a total of 10 optimal results were selected to form a dataset.

Implementation. The solar radiation calculation was performed using the open-source plugin Ladybug [57], inside the visual programming platform Grasshopper3d in McNeel's Rhinoceros 3d modeling software. A communication module for Grasshopper was developed using web sockets, connecting the parametric model and the solar simulations to an external optimization algorithm. The optimization algorithm was a customized implementation of Simulated Annealing (SA). The whole workflow was controlled by a command-line program that called the optimization algorithm and obtained the performance results from Grasshopper.

The 342 optimizations were run on a desktop computer for a fixed number of optimization steps (n = 3000). Each optimization required an average of approximately 20 min to complete. After selecting the ten best solutions for each boundary condition (Fig. 3), a dataset of 3420 pairs of boundary condition – optimal geometry was created.

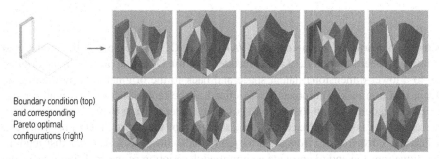

Boundary condition (top) and corresponding Pareto optimal configurations (right)

Fig. 3. Example of a boundary condition and the selected Pareto optimal geometries.

4.3 Learning

Generative Models. Two popular types of generative models in ML are the Generative Adversarial Networks (GAN) and the Variational Auto-encoders (VAE). While either type could be used with the suggested method, this case study focuses on the use of a VAE because of the simpler setup and its natural ability to model a latent space. This last feature is essential since it is the navigation of this latent space that enables sampling optimal geometries for specific boundary conditions.

Data Format. In order to use the generated data with an ML model they first had to be converted to vectors. Since the boundary conditions in this problem are geometric, both boundary conditions and optimal geometries were incorporated into a single geometric representation. In ML, there are three primary ways to describe geometric data [58]: i) image-based (single or multi-view), ii) voxel-based, and iii) point clouds. Image-based methods are currently the most robust and well-developed methods and compared to the original parametric description of the geometries, they allow better modeling of the spatial relationships between the individual parameters of the vector representation through the use of convolutions in the learning model.

Since all geometries in the dataset were created using a heightmap, a single depth map from a top view was used to describe the data (Fig. 4). Multiple different image resolutions were considered for the depth map before a resolution of 16X16 pixels was selected based on initial results when using the VAE.

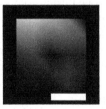

Fig. 4. From left to right: i) optimized geometry with solar radiation colors – SE Isometric, ii) optimized geometry with depth map – SE Isometric, iii) top view projection of (ii) – final format of data.

Using the Variational Auto-encoder (VAE). A VAE is a probabilistic model that learns an encoder function E(x) and a decoder function D(z), mapping from the original data to a lower-dimensional latent space and back to the original data by training on a reconstruction task. The objective is defined as the reconstruction loss with a regularizer. The reconstruction loss encourages the decoder to learn to reconstruct the data. The regularizer is the Kullback-Leibler divergence of the approximate posterior (i.e., the encoder's distribution) from the prior (commonly chosen as a Gaussian distribution) [41]. Equations (2) and (3) provide a simplified description of the VAE in terms of the encoder and decoder functions, where x is an input vector, z is the mapping of x in the latent space, g_φ is a reparametrization function[3], and y is the reconstruction of x.

$$z = E(x), z' = g_\varphi(z), y = D(z') \tag{2}$$

$$y = D(g_\varphi(E(x))) \tag{3}$$

After a VAE has been trained, the decoder function can be isolated and used as a generative model that samples the latent space and generates data instances from the original distribution. In this work, the decoder is used to generate images (depth maps) of boundary conditions and corresponding optimal geometries.

The retrieval of data instances for a specific boundary condition was achieved as follows. First, a loss function Lb was defined as the distance of a generated instance's boundary condition from the desired boundary condition, described in (4). Next, the boundary condition – i.e., the surrounding buildings abstracted to simple parallelepipeds – was translated to a depth map, following the same data format on which the VAE was trained, but without any corresponding optimal geometry. The desired geometries were found by solving the optimization problem (5) of finding the sample z in the latent space, for which the decoder produces a depth map that minimizes the boundary condition loss L_b.

$$L_b(target\_boundary, y) = Distance(target\_boundary, Boundary\_Condition(y)) \tag{4}$$

$$J(z) = L_b(target\_boundary, D(z)) \tag{5}$$

Since the decoder – and consequently the loss L_b – is a differentiable function, problem (5) can be solved using gradient descent. Vector z is initialized as a random sample of the latent space. The loss L_b is calculated, and its gradient is backpropagated to the decoder's input, resulting in an update of z. Several updates are performed, until convergence. Using gradient descent in this process is of particular importance because it enables high-speed retrieval of the appropriate latent space vector, in contrast to alternative search methods such as stochastic sampling.

Model Architecture. *Training.* The VAE was implemented as a convolutional neural net (Fig. 5). The encoder consists of two convolutional layers followed by a fully connected layer with output size 32. This output corresponds to the mean and standard deviation of a normal distribution of dimension 16, so the latent space is 16-dimensional. The decoder – or generative model – follows a mirrored structure of the encoder. The input-output of the VAE is a 16 × 16 grayscale image. The reconstruction loss was defined as the L2 distance (squared difference) of the input-output images.

Inference. For the process of finding appropriate boundary condition – optimal geometry instances for a specific boundary condition, a boundary condition loss function was defined. This function calculates how close a generated image's boundary condition is to the desired boundary condition. The generated image is first masked to only leave the boundary condition visible. Then, the masked image is compared to the desired boundary condition image (Fig. 6). The distance of the two images was calculated using the sigmoid cross-entropy.

---

[3] The reparametrization is an essential component of the VAE, but only mentioned here for reasons of completeness. For details we direct the interested reader to [41].

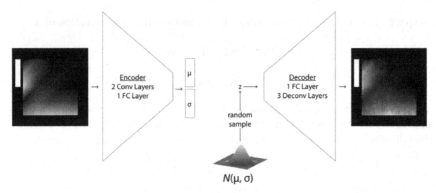

Fig. 5. VAE architecture used for training.

Training. The total dataset was split into two parts, a training set containing 90% of the data (3,080 data points that belong to 308 boundary conditions) and a test set containing the rest 10% of the data (340 data points that belong to 34 unique boundary conditions). The split was done through random selection and care to place all ten data points from the same boundary condition in the same group, ensuring that the boundary conditions found in the test dataset have not been encountered during the training. The VAE model was implemented using the Python library TensorFlow [59] and trained for 1000 epochs, using the Adam optimizer [60] and batch size 32. The loss function was implemented as the single sample Monte Carlo estimate of the expectation [61], where the reconstruction loss is the squared difference of the input-output images. Only minor improvements in the loss were gained between 200 and 1000 epochs. At 1000 epochs, a validation loss of 9.3 was achieved.

Inference. Inferred geometries were generated for the 34 unique boundary conditions of the test set using gradient descent. Due to the random initialization of the process and the non-convex shape of the latent space, different geometries can be obtained for the same boundary condition through repeated optimizations. For each of the boundary conditions, 100 geometries were generated. The optimization algorithm Adam was used with a learning rate of 0.02 for 400 iterations. Convergence was typically observed in less than 200 iterations.

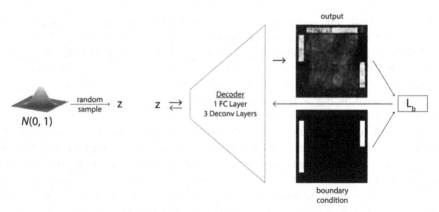

Fig. 6. Inference using the decoder part of the VAE.

5 Results

The method introduced for predicting optimal geometries for specific boundary conditions relies on the two underlying processes: i) a high-quality mapping of the data to a latent space, and ii) the successful navigation of this low-dimensional space.

The first process requires a mapping function that can encode all the critical information, as well as a well-structured latent space that enables the generation of new data through successful interpolations. Both processes are evaluated by assessing the quality of reconstructions that the VAE produces for the test set. If the VAE was used for a visual task, the reconstruction quality would refer to the similarity of the input and output images. However, since the overall goal of this problem relates to building performance, the evaluation was performed with respect to the specific performance goals of solar radiation and volume compliance, as they have been detailed in Sect. 4.2. The geometries derived from the VAE-reconstructed depth maps are expected to perform close to the optimization-derived geometries.

Similarly, the navigation of the latent space is evaluated based on the performance of the geometries generated, constrained by the boundary conditions in the test set, using the process described in Sect. 4.3.

Potential inaccuracies in the actual performance metrics of the dataset may have been introduced during the resampling process, when meshes based on a 5×5 heightmap were encoded to 16×16 depth map images. To avoid this issue when comparing SA-optimized with VAE-generated geometries, the reported performance of both the test set ground truth and the test set inferences was calculated following a common process based on the 16×16 depth map encodings of the geometry.

5.1 Reconstruction Performance

The reconstruction results for all 34 boundary conditions of the test set were coded into three categories after careful observation of the per-boundary condition scatterplots and a comparison of the mean performances. The results for 5 boundary conditions were coded as type a: performance very close to the test set, 27 were coded as type b:

performance on one axis similar to the test set and the other axis better than the test set, and 2 were coded as type c: performance better than the test set. Diagrams a, b, and c of Fig. 7 show a representative sample from each type. While some individual geometries with bad overall performance were generated, for each boundary condition, the mean performance of the generated geometries was similar or better than that of the test set.

In more detail, in Fig. 7, the performance of the test set samples is plotted against the performance of their VAE-reconstructions. The diagrams a, b, and c each correspond to a different boundary condition. In the top row, each geometry instance corresponds to a point on the scatterplot. In the bottom row, the mean and standard deviation of each group of geometries are plotted. The best overall performance would be located in the bottom left corner of the plot. A well-trained VAE should produce reconstructions with performances close to those of the test set. Because the VAE is a probabilistic model, multiple reconstructions were sampled for each instance of the test set ($n = 100$). Additionally, the scatterplot includes the performance of two random geometry generators as baselines for comparison: one uniform random and one Gaussian ($\mu = 5$ m, $\sigma = 1.5$ m). Two more baselines are included, coming from simple heuristics: a geometry with a flat horizontal roof and volume equal to the target (optimal volume deviation) and a geometry with a tilted roof at 42° facing south (optimal solar gain).

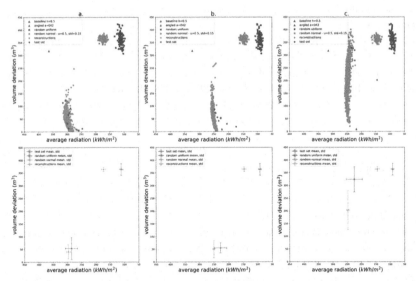

Fig. 7. Performance of geometries for three representative cases. Each column a, b, c includes results for a single boundary condition. Geometries from the test set, VAE-reconstructions, baselines, and random generators.

Figure 7a indicates that most reconstructions perform close to the test set or along a curve close to the Pareto front line as implied by the test set samples. The mean performance of the reconstructed geometries is very close to that of the test set geometries. However, in Fig. 7b, the distribution of the reconstructions does not follow the one of the test set. The mean solar radiation performance of the reconstructions is better than that of the test set, while the mean volume deviation is approximately the same. Last, in Fig. 7c, the performance of the reconstructions is superior in both axes.

5.2 Inference Performance

To evaluate the process of navigating the latent space, the performance of the inferred geometries is compared against the performance of the reconstructed geometries. Figure 8 shows representative examples of optimal geometry inference for three different boundary conditions, with varying success. In Fig. 8a, the inferred geometries overlap with the reconstructed ones, which means that the optimal geometries as encoded through the VAE were successfully found. For other boundary conditions, such as the one in Fig. 8b, the inference is not successful: there is a wide spread of performance for the inferred geometries, with their mean performance located far from that of the reconstructions. Last, in Fig. 8c, the performance of many inferred geometries is better than the one of the reconstructions.

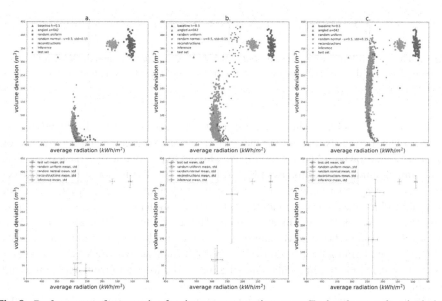

Fig. 8. Performance of geometries for three representative cases. Each column a, b, c includes results for a single boundary condition. Geometries from the test set, VAE-inferences, VAE-reconstructions, baselines, and random generators.

The results for all 34 boundary conditions of the test set were coded in the three representative types. Results for 28 boundary conditions were similar to Fig. 8a, i.e., successful, 5 were found to be similar to Fig. 8b, i.e., not successful, and Fig. 8c is the only case of this type.

5.3 Hypervolumes

To further evaluate the generated geometries' performances for both the volume and the solar radiation objectives, the hypervolumes of the Pareto fronts were calculated, as shown in Fig. 9. For each boundary condition, three Pareto fronts are compared: the ground truth (test set), the Pareto front of the reconstructed geometries, and the Pareto front of the inferred geometries. The hypervolumes of all three Pareto fronts were calculated using the Python library pymoo [62], using a common reference point for each boundary condition.

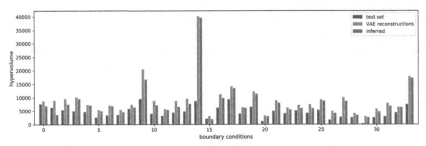

Fig. 9. The hypervolumes of the Pareto front of the VAE-reconstructions and the VAE-inferences are plotted against the hypervolume of the test set for each of the 34 unique boundary conditions in the test set.

The bar graph in Fig. 9 shows that for most boundary conditions, the hypervolume of the reconstructions is higher than that of the test set. VAE-generated geometries for boundary condition no. 14 have a significantly higher hypervolume than the test set. This is the same boundary condition as in Fig. 7c and Fig. 8c. The reason is that the test set geometries for boundary condition 14 are far from optimal. The mean performance of the volume objective is close to random, as the optimization process that generated these geometries got stuck to some local optimum. However, the VAE was able to generalize correctly from higher-quality examples and generated better performing geometries than those found through optimization. These results indicate that the VAE has successfully interpolated the training data, allowing the model to generalize from the training examples to new cases.

Finally, the inferred geometries Pareto front has a hypervolume close to that of the reconstructed ones for most boundary conditions. This confirms that the latent space navigation method, using gradient descent, can successfully find the latent vectors that generate optimally performing geometries.

6 Discussion

6.1 VAE Sampling for Optimal Solar Gain Performance

The results suggest that a VAE trained on optimally performing geometries can generate geometries with similar performance properties for new simulation contexts. To that extent, the VAE can potentially replace a computationally expensive optimization process, offering a drastic speed improvement. For reference, a single geometry in the dataset typically took approximately 20 min to get optimized, while the introduced method used less than 5 s.

In the case study, the VAE-generated geometries through the latent space search had a higher hypervolume than those generated through simulated annealing optimization for most boundary conditions. This suggests that, in general, the proposed ML-based method generated higher and more consistent quality results than the optimization method. A potential reason is that during training, as the VAE learns to compress the inputs to a lower-dimensional space, it identifies the essential features to prioritize during the compression. As a result, the higher frequencies – or the noisy information – tend to get lost. On the other hand, a stochastic optimization process such as the SA tends to generate noisy results. Through this process, and by generalizing from all of the training data, the VAE may have filtered out noisy geometric features that were decreasing the solar performances. Finally, the results of the VAE model could be further improved using more extensive and higher quality datasets and hyperparameter tuning.

6.2 Beyond Quick Optimization

Apart from quick automatic optimization, the suggested method opens up the potential for optimizing interactively, directly inside the 3d modeling environment. The way that the generative model has been used frees it from any tie to a specific parametric model and any associated limitations. Designer intentions regarding geometric form can be indicated through modeling and used to guide the generation process. For example, a user-designed geometry can easily be encoded as a depth map and guide the latent space search with an appropriate modification of the loss function L_b.

6.3 Generalizability

The case study demonstrated how a VAE can generate geometries with optimal solar gain and predefined size. However, the suggested method of optimal geometry generation through latent space navigation of a generative model can be used with any performance metric of interest. The optimization workflow described in Sect. 4.2 could be followed, but individual components such as the geometry representation or the optimization algorithm may be updated to match the needs of each specific problem.

Furthermore, the suggested method is not limited to a single performance objective. The case study already hints at the use of multiple objectives, using the volume target.

6.4 Limitations

Concerning the case study, the problem has been simplified and limited in scope in order to facilitate the evaluation of the overall method as well as its individual components. In order to address problems of real-world complexity, a different geometry representation and a more extensive dataset may be needed.

One overall limitation of the suggested method is that the generative model may generate unpredictable results for simulation contexts that are entirely outside of the range of the training set. Appropriate coding of the boundary conditions may alleviate this issue. Additionally, similar to an optimization process, the performance objectives must be specified in advance, i.e., during the training process. The adaptability of the suggested method to changing objectives remains an open question.

7 Conclusion

This research introduced a novel method for optimal geometry generation that does not require the designer to use a parametric model. The method aims to provide a more intuitive and interactive alternative for guiding the early phase of performance-driven design than currently available tools. The case study demonstrated the feasibility of using a VAE as a generative model for optimally performing geometries. Future work can focus on expanding the range of the problem variables with real-world problem definition complexity and datasets.

In order to take advantage of the full potential of the suggested method and meet the promise for early-phase design support, future work will also focus on different ways of presenting the results and modes of interactivity inside the modeling environment.

Acknowledgements. The first author is partially funded by an Onassis Scholarship (Scholarship ID: F ZO 002/1 – 2018/2019).

References

1. Paulson, B.C., Jr.: Designing to reduce construction costs. J. Constr. Div. **102**, 587–592 (1976)
2. Collaboration, Integrated Information and the Project Lifecycle in Building Design, Construction and Operation. Construction Users Roundtable (2004)
3. Morbitzer, C.A.: Towards the Integration of Simulation into the Building Design Process (2003). http://www.esru.strath.ac.uk/Documents/PhD/morbitzer_thesis.pdf
4. Bradner, E., Iorio, F., Davis, M.: Parameters tell the design story: ideation and abstraction in design optimization. In: 2014 Proceedings of the Symposium on Simulation for Architecture and Urban Design, p. 26. Society for Computer Simulation International, Tampa, FL, USA (2014)
5. Evins, R.: A review of computational optimisation methods applied to sustainable building design. Renew. Sustain. Energy Rev. **22**, 230–245 (2013). https://doi.org/10.1016/j.rser.2013.02.004
6. Tian, Z., Zhang, X., Jin, X., Zhou, X., Si, B., Shi, X.: Towards adoption of building energy simulation and optimization for passive building design: a survey and a review. Energy Build. **158**, 1306–1316 (2018). https://doi.org/10.1016/j.enbuild.2017.11.022
7. Caldas, L.: An evolution-based generative design system : using adaptation to shape architectural form (2001). http://dspace.mit.edu/handle/1721.1/8188
8. Mueller, C.T., Ochsendorf, J.A.: Combining structural performance and designer preferences in evolutionary design space exploration. Autom. Constr. **52**, 70–82 (2015). https://doi.org/10.1016/j.autcon.2015.02.011
9. Turrin, M., von Buelow, P., Stouffs, R.: Design explorations of performance driven geometry in architectural design using parametric modeling and genetic algorithms. Adv. Eng. Inform. **25**, 656–675 (2011). https://doi.org/10.1016/j.aei.2011.07.009

10. Nagy, D., et al.: Project discover: an application of generative design for architectural space planning. In: 2017 Proceedings of the Symposium on Simulation for Architecture & Urban Design, p. 8, Toronto, Canada (2017)

11. Attia, S., Hamdy, M., O'Brien, W., Carlucci, S.: Assessing gaps and needs for integrating building performance optimization tools in net zero energy buildings design. Energy Build. **60**, 110–124 (2013). https://doi.org/10.1016/j.enbuild.2013.01.016

12. Wortmann, T., Costa, A., Nannicini, G., Schroepfer, T.: Advantages of surrogate models for architectural design optimization. AI EDAM. **29**, 471–481 (2015). https://doi.org/10.1017/S0890060415000451

13. Lin, S.-H.E., Gerber, D.J.: Designing-in performance: A framework for evolutionary energy performance feedback in early stage design. Autom. Constr. **38**, 59–73 (2014). https://doi.org/10.1016/j.autcon.2013.10.007

14. Soares, N., et al.: A review on current advances in the energy and environmental performance of buildings towards a more sustainable built environment. Renew. Sustain. Energy Rev. **77**, 845–860 (2017). https://doi.org/10.1016/j.rser.2017.04.027

15. Ashour, Y.S.E.-D.: Optimizing Creatively in Multi-Objective Optimization (2015). http://search.proquest.com/docview/1759161420/abstract/E1E30D170B2E4A04PQ/1

16. Wortmann, T., Schroepfer, T.: From optimization to performance-informed design. In: 2019 Proceedings of the Symposium on Simulation for Architecture & Urban Design, p. 8, Georgia Tech, College of Design, School of Architecture, Atlanta, GA, USA (2019)

17. Brown, N.C., Mueller, C.T.: Design variable analysis and generation for performance-based parametric modeling in architecture. Int. J. Archit. Comput. **17**, 36–52 (2019). https://doi.org/10.1177/1478077118799491

18. Harding, J., Joyce, S., Shepherd, P., Williams, C.: Thinking topologically at early stage parametric design. In: Hesselgren, L., Sharma, S., Wallner, J., Baldassini, N., Bompas, P., Raynaud, J. (eds.) Advances in Architectural Geometry 2012, pp. 67–76. Springer, Vienna (2013). https://doi.org/10.1007/978-3-7091-1251-9_5

19. Harding, J.E., Shepherd, P.: Meta-parametric design. Des. Stud. **52**, 73–95 (2017). https://doi.org/10.1016/j.destud.2016.09.005

20. Davis, D.: Modelled on software engineering: flexible parametric models in the practice of architecture (2013). https://researchbank.rmit.edu.au/view/rmit:161769

21. Holzer, D., Hough, R., Burry, M.: Parametric design and structural optimisation for early design exploration. Int. J. Archit. Comput. **5**, 625–643 (2007). https://doi.org/10.1260/147807707783600780

22. Toulkeridou, V.: Steps towards AI augmented parametric modeling systems for supporting design exploration. In: Blucher Design Proceedings, pp. 81–92. Editora Blucher, Porto, Portugal (2019). https://doi.org/10.5151/proceedings-ecaadesigradi2019_602

23. Carrara, G., Kalay, Y.E., Novembri, G.: A computational framework for supporting creative architectural design. In: Evaluating and Predicting Design Performance. Wiley, New York, N.Y (1991)

24. Dorst, K., Cross, N.: Creativity in the design process: co-evolution of problem–solution. Des. Stud. **22**, 425–437 (2001). https://doi.org/10.1016/S0142-694X(01)00009-6

25. Singh, V., Gu, N.: Towards an integrated generative design framework. Des. Stud. **33**, 185–207 (2012). https://doi.org/10.1016/j.destud.2011.06.001

26. Shaviv, E.: Integrating energy consciousness in the design process. Autom. Constr. **8**, 463–472 (1999). https://doi.org/10.1016/S0926-5805(98)00101-0

27. Attia, S., Gratia, E., De Herde, A., Hensen, J.L.M.: Simulation-based decision support tool for early stages of zero-energy building design. Energy Build. **49**, 2–15 (2012). https://doi.org/10.1016/j.enbuild.2012.01.028

28. Ochoa, C.E., Capeluto, I.G.: Advice tool for early design stages of intelligent facades based on energy and visual comfort approach. Energy Build. **41**, 480–488 (2009). https://doi.org/10.1016/j.enbuild.2008.11.015

29. Jones, N.L., Reinhart, C.F.: Effects of real-time simulation feedback on design for visual comfort. J. Build. Perform. Simul. 1–19 (2018). https://doi.org/10.1080/19401493.2018.1449889

30. Wortmann, T.: Surveying design spaces with performance maps: a multivariate visualization method for parametric design and architectural design optimization. Int. J. Archit. Comput. **15**, 38–53 (2017). https://doi.org/10.1177/1478077117691600

31. Østergård, T., Jensen, R.L., Maagaard, S.E.: Building simulations supporting decision making in early design – a review. Renew. Sustain. Energy Rev. **61**, 187–201 (2016). https://doi.org/10.1016/j.rser.2016.03.045

32. Tian, W.: A review of sensitivity analysis methods in building energy analysis. Renew. Sustain. Energy Rev. **20**, 411–419 (2013). https://doi.org/10.1016/j.rser.2012.12.014

33. Sileryte, R., D'Aquilio, A., Stefano, D.D., Yang, D., Turrin, M.: Supporting exploration of design alternatives using multivariate analysis algorithms. In: 2016 Proceedings of the Symposium on Simulation for Architecture and Urban Design, p. 8, London, United Kingdom (2016)

34. Yang, D., Sun, Y., di Stefano, D., Turrin, M.: A computational design exploration platform supporting the formulation of design concepts. In: 2017 Proceedings of the Symposium on Simulation for Architecture & Urban Design, p. 8, Toronto, Canada (2017)

35. Yang, D., Ren, S., Turrin, M., Sariyildiz, S., Sun, Y.: Multi-disciplinary and multi-objective optimization problem re-formulation in computational design exploration: a case of conceptual sports building design. Autom. Constr. **92**, 242–269 (2018). https://doi.org/10.1016/j.autcon.2018.03.023

36. Pantazis, E., Gerber, D.: A framework for generating and evaluating façade designs using a multi-agent system approach. Int. J. Archit. Comput. **16**, 248–270 (2018). https://doi.org/10.1177/1478077118805874

37. Nguyen, A.-T., Reiter, S., Rigo, P.: A review on simulation-based optimization methods applied to building performance analysis. Appl. Energy **113**, 1043–1058 (2014). https://doi.org/10.1016/j.apenergy.2013.08.061

38. Goodfellow, I.: NIPS 2016 Tutorial: Generative Adversarial Networks. arXiv:1701.00160 [cs]. (2017)

39. Karras, T., Aila, T., Laine, S., Lehtinen, J.: Progressive Growing of GANs for Improved Quality, Stability, and Variation. arXiv:1710.10196 [cs, stat]. (2018)

40. Liu, Z., Luo, P., Wang, X., Tang, X.: Deep learning face attributes in the wild. In: Proceedings of International Conference on Computer Vision (ICCV) (2015)

41. Kingma, D.P., Welling, M.: Auto-Encoding Variational Bayes. arXiv:1312.6114 [cs, stat]. (2014)

42. Wu, J., Zhang, C., Xue, T., Freeman, B., Tenenbaum, J.: Learning a probabilistic latent space of object shapes via 3d generative-adversarial modeling. In: Advances in Neural Information Processing Systems, pp. 82–90 (2016)

43. Huang, W., Zheng, H.: Architectural drawings recognition and generation through machine learning. In: Proceedings of the 38th Annual Conference of the Association for Computer Aided Design in Architecture (ACADIA), p. 10, Mexico City, Mexico (2018)

44. Liu, H.L.: An anonymous composition. In: ACADIA 19: Ubiquity and Autonomy [Proceedings of the 39th Annual Conference of the Association for Computer Aided Design in Architecture (ACADIA) ISBN 978-0-578-59179-7] (The University of Texas at Austin School of Architecture, Austin, Texas 21–26 October, 2019), pp. 404–411. CUMINCAD (2019)

45. Mohammad, A.B.: Hybrid elevations using GAN Networks. In: ACADIA 19:UBIQUITY AND AUTONOMY [Proceedings of the 39th Annual Conference of the Association for Computer Aided Design in Architecture (ACADIA) ISBN 978-0-578-59179-7] (The University of Texas at Austin School of Architecture, Austin, Texas 21–26 October, 2019) pp. 370–379. CUMINCAD (2019)

46. Zhang, H., Blasetti, E.: 3D architectural form style transfer through machine learning. In: CAADRIA 2020, p. 10 (2020)

47. Goodfellow, I., et al.: Generative adversarial nets. In: Ghahramani, Z., Welling, M., Cortes, C., Lawrence, N.D., and Weinberger, K.Q. (eds.) Advances in Neural Information Processing Systems 27, pp. 2672–2680. Curran Associates, Inc. (2014)

48. Brock, A., Donahue, J., Simonyan, K.: Large Scale GAN Training for High Fidelity Natural Image Synthesis. arXiv:1809.11096 [cs, stat]. (2019)

49. Isola, P., Zhu, J.-Y., Zhou, T., Efros, A.A.: Image-To-Image translation with conditional adversarial networks. In: Presented at the Proceedings of the IEEE Conference on Computer Vision and Pattern Recognition (2017)

50. Caldas, L.G., Norford, L.K.: Shape generation using pareto genetic algorithms: integrating conflicting design objectives in low-energy architecture. Int. J. Archit. Comput. 1, 503–515 (2003). https://doi.org/10.1260/147807703773633509

51. Lin, S.-H., Gerber, D.J.: Evolutionary energy performance feedback for design: multidisciplinary design optimization and performance boundaries for design decision support. Energy Build. 84, 426–441 (2014). https://doi.org/10.1016/j.enbuild.2014.08.034

52. Magnier, L., Haghighat, F.: Multiobjective optimization of building design using TRNSYS simulations, genetic algorithm, and Artificial Neural Network. Build. Environ. 45, 739–746 (2010). https://doi.org/10.1016/j.buildenv.2009.08.016

53. Méndez Echenagucia, T., Capozzoli, A., Cascone, Y., Sassone, M.: The early design stage of a building envelope: multi-objective search through heating, cooling and lighting energy performance analysis. Appl. Energy 154, 577–591 (2015). https://doi.org/10.1016/j.apenergy.2015.04.090

54. Wang, W., Zmeureanu, R., Rivard, H.: Applying multi-objective genetic algorithms in green building design optimization. Build. Environ. 40, 1512–1525 (2005). https://doi.org/10.1016/j.buildenv.2004.11.017

55. D'Cruz, N., Radford, A.D., Gero, J.S.: A pareto optimization problem formulation for building performance and design. Eng. Optim. 7, 17–33 (1983). https://doi.org/10.1080/03052158308960626

56. Wortmann, T.: Opossum: introducing and evaluating a model-based optimization tool for grasshopper. In: Protocols, Flows and Glitches, Proceedings of the 22nd International Conference of the Association for Computer-Aided Architectural Design Research in Asia (CAADRIA), pp. 283–293. The Association for Computer-Aided Architectural Design Researchin Asia (CAADRIA), Hong Kong (2017)

57. Roudsari, M.S., Pak, M., Smith, A., others: Ladybug: a parametric environmental plugin for grasshopper to help designers create an environmentally-conscious design. In: Proceedings of the 13th international IBPSA conference held in Lyon, France Aug, pp. 3128–3135 (2013)

58. Lun, Z., Gadelha, M., Kalogerakis, E., Maji, S., Wang, R.: 3D Shape Reconstruction from Sketches via Multi-view Convolutional Networks. arXiv:1707.06375 [cs] (2017)

59. Abadi, M., ET AL.: Tensorflow: A system for large-scale machine learning. In: 12th Symposium on Operating Systems Design and Implementation, pp. 265–283 (2016)

60. Kingma, D.P., Ba, J.: Adam: A Method for Stochastic Optimization. arXiv:1412.6980 [cs]. (2017)

61. Convolutional Variational Autoencoder|TensorFlow Core, https://www.tensorflow.org/tutori
    als/generative/cvae. Accessed 27 Feb 2021
62. Blank, J., Deb, K.: Pymoo: multi-objective optimization in Python. IEEE Access. **8**, 89497–
    89509 (2020)

Art Places and Their Impact on Property Prices of Condominiums in Singapore

Xinyu Zeng$^{(\boxtimes)}$ ⓘ and Bige Tunçer$^{(\boxtimes)}$ ⓘ

Singapore University of Technology and Design, 8 Somapah Road, Singapore 487372, Singapore
xinyu_zeng@mymail.sutd.edu.sg, Bige_tuncer@sutd.edu.sg

Abstract. During its formative years, the designation of spaces for arts was often deemed to be a luxury in Singapore, which is a land-scarce country. As her economy progresses to one of the world's highest GDP per capita, art development is now deemed as a part of the greater service sector. The idea of how arts have any actual positive impact on the economy is often based on a theoretical discussion but with little data-driven evidence. As art development in Singapore is in a rather early stage, it relies strongly on government policies and support. This paper explores if arts can have a positive impact on Singapore's condominium projects to quantify the economic benefits rather than to tackle the macro-economic benefits, which might be compounded with other factors. The quantitative methods used in this paper include comparative methods using GIS, single factor regression, and two-factor regression.

Keywords: Art places · Condominium property price · Quantitative methods · GIS

1 Introduction

Singapore is a major global city with one of the highest GDP per capita and a financial hub in the world. Kong (2015) highlighted that a global city is not only a site of economic activities but also has increasing needs for creative production and consumption of culture and arts. However, the art and culture developments in Singapore seem not to match its economic growth, as Chang (2000) stated that Singapore still has challenges to achieve its ambitious goal of being the "Global City for the Arts".

The gap between arts and economic development in Singapore has a historical reason. In the early days of Singapore's independence, faced with a complex international situation, and scarce land area and resources, survival was the most important issue. Arts were regarded as a luxury and were encouraged not to have. At the start of Singapore's independence, her immediate concerns were mostly pragmatic ones. As highlighted by Lim (2019), Singapore needed a "rugged society" that can help to contribute to Singapore's economic growth instead of chasing "rock idols and pop dreams" during the formative years. As Singapore's economy began to take off in the 1980s, arts were planned as a creative field that would contribute to the economy and be supported by the state. According to Kong (2012), Singapore started to view arts as part of the greater

© Springer Nature Singapore Pte Ltd. 2022
D. Gerber et al. (Eds.): CAAD Futures 2021, CCIS 1465, pp. 107–124, 2022.
https://doi.org/10.1007/978-981-19-1280-1_7

service sector to generate economic benefits as Singapore progressed from a manufacturing economy to a service-oriented economy in the late 1980s to 1990s and as living conditions improved.

Singapore's arts rely heavily on government policies, and government policies are selective and cannot be supported without supervision. Chong (2014) summarized the art development in Singapore as "bureaucratic imagination of the art". In other words, art development in Singapore is top-down and planners only support the development of arts and its generic that the planners deem to support Singapore's progress towards a global city. This may on the other hand hinder the diversity of Singapore's artworks.

Compared with other disciplines, the definition of arts is inherently vague and fuzzy. It is not difficult to understand that in most studies on the importance of arts, there seems to be a lack of quantitative studies.

This research primarily aims to quantify the economic impact of art development in Singapore and can hopefully create a dialogue with government authorities for their cultural planning processes. This study intends to quantify the impact of art by refining the questions to more specific areas, such as focusing on the art places and their immediate impact on the property price of condominiums within the catchment area of art places.

The economic impact from the creative sectors can be classified at different levels, such as GDP and employment rate, innovation industry, and cultural identity. Due to the numerous and intricate factors affecting the economy, it is difficult to quantify the contribution of arts to the economy alone. For example, it is understood that art places might improve tourism revenue. However, the actual impact of art places on tourism revenue is difficult to quantify, as there are many spillover effects. As such, in this research, property price is singled out and regarded as a projection of economic impact to be studied. Quantitative studies are done by comparing the property prices within and outside the catchment area of art places, as well as studying the impact on the surrounding property prices before and after the opening of the art places.

The selection of 'art places' for this study is not straightforward. The definition of arts is broad, as it can be regarded as the unique creation from artists and expression of identity. This definition encompasses a diversity of genres, such as theatre, dance, music, traditional arts, and literary arts. Arts could also be seen as part of the creative cluster where individual creativity, skill, and talent can generate wealth and create jobs. Hence, it is difficult to conduct a comprehensive study that covers every aspect of arts. Therefore, art places in this study are selected as an entry point of the artistic concept, as this is more tangible and understandable compared to other art forms.

Chanuki (2016) also highlighted that there is a significant lack of quantitative evidence to support arts help to improve economic conditions of urban neighborhoods despite many inherently believe it. The approach adopted in his study uses the metadata of geotagged photographs of arts in London neighborhoods and compared them to test the relative gain in property prices.

Fabiana and Pierfrancesco (2019) believe that street arts can add socio-economic value to properties, and used Naples, Italy as a city for a case study. In their research, theoretical quantitative methods are discussed but the actual numbers are not presented. This again illustrated the actual challenges faced by researchers on quantifying art places' impact on the economy and urban environment.

My research seeks to use advanced computer-aided methodologies to address the gaps between the quantitative evidence and common understanding of how art places add value to the urban environment and help to improve the quality of the built environment.

2 Methodology

The key variables used in this study are (1) property prices of all new apartments sold and (2) identified art places or art clusters. The property prices are then adjusted using the property price index collected by Singapore authorities to remove potential inflation and gentrification effects.

Data Processing is also conducted to aggregate the time-series of the property price of each town after removal of the inflation and gentrification effects.

The data sampling used in this study is intended to be comprehensive and covers all possible data points available from the public domain.

The methodologies used are from a (1) observational research method and (2) statistical regression approach.

For the observational research method, the property price of apartments within the catchment area of the art places are compared to the property price of their respective towns in the same period. Another observational comparison is to compare the property prices in the town before and after the art places are opened in the affected townships; furthermore, the difference of the property prices observed has already accounted for the inflation and gentrification effects by using the local property price index compiled by Singapore authorities for each region.

As for the statistical regression approach, there are a simple single-factor approach and a two-factor approach to remove the potential causality of the Central Business District (CBD) effect on property price.

2.1 Data Collection and Processing

Property Price Data. In this study, property price refers to the median price of condominium prices sold by developers from June 2007 to December 2019 and adjusted by the specific Property Price Index (PPI) for each locality, namely Core Central Region (CCR), Rest of Central Region (RCR) and the Outside Central Region (OCR) (Fig. 1). The earliest start time available from the public domain starts from June 2007.

PPI is computed by Urban Redevelopment Authority (URA) to track housing price movement over each quarter and there are specified PPIs for different localities. This PPI is used to adjust and rebase the transaction to remove the effects of inflation and price movements due to time differences. The base level is set at 100, as of Q1 2009. The data is obtained from the official website www.ura.gov.sg, and www.data.gov.sg. The inflation for each period is adjusted, to compare apples to apples. The impact from potential gentrification or improved quality of the neighborhood and the property are indirectly adjusted and accounted for, as the PPI index is only tracking the general price movement and is not a quality constant index. In other words, the increase in the property price over the years due to inflation or gentrification is reflected in the index. Hence, by reversing and rebasing the index to a single date (Q1 2009), these improved property prices due to inflation or improved quality in neighborhoods are eliminated.

The data excludes landed properties, meaning single-family houses, as these are less homogenous, such as the land area, design, and their built-up area. Similarly, public housing apartments, such as (Housing Development Board) HDB flats and executive condominiums are excluded as they are subsidized by the government or with a pricing ceiling, making the price is less market-driven and hence difficult to make a reasonable comparison. The secondary sales are also excluded as there could be different maintenance conditions.

There are 899 properties in the database. The GPS coordinate of each property is collected from www.onemap.sg and transferred into excel. This data is further imported into QGIS to measure the distance to art places.

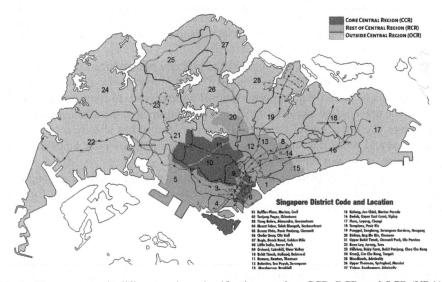

Fig. 1. Singapore map in different regions classifications, such as CCR, RCR, and OCR (URA)

55 Town Planning Areas in Singapore. In order to do a comparison study on the impact of art places on property prices, Singapore city needs to be studied in more detail, as individual towns. This will allow a more meaningful study of how art places affect the property price in each town rather than the whole city of Singapore. For simplicity and for the purpose of this research, the property price of each town is assumed to be homogenous and similar in nature.

The selection of Singapore's towns in this study is based on the Planning Area Census 2010 which was obtained from data.gov.sg. In this Planning Area Census 2010, there are 55 planning areas (towns) in Singapore. This is used in the Census 2010 published by the Department of Statistics, Singapore. The map boundaries used in the Planning Area are also based on URA's Masterplan 2008.

Art Places Data. The list of art places in this study was collected from the Ministry of Culture, Community and Youth (MCCY), National Art Council (NAC), National Heritage Board (NHB), and visitsingapore.com by Singapore Tourism Board (STB). First, 193 art places were collected. After the removal of repetitions, a total of 143 art places, including 89 art places and 54 public arts (sculpture and graffiti) remained. Public arts were also removed as they were too scattered and not a social focal point.

The distances between the property and the art places were calculated through QGIS. Similarly, the distance between properties and the CBD, art places (including 2 art places clusters), and the CBD were also calculated through QGIS. In this study, the CBD will be represented by Raffles Place MRT Station. The years of the opening time of each art place were also collected and recorded.

Art Places Clusters. The art places clusters were formed to reduce the potential cross effect due to the close of the proximity of the major art places. In other words, it is hard to quantify the effect of each art place as there could be spillover effects by other major art places. Two clusters are with similar characteristics are considered in this study (Fig. 2).

Cluster A comprises Sands Theatre/ArtScience Museum, Asian Civilisations Museum, Esplanade, National Gallery Singapore, National Museum of Singapore, Victoria Theatre. These are the highly recognizable art places.

Cluster B comprises of School of the Arts, LASALLE College of Arts, Nanyang Academy of Fine Arts, National Design Centre, Singapore Art Museum, Stamford Arts Centre, Selegie Arts Centre. These are the art places that focus more on arts education.

The centroids of Cluster A and Cluster B were computed to represent the clusters for further analysis.

Fig. 2. Art cluster A and cluster B

2.2 Comparative Method (with 3 km Radius vs Town)

The research question of this study is whether art places have a positive influence on property prices. The assumption is that a property's proximity to an art place or cluster increases its price per square foot compared to the properties that are further away from the art places.

In order to quantify the impact of art places on property prices, a catchment area with a 3 km radius from the art places was set. The 3 km radius was chosen as it takes approximately 10 min to drive or within 30 min to travel via public transportation, making it a reasonable distance to test the influence of art places on the neighborhood. The average medium property price in the catchment area is compared with the property price of all towns that are included or partially included in the catchment area.

To be more specific about each town's comparison, the average medium price of the property that falls into the catchment area of a particular town is compared with the average medium property price of the corresponding town.

Below is a diagram to illustrate how the comparison is being done. For example, the average of Town A property prices (P1 to P5) is compared to the average of properties that are within town A and 3 km radium of the art places (P4 to P5). Similarly, town B (P6 to P11) is compared to P6 to P8. On the overall basis, the average of properties within all towns that are intersected by 3 km radium, namely Town A and Town B for this instance (P1 to P11), are compared to the average of all properties within the radius (P4 to P8) (Fig. 3).

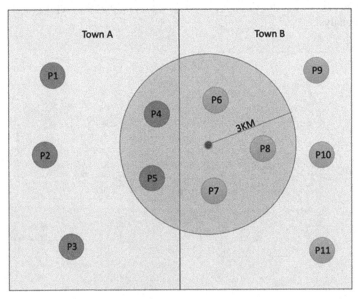

Fig. 3. Illustration of catchment area vs town in the analysis of the comparative table method

After the GPS coordinate of all art places was compiled into QGIS, 7 representative art places were chosen as the object of comparison. The selection of the 7 representative art places was based on general principles described below and experience.

One of the general principles is that the catchment art places should not be overlapping each other, as the overlapping effect will make it difficult to quantify the specific impact from a particular art place. Hence, most of the art places within the Museum Planning Area, which is located in the Central Area of Singapore with many art places and historical buildings, are not selected. Similarly, Clusters A and B are not selected.

However, Holland Village and Singapore Botanic Garden are selected despite that they are slightly overlapped because these two art places are significantly important. Holland Village is a creative community and incubator for local artists given its strong cultural and community background that fuse the European culture and Singapore's local lifestyle. Botanic Gardens, on another hand, is the only tropical garden on the UNESCO's World Heritage List and is also an important area for Singapore residents to gather and to enjoy free musical performances.

The second general principle, with its reason similar to the first principle, is that art places within large institutions are not considered as such institutions could also have an impact on property price, making it difficult to test the influence of art place on its own merit. For example, the NUS museum and NTU Centre for Contemporary Art are excluded in this test as the institution of higher learnings such as NUS might have a positive impact on the property price. The value brought forth by the NUS museum which is being part of NUS could not be individually accounted for and separated from the overall positive impact.

Lastly, art places of less significance, such as small studios are not considered (Figs. 4 and 5).

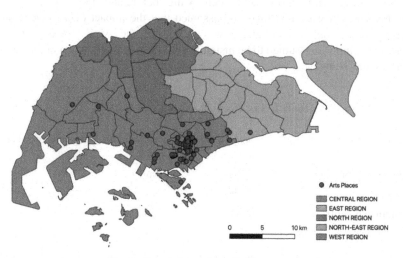

Fig. 4. Selected art places in Singapore

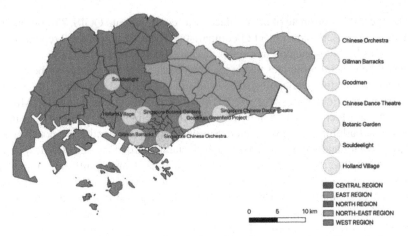

Fig. 5. 7 Representative art places selected for the comparative table analysis

2.3 Comparative Method (Before and After Years of Opening)

One of the methods to test out whether art places have any positive impact on property prices is to compare the property prices before and after the opening date within a certain catchment area.

In order to do so, art places that opened before 2007 have to be filtered out, as the property data collected from URA is after 2007. There are a total of 11 art places that opened after 2007, which is also the start of the property data collection provided by URA (Table 1). 3 art places are studied as the remaining 8 art places are all overlapping with one another, making it difficult to identify the "before and after" effects.

In this study, it was simplified and assumed that the impact of art places on the property price was only evident after it was opened. In other words, the expectation of art places opening was assumed to be insignificant to homeowners when purchasing the property in this study.

Table 1. List of art places opened after 2007

Name of art places	Year of Opening	Location not overlapped by other art places
The peranakan museum	2008	No
School of the arts	2008	No
Goodman arts centre	**2011**	**Yes**
ArtScience museum/red dot/sands theatre	2011	No
Gillman barracks	**2012**	**Yes**
Singapore dance theatre (SDT)	2013	No

(continued)

Table 1. (*continued*)

Name of art places	Year of Opening	Location not overlapped by other art places
Singapore Chinese dance theatre	**2014**	**Yes**
National gallery Singapore	2015	No
Indian heritage centre	2015	No
Arts equator	2016	No
Stamford arts centre	2019	No

2.4 Regression Methods

Comparative tables focus on the immediate catchment areas and the corresponding towns. As the number of property price data points within the catchment areas are relatively small, it is unable to provide statistically significant findings. To improve the data reliance, regression methodology is introduced. There are two regressions tests, namely single-factor regression and two-factor regression.

The regression tests out the relationship between art places and property prices. The hypothesis is that the closer proximity to art places the higher property prices.

There are two types of regressions to be examined in this study. One is Single Factor Regression and the other is the Two-Factor Regression.

For the Single Factor Regressions, the dependent variables are the property price adjusted by PPI for inflation and potential gentrification effects. The independent variables will be the distance between the property and the art places (clusters).

For the Two-Factor Regressions, the independent variable will be the distance to the CBD and whether the properties are within the 3 km radius of the art places(clusters). The reason for adding the new independent variables is to remove the CBD compound effect on property prices.

Single Factor Regression. In Single Variable Regression, Variable "y" equals the property price adjusted for inflation (using PPI index). Variable "x1" means the distance to a certain art place. Coefficient "a" means that the change of the property price for every 1 m away from a certain art place. Coefficient "b" equals the intercept of the property price, thus, hypothetically if the property is located exactly in a certain art place, the property price value is "b".

$$Y = ax1 + b$$

Where,
y = property price adjusted for Inflation (using PPI index),
x1 = distance to the art place
b = intercept

Two-Factor Regression. In Two-Factor Regression, Variable "y" equals the property price adjusted for inflation (using PPI index). Variable "x1" represents the distance to CBD (Raffles Place MRT) while Variable "x2" represents if the property is within a 3 km radius of a certain art place. Coefficient "a" means that the change of the property price for every 1 m away from CBD, of which Raffles Place MRT is used as a proxy. Coefficient "b" means the change to the property price if the property falls into the 3 km radius of a certain art place. Coefficient "c" equals the intercept of the property price, thus, hypothetically if the property is located exactly in a certain art place, the property price value is "c".

$$Y = ax1 + bx2 + c$$

Where,
y = property price adjusted for Inflation (using PPI index),
x1 = distance to CBD (Raffles Place MRT)
x2 = 3 km Radius of the art place
c = intercept

3   Results

The first observation and comparative test are to set a 3 km catchment area and compare its property prices with the average price of corresponding towns. The results are generally positive except for a few towns that have other stronger or unique factors that result in a higher property price compared to the catchment area of the art places within the town. Towns such as Downtown Core, Tanglin, and Geylang do not exhibit higher property prices for the specific catchment area from art places. Similarly, Gillman Barracks based on this test also did not exhibit higher property prices for its catchment area compared to the rest of the towns. This is believed to be due to the lack of amenities for its immediate catchment area compared to the rest of town. In short, this test will show positive results only if the art places have stronger attributing factors compared to the other amenities within the town.

The second test is to determine the property price of the catchment area and the corresponding towns on a "before and after" basis. As the property price is collected from the year 2007, there are only 3 art places that fulfilled the criteria to do the "before and after" test. In all 3 art places, the test showed that not only do the art places brought higher value to the catchment areas and but also to all the corresponding towns.

In the single-factor regression, the distance to the art places is used to explain the variations to the property prices. The test results for most art places (namely Cluster A, Cluster B, Botanic Garden, Gillman Barracks, and Holland Village) are statistically significant and showed that the closer to art places, the higher their property prices. In this single factor regression, only Goodman Arts Centre did not have a statistically significant result. As such, it was believed that one key variable, namely distance to CBD, might be confounded in the analysis and not accounted for.

Two-factor regression is further introduced to remove the compounding effect of CBD on property price. The distance to CBD and if properties are within the 3 km catchment area are the two factors tested for this regression. These results are statistically significant and they showed that if properties are within a 3 km catchment area of art places, they have higher property price like-for-like. The results also showed an increase of the R-square to 32.89% to 56.65% from the previous 28.6% to 40.7%, implying that the higher percentage in changes in property price can be accounted for by the introduction of the new variable. However, both Goodman Arts Centre and Gillman Barracks do not have statistically significant results. The reasons for the non-conclusion results are (1) low sample size for the properties sold in the catchment area of Gillman Barrack and (2) Goodman Arts Centre is close to Geylang area which is a 'red-light district'.

3.1 Comparative Method (with 3 km Radius vs Town)

The overall comparison shows that art places have a positive impact on property prices (Table 2). The average property prices within the 3 km radius of art places are 1.60% to 19% higher than the average with all towns touched by the catchment area. The property prices within some specific towns even have an increase as high as 37.7%.

However, 2 out of the 7 selected art places are showing negative results. This is because of a few towns that have other stronger or unique factors that result in higher property prices compared to the catchment area.

Table 2. Summary of the results of the comparative methods for 7 representative art places

	Goodman Art Centre			Singapore Chinese Orchestra			Gillman Barracks			Singapore Chinese Dance Theatre			Botanic Garden			Souleelight			Holland Village		
		Adjusted Price	Difference %		Adjusted Price	%		Adjusted Price	%		Adjusted Price	Difference%		Adjusted Price	Difference %		Adjusted Price	%		Adjusted Price	%
Average within Circle (3Km)		904.19			1733.91			1051.9			838.09			1761.28			819.25			1423.41	
Average with all towns touched by the Circle	All Towns	916.76	-1.40%	All Towns	1552.35	11.70%	All Towns	1143.3	-8.00%	All Towns	811.61	3.30%	All Towns	1480.61	19.00%	All Towns	805.98	1.60%	All Towns	1397.88	1.80%
Average within Circle and within Town 1		1337.78			1852.8			1101.63			838.09			1829.89			918.8			1567.89	
Average within Town 1	Kallang	971.64	37.70%	Downtown Core	1884.86	-1.70%	Bukit Merah	1262.59	-12.70%	Bedok	811.61	3.30%	Tanglin	1703.86	7.40%	Bukit Batok	878.8	4.60%	Tanglin	1703.86	-8.00%
Average within Circle and within Town 2		828.73			1503.3			985.6						1447.29			744.59			972.24	
Average within Town 2	Geylang	829.94	-0.10%	Outram	1503.3	0.00%	Queenstown	1059.35	-7.00%				Bukit Timah	1261.76	14.70%	Bukit Panjang	733.88	1.90%	Queenstown	1059.35	-8.20%
Average within Circle and within Town 3		1335.07			1490.79									1439.26						1441.41	
Average within Town 3	Marina Parade	1033.04	29.20%	Bukit Merah	1262.59	18.10%							Novena	1173.49	22.60%				Bukit Timah	1261.76	14.20%
Average within Circle and within Town 4														2329.76							
Average within Town 4													Newton	2125.81	9.60%						

Goodman Arts Centre: In this instance, the impact was negative. The reason for the negative result is mainly because of the uniqueness of Geylang town as a 'red-light district' or pleasure district that is catering to prostitution and sex-oriented business and there is a higher number of property transactions in Geylang that resulted in a tilt in the results.

The average price within 3 km of Goodman Arts Centre and Geylang town is 0.1% lower than the average price within the same town. There is only a marginal difference between the town and properties within a 3 km radius of Goodman Arts Centre.

However, the results of Kallang and Marine Parade, also within the catchment area of Goodman Arts Centre, show a positive influence on property prices (37.70% and 29.20% increase respectively).

The expected positive impact of art places is not found in Geylang given the uniqueness of its locality. Geylang is highly associated with the red-light district of Singapore. There is also very limited or even non-existent urban rejuvenation within the town, despite its relative closeness to the CBD area. The other neighboring towns to Geylang, such as Marine Parade and Kallang, are deemed to be 'better' towns. Marine Parade is even one of the better residential enclaves in Singapore with mid-high-end condominiums. Kallang is also deemed to be good as it has the National Stadium and the Kallang River that is used for water sports, such as canoeing.

Gillman Barracks: The catchment area of Gillman Barracks encompasses 2 towns, Bukit Merah and Queenstown. The average price within the 3 km radius is $1051.9 psf, while the average price within all towns is $1143.3, which is 8.0% higher.

The results show that Gillman Barracks has a negative impact on property prices. One of the reasons for this is that Gillman Barracks might not be deemed as a successful art place, as it was also reported that many art owners/tenants in Gillman Barracks were out of business. Shetty (2015) reported in a local newspaper Straits Times that nearly a third of Gillman Barracks galleries have decided not to renew their leases because they feel that "the infrastructural additions to the barracks, such as better signage, covered walkways, and more food and beverage options, and programming of pop-up events, came too late. They were introduced about a year after the opening of the barracks."

Other than a lack of amenities for its immediate catchment area compared to the rest of the town, another reason is that most properties outside of the circles in towns are near the southern waterfront, which is the luxury residence cluster in Singapore.

Similarly, towns such as Downtown Core, Tanglin, and Queenstown do not exhibit higher property prices for the specific catchment areas from art places.

Downtown Core: The average price within a 3 km radius of the Chinese Orchestra is 1.7% lower than the average price within Downtown Core. The reason for the impact of art places in Downtown Core is less obvious because there are other attractions and major amenities, such as integrated resort (Marina Bay Sands), Clarke Quay, Boat Quay, and CBD, are in Downtown Core.

Tanglin: The average price within the catchment area is 8% lower than the average price in Tanglin. The reason for the negative result is because only a small amount of the properties in Tanglin are in the 3 km circle of Holland Village, the rest are near Orchard, which is the upscale shopping area and tourist attraction of Singapore.

Queenstown: The average price within the catchment area is 8.2% lower than the average price in Queenstown. The reason for the negative result is due to the property locations within the 3 km circle are considered not as prime, as other properties are either close to Orchard or near to Southern Waterfront, which is the luxury residential cluster in Singapore.

In summary, this method shows positive results only if the art places have a strong attributing factor compared to the other amenities within the town.

3.2 Comparative Method (Before and After Years of Opening)

The impact of art places is assessed based on the before and after opening year. For example, Gillman Barracks opened in 2012, and the average property price within the catchment area, defined as within the 3 km radius, is compared to between those in and before 2012 versus those in 2013 and after (Table 3).

Table 3. Summary of the results of the 3 art places under "before and after" analysis

	Gillman Barracks			Goodman Art Centre			Singapore Chinese Dance Theatre		
		Adjusted Price	Difference%		Adjusted Price	Difference%		Adjusted Price	Difference%
Year of Opening		2012			2011			2014	
Average within Circle (3Km) before and on year of opening		1004.76			814.82			811.64	
Average within Circle (3Km) after year of opening	All Towns	1169.76	16.40%	All Towns	1009.6	23.90%	All Towns	956.76	17.90%
Average within Town 1 before and on year of opening		1186			876.1			796.64	
Average within Town 1 after year of opening	Bukit Merah	1331.53	12.30%	Kallang	1157.9	32.20%	Bedok	1002.4	25.80%
Average within Town 2 before and on year of opening		876.49			738.88				
Average within Town 2 after year of opening	Queenstown	1229.14	40.20%	Geylang	951.35	28.80%			
Average within Town 3 before and on year of opening					923.18				
Average within Town 3 after year of opening				Marina Parade	1291.54	39.90%			

Gillman Barracks. The average property price within the catchment area of Gillman Barracks after the year of opening 2012 (from 2013 onwards) was $1,331.53 psf while the average property price within the same catchment area before the year 2012 was $1,186 psf, reflecting a 16.4% premium. This implied that the Gillman Barracks might have positively contributed to the increase of the property within the catchment area, as these prices were already adjusted for inflation, using the PPI for its region to adjust back the price to the Q1 year 2019 (index rebased to 100).

Goodman Arts Centre. The average property price within the catchment area of Goodman Arts Centre after the year of opening 2011 (from 2012 onwards) was $1,009.6 psf while the average property price within the same catchment area before and on the year 2011 was $814 psf, reflecting 23.9% premium. This implied that the Goodman Arts Centre has a positive influence on property prices within the catchment area.

Singapore Chinese Dance Theatre. The 3 km radius catchment area of Singapore Chinese Dance Theatre only encompasses one town: Bedok. The average property price within the catchment area of Singapore Chinese Dance Theatre after the year of opening 2014 (from 2015 onwards) was $956.76 psf while the average property price within the same catchment area before and on the year 2014 was $811.64 psf, reflecting 17.9% premium. This implied that the Singapore Chinese Dance Theatre has a positive influence on property price within the catchment area.

The results all showed that not only have the art places brought higher property values to the 3 km radius catchment areas but also all the corresponding towns in this "before and after" analysis.

3.3 Single Variable Regression

Table 4. Summary of results for single factor regression

Single Factor Regression	Cluster A	Cluster B	Botanic Garden	Goodman Arts Centre	Gillman Barracks	Holland Village
R-square	28.61%	27.60%	39.92%	0.10%	40.67%	32.70%
Coefficient "a"	-0.078	-0.077	-0.086	-0.055	-0.078	-0.071
Intercept "b"	1603.59	1559.21	1718.78	697.58	1818.08	1733.41
P-Value for "a"	0.00%	0.00%	0.00%	24.20%	0.00%	0.00%
P-Value for "b"	0.00%	0.00%	0.00%	0.00%	0.00%	0.00%

Most of the results are statistically significant, except for Goodman Arts Centre. These results showed that the nearer to Art Places, the higher the property price. Every decrease of 1 m resulted in a decrease of between $0.071 to $0.086 psf to the property price (Table 4).

Using Cluster A for example, the coefficient "a" has a value of −0.078, it implied that for every 1 m away from Cluster A, the property price will decrease by $0.078 psf. Intercept "b" has a value of 1,603.59, implying property price will be highest and at $1,603.59 psf if it is located at the hypothetical point at Cluster A.

The R-Square value for this case is 28.61%, implying that other variables were not accorded for in this study. In other words, there are about 71.39% of the property price could not be explained by the distance to the art place Cluster A. This regression is statistically significant as the p-value for Coefficients "a" and Intercept "b" is 0.00% and 0.00% respectively, both are below 0.05 or 5% probability that they occurred by chance. In other words, the probability of coefficients "a" and "b" occurred by chance is only 0.00% and 0.00% respectively.

In this test, only the result of Goodman Arts Centre is statistically not significant as the p-value for Coefficient "a" is 24.2%, which is high above 0.05 or 5% probability. The reason for the high P-Value for coefficient "a" may be due to the fact this is only a single variable and the distance to CBD was not accounted for, resulting in an inconclusive finding for Goodman Art Centre.

Overall, most of the art places are statistically significant and showed that the closer to art place has a positive effect on property prices. Coefficient "a" has a value between −0.071 to −0.086, which means 1 m away from a certain art place, the property prices drop from $0.071 to $0.086 psf respectively. R-square values range from 27.60% to 40.67% suggesting that other variables were not ac-corded for in this study. In other words, there are about 59.33% to 72.40% of the property prices that could not be explained by solely the distance to a certain art place. As property price is a function of many inputs, it is impossible to obtain a high r-square value in this single variable regression.

3.4   Two-Factors Regression

Table 5. Summary of results for single factor regression

Two-Factor Regression	Cluster A	Cluster B	Botanic Garden	Goodman Arts Centre	Gillman Barracks	Holland Village	National Gallery
Adjusted R-square	33.81%	39.11%	54.65%	39.15%	31.08%	32.89%	35.86%
Coefficient "a"	-0.064	-0.048	-0.055	-0.086	-0.08	-0.08	-0.058
Coefficient "b"	268.81	424.48	697.58	-368.76	-39.26	325.13	347.96
Intercept "c"	1495.24	1321.72	1355.91	1764.98	1651.19	1631.34	1434.04
P-Value for "a"	0.00%	0.00%	0.00%	0.00%	0.00%	0.00%	0.00%
P-Value for "b"	0.00%	0.00%	0.00%	0.00%	58.37%	0.00%	0.00%
P-Value for "c"	0.00%	0.00%	0.00%	0.00%	0.00%	0.00%	0.00%

Most of the results are statistically significant, except for Gillman Barracks. These results showed that being within the 3 km catchment area of Art Places, properties can have higher prices in a like-for-like situation. The art places are found to provide to increase the catchment area property price by $268.81 to $697.58 psf compared to others with the same distance to CBD (Table 5).

Using Botanic Garden as an example, its coefficient "a" has a value of −0.055, it implied that for every 1 m away from Raffles Place, the property price will decrease by $0.055 psf. Its coefficient "b" has a value of 697.58, implying that the property within the 3 km radius catchment area of Botanic Garden, the price is $697.58 psf higher than those not. The coefficient "c" has a value of 1355.97, implying property price will be highest and at $1,355.97 psf if it is located at the hypothetical point at Raffles Place MRT. This regression is statistically significant as the p-value for coefficients "a", "b", and "c" are all 0.00%, below 0.05 or 5% probability that they occurred by chance. In other words, the probability of "a", "b", and "c" occurred by chance is only 0.00%.

The Adjusted R-Square value for this case is 54.65%, which implies that other variables were not accorded for in this study. In other words, about 45.35% of the property prices could not be explained by the distance to CBD and whether it is within the 3 km radius catchment area of Botanic Garden.

In this test, only the result of Gillman Barracks is statistically not significant as the p-value for Coefficient "b" is 58.37%, which is high above 0.05 or 5% probability that they occurred by chance. In other words, the probability of "b" occurred by chance is 58.37%. The reason for the high P-value is because the properties within the 3 km catchment area of Gillman Barracks are too few (only 7 properties), thus the sample size is too small to be statistically meaningful.

Overall, the rest of the art places have statistically significant results. All results showed a negative number for Coefficient "a", implying that the further from CBD, the lower the property price. Most art places have positive values for Coefficient "b", between 268.81 to 697.58, implying that the presence of an art place within the 3 km added about $268 psf to $697.58 psf to the property price. Goodman Arts Centre is the only exception in this study, with a negative value for Coefficient "b". As discussed earlier on the comparative table, the reason for the negative result is mainly because of the uniqueness of Geylang and the higher number of transactions in Geylang that resulted in a tilt in the results for Goodman Arts Centre.

The R-square values have improved in the two-factor regression compared to the Single Factor regression. The R-Square Values for the Single Variable Regression are between 28.61% to 40.67%, while the R-Square Values for the Two Variables Regression are between 32.89% to 54.65%. The introduction of one additional variable, namely the distance to CBD, improves the value of R-Square for the Single Variable Regression. Since property prices are influenced by multiple factors, it is not unusual for the R-square for even the two variable regressions to be below 70%.

4 Discussion

4.1 Summary of Major Findings

In this paper, we have adapted quantitative research methods to test out whether art places have a positive influence on property prices and the community. We have managed to analyze the influence art places on property prices as a reflection of how art impacts the economy.

As discussed earlier, the property prices are adjusted and rebased using the Property Price Index computed by Singapore authority, the effects of inflation and potential gentrification or improved quality are already accounted for and removed from the comparison. PPI is not a quality constant index but only tracks general price movements. In other words, the increase in the property price over the years due to inflation or gentrification is reflected in the index. Hence, by reversing and rebasing the index to a single date (Q1 2009), these improved property prices due to inflation or improved quality in neighborhoods are eliminated.

In the first comparative method test, the results are generally positive, showing that the average property price within the 3 km radius of art places is higher than the average with all towns touched by the catchment area.

In the second comparative test of "before and after" analysis, all catchment areas and corresponding towns have higher property prices after the art places were established. This further suggests that art places do have a positive impact on both the catchment areas as well as the corresponding towns.

As there are externalities other than art places that might have a stronger influence on property prices, the first test of 3 km catchment areas might not fully explain if art places have a positive impact on property prices. For example, if there were two properties in the same town but one is nearer to an art place while another is nearer to a Shopping Belt or CBD, it is more likely that the one near to Shopping Belt or CBD has a higher property price. This does not imply that art places do not have a positive impact on property prices.

The comparative table focuses on the immediate catchment area and the corresponding towns. As the number of property price data points within the catchment areas are relatively small, it is unable to provide statistically significant findings.

In order to improve the data reliance, regression methodology is introduced. There are two regressions tests, namely single-factor regression and two-factor regression.

In the single-variable regression test, most of the art places are statistically significant and showed that the closer to art place has a positive effect on property prices. As property price is a function of many inputs, it is impossible to obtain a high R-square value in this single variable regression. The causality of art place to property price might be weak due to the low R-square of between 28% to 41%.

In the two-factor regression, there are two improvements from the single variable regression. First, the potential compounding effect from Central Business District (CBD) is removed as the distance to CBD is included as one of the factors. Second, the R-square value has increased significantly from 32% to 55%. Coefficient "b" value between 268.81 to 697.58, in this test, means that the presence of an art place within 3 km catchment areas added \$268 psf to \$697 psf to the property price.

Summing up the results from comparative tables and regression tests, the art places do provide higher property prices from different analysis approaches.

4.2   Study Strength and Limitations

This study has managed to provide plausible and statistical observations that art places improved property prices, even after removing the inflation and gentrification effects. Most of the past research did not provide quantitative evidence of how art places help to improve property prices, especially in Singapore's context.

However, there are several limitations to this study. Property price is also influenced by other factors that are not accounted for in this study. First, the study of property prices is solely based on two key factors, which are distance to CBD and distance to art places. This resulted in only an R-square of between 33% to 57%. If the number of factors could be expanded in a further study, the reliance of the regression could be further strengthened. It is believed that the other factors that are not considered in this research and might have an impact on property price are factors such as amenities around the neighborhood, different facilities within the condominium project, distance to prominent schools, the total gross floor area, the floor level, etc. If these variables are introduced in a future study, there could be an improvement to the R Square value.

Second, the regression can be further refined to test the nonlinear relationship between art places and property prices. The basic assumption used in the model is that the relationship is linear. In a complex environment, the actual relationship might be nonlinear and therefore this research might have over-simplified the current assumptions.

Third, there could be other socio-economic benefits that were not tested and examined in this study. Hence, the socio-economic benefits from art places might not have been fully accounted for.

4.3   Implications on Practice

One key recommendation is that there is a need to have decentralized art places. It was observed that most of the art places are in the central region.

The introduction of art places in the neighborhood will improve the property prices in its corresponding town. For example, under the "before and after" analysis, the results all showed that not only have the art places brought higher property values to the 3 km radius catchment areas but also all the corresponding towns. This is a shred of strong evidence that art places help to drive prosperity to their neighboring area. Do note that all property prices in this study are already adjusted for the inflation and other potential effects using URA Property Price Index to rebase the property price to Q1 2009. Hence, for equality basis, the need to have decentralized art places is crucial.

4.4 Future Research

Given the limitations discussed, it is interesting to also examine the socio-economic benefits of art places. For example, in future research, a survey can be conducted to understand the community perception of the art places and to understand their behavior towards art places.

There could be a further refinement to examine the economic benefits, including expanding the number of factors and using more advanced methodologies, including deep learning to unravel the economic linkages between art places and the economy.

References

Chanuki, I.S., Tobias, P., Helen, S.M.: Quantifying the link between art and property prices in urban neighbourhoods. Roy. Soc. Open Sci. **3**(4), 160146 (2016)

Chang, T.C.: Renaissance revisited: singapore as a 'Global City for the Arts.' Int. J. Urban Reg. Res. **24**(4), 818–831 (2000)

Chong, T.: Bureaucratic imaginations in the global city: arts and culture in singapore. In: Lee, H.-K., Lim, L. (eds.) Cultural Policies in East Asia, pp. 17–34. Palgrave, London (2014). https://doi.org/10.1057/9781137327772_2

Fabiana, F., Pierfrancesco, D.P.: How can street art have economic value? Sustainability **11**, 580 (2019)

Lim, C.T.: Chapter 2: The Anti-Yellow Culture Campaign in Singapore: 1953–1979. In The State and The Arts In Singapore Policies and Institutions, pp. 31–48. World Scientific Publishing Co. Pte. Ltd, Singapore (2019)

Kong, L.: Ambitions of a global city: arts, culture and creative economy in 'Post-Crisis' Singapore. Int. J. Cult. Policy **18**(3), 279–294 (2012)

Kong, L., Ching, C., Chou, T.: Arts, Culture and the Making of Global Cities: Creating New Urban Landscapes in Asia. Edward Elgar, Cheltenham (2015)

Shetty, D.: Nearly a third of Gillman Barracks galleries have decided not to renew their leases. In: Straits Times (2015). https://www.straitstimes.com/lifestyle/arts/nearly-a-third-of-gillman-barracks-galleries-have-decided-not-to-renew-their-leases

Social Distancing and Behavior Modeling with Agent-Based Simulation

Ming Tang$^{(\boxtimes)}$ (iD)

School of Architecture and Interior Design, College of DAAP, University of Cincinnati, Cincinnati, OH 45221-0016, USA
tangmg@ucmail.uc.edu

Abstract. The research discusses applying agent-based simulation (ABS) technology to analyze the social distancing in public space during the COVID-19 pandemic to facilitate design and planning decisions. The ABS is used to simulate pedestrian flow and construct the micro-level complexity within a simulated environment. This paper describes the various computational methods related to the ABS and design space under the new social distancing guidelines. We focus on the linear phases of agent activities, including (1) environmental query, (2) waiting in a zone, (3) waiting in a queue, and (4) tasks (E-Z-Q-T) in response to design iterations related to crowd control and safety distance. The design project is extended to the agents' interactions driven by a set of tasks in a simulated grocery store, restaurant, and public restroom. We applied a quantitative analysis method and proximity analysis to evaluate architectural layouts and crowd control strategies. We discussed social distancing, pedestrian flow efficiency, public accessibility, and ways of reducing congestion through the intervention of the E-Z-Q-T phases.

Keywords: Agent-based simulation · Social distancing · Crowd control

1 Introduction ABS for Crowd Behavior Simulation

An Agent-based simulation (ABS) consists of multiple agents controlled by rules to interact with each other within a virtual environment, thereby formulating a bottom-up system. The ABS concept has been widely used in computer science, biology, and social science to simulate swarm intelligence, dynamic social behavior, and fire evacuation. The simulation consists of interacting agents who can create various complexity. Agents can be defined as autonomous "physical or social" entities or objects that act independently of one another. An ABS consists of numerous agents, which follow localized rules to interact with a simulated environment, thereby formulating a dynamic system. Since Craig Reynolds' artificial "bodies" and flock simulation, the ABS concept has been widely used to study de-centralized systems, including human social interaction.

Many computational methods were applied to simulate agents involving movement, including *"the simple statistical regression, spatial interaction theory, accessibility approach, space syntax approach and fluid-flow analysis"* (Batty [1]). Michael Batty described the property of *"autonomy"* and *"the embedding of the agent into the*

© Springer Nature Singapore Pte Ltd. 2022
D. Gerber et al. (Eds.): CAAD Futures 2021, CCIS 1465, pp. 125–134, 2022.
https://doi.org/10.1007/978-981-19-1280-1_8

environment" as the two fundamental properties of agents in an ABS. ABS focuses on the agent's properties and processes to respond to external changes, specifically how the agents can "sense" and "act" to form a complex system. The movements are usually based on simple rules such as separation, alignment, and cohesion. Computer scripts can be used to control an agent's velocity, maximum force, range of vision, and other properties. Many research projects have been done to examine how agents "sense" the landscape and "walk" through it, such as studying train station crowd control (Tang [2]), building-occupant relations with discrete-event trigger (Schaumann [3]), dynamic coupling for flood evacuation simulation (Shirvani et al. [4]), and Clayton and Yan's panic evacuation simulation (Clayton, Yan [5]). Several scholars have discussed the research related to context awareness of the multi-agent system to assist space planning. (Gerber [6], Buš [7], Hua [8], Aschwanden[9] and Keifer [10].)

The ABS for pedestrian simulation can be traced back to the "level of services" by John Fruin in the 1960s and later the "social forces model" in the 1990s by Dirk Helbing and Peter Molnar. (Oasys [11]) The ABS approach has a significant advantage in crowd behavior simulation compared with other computation methods such as cellular automation and space syntax[1]. (Tang [2]) An agent can make decisions while evaluating the result generated in a real-time environment in an autonomous, bottom-up ABS approach. ABS allows a complex movement pattern to emerge from the simple interaction among agents. Each agent can "sense" its neighbors and "react" to them by modifying its location, velocity, shape, or other attributes. ABS for crowd behavior simulation is established in the same relational and computational strategies from the early physiological field. Some of the emerging methods in the crowd simulation involve utilizing AI to generate realistic crowd dynamics that respond to the crowd's perception of the surroundings, pre-programmed tasks, and goals. Unlike the "reactive" agent in Reynolds' flock simulation [12], these "cognitive agents" sometimes act as the non-player controlled characters (NPC) found in video games, which have their own decision trees. As a result, the agents can respond to the various "social distancing rules" and other agents' movement in real-time and adjust their behavioral parameters.

2 Methodology to Simulate Social Distancing in ABS

We applied the ABS in research titled "Return to the Third Places: Architectural intervention at the Price Hill, Cincinnati during the COVID-19," which focuses on the strategies responding to the design challenges of the social distancing of public space. "Third place" is a term coined by sociologist Ray Oldenburg and refers to places where people spend time between home ('first' place) and work ('second' place). They are locations where people exchange ideas, have a good time, and build relationships. These places can be described as "a community's living room", and provide social association, community identity, and civic engagement[2]. Social distancing with a minimum of six feet

---

[1] "The simulated behaviors of cellular automation are often unpredictable and lack purposive planning goals….and the interactions among agents, complex social behavior cannot be simulated through space syntax." (Tang, 2018).

[2] These semipublic, semiprivate places such as restaurants, bars, gyms, houses of worship, barbershops, coffee shops, post offices, main streets, beer gardens, bookstores, parks, community centers, and gift shops—inexpensive places where people come together, and life happens.

(two meters) during the COVID-19 pandemic implies many challenges of everyday life in work and home, particularly to the "third places." The research examined several existing "third places" at Price Hill, Cincinnati, and evaluated architectural solutions to create resilient places allowing social distancing during the COVID-19 pandemic. Specifically, the research investigated the (1) Proposal for design and renovation of a grocery store, restaurant, and restroom to maintain social distancing and crowd control during COVID-19, with the addition of (2) creating an ABS allowing the proposed design to be evaluated under various occupancy and program scenarios.

We started this project by researching existing design strategies responding to the COVID-19 pandemic. We found that many people are looking for environmental stimulus to help them maintain their social distancing. Some apparent solutions currently being used are markers installed on the ground, plastic shield installed in the checkout line, and outdoor waiting areas. Some generative design strategies were available for automatically creating furniture layout with social distancing. While these all help architects propose a spatial layout, they do not completely guarantee effective use of space while maintaining safety. With that in mind, we start to model our ABS to simulate people's movement through space with particular goals, including checking in a restaurant, checking out a grocery store, and using a restroom. These procedurals are simulated into four phases: start with environment query, walk into a waiting area, walk into a waiting queue, and finally fulfill specific tasks such as taking an order, eating, checkout, or using a toilet. ABS became an essential tool to evaluate whether the proposed design would impact safety and social distancing during these activities (Fig. 1).

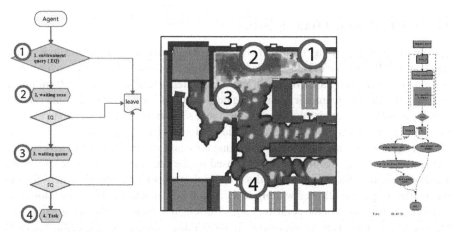

Fig. 1. Left and middle. The four phases in a sequence. 1. Environment query. 2. Waiting zone, colored based on agent proximity. 3. Waiting queue. 4. Task. Right: decision tree of an agent. The dynamics of crowd movement are not modeled at the global level but instead simulate the local interactions among the decision trees, events, and tasks.

2.1 Environmental Query

Based on the commercial ABS software called Oasys Massmotion[3], our research team applied behavior trees and the environmental query system (EQ) concept, a standard method for building NPCs in game design (Epic Game [13]). A behavior tree asset can be used to execute branches containing logic and serves as the "brain" for an agent. An AI Perception System provides a way for an NPC to receive data from the environment. The data includes where noises are coming from or if the NPC sees something. This is accomplished with the agent's Perception Component acting as a stimuli listener and gathering registered stimuli sources. Then, the behavior trees can be used to make decisions on which logic to execute.

In this process, the EQ is used to retrieve information about the environment. EQ is primarily made up of locations or Agents and the surrounding environment. EQ can instruct NPC characters to find the best possible place to provide a line of sight to a player to attack, the nearest health or ammo pickup, or the closest cover point in a computer game. In the context of crowd behavior related to environmental awareness, we collect data from EQ on the subsequent three phases (Z-Q-T), such as the current population density in the waiting area, the number of agents in the checkout line, or the dining table or toilet occupancy. The EQ gathers data as universal knowledge rather than an individual agent's sensory experience. For instance, once the agent enters the restroom, its perception system will automatically be given the number of people waiting in line from EQ. This intelligence is not based on an individual agent's vision or hearing, but universal knowledge gathered from a global EQ.

2.2 Waiting Zone and Waiting Queue

The environment in ABS is composed of various objects, including the static floor, stairs, ramp, and wall, which can be computed as a navigation mesh. The virtual environment is formed by importing a BIM model, including interior walls, stairs, and furniture. It also includes "smart geometries" such as gates, elevators, vehicles, portals, and servers, which can form impact circulations and the agent's behavior. The smart geometries are used to trigger certain events or set capacities and goals for the agents.

Among those environment objects, waiting zone and waiting queue are two essential elements dictating an agent's movement based on the predefined rules. Because of the social distancing rule, 2 m is defined as the minimum personal space, which dictates the number of people allowed waiting in a line and how many agents can be allowed to enter the waiting area. In the context of a goal-oriented movement, the agent will use an EQ to evaluate how many agents already in the waiting zone against a specific threshold value. If the zone is too crowded, the agent will choose not to enter the zone and leave the building. The same rule will apply when the agent moves from the waiting zone to the waiting queue. The following diagram highlighted the four phases (E-Z-Q-T) in the

---

[3] Oasys MassMotion is an advanced crowd simulation software that uses crowd modelling technology to provide leading technology to designers, operators and owners with clear information about crowding, usage patterns and occupant safety in a facility. https://www.oasys-software.com/products/pedestrian-simulation/massmotion/.

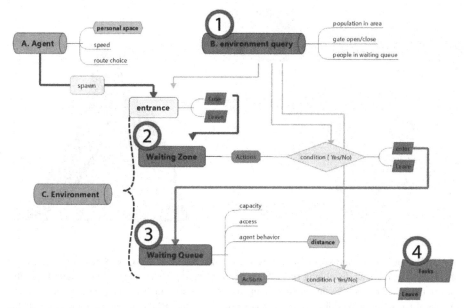

Fig. 2. Agent, environment, and environmental query system overlay with four phases. E-Z-Q-T as 1-2-3-4.

ABS. The system includes Agent profile (A), Environment (B), and the Environment query (C). (Fig. 2).

The real-time status of the waiting zone and waiting queue can be captured and broadcasted to the agent at a certain distance so it will be aware of the situation when it gets close. When the simulation is cached, a baked proximity map is also generated to indicate the number of agents close together in an area within a specific time.

The color map representing the agent proximity is used to study the social distancing and its impact on crowd behavior. We used various EQ metrics to update the agent's decision tree based on their proximity, density, and clocked time to complete specific tasks. As an independent entity, every agent in the environment constantly analyzed its proximity to other agents within the waiting zone and queue and made its own decisions. (Fig. 3 and 4).

2.3 Tasks

Besides enlarging waiting areas to help with the capacity limitations due to social distancing, we also evaluated strategies to reduce the time an agent took to complete the specific task. For instance, some stores use a Scan & Go Mobile service that allows customers to scan and pay for groceries on their smartphones to reduce the checkout time. Some stores such as Kroger provided customers self-scan shopping carts to eliminate the human cashier service. After simulating these two scenarios (regular checkout vs. smart scan without checkout), we concluded that Scan & Go Mobile service would reduce space needed near the checkout area and minimize the queue line in the store. We compared the current and projected customer flow and generated several scenarios

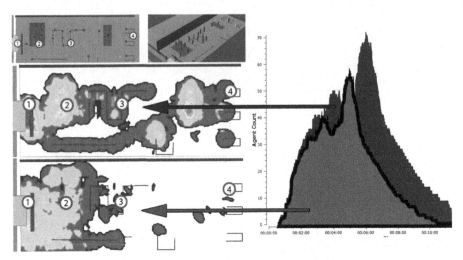

Fig. 3. Agent density and space proximity map. E-Z-Q-T as 1-2-3-4. ABS without social distancing vs. with social distancing rules. Each agent's autonomous "action" lies in modifying its movement based on its rules and environment. Top. Floor plan and interior perspective of a check-in area of a restaurant. Middle: proximity map without social distancing. Bottom: proximity map with 2-m social distancing with the same number of agents in the same given time. Notice the hot waiting areas' issues are replaced with a larger waiting area, while some agents choose not to walk in the restaurant after EQ. Right. Compare the number of occupancies. Red: agents with social distancing. Blue: agents without social distancing. (Color figure online)

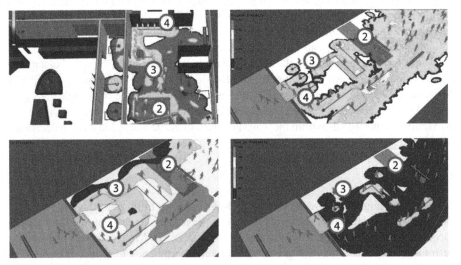

Fig. 4. The shape of queue lines, such as straight lines, L-shape lines, or U-shape lines, impacts the agent's proximity. Bottom left: the amount of time within five meters proximity. Bottom Right: the amount of time within two meters proximity. Notice all agent has less than 60 s within social distance. (blue color). E-Z-Q-T as 1-2-3-4. (Color figure online)

for crowd simulation. (Fig. 5) A fluent crowd movement pattern emerged based on the faster checkout time and fewer micro-scale interactions among agents. The impact of a speedier task on the spatial layout can be evaluated and presented through heatmaps and quantitative datasets.

1 Check-Out: 1 Check-Out + Scan & Go Mobile:

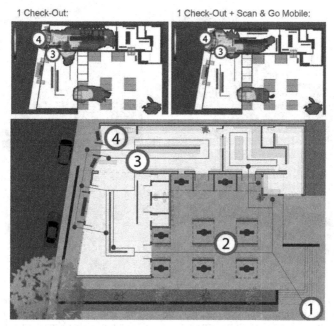

Fig. 5. Grocery store renovation. To address capacity limitations during COVID-19, we created an outdoor waiting space for the customers to sit. The waiting area includes separated personal space, each with a pergola and seating. Customers may choose to wait outdoors with social distancing before going into the grocery store or food bank. Top left: Conventional checkout methods. ABS shows that congestion areas are mainly located in the checkout area. Top right: Simulation results show that Scan & Go mobile will result in less congestion and a safe social distancing, while the conventional checkout will result in a sizeable waiting zone. These areas can be used to accommodate more customers and larger shopping areas. E-Z-Q-T as 1-2-3-4.

3   ABS for Public Restroom Design

The research on social distancing simulation in a restroom focused on the waiting area and queue outside the toilet area. We estimated the occupancy increase during the public event at the price landing park, where the proposed restroom is located. We evaluate the crowd flow through the total travel time, density, and public accessibility in different scenarios. Based on the result of ABS, we analyzed whether various waiting areas, utilities, and spatial layout can improve pedestrian flow efficiency and shorten waiting time and reduce congestion. (Fig. 6) After several iterations of the public restroom design, the E-Z-Q-T sequence was applied to analyze how agents would move through space as they go through the necessary programmatic areas.

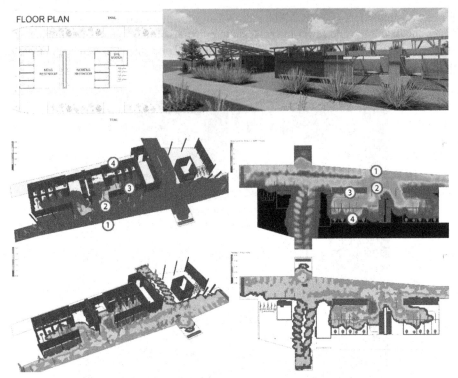

Fig. 6. E-Z-Q-T as 1-2-3-4. Design iterations were simulated with projected pedestrian flow. According to this simulation analysis, all areas are either blue or green, proving that the proposed design can handle many people while still maintaining social distancing. The red color area illustrates the high proximity waiting zone, which has direct natural ventilation. Top-left: occupancy time. Top-right: population density. Bottom-left and bottom-right: Proximity map. (Color figure online)

4   Conclusion

The crowd simulation method relies on the emergent properties and local interactions among agents. The project applied behavior simulation systems and investigated how to integrate E-Z-Q-T in space design. We observed that social distancing affected spatial organization and circulations in various ways. The social distancing requires a larger waiting area and results in a more extended connection between programmed spaces. With EQ and behavior trees, we simulated how agents enter/exit a given room, what happens when they meet agents coming the other way around a corner, the way they walk crossflows and counter-flows, and how they react to the dynamically changing crowd density in the waiting zone or queue, respecting the social distancing. Data visualization such as the proximity map became a valuable tool for us to study the capacity of a given space while maintaining a set physical distance. For instance, the blue and green areas show that the design ensures the social distancing requirements. The warmer tones mark areas that need to be improved (Fig. 7).

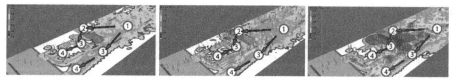

Fig. 7. Proximity map of 150, 250, and 500 people in a store. With the "travel cost" function, more agents start to choose the alternative checkout queue when the waiting zone is crowded. E-Z-Q-T as 1-2-3-4. (Color figure online)

A proposed spatial layout can be evaluated within ABS by analyzing the interaction between the simulated crowd and the surrounding environment. Designers can observe agents' changing behaviors by testing different spatial layouts and various behavior rules. We believe the crowd simulation can produce measurable improvement in the design by predicting specific "high-risk" areas with potential congestion issues. The simulation results were used to suggest alternative pedestrian paths and compare the different crowd control strategies.

We are currently investigating the process to reinforce EQ with digital billboard placement. The design proposal is to create a visible billboard allowing the real-time occupancy data to be broadcasted to people. The new ABS will count each agent's vision and perception system and trigger their decision tree. We have computed the most visible vertical surfaces through the environment (Fig. 8). The digital billboard will inform agents of the waiting zone and waiting queue status and create a more precise EQ for crowd simulation.

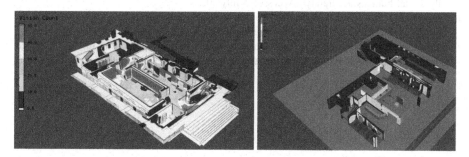

Fig. 8. Visibility analysis on the vertical surfaces.

Acknowledgments. Thanks to the students Sarah Auger, Maddison DeWitt, Brittany Ellis, Andy Failor, Lisa Garcia, Ashley Kasel, An Le, Hannah Loftspring, Kyle Munn, Deborah Park, Sabrina Ramsay, Haley Schulte, Brayden Templeton, Pwint Wati Oo (Audrey) participated the project in Fall 2021 at the University of Cincinnati. Thanks to UC Forward, Price Hill Will, and Meiser's Fresh Grocery & Deli provided advice and support. Thanks to Nathan Deininger provided proofread and editing.

References

1. Batty, M.: Cities and Complexity, Understanding Cities with Cellular Automata, Agent-Based Models, and Fractals. MIT Press, Cambridge (2007)
2. Tang, M., Auffrey, C.: Advanced digital tools for updating overcrowded rail stations: using eye tracking, virtual reality, and crowd simulation to support design decision-making. Urban Rail Transit **4**, 249–256 (2018)
3. Schaumann, D., et al.: Toward a multi-level and multi-paradigm platform for building occupant simulation. In: Symposium on Simulation for Architecture and Urban Design (2019)
4. Shirvani, M., Georges K., Paul R.: Agent-based simulator of dynamic flood-people interactions. J. Flood Risk Manag. **14**(2) (2021). https://doi.org/10.1111/jfr3.12695
5. Zarrinmehr, S., Asl, M.R., Yan, W., Clayton, M.: Optimizing building layout to minimize the level of danger in panic evacuation using genetic algorithm and agent-based crowd simulation. In: BIMSIM Group (2021). Accessed 15 Mar 2021
6. David, G., Rorigo, L.: Context-aware multi-agent systems. In: ACADIA Conference Proceeding (2014)
7. Peter, B.: Emergence as a design strategy in urban development using agent-oriented modeling in simulation of reconfiguration of the urban structure. In: ECAADE 2012 Conference Proceeding (2012)
8. Hao, H., Ting, L.: Floating bubbles, an agent-based system for layout planning. In: CAADRIA 2010 Conference Proceeding (2010)
9. Gideon, A.: Agent-based social pedestrian simulation for the validation of urban planning recommendations. In: SIGRADI 2012 Conference Proceeding (2012)
10. Kiefer, A.W., Bonneaud, S., Rio, K., Warren, W.H.: Quantifying the coherence of pedestrian groups. In: Proceedings of the Cognitive Science Society, Berlin, Germany (2013)
11. Oasys webpage. https://www.oasys-software.com/news/can-pedestrian-simulation-tools-really-model-humans-accurately/, Accessed 5 July 2021
12. Reynolds, C.W.: Flocks, herds, and schools: a distributed behavioral model. In: Computer Graphics, SIGGRAPH 1987 Conference Proceedings, pp. 25–34 (1987)
13. Unreal engine webpage: Environment Query System: Overview. Accessed 5 July 2021, https://docs.unrealengine.com/en-US/Engine/ArtificialIntelligence/EQS/EQSOverview/index.html

Interactive Media in Public Spaces:

Ethics of Surveillance, Privacy and Bias

Ana Herruzo[1,2(✉)] 🅘

[1] Universidad Politécnica de Madrid, P Juan XXIII, 11, 28040 Madrid, Spain
Ae.herruzo@alumnos.upm.es
[2] Arizona State University, Tempe, AZ 85281, USA

Abstract. When designing public interactive environments, new advances in computing and data collection techniques from the users can enhance the public's engagement and interaction with the designed space. Consequently, ethical questions arise as the ambiguity surrounding user information extraction and analysis may lead to privacy issues and biases. This paper examines topics of surveillance, privacy, and bias through two interactive media projects exhibited in public spaces. In particular, this paper focuses on analyzing the curation and manipulation of a database of an AI-powered and computer vision-based interactive installation that uses advances in computing such as DeepLearning and Natural Language Processing but runs into privacy and gender bias issues. This paper aims to showcase how decision-making, data curation, and algorithmic processes can directly impact and reinforce surveillance, privacy, and bias in public spaces.

Keywords: Media architecture · Artificial intelligence · Ethics

1 The Shifting Paradigm

We live in unprecedented times, where the rise of technology and artificial intelligence are now contributing to unique new possibilities in art and architecture and contributing to society's advances in many ways. In a similar manner that digitalization has impacted society, creating new economic models, it is also essential to reflect on its influence at cultural and artistic levels and the consequences of implementing new technological advances in our practices, the users, the cities, or today's society. As more artists and architects begin to integrate digital media into architectural and public spaces, benefiting from using the latest technologies to create interactive experiences, some ethical questions can arise.

Due to the rapid evolution in hardware and computing power, machine learning is becoming more accessible, affordable, and widespread; a whole new range of tools are available to artists, architects, and designers to explore human-computer interactions and digital art. When bridging art with new technological advances such as surveillance systems and machine learning algorithms, we begin working and manipulating data from users. We encounter an interesting space, an interstitial bubble, where questions and ethics of surveillance and privacy should be accounted for, especially now that it has

© Springer Nature Singapore Pte Ltd. 2022
D. Gerber et al. (Eds.): CAAD Futures 2021, CCIS 1465, pp. 135–150, 2022.
https://doi.org/10.1007/978-981-19-1280-1_9

been demonstrated that some machine learning algorithms can contribute to perpetuating cultural and stereotypical biases [1].

When designing public interactive environments, new advances in computing and data collection techniques from the users can enhance the public's engagement and interaction with the designed space. Consequently, ethical questions arise as the ambiguity surrounding user information extraction and analysis may lead to privacy issues and biases. In this paper, we examine data collection, privacy, and gender bias by studying a couple of digital media interactive projects in the public space of two different museums. We examine these projects by looking at their design process, the technology implemented, and their impact on Surveillance, Privacy, and Bias. We pay special attention to analyzing the creation and manipulation of a database of an AI-powered and computer vision-based interactive installation deployed in the public space of a museum in Los Angeles, CA. This project uses advances in computing such as DeepLearning and Natural Language Processing to create deeper engagements with users and add new layers of interactivity, but as a consequence, runs into issues of privacy and gender bias.

This paper aims to exemplify how when integrating new technologies into interactive art pieces. When deploying them into public spaces, ethics, trust, and transparency towards users' data in contemporary digital culture should be accounted for during the design process. Also, ultimately, it aims to showcase how decision-making, data curation, and algorithmic processes can directly impact and reinforce surveillance, privacy, and bias in public spaces.

2   Interactive Environments and Architecture

In his book New Media Art, Mark Tribe [2] describes the term New Media art as "projects that make use of emerging media technologies and are concerned with the cultural, political, and aesthetic possibilities of these tools". He locates New Media Art as a subset of two broader categories: Art and Technology and Media Art. Tribe also explains how the term is used interchangeably with categorical names like "Digital art," "Computer art," "Multimedia art," and "Interactive art".

The work discussed in this paper falls within that same intersection, studying and utilizing scientific and technological disciplines with an overall application to different media environments, not just digital but also physical and human-scaled spaces.

The concepts of interactive art and environments have been present for many years now, starting with pioneering experiments such as in 1977 Myron Krueger's "Responsive Environments" [3], which explores different types of human-machine interaction, and the potentials of interactive art, and its implications in a number of fields. Since then, there has been much experimentation with different human-computer interactivity scenarios, such as computer vision, voice-activation, body and skeleton tracking, EEG brain waves, thermal cameras, and many types of sensors and controllers.

Technology can allow us to create interactive pieces where the user can be an active agent of the story and design media in a non-linear narrative, opposed to traditional time-based media such as cinema. These new technologies have also allowed us to take the media out of the box (the traditional fixed display), break with linear narrative storytelling, and make the user, with its interactive experience, the active driver of the story.

Throughout the cinematic century, the dominance of the moving image has been a single image on a single frame. Anne Friedberg, in her book The Virtual Window [4], does a quick run through the history of the cinematic century and the evolution of digital media formats, explaining how the traditional "canvas/frame/window" used in film has now multiple forms and shapes and is made of all kinds of hardware or materials or tridimensional physical environments. Andy Warhol's multimedia experiments, such as the expanded sensorium of performative "happenings" on "Exploding plastics inevitable shows" with the Velvet Underground or "The Eameses' Multimedia Architecture," are examples where artists and architects began to develop successful installations that broke with the single frame in the 1960s. However, overall, breaking the frame, and transitioning into new immersive, experiential, interactive environments, is a phenomenon expanding highly in the last decades with the rise of New Media Art (Fig. 1).

Fig. 1. Projection-mapping project for the opening of the exploratorium museum, by obscura digital.

Projection-Mapping is an excellent example of taking media out of the box, and creating custom formats, and using the building as a canvas to paint, design or interact. This technique has become widely used in the last years to augment media, create experiences on the facades of buildings, and turn buildings into fantasies and interactive spaces. Other examples of taking media out of the box and integrating it into the built environment are Gensler's new Digital Experience Design projects [5], having teams specializing in this discipline opening in several offices across the United States. These teams focus on integrating digital media into architecture and incorporating the user experience design into their projects. These teams that work with multiple applications in digital environments (ubiquitous connectivity, touchless solutions, mobile access, and data intelligence) are made of multidisciplinary roles, experience designers, UX designers, computer scientists, systems integrators, motion graphic artists, animators, media producers, and creative technologists; is essence hybrid art and technology teams.

Artist Refik Anadol [6] creates large immersive digital installations that interact with the architecture of the space, and he creates mesmerizing animations of particle systems using neurological, environmental, and geographic data, with the help of artificial intelligence; more specifically, Machine Learning. His work is an excellent example of how, along with the rapid evolution of real-time rendering engines, programmable shaders, and new algorithms, it is now possible to effectively create instant real-time data-driven media at large resolutions and excellent rendering quality. Due to this evolution in hardware and computing power, machine learning is also becoming more accessible, affordable, and widespread; therefore, an entirely new range of tools is available to explore user interactions and digital art, as exemplified in the case studies described in this paper.

2.1 The Shifting Context: Public Space and Its Audience

Interactive experiences using machine learning are often developed in closed environments such as labs and have a small pool of user testing during the project's development. In interactive art, they are often exhibited in private spaces, where issues of privacy or bias will not stand out as much, given the exposure of the piece and the number of observers.

This changes when we work in a public space, the user pool, becomes the general population, and the exposure of the media can reach anyone passing through the space, and it can be filmed, photographed and shared. This shift from private or semi-public to exhibiting in public space detonated this paper's findings. Along with the realization of the ethical issues that can arise when using interactive experiences, computer vision, and user analysis based generated art in public environments. The boundaries of consent, privacy and biases become blurred.

In these case studies, we describe two projects that create interactive digital canvases in public environments in the courtyards of two museums. We perform an analysis looking at topics of surveillance, privacy and bias, and methodologies implemented to avoid the ethical issues that could be derived from them.

3 Surveillance, Privacy and Bias

Nowadays, as part of the digital age, advances in Computer Science -while becoming part of our daily lives- are triggering a series of unprecedented models in today's society. Shoshana Zuboff [7], in her book "The Age of Surveillance Capitalism", describes the shift that has occurred from traditional mass production lines in factories from the 19th century to a new business model in digital form that is fully reliable in users data, predictions models, and behavior manipulation to produce revenue. She performs an exciting walkthrough of the evolution of the history of Silicon Valley tech giants and digital corporations such as Google, Microsoft, Apple to describe how this new business model has emerged in the last 20 years. This model relies highly on user data and its analysis using machine learning algorithms to extract accurate predictions. She argues that these predictions are ultimately able to impact the behavior of millions of users worldwide.

Topics of users' data gathering, and privacy are increasingly becoming of public interest these days, for example, on Netflix's documentary "The Social Dilemma". In an interview with a Silicon Valley software developer, he describes how many young programmers developed groundbreaking technologies and software tools such as Google Drive, Google Maps, but are now highly concerned about how design decisions that were made by 25/35 year-olds are impacting the life and behaviors of billions of people. The industry was not prepared and unaware of the impact these advances and tools would have worldwide as Surveillance Capitalism becomes the dominating form of digital life. The importance of decision-making is a crucial topic to be revisited further in this paper.

During the COVID-19 pandemic, it has been proved that forms of surveillance are beneficial, such as tracking the spread of the virus with the pandemic applications, Q.R. codes scanning, radarCovid in Europe and others implemented in Asia. The European app does not serve the goal of surveillance capitalism and behavior surplus, or engage in prediction practices, as stated on their privacy terms [8] but instead is solely meant to help stop the spread of a virus that has dramatically disrupted the world as we have known it up to now. Nevertheless, in many other cases, these applications, as a consequence of surveillance capitalism, incur on instrumentalism [7]. These prediction paradigms are often offered to us as the technological dream of public safety, national security, fraud detection, and even disease control and diagnosis. They are often accepted as valid and objective, but instead, they can naturalize and amplify discriminatory outcomes [9].

3.1 Machine Learning Biases

An algorithm bias or A.I. bias is a phenomenon that happens when an algorithm generates results that are "prejudiced" due to wrong assumptions during the process [10].

Machine learning derived biases are essential topics discussed and highly studied nowadays in computer science due to their possible impact and potential to harm groups or individuals. It is a difficult task since it also has social, cultural, and legal consequences in their applications beyond their mere scientific development.

There are severe concerns about how A.I. and Machine learning can perpetuate cultural stereotypes, resulting in biases regarding gender and race. Some studies measure attitudes towards, and stereotypes of social groups [11], and these cultural stereotypes have been proven to propagate in some machine learning algorithms when using text-based data sets. In a study on "Semantics derived automatically from language corpora contain human-like biases" [1], scientists replicated findings where female names are more associated with family than career words, compared with male names. To avoid biases being perpetuated by automated machine learning algorithms, scientists are beginning to propose different methods for algorithms to avoid these biases. For example, [12] provides a methodology for modifying an embedding to remove gender stereotypes and developed algorithms to "de-bias" the embedding.

In the city's environment, we are also beginning to acknowledge issues relating to the application of these algorithms. Journalist J. Fasman [13] talks about the "predictive policing programs" implemented in the U.S. to deploy more police officers in specific spaces by using historical crime data, but this data is not objective since neighborhoods can change over time. He highlights how these programs can run the risk of essentially calcifying past racial biases into current practices.

Machine learning bias often stems from problems introduced by the individuals who design or train the machine learning systems. While creating machine learning algorithms and training them with data sets, data scientists should try to work with the data in ways that will minimize these biases. Nevertheless, decision-makers are the ones who need to evaluate when machine learning technology should or should not be applied. In the case studies analyzed in this paper, the designers/developers are both scientists and decision-makers; it is an interesting contemporary scenario that will allow us to make decisions and curate the database to avoid creating biases.

4   Case Studies

4.1   Emergence

"Emergence" is an interactive and audiovisual project, for the opening of the Exploratorium museum. It was created by Obscura Digital, and the author was the Lead Interactive Engineer. The main facade was displaying media created in the studio by filming analog experiments and were designed to fit the front facade of the Pier. The second façade, located in the museum's courtyard, displayed an interactive experience, and it is the part we will focus on in this paper (Fig. 2).

Fig. 2. Interactive experience consisting of thermal imaging surveillance of public space users and projection-mapped onto the museum's façade.

This interactive experience used thermal imaging cameras to record the attendees and project them onto the building, to make the attendees part of the project itself. In this way, they saw themselves augmented onto the building but transformed into a different colour palette since the thermal imaging cameras can capture and analyze temperatures of objects and return media by applying different color ranges to temperatures. Thermal imaging systems function similarly to conventional cameras and are used to capture an image of an environment, but instead of using visible light to construct an image, a thermal imaging system utilizes I.R. (infrared) light to form an image.

The two thermal imaging cameras were placed in public space, and any person utilizing the space could easily enter the field of view of these cameras and suddenly be projected and augmented to a canvas ten times their size. Their images could be seen from afar and several streets away due to the scale of the building and the projection.

The footage from the cameras was live-streamed to the severs and projection-mapped onto the Museum's facade.

Analysis

Surveillance. No surveillance or privacy issues were raised during the design process or after the project. citizens attending the show enthusiastically engaged with the installation, bringing different objects up to the cameras and playing with them to explore how temperature changes were reflected in the footage.

Privacy. Thermographic cameras are widely used in veterinary testing [14] to diagnose animals without carrying out exploratory procedures. Perhaps if this project had been developed in a post-pandemic future, now that our temperature is being taken prior to entering many establishments, flights and restaurants, privacy issues could be raised. If a person walking through the courtyard were to have a fever, this information could be revealed.

Bias. Publicly displaying users' corporeal thermal images could also lead to bias, for example, viewers assuming that someone with a high temperature could automatically be infected by COVID-19.

4.2 WISIWYG: What I See is What You Get

The second case study that we look at in this paper goes much deeper into surveillance, privacy and bias since during this project, these issues were raised during the design process and user testing, which allowed us to correct them. We will detail a complete description of the installation, the design process, and how we managed to rectify the possibility of incurring issues of privacy and bias.

This project was developed in an academic environment, as part of the Applied Computer Science-Media Arts program at Woodbury University and led by professors Ana Herruzo and Nikita Pashenkov. It was exhibited at a public open space of the J. Paul Getty Museum. It consists of a large interface that essentially serves as a mirror to the audience and incorporates live interactive visuals based on data extracted from the user's movements and facial expressions, age, and gender. These visuals are accompanied by synthetic texts generated using machine learning algorithms trained on the Museum's art collection (Fig. 3).

Fig. 3. Users while experiencing the installation and playing with their silhouettes on the screen.

As part of an academic project, in conversations with educational specialists at the museum, interest was shown in exploring human emotions as a thematic element in a project that aimed to merge art and technology. The theme of working with human emotions led us to explore the state-of-the-art scenarios and algorithms that could help us detect facial expressions in humans. Later, we explored how we could connect human emotions depicted in the art pieces from the Museum's art collection with an installation that would read the attendees facial expressions. Part of the project included guided visits to the Museum to study some of the artworks in its collection, learning about the artists' intentions and what the works are communicating in their portrayal of emotions and facial expressions.

While studying the Museum's art collection [15] that comprises Greek, Roman, and Etruscan art from the Neolithic Age to Late Antiquity; and European art from the Middle Ages to the early 20th century; with a contemporary lens, questions arose regarding static and finished pieces of art, in contrast to interactive and responsive artworks [3]. A key observation guiding our concept was that artworks in exhibitions and museums are typically accompanied by a title and a brief description. In our case, unique visuals would be generated based on user interaction, and we proposed creating new synthetic titles and descriptions to accompany each user engagement. It seemed fitting that the textual output would be generated by machine learning algorithms trained on existing texts from the Museum's art collection to create a unique connection bridging the Museum's carefully curated static content with new dynamically generated visuals. The resulting installation utilizes a Kinect sensor to analyze and mirror the users' movements and a separate camera to read facial expressions via computer vision and Deep Learning algorithms, using their outputs in the next stage as a basis for text generation based

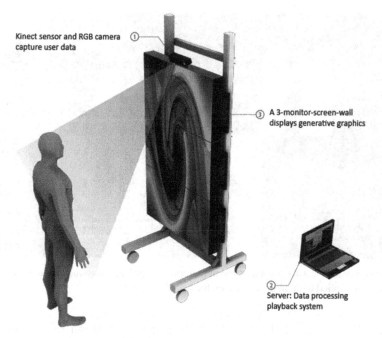

Kinect sensor and RGB camera ①
capture user data

③ A 3-monitor-screen-wall
displays generative graphics

② Server: Data processing
playback system

Fig. 4. Installation diagram

on natural language models trained on the text descriptions of the Getty Museum's art collection (Fig. 4).

Our goal was to create a digital art interface with an intimate connection with the users while simultaneously generating new periodic content that is constantly evolving, changing, never the same. This is in direct contrast to architectural buildings, or the art exhibited by the Museum, both consisting of finished, static objects. The project called "WISIWYG" is a play on the popular acronym "What You See Is What You Get" (WYSIWYG), based on the idea that the installation incorporates computer vision processing and machine learning algorithms to generate outputs according to what it sees from its own perspective. In some way, it is an "intelligent" facade that serves as a mirror that reads you and returns media generated with your data, movements, and interactions, and also outputs a text description of the new art piece created with all of the information extracted (Fig. 5).

Features that make this project unique include the combination of real-time generative graphics with exciting new machine learning models.

Computer Vision. The primary driver of the project is the camera input, processed through computer vision and machine learning algorithms. In the course of users' engagement, computer vision is first used to isolate faces and determine the number of people in the camera's field of view. This step is accomplished via a traditional computer vision face tracking method using the popular Open Computer Vision (OpenCV) library [16]. The second step uses a Deep Learning model based on a Convolutional Neural Network

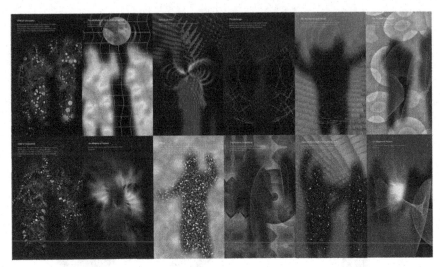

Fig. 5. Gallery showcasing several of the animations designed by the students and the different color palettes applied to them depending on the users' emotions.

(CNN) [17] constructed in Python with the help of Keras [15] and TensorFlow frameworks. The Python code to construct the facial expression detection models is available in an open GitHub code repository [19].

In addition to facial expression detection, the project incorporated Deep Learning models for age and gender detection. The age detection was built on a successful CNN architecture based on Oxford Visual Group's (VGG) model [20].

Data Curation. This portion of the project began by screening Getty's art collection online and selecting artwork depicting people. The focus was on artefacts containing people, based on the total number of pieces on display at the Getty Center at the time of data collection (1,276 results according to the website). As the next step, a database was created by recording the artworks' titles and descriptions and subjectively estimating the number and ages [23] of people featured.

The database was created by analyzing the 1276 artworks and collecting the following data for each piece: Title, Artist/Maker, Date, Description, Primary Sentence, Number of People, Gender, Age, Emotion and reference image (Fig. 6).

In analyzing the Getty's Museum's art collection, students experimented with the Deep Learning natural language model GPT-2 [24], an acronym for Generative Pre-Trained Transformer, released by the non-profit foundation OpenAI in February 2019. The language model generates convincing responses to textual prompts based on set parameters such as the maximum length of response and 'temperature' indicating the relative degree to which the output conforms to the features resembling training data.

Initially, the text prompts were pre-generated by the students based on their own analysis of the following: artworks and consisted of short singular and plural descriptions like "sad young boy", "two happy women", "old, scared people". The text prompts were interactively inputted to the GPT-2 model to generate responses entered into a database

Title: Rembrandt Laughing
Artist/Maker: <u>Rembrandt Harmensz. van Rijn</u> (Dutch, 1606 - 1669)
Date: 1628
Description: Intently interested in the expression of human emotion, Rembrandt often used himself as his own model in his early years as an independent master in Leiden. Here, in a small and freely painted work, he appears in the guise of a soldier, relaxed and engaging the viewer with a laugh. For this sophisticated self-portrait, painted at age twenty-one or twenty-two, Rembrandt combines a study of character and emotion (known in Dutch as a *tronie*) with a rare jovial self-presentation. The lively, short brushwork in the face and brisk handling of the neutral background convey a sense of spontaneity and immediacy. This is one of a small number of paintings by Rembrandt from the late 1620s executed on copper. He signed it in the upper-left corner with his monogram of interlocking letters, "RHL" which he used only briefly, from late 1627 to early 1629.
Primary Sentence: He appears in the guise of a soldier, relaxed and engaging the viewer with a laugh.
Number of People: 1
Gender: Male
Age: 20s
Emotion: happy
Image:<u>https://media.getty.edu/museum/images/web/enlarge/3447160 1.jpg</u>

Fig. 6. Item from the database created for the 1276 artworks from the Getty museum's collection.

and associated by rows with tagged columns for age, the number of people, gender, and facial expressions. The database content was then programmatically correlated with the outputs of computer vision processing and Deep Learning classification using the Pandas library in Python by selecting a random database cell containing a response that matches detected facial expression tags. Finally, the selected response was rendered as the text description that accompanied visual output onscreen (Fig. 7).

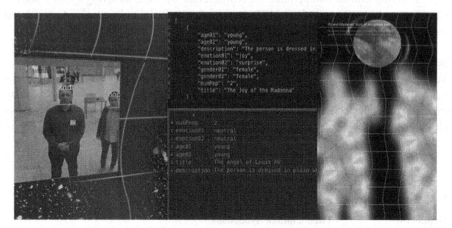

Fig. 7. Image of the students testing the face recognition algorithms at the Getty on the left. On the right, examples of user data incoming into the system and the generated graphic content.

Analysis

Surveillance. Computer vision, the installation uses camera-based computer vision to analyze the facial expressions of the users. The piece is surveilling the attendees.

Privacy. The installation creates visuals, using the data from the users, performing an aesthetic and artistic interpretation of the data, an abstraction. Therefore, this data is not transparently displayed. However, the installation created text descriptions to accompany these graphics, and these descriptions use gender pronouns and allegations of emotions and age (Table 1).

Table 1. Example of a prompt generated by the machine learning algorithms. Data fed is on the left, and A.I. generated prompt on the right.

Input data from users	Machine learning generated prompts
Gender: "Woman" **Emotion:** Fear **Age:** Young	**Title**: Child of Sybessems **Description:** To her horror, she thought only of demons. To explain her blindness, the Christian artist painted a detailed representation of himself in front of the blind girl, with Her blind left eye partially obscured by an opaque, tear-bald mask

Bias. At the beginning of user testing, some of the students raised concerns about using gender in the camera-based analysis and generated texts, implying that it could potentially make some of the attendees uncomfortable. This also raised important questions regarding the role of gender in today's society and opened the door for further discussions involving human interactions. We realized that we were incurring a bias since the algorithm we were using performed a binary gender analysis, and we had not considered non-binary gender identification.

Decision Making: De-gendering
We attempted to address this issue in part by avoiding the use of gender and programmatically manipulating the generated texts, as well as manually editing our database by screening for male and female pronouns, attempting to "de-gender" it by replacing those with the plural "they, "or neutral "person", wherever appropriate (Table 2).

Table 2. Example of the de-gendered A.I. prompts, ones the databases was revisited.

Input data from users	Degendered prompts
Gender: -- **Emotion:** Fear **Age:** Young	**Title**: Child of Sybessems **Description:** To their horror, the person thought only of demons. To explain their blindness, the Christian artist painted a detailed representation of themselves in front of the blind person, with their blind left eye partially obscured by an opaque, tear-bald mask

In this particular case, we purposely avoided gender biases by ultimately deciding not to use the data from gender recognition and by carefully manipulating and re-formatting our database. For a more detailed description of this project's creative and technical processes, see the following references [25] and [26].

5 Discussion

With the intent to bring art and technology closer and explore the boundaries of interactive art, users' interaction, and the levels of communication that the art piece can achieve with its viewer and user, we encounter an interesting space where questions and ethics of surveillance and privacy become relevant. A practice involving surveillance might be legal in some contexts but could be considered unethical in terms of human rights [27]. This adds to concerns about machine learning perpetuating cultural and stereotypical biases; or new technological advances impacting today's society. This phenomenon also begins to be present in artistic environments as well, especially when working with interactive experiences in public spaces.

Wired magazine published an article recently interviewing the authors of "Face of the riot" [28], who created a web page [29] where they duplicated every face from the 827 videos taken from the insurrection at the Capitol on January 6, 2021. They created the web page using open-source software and machine learning algorithms similar to those presented in this paper. It is crucial to understand how these databases are curated and later used to train the A.I. algorithms. Faces of the riot use 827 videos from the riots, but, what if this database could be manipulated? Or What if some of the videos were from people who were not present at the actual riot? These questions showcase how nowadays, invasions of users' privacy can occur quite easily, as well as exposure to biases. Therefore, implementing new policies and regulations regarding users' data and privacy is beginning to be highly discussed these days [30].

In his book Automated Media [31], Andrejvic brings up some interesting points related to this topic, highlighting that if automated processes were developed by decentralized systems and by independent people, and if we all had the skills to program our algorithms, we could avoid biases. Nevertheless, he explains how we need an economic investment to support the research in our current societal model to build the infrastructure and the code to generate this automation, maintain it, and debug it. Moreover, within the current economic models, this implies that the organism in charge of funding and creating the automation will inevitably make choices that will influence the level of biases that the automation will carry.

However, in this paper, we try to go further, since we showcase how in our case study while using a decentralized system, crafting our database, and having the skills to train the algorithm, we also run into the same ethical questions as if an external socioeconomic institution were to take care of financing and servicing the technical aspects of the project. There are decisions to be made when choosing and curating a database to train algorithms. That decision-making process is not part of the automation but is done prior and is done by us. We are the decision-makers.

6 Conclusions

When designing public interactive environments, new advances in computing and data collection techniques from the users can enhance the public's engagement and interaction with the designed space. Consequently, ethical questions arise as the ambiguity surrounding user information extraction and analysis may lead to privacy issues and biases.

Due to the rapid evolution in hardware and computing power, machine learning is becoming more accessible, affordable, and widespread; a whole new range of tools are available to artists, architects, and designers to explore human-computer interactions and digital art. In interactive art, they are often exhibited in private spaces, where issues of privacy or bias will not stand out as much, given the exposure of the piece and the number of observers. This changes when we work in a public space, the user pool, becomes the general population, and the exposure of the media can reach anyone passing through the public space, and the content can be filmed, photographed, and shared.

This shift from working in a private or semi-public environment to exhibiting in public space detonated ethical issues using interactive experiences, computer vision, and user analysis based generated art. The boundaries of consent, privacy and biases became blurred.

This paper brings transparency to a methodology involved in curating a database fed into an algorithm and how it was manipulated to avoid biases. Aiming to communicate and help designers be aware and more conscious of the data they are using and how their decisions can help avoid incurring biases and privacy issues.

The findings presented in this paper contribute to an ongoing research line that seeks to showcase how new methodologies that benefit from new technological advances could be implemented in the arts to contribute to more ethical frameworks. As technology moves at a faster pace than policy regulations at governmental and institutional levels, we as designers can begin to acknowledge the ethical realities that we may incur while implementing new advances in computing and technology into our works and come up with solutions. Artists and architects can have complete autonomy towards the work that they develop and are usually connected in networks where their work will be displayed, exhibited, and shared with the public, reaching an audience. Therefore, with the use of computational means and decision making, we can begin to tackle and address issues of privacy and bias, propose solutions, and come up with innovative and contemporary practices that can serve as examples to follow.

References

1. Caliskan, A., Bryson, J.J., Narayanan, A.: Semantics derived automatically from language corpora contain human-like biases. Science **356**, 183–186 (2017). https://doi.org/10.1126/science.aal4230
2. Tribe, M., Jana, R., Grosenick, U.: New Media Art (Taschen Basic Art). Taschen America, LLC (2006)
3. Krueger, M.W.: Responsive environments. In: Proceedings of the June 13–16, 1977, National Computer Conference, pp. 423–433. Association for Computing Machinery, New York, NY, USA (1977). https://doi.org/10.1145/1499402.1499476

4. Press, T.M.: The Virtual Window, https://mitpress.mit.edu/books/virtual-window. Accessed 1 Oct 2019

5. ProjectsIDigital Experience Design I Expertise, https://www.gensler.com/expertise/digital-experience-design/projects. Accessed 20 Mar 2021

6. Anadol, R.: Synaesthetic architecture: a building dreams. Archit. Des. **90**, 76–85 (2020). https://doi.org/10.1002/ad.2572

7. Zuboff, S.: The Age of Surveillance Capitalism: The Fight for a Human Future at the New Frontier of Power. Public Affairs, New York (2019)

8. App RadarCOVID. https://radarcovid.gob.es/. Accessed 16 Mar 2021

9. Gill, K.S.: Prediction paradigm: the human price of instrumentalism. AI Soc. **35**(3), 509–517 (2020). https://doi.org/10.1007/s00146-020-01035-6

10. What is Machine Learning Bias (AI Bias)?. https://searchenterpriseai.techtarget.com/definition/machine-learning-bias-algorithm-bias-or-AI-bias. Accessed 9 Mar 2021

11. Nosek, B.A., Banaji, M.R., Greenwald, A.G.: Harvesting implicit group attitudes and beliefs from a demonstration web site. Group Dyn. Theory Res. Pract. **6**, 101–115 (2002). https://doi.org/10.1037/1089-2699.6.1.101

12. Bolukbasi, T., Chang, K.-W., Zou, J., Saligrama, V., Kalai, A.: Man is to Computer Programmer as Woman is to Homemaker? Debiasing Word Embeddings. ArXiv160706520 Cs Stat. (2016)

13. Surveillance and Local Police: How Technology Is Evolving Faster Than Regulation. https://www.npr.org/2021/01/27/961103187/surveillance-and-local-police-how-technology-is-evolving-faster-than-regulation. Accessed 1 Feb 2021

14. Thermal Camera Applications - Veterinary I Non-Invasive Testing of Animals. https://www.pass-thermal.co.uk/thermal-camera-applications-veterinary. Accessed 20 Mar 2021

15. Collection (Getty Museum). http://www.getty.edu/art/collection/. Accessed 20 Mar 2021

16. OpenCV: https://opencv.org/. Accessed 7 July 2019

17. Lopes, A.T., de Aguiar, E., De Souza, A.F., Oliveira-Santos, T.: Facial expression recognition with convolutional neural networks: coping with few data and the training sample order. Patt. Recognit. **61**, 610–628 (2017). https://doi.org/10.1016/j.patcog.2016.07.026

18. Home - Keras Documentation, https://keras.io/. Accessed 7 July 2019

19. ACS-Woodbury: ACS-Woodbury/WISIWYG (2019)

20. Parkhi, O.M., Vedaldi, A., Zisserman, A.: Deep face recognition. In: Proceedings of the British Machine Vision Conference 2015, pp. 41.1–41.12. British Machine Vision Association, Swansea (2015). https://doi.org/10.5244/C.29.41

21. IMDB-WIKI - 500k+ face images with age and gender labels, https://data.vision.ee.ethz.ch/cvl/rrothe/imdb-wiki/. Accessed 7 July 2019

22. Serengil, S.: Apparent Age and Gender Prediction in Keras. https://sefiks.com/2019/02/13/apparent-age-and-gender-prediction-in-keras/. Accessed 20 Mar 2021

23. Rothe, R., Timofte, R., Gool, L.V.: DEX: Deep EXpectation of apparent age from a single image. In: 2015 IEEE International Conference on Computer Vision Workshop (ICCVW), pp. 252–257. IEEE, Santiago, Chile (2015). https://doi.org/10.1109/ICCVW.2015.41

24. Radford, A., Wu, J., Child, R., Luan, D., Amodei, D., Sutskever, I.: Language Models are Unsupervised Multitask Learners. 24

25. Herruzo, A., Pashenkov, N.: Collection to creation: playfully interpreting the classics with contemporary tools. In: Yuan, P.F., Yao, J., Yan, C., Wang, X., Leach, N. (eds.) Proceedings of the 2020 DigitalFUTURES, pp. 199–207. Springer, Singapore (2021). https://doi.org/10.1007/978-981-33-4400-6_19

26. Herruzo, A., Pashenkov, N.: "What I See Is What You Get" explorations of live artwork generation, artificial intelligence, and human interaction in a pedagogical environment. In: Brooks, A., Brooks, E.I. (eds.) Interactivity, Game Creation, Design, Learning, and Innovation. pp. 343–359. Springer International Publishing, Cham (2020). https://doi.org/10.1007/978-3-030-53294-9_23

27. Lyon, D., Haggerty, K.D., Ball, K. (eds.): Routledge handbook of surveillance studies. Routledge, Abingdon, Oxon ; New York (2012)

28. A Site Published Every Face From Parler's Capitol Riot Videos. https://www.wired.com/story/faces-of-the-riot-capitol-insurrection-facial-recognition/

29. Faces of the Riot. https://facesoftheriot.com/. Accessed 3 Mar 2021

30. Williams, S.: Data Action: Using Data for Public Good. MIT Press, Cambridge (2020)

31. Andrejevic, M.: Automated Media. Routledge, Milton Park (2019)

Algorithms for Floor Planning with Proximity Requirements

Jonathan Klawitter$^{(\boxtimes)}$ ⓘ, Felix Klesen ⓘ, and Alexander Wolff ⓘ

Universität Würzburg, Würzburg, Germany
Jonathan.klawitter@uni-wuerzburg.de

Abstract. Floor planning is an important and difficult task in architecture. When planning office buildings, rooms that belong to the same organisational unit should be placed close to each other. This leads to the following NP-hard mathematical optimization problem. Given the outline of each floor, a list of room sizes, and, for each room, the unit to which it belongs, the aim is to compute floor plans such that each room is placed on some floor and the total distance of the rooms within each unit is minimized.

The problem can be formulated as an integer linear program (ILP). Commercial ILP solvers exist, but due to the difficulty of the problem, only small to medium instances can be solved to (near-) optimality. For solving larger instances, we propose to split the problem into two subproblems; floor assignment and planning single floors. We formulate both subproblems as ILPs and solve realistic problem instances. Our experimental study shows that splitting the problem helps to reduce the computation time considerably. Where we were able to compute the global optimum, the solution cost of the combined approach increased very little.

Keywords: Floor planning · Proximity requirements · Integer linear programming · NP-hard

1   Introduction

Designing architectural floor plans of an office building is a challenging endeavor involving a multitude of tasks. Among other things, one has to draft a building outline, decide on the number of floors, and place rooms, stairs and elevators. At every step of the process, the planner must meet some predefined or implicitly understood requirements. Therefore, floor planning is a cumbersome process of trial and error requiring a significant amount of human labour and time. It is thus of interest to support such manual processes with (semi-) automated approaches [17].

We consider a simpler problem. In our variant of the problem, the planner has already fixed the outline of each floor, and the position and width of the corridors. This is often the case due to fixed lot sizes as well as distance and construction rules. We further assume that the planner has a list of rooms and, for each room, a minimum size.

© Springer Nature Singapore Pte Ltd. 2022
D. Gerber et al. (Eds.): CAAD Futures 2021, CCIS 1465, pp. 151–171, 2022.
https://doi.org/10.1007/978-981-19-1280-1_10

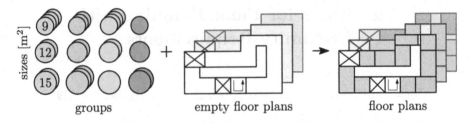

Fig. 1. Given a set of rooms, each with its size and group, and a set of empty floor plans, the FLOOR PLANNING WITH GROUP PROXIMITY problem asks for a placement of the rooms such that rooms belonging to the same group are close together.

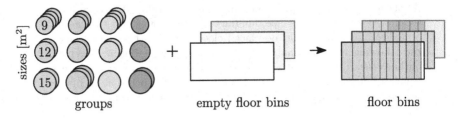

Fig. 2. The FLOOR ASSIGNMENT problem asks for a mapping of rooms to floors such that rooms of the same group appear, preferably, on only few and neighboring floors.

Allen and Fustfeld [2] highlighted the importance of the architectural layout for communication. Based on their observation, some works assume that the planner has exact information about which pairs of rooms should be placed next to each other [23]. We model proximity relations differently. In the spirit of Allen and Fustfeld, we aim to arrange rooms with respect to the *organisational units* that will use them later. Technically, the task is to map rooms to location within floors of the building such that the rooms that belong to the same unit or *group* are placed close together; see Fig. 1. We call this problem FLOOR PLANNING WITH GROUP PROXIMITY and define it formally in Sect. 3.

Most variants of floor planning (including ours) are variants of basic combinatorial *packing problems* such as KNAPSACK or BIN PACKING; hence they are usually NP-hard. For this reason it is unlikely that efficient algorithms for floor planning exist. Still, algorithms that support architects in this phase of the planning process are needed since computers are usually faster than humans in solving NP-hard optimization problems to (near-) optimality.

We show that FLOOR PLANNING WITH GROUP PROXIMITY can be formulated as an integer linear program (ILP). Commercial ILP solvers exist, but due to the complexity of our problem, only small to medium problem instances can be solved to (near-) optimality. For large problem instances, we propose to split the problem into two independent subproblems. First, we map the rooms to floor "bins" considering the proximity of rooms within their groups; see Fig. 2. For example, it is preferable to map all rooms of one group to the same floor. If this is not possible, they should be mapped to only few neighboring floors [2]. We call this problem FLOOR ASSIGNMENT WITH GROUP PROXIMITY. Second, we solve FLOOR PLANNING WITH GROUP PROXIMITY for each floor separately.

Contribution. We introduce a new type of floor planning problem that occurs when planning office buildings. In our model, we assume that for each floor we are given its outline, the corridor, and the stairs. Our objective is to place the rooms into the empty floor plans such that rooms that belong to the same group are close together. For the precise problem definition, see Sect. 3.

We formulate our floor planning problem as an ILP; see Sect. 4. According to our computational experiments, it takes too long to find optimal solutions of this global formulation for large instances of the problem. Thus we split the global problem into two independent subproblems; floor assignment and (single) floor planning, which we then solve by separate ILPs (see Sect. 5). While optimal solutions for the subproblems do not necessarily lead to optimal solutions of the global problem, our computational experiments show that the loss in the global objective is acceptable and the runtime improvements are considerable. We also test a simple greedy heuristic for the floor assignment subproblem; see Sect. 8.

While the runtime of ILP solvers is difficult to predict, linear optimization is quite powerful and allows the user to add additional constraints (such as "these two groups must go to the basement") easily. If a user is dissatisfied with a solution that is provably optimal, then it is not the algorithm to be blamed – but the model. In other words, the objective must be changed or further constraints must be added in order to exclude solutions with certain undesired features. Once the model has been settled, it is possible to drop the requirement of optimal solutions; quite often ILP solvers quickly find near-optimal or even optimal solutions, but then need very long to *prove* their optimality; see Table 5.

We think that our work has the potential to support architects in finding sustainable solutions both for themselves and the users of their designs.

2   Related Work

The literature on floor planning problems is quite diverse. On the one hand, a panoply of different algorithmic methods have been used to tackle such problems. They range from logic programming [16], constraint programming [7], quadratic programming [21], over using shape rules and grammars[10,12,28] or graph-theoretic tools [18,22,23] to evolutionary algorithms [12,15] to name just a few. We refer the interested reader to the surveys by Del Río-Cidoncha et al. [8] and by Lobons and Donath [17]. On the other hand, nearly every paper considers a different problem definition. We can group these variants as being purely combinatorial (where only room sizes matter), more geometric (where the actual shapes and their aspect ratios matter), or purely topological (where only adjacencies matter). Our problem variant includes both geometric and, with the group proximity constraints, topological aspects.

A purely combinatorial version of floor planning is the well-known BIN PACK-ING problem, where items of different sizes must be packed into bins, each of a fixed capacity, in a way that minimizes the number of bins used. Bergman et al. [5] consider a variant called BIN PACKING WITH MINIMUM COLOR FRAGMENTATION, where each item is associated with a color. Then the goal is to find a bin packing where items of a common color are placed in the fewest number of bins possible. This problem is closely related to FLOOR ASSIGNMENT WITH GROUP PROXIMITY though the quality of a solution is measured differently. If rooms are set to be rectangular and their sizes are prescribed (with aspect ratios of rooms either fixed or bounded), we are in the range of RECTANGLE PACKING problems [6,13].

Marson and Musse [18] showed how to generate floor plans for residential houses where just a few adjacencies are specified. They prescribed room sizes and used squarified treemaps to subdivide the fixed building outline. Knecht and König [15] used an evolutionary approach to generate subdivisions of a given rectangle into a given number of smaller rectangles (rooms). In a second step, they used a genetic algorithm to change the topology of the resulting subdivision and to obtain the desired adjacencies. More recently, Shi et al. [24] used a Monte-Carlo tree search to evaluate and select promising candidates among many floor plans that they build room by room. Due to their runtime, some of these approaches only work for small houses, but not for large office buildings.

If all allowed room adjacencies are already prescribed, the input of a floor planning problem takes the form of a (well-behaved) triangulated embedded planar graph. The corresponding floor plan is then called a *rectangular dual* [11]. Formally, a rectangular dual is a dissection of a rectangle into smaller rectangles such that the adjacency graph of the smaller rectangles is the given embedded graph. Upasani et al. [27] presented an iterative procedure that takes a rectangular dual and lower and upper bounds on the room dimensions (in x- and y-direction) as input. The algorithm then optimizes the layout by alternatingly computing network flows in the graphs that represent the horizontal and the vertical contacts between the rectangular rooms. Instead of expecting an adjacency matrix as input, Simon [26] generated floor plans with a genetic algorithm that minimizes traffic flow between rooms, e.g., class rooms in schools.

A problem related to floor planning is the *facility layout* problem where facilities have to be arranged efficiently within an organization. In contrast to floor planning, facility layout is less about subdividing a given building, but about the placement of the facilities and the resulting paths between them. The aim is to place facilities such that the paths allow for low material handling costs, short lead time, and high productivity. We refer interested readers to the surveys by Meller and Gau [20] and by Drira et al. [9]. Like for floor planning, there also exist multi-floor variants where departments have to be placed on floors and convenient positions for lifts have to be found; see a recent survey by Ahmadi et al. [1]. Interestingly, Meller and Bozer [19] also suggest a two-stage approach for the multi-floor facility layout problem.

Barth et al. [3] and Bekos et al. [4] considered a related problem that was motivated by drawing semantic word clouds. Given a set of rectangular shapes with fixed sizes and an adjacency graph defined on these shapes, place the shapes such that no two shapes overlap. Their objective was to realize the maximum number of desired adjacencies as side contacts of the rectangular shapes.

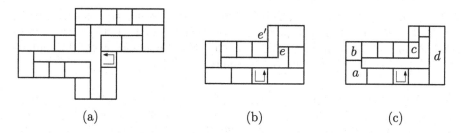

Fig. 3. Three orthogonal floor plans: (a) is valid; (b) is invalid since the corridor wall e does not overlap with the outer wall e' (when projected to the y-axis); and (c) is invalid since room a is not rectangular, room b shares not enough wall with the corridor for a door, room c has no window, and room d occupies two building corners.

3   Our Model and Problem Definitions

In this section, we describe a model for floor plans, allowed room placement, and group proximity that is tailored to our ILP approach. We then make the problem statements precise.

Floor Plans. A *floor plan* (of a single floor) describes a subdivision of the floor's *outline*, a simple polygon, into smaller polygons by inner walls. Each of the smaller polygons represents a room, the corridor, stairs, etc. Here we consider only *orthogonal floor plans*, that is, each wall is drawn either horizontally or vertically. We require that the *corridor* runs parallel to the outline, that is, the polygon describing the corridor is combinatorially the same as the polygon describing the outline; see Fig. 3(a). In particular, to keep our model simple, each vertical or horizontal segment of the corridor must overlap vertically or horizontally, respectively, with its combinatorial counterpart of the outline; see Fig. 3(b) for a counterexample. For the following requirements, see Fig. 3(c). We insist that all *rooms* are rectangular and stretch from the corridor to the outline. To ensure that each room gets a door and a window, we require that the room has a certain minimum overlap with the corridor and with the outline. To keep the model simple a room may occupy at most one corner (but see Sect. 9).

Empty Floor Plans. Part of the input of the floor planning problem are *empty floor plans*. Geometrically, we require that an empty floor plan consists of the outline, the corridor, at least one room that represents *stairs* or elevators, and potentially other *blocked areas*. Stairs connect the individual floors of a building with each other. Blocked areas represent, for example, sanitary facilities, which are often located at the same place on each floor. The *capacity* of an empty floor plan is the size of the area not covered by the corridor and blocked rooms.

We model an empty floor plan as follows. We subdivide the unoccupied area between the outline and the corridor into rectangles of two types; see Fig. 5. Each pair that consists of a corner of the outline and the corresponding corner of the corridor spans a *corner rectangle*. The rectangles that form the remaining area are *edge rectangles*. For the sake of brevity, we call corner and edge rectangles simply *corners* and *edges*, respectively. For each corner v, its capacity κ_v is its area. Similarly, for each edge e, the capacity of e is denoted by κ_e. The sum of these capacities is the capacity of the empty floor plan.

Room Placement. We now describe a set of rules that define how rooms can be placed into empty floor plans to obtain valid floor plans. Recall that each room comes with a (minimum) *size*. A *room placement* is a mapping of rooms into the unoccupied area, defined by the corners and edges, and their capacities. A valid room placement satisfies the following conditions.

(P1) Each room is either mapped to an edge or to a pair consisting of a corner and an adjacent edge.
(P2) For each corner v, at most one room may be mapped to v and this room must occupy v fully. Moreover, the size of the room must exceed the capacity of v. This excess must be enough for the room to admit a door and a window.
(P3) For each edge e with adjacent corners v and v', the sizes of rooms mapped to e plus the total excess of the rooms mapped to (e, v) and (e, v') may not exceed the capacity of e.

Note that Item (P3) implies that a room may occupy at most one corner. Further restrictions on room placements are possible, for example, we could forbid rooms to be placed at an edge where its aspect ratio becomes undesirably large.

We want to point out that a room placement does not fix the positions of rooms along an edge, but only allocates the necessary space. Hence, the exact positions of such rooms need to be computed in a post-processing step. We return to this matter in Sect. 8.

Group Proximity. Given a room placement, we describe how to measure the proximity of the rooms within a group, for short, the *group proximity*. To this end, let V be the set of corners and E the set of edges. Let $O = V \cup E$ be the set of *objects* (that is, vertices and edges). For two objects $o, o' \in O$, let $\delta_{o,o'}$ denote the distance of o and o'. For example, $\delta_{o,o'}$ could be the length of a path from o to o' along the corridor (possibly using stairs). In general, however, the planner can set the distances as they see fit provided that, for $o = o'$, $\delta_{o,o'} = 0$. For a group g, we then define the proximity of g as the sum of distances $\delta_{o,o'}$ over all pairs $o, o' \in O$, where both o and o' contain at least one room of g.

Problem Definitions. We now define our two problems. To this end, let G be the set of groups, and let S be the set of room sizes. For each $g \in G$ and each $s \in S$, let $\rho_{g,s}$ denote the number of rooms of size s that belong to group g.

First, we define the problem FLOOR PLANNING WITH GROUP PROXIMITY. Let V and E be the sets of corners and edges, respectively, that together with their distance matrix δ and capacities κ describe the available empty floor plans. A feasible solution for the problem is a valid placement of all rooms into V and E

with respect to κ, as defined above. We say that a feasible solution is optimal if it minimizes the sum of proximities over all groups in G with respect to δ.

Next, we define the problem FLOOR ASSIGNMENT WITH GROUP PROXIMITY. Let F be the set of available floors. Each floor f in F has a capacity κ_f, and for two floors f, f' in F, $\delta_{f,f'}$ denotes their distance. For example, if all floors belong to a single building, the ith floor and jth floor could have distance $|j - i|$. A feasible solution for the problem is an assignment of all rooms to the floors in F such that no floor $f \in F$ is overfilled with respect to κ_f. Given a feasible solution and a group $g \in G$, let $F_g \subseteq F$ be the set of floors that contain a room of group g. We say that a feasible solution is optimal if it minimizes $\sum_{g \in G} \sum_{f,f' \in F_g} \delta_{f,f'}$.

For ease of reading, we refer to these two problems from now on simply with FLOOR PLANNING and FLOOR ASSIGNMENT.

4 An ILP for Floor Planning

Linear programming is a popular tool to solve combinatorial optimization problems. A *linear program (LP)* consists of (i) real-valued variables $x_1, \ldots, x_n$, (ii) a target function that is restricted to be linear in the variables (e.g., minimize $c_1 x_1 + \cdots + c_n x_n$ for some constants $c_1, \ldots, c_n$), and (iii) a set of linear constraints $(a_{i,1} x_1 + \cdots + a_{i,n} x_n \geq b_i$ for $i = 1, \ldots, m,)$. Linear programs can be solved efficiently [14]. A *(mixed-)integer linear program (ILP)* is a generalization of a linear program where some variables can be restricted to integer values. In particular, binary "decision" (that is, 0–1) variables can be used. This makes it possible to encode NP-hard combinatorial optimization problems. Consequently, ILPs cannot be solved efficiently in general. In practice, however, small and medium-sized instances of such problems can often be solved relatively fast [25]. For example, we can solve the below ILP formulation for FLOOR PLANNING for a single floor with 40 rooms and three groups in under one second.

We now describe how to formulate FLOOR PLANNING as an ILP. The input of the ILP consists of the empty floor plans given by the sets V and E of corners and edges, respectively, their adjacency relations, their distances δ, and their capacities κ. Let $O = V \cup E$ be the set of *objects*, that is, the corners and edges. The ILP further gets the set G of groups, the set S of room sizes, and the room quantities ρ as input. Note that the ILP views all numbers in the input as constants. (Since the distances are part of the input, they hide the number of floors from the ILP. The distances are also not restrained to the actual geometry of the floor plans and can thus be set as desired by the planners.)

We need the following variables, all of which are binary except for the first one, which is an integer. To help intuition, we specify the intended meaning of the variables.

$x_{g,s,e} \geq 0$ denotes the number of group-g, size-s rooms placed at edge e.

$y_{g,s,v,e} = 1 \Leftrightarrow$ a group-g, size-s room occupies corner v and extends into edge e.

$z_{g,o} = 1 \Leftrightarrow$ a room of group g is at object o (which is an edge or a vertex).

$u_{g,o,o'} = 1 \Leftrightarrow$ rooms of group g are at objects o and o'.

If placing a room of size s into corner v along edge e would not allow this room to have a door or a window, we set $y_{g,s,v,e} = 0$ for every group g. (This partially enforces room placement condition (P2).) Similarly, if placing a room of size s at edge e would make the room's aspect ratio too extreme, we could set $x_{g,s,e} = 0$ for every group g.

Recall that the problem asks to minimize the sum of distances between pairs o, o' of objects that contain rooms from the same group g. The triples g, o, o' that contribute to this sum are those with $u_{g,o,o'} = 1$. Hence, the objective function of our ILP for FLOOR PLANNING (FP$_{\text{ilp}}$) is:

$$\text{minimize} \sum_{g \in G, o, o' \in O} u_{g,o,o'} \cdot \delta_{o,o'},$$

which is subject to the following constraints. The first three constraints enforce the room placement conditions (P1) to (P3).

- Place all rooms (P1):

$$\sum_{e \in E} x_{g,s,e} + \sum_{\substack{v \in V \\ e \text{ adjacent to } v}} y_{g,s,v,e} = \rho_{g,s} \quad \text{for } g \in G, s \in S$$

- Place at most one room in a corner v (P2):

$$\sum_{\substack{g \in G, s \in S, \\ e \text{ adjacent to } v}} y_{g,s,v,e} \leq 1 \quad \text{for } v \in V$$

- Do not overfill an edge e (P2)–(P3); that is, the sum of the sizes of all rooms fully placed at e and those extending into e from a corner must not exceed κ_e:

$$\sum_{g \in G, s \in S} x_{g,s,e} \cdot s + \sum_{\substack{g \in G, s \in S, \\ v \text{ adjacent to } e}} y_{g,s,v,e} \cdot (s - \kappa_v) \leq \kappa_e \quad \text{for } e \in E$$

Note that the room sizes in S are constants from the point of view of the ILP. We need the following constraints to set the binary variables of types u and z.

- Force $z_{g,e}$ to 1 if a room of group g is placed at edge e:

$$x_{g,s,e}/\rho_{g,s} \leq z_{g,e} \quad \text{for } e \in E, g \in G, s \in S$$

- Force $z_{g,v}$ to 1 if a room of group g is placed at corner v:

$$y_{g,s,e,v} \leq z_{g,v} \quad \text{for } v \in V, g \in G, s \in S, \ e \text{ incident to } v$$

- Force $u_{g,o,o'}$ to 1 if rooms of group g are placed at objects o and o':

$$z_{g,o} + z_{g,o'} - 1 \leq u_{g,o,o'} \quad \text{for } o, o' \in O, g \in G$$

Note that we define only lower bounds for variables of types u and z here. However, setting, for example, $u_{g,o,o'}$ to 1 *without* rooms of group g being placed at both objects o and o' would increase our objective function. Hence, the ILP solver sets $u_{g,o,o'}$ to 0 in this case.

Dealing with Unsolvable Instances. Consider an instance of FLOOR PLANNING where the empty floor plans have a total capacity K and all rooms together require an area of A. If the rooms require more area than available (that is, $A > K$), then clearly, there is no solution. However, even if enough area is available (that is, $A \leq K$) there may not necessarily exist a solution, because the rooms placed at the same edge e may not fill up e completely and hence available area remains unused. Moreover, there might not be enough large rooms too occupy all corners. In general, this is more likely to happen when there is not much spare area available. One way to deal with such unsolvable instances is to scale down room sizes, which effectively decreases A.

5   An ILP for Floor Assignment

Due to the complexity of the FLOOR PLANNING problem, our ILP formulation from the previous section can only be solved for instances of moderate size. We thus propose to split large instances into smaller ones. More precisely, we solve the respective FLOOR ASSIGNMENT problem that splits a large multi-floor instance into individual floors. As a result, we get single-floor instances of FLOOR PLANNING that can usually be solved within a reasonable amount of time.

   We now describe an ILP for the FLOOR ASSIGNMENT problem. Recall that the problem asks us to assign every room to one of the floors such that the rooms of each group are assigned only to few floors that are close together. The input is given by the set F of floors, their distances δ, their capacities κ, the set G of groups, the set S of room sizes, and the room quantities ρ. Note that distances of floors set in δ do not need to grow linearly. In particular, one may set distances to model that people take the stairs to go up one or two floors, but take the elevator for more floors.

   We need the following variables all of which are binary except for the first one, which is an integer.

   $x_{g,s,f} \geq 0$ denotes the number of group-g, size-s rooms assigned to floor f.

   $z_{g,f} = 1 \Leftrightarrow$ a room of group g is assigned to floor f.

   $u_{g,f,f'} = 1 \Leftrightarrow$ rooms of group g are assigned to floors f and f'.

Then our ILP for FLOOR ASSIGNMENT (FA_{ilp}) is as follows.

$$\text{Minimize} \sum_{g \in G, f, f' \in F} u_{g,f,f'} \cdot \delta_{f,f'}$$

subject to the following constraints.

- Assign all rooms:

$$\sum_{f \in F} x_{g,s,f} = \rho_{g,s} \quad \text{for } g \in G, s \in S$$

- Do not overfill any floor:

$$\sum_{g \in G, s \in S} x_{g,s,f} \cdot s \leq \kappa_f \quad \text{for } f \in F$$

- Force $z_{g,f}$ to 1 if a room of group g is assigned to floor f:

$$x_{g,s,f}/\rho_{g,s} \leq z_{g,f} \quad \text{for } f \in F, g \in G, s \in S$$

- Force $u_{g,f,f'}$ to 1 if rooms of group g are assigned to floor f and f':

$$z_{g,f} + z_{g,f'} - 1 \leq u_{g,f,f'} \quad \text{for } f, f' \in F, g \in G$$

6   A Heuristic for Floor Assignment

In this section, we propose a heuristic for the FLOOR ASSIGNMENT problem. Our heuristic, which we call FA$_{\text{heu}}$, tries to distribute the rooms evenly among the floors. This is motivated by the observation that an instance of FLOOR PLANNING with a nearly full floor is less likely to have a solution. Roughly speaking, FA$_{\text{heu}}$ consists of three steps, namely, (i) reserving area on each floor, (ii) allocating space to the groups, and (iii) distributing the rooms of each group to the corresponding allocated space. We explain these steps now more precisely.

The first step of FA$_{\text{heu}}$ works as follows. Recall that F is the set of floors. Now let K be the sum of the capacities of all floors, let A be the sum of sizes of all rooms, and thus $K - A$ is the excess area of the building. Then FA$_{\text{heu}}$ reserves on each floor an area of size $(K - A)/|F|$, that is, an equal proportion of the excess area.

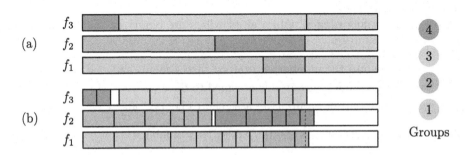

Fig. 4. FA$_{\text{heu}}$ applied to a small problem instance (sM-3M, see Sect. 7). (a) Free space (grey) is evenly distributed to the three floors and the four groups get area allocated one after the other. (b) Allocated space is converted into rooms whereat at most one room per floor uses reserved space.

In the second step, the remaining area is allocated to the groups. To this end, FA_{heu} iterates through groups and floors concurrently. More precisely, suppose that currently group g and floor f are handled. Then as much area of f is allocated to g as either f has available or as g still requires. Accordingly, FA_{heu} either proceeds with the next floor or the next group. For an example, consider Fig. 4(a), where group 1 requires $105\,\text{m}^2$ of the unreserved $129\,\text{m}^2$ on floor f_1. Proceeding with group 2, which requires $101\,\text{m}^2$, the heuristic first allocates the remaining $24\,\text{m}^2$ of f_1 to group 2 and then continues with floor f_2.

In the last step, FA_{heu} converts the allocated areas of the second step into an assignment of rooms to floors as follows. If the whole allocated area of a group g is on one floor, then the area can be straightforwardly partitioned into the rooms of g (like for group 1 in Fig. 4(a–b)). Otherwise, if g is the last group that got area allocated on a floor f (like group 2 on floor f_1 in the example), then FA_{heu} repeatedly assigns the largest remaining (that is, unplaced) room of g on floor f that fits inside the allocated area. If this is no longer possible, but the allocated area of g on f is not fully used up yet, FA_{heu} assigns the smallest remaining room of g to f. In the example, first the largest room and then the smallest room of group 2 are assigned to f_1. In this way, FA_{heu} processes floor after floor and group after group.

Note that as long as the reserved area on each floor is at least as large as the largest room size, FA_{heu} always returns a valid solution. We always place the smallest room at the end of a floor in order to equally distribute the excess area.

The first step takes $\mathcal{O}(|G| \cdot |S| + |F|)$ time since we need to sum the sizes of all rooms and the capacities of all floors. For the second step, we need $\mathcal{O}(|G| + |F|)$ time to greedily allocate the area of the floors to groups. The third step takes $\mathcal{O}((|G| + |F|) \cdot |S|)$ time since every room is assigned successfully once and each size may be tried unsuccessfully once per floor. Therefore, the total runtime of FA_{heu} is also $\mathcal{O}((|G| + |F|) \cdot |S|)$.

7   Test Data

To get a rough idea of realistic problem instances, we considered the situation at two institutes at the University of Würzburg; Mathematics and Computer Science. The actual floor plans of the mathematics buildings inspired the room sizes of our instances. We mapped each type of staff, such as professors and research assistants, to a different room size. Accordingly, we modeled the institutes' chairs as groups, with their respective staff. Based on this data, we built problem instances that vary in the number of groups, in the number of floors, in the floor sizes, and in the number of different room sizes.

We first treat floor plans. We designed four different empty floor plans, f_S, f_M, f_L, f_{XL} in order of increasing size; see Fig. 5. As an example, Table 1 shows the distance matrix of f_S. For all instances, we set the distance of the ith and the jth floor to $|j - i| \cdot 20\,\text{m}$ (seen as a penalty) and calculate the distances of edges and vertices of different floors with their distances to the stairs.

Note that the floor capacities correlate with the floor complexities in terms of numbers of edges and corners. We want to point out that FP_{ilp} does not

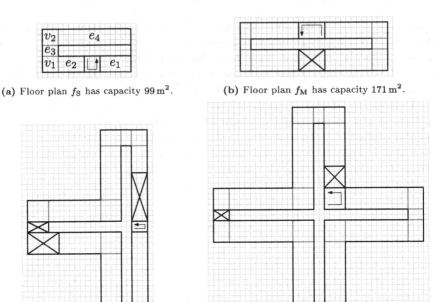

(a) Floor plan f_S has capacity $99\,\mathrm{m}^2$.

(b) Floor plan f_M has capacity $171\,\mathrm{m}^2$.

(c) Floor plan f_L has capacity $318\,\mathrm{m}^2$.

(d) Floor plan f_{XL} has capacity $512\,\mathrm{m}^2$.

Fig. 5. The four empty floor plans used in our problem instances.

take the actual geometry of a floor into account but only its complexity, the capacities, and the distance matrix. In particular, our problem definition does not insist that the same empty floor plan is used for every floor. For simplicity and comparability of the results, in each of our test instances, we do use the same empty floor plan for every floor.

We define the following sets of groups.

M: A set of groups based on the chairs of the Institute of Mathematics. It contains 11 groups with a total of 122 rooms of sizes 8, 15, or $18\,\mathrm{m}^2$. The exact groups are shown in Table 2.

C: A set of groups based on the chairs of the Institute of Computer Science that contains 9 groups with a total of 177 rooms of sizes 10, 15, 20, and $25\,\mathrm{m}^2$; see again Table 2.

sM: A small set of groups that contains the math groups 1, 2, 4, and 11.

MC: A large set of groups; the union of the sets M and C. It thus contains 20 groups with a total of 299 rooms of six different sizes.

Finally, we can define our six problem instances. We have three medium-sized instances that use the set M of math groups but floors of different sizes. Furthermore, we have a small, a large, and a very large instance.

M-18S: Math groups M and 18 copies of f_S.

M-9M: Math groups M and 9 copies of f_M. The problem instance is closest to the actual situation at the Institute of Mathematics.

M-3XL: Math groups M and 3 copies of f_{XL}.

sM-3M: Subset of math groups sM and 3 copies of f_M.

C-11L: Computer science groups C and 11 copies of f_L. This instance is larger and more complex than the math instances as it has more rooms and different room sizes.

MC-15L: Combined groups MC and 15 copies of f_L.

Table 1. Distances in the empty floor plan f_S.

		e_1	e_2	e_3	e_4	v_1	v_2
Edges	e_1	0	5	7	2	8	8
	e_2	5	0	1	2	0	2
	e_3	7	1	0	1	0	0
	e_4	2	2	1	0	2	0
Corners	v_1	8	0	0	2	0	1
	v_2	8	2	0	0	1	0
Stairs		2	2	5	2	5	5

Table 2. The mathematics and computer science groups.

Institute	Math			CS			
Sizes [m²]	8	15	18	10	15	20	25
Groups 1	3	3	2	6	8	2	1
2	4	1	3	2	19	2	2
3	5	1	3	3	10	1	1
4	5	1	1	0	3	1	1
5	9	1	2	2	14	3	2
6	11	1	5	3	17	2	2
7	16	1	3	0	15	2	2
8	8	1	3	3	28	1	2
9	4	1	3	1	14	1	1
10	7	1	4				
11	8	1	3				

8 Experimental Evaluation

In this section, we evaluate our three approaches for the FLOOR PLANNING problem on our problem instances from Sect. 7. First, we examine the performance of FP_{ilp} on its own. Then we combine FP_{ilp} with either FA_{ilp} or FA_{heu}, and discuss the strengths and limitations of these approaches. Finally, we detail how we post-process FP_{ilp} solutions.

Using an ILP for Floor Planning. All ILP tests were run using CPLEX with OPL on a virtual machine that runs Ubuntu 18.04 with 16 cores and 96 GB of RAM. After preliminary experiments, we decided to use settings of CPLEX that resemble a depth-first search with a focus on finding feasible solutions. We tested FP_{ilp} twice on each of the six problem instances, once with a realistic time limit of one hour and once with a time limit of 12 h for comparison. The bare results are shown in Table 3.

FP$_{\text{ilp}}$ found solutions for small to medium-sized instances but it managed to solve only the smallest instance sM-3M to optimality. The resulting floor plans, found within five minutes, are shown in Fig. 6a. Knowing that at least one group needs to be split to two floors, the results looks satisfactory.

Concerning the medium-sized instances (M-XX), FP$_{\text{ilp}}$ performed significantly better with the higher time limit. For example, for the instance M-9M, the floor plans after 1 h, 12 h, and 66 h of computation time are shown in Fig. 7. We observe that the floor plans after 12 h and especially after 66 h appear considerably tidier than after 1 h, though still not fully satisfactory.

The situation is worse for larger problem instances. Namely, for MC-15L, no solution was found even after 12 h. For C-11L, FP$_{\text{ilp}}$ found solutions but ran out of memory after roughly 8 h. The RAM usage shown in Table 4 suggests that this is a systematic problem. Hence, these results indicate that it is necessary to break down large problem instances into smaller ones by computing a floor assignment first.

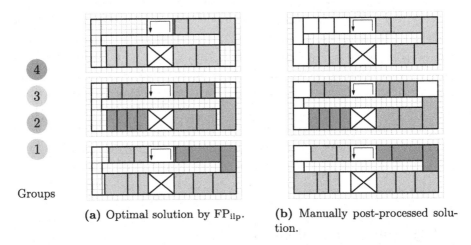

(a) Optimal solution by FP$_{\text{ilp}}$.   (b) Manually post-processed solution.

Fig. 6. The solutions found by FP$_{\text{ilp}}$, like the optimal solution for the problem instance sM-3M here, should not be interpreted as final floor plans.

Table 3. Best solutions found for different approaches and instances. The two versions of FP$_{\text{ilp}}$ differ only in their time limit. Results written in *red italic* required scaling of room sizes to admit solutions.

Instance	sM-3M	M-3XL	M-9M	M-18S	C-11L	MC-15L
FP$_{\text{ilp}}$ (12 h)		848	6754	3398	*ooMem*	–
FP$_{\text{ilp}}$ (1 h)	**241**	1516	13816	3877	85142	–
FA$_{\text{ilp}}$ (1 h) + FP$_{\text{ilp}}$	*458*	642	*5207*	*5090*	*19934*	*15737*
FA$_{\text{heu}}$ + FP$_{\text{ilp}}$	512	1398	*6436*	*7538*	*23195*	*26300*

Floor Assignment as a Preprocessing Step. The two floor assignment algorithms FA_{ilp} and FA_{heu} distribute rooms to floors such that FP_{ilp} has to solve only single-floor instances. We compare the solutions found by either combination with those found by FP_{ilp} alone. To this end, we combined the solutions for the single-floor instances to a solution of the original multi-floor instance and computed the quality according to the objective function of FP_{ilp}; see Table 3. FA_{ilp} instances were again solved with CPLEX and DFS-like settings.

Both floor assignment approaches made it possible to find solutions for the two largest problem instances where FP_{ilp} on its own did not succeed. Furthermore, the combination of FA_{ilp} and FP_{ilp} found a better solution to M-3XL than FP_{ilp} alone and this in a total of only 14 min. In general, the combination of FA_{ilp} and FP_{ilp} achieved better results than the combination of FA_{heu} and FP_{ilp}. However, there is a caveat concerning these results. For most problem instances, it was necessary to downscale the room sizes for several floors such that they would admit solutions. In this regard, FA_{heu} performed better than FA_{ilp}; we discuss this in more detail below.

Table 4. Maximum amount of memory used by FP_{ilp} (excluding overhead).

Instance	sM-3M	M-3XL	M-9M	M-18S	C-11L	MC-15L
FP_{ilp} (12 h)	–	900 MB	16 GB	3 GB	>79 GB	66 GB
FP_{ilp} (1 h)	10 MB	900 MB	15 GB	3 GB	13 GB	6 GB

Table 5. Runtimes of FA_{ilp} in seconds for finding and for proving solutions optimal (rows 1 and 2); aggregated runtimes for the FP_{ilp} runs over all floors (rows 3 and 4).

Instance	sM-3M	M-3XL	M-9M	M-18S	C-11L	MC-15L
FA_{ilp} finding OPT	0.01	0.35	1.09	5.47	2.36	4.70
FA_{ilp} proving OPT	0.18	0.35	2134.32	–	34.84	1067.23
FP_{ilp} (FA_{ilp})	1.33	808.69	3.10	0.37	2.32	8.97
FP_{ilp} (FA_{heu})	1.22	170.44	4.06	0.63	4.27	10.71

Figures 8a and 8b show the floor plans computed for M-9M with the help of FA_{ilp} and FA_{heu}, respectively. We find that both solutions appear tidier than those found by FP_{ilp} alone (Fig. 7). Moreover, since each group appears on at most two floors, these solutions appear closer to what a human planner would construct. We further want to point out that, for FA_{heu}, only one floor required downscaling of room sizes.

Next, we consider the runtime of the floor assignment approaches. While the runtime of FA_{heu} is negligible, we gave FA_{ilp} a time limit of one hour. Table 5 shows the time required by FA_{ilp} to find optimal solutions and to prove that they are optimal. Often, FA_{ilp} found the optimal floor assignment solution within a few seconds for all instances, but required a few minutes to prove their optimality; for M-18S it was not able to prove optimality even after three days.

For both FA_{heu} and FA_{ilp} as preprocessing steps, Table 5 shows the runtime required by FP_{ilp} to find and prove optimal solutions (aggregated over all floors). FP_{ilp} solved most buildings in less than 10 s. Only the instances with the large empty floor plan f_{XL} still took a few minutes to be solved. Nevertheless, we can conclude that first computing a floor assignment and then using FP_{ilp} for single floors yields a tremendous speed up.

Scaling to the Rescue. Since the floor assignment approaches do not take the empty floor plans into account when distributing the rooms it may happen that a floor assignment solution results in unsolvable floor planning instances. For such an unsolvable instance, we scaled down all room sizes with the same scaling factor such that the instance became solvable. For the different problem instances, Table 6 shows the worst needed scaling factor among all floors and the number of floors that required scaling.

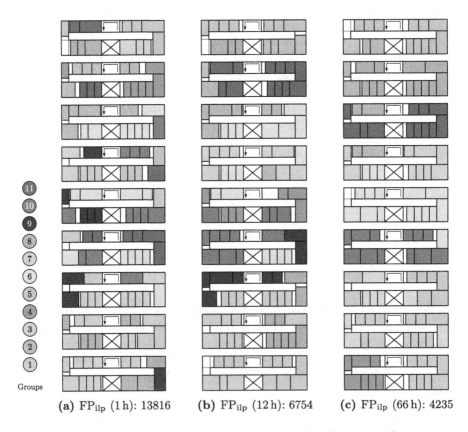

(a) FP_{ilp} (1 h): 13816   (b) FP_{ilp} (12 h): 6754   (c) FP_{ilp} (66 h): 4235

Fig. 7. Solutions for M-9M computed by FP_{ilp} with different time limits.

Taking a closer look at the solutions in Fig. 8, we observe that downscaling was not only necessary because of a lack of empty space, but also because of a bad distribution of room sizes. For example, the sixth floor in the FA_{heu} solution has many small rooms but only one large room and therefore not all corners can be occupied. The same problem occurred on the second floor in the FA_{ilp} solution. We further find the FA_{ilp} solutions mostly required downscaling because of tightly packed floors. By design, this is less likely to happen with FA_{heu}.

Algorithms vs. Human. For the problem instance M-9M we also constructed a floor plan manually; see Fig. 8c. Compared to the floor plans computed by the three different algorithmic approaches (see again Figs. 7 and 8), the manual floor plan appears more coherent in the following sense. Groups form only few clusters and do not alternate in the cyclic order around the corridors (whereas they do in the ground-floor of Fig. 7b). Interestingly, the floor plan computed

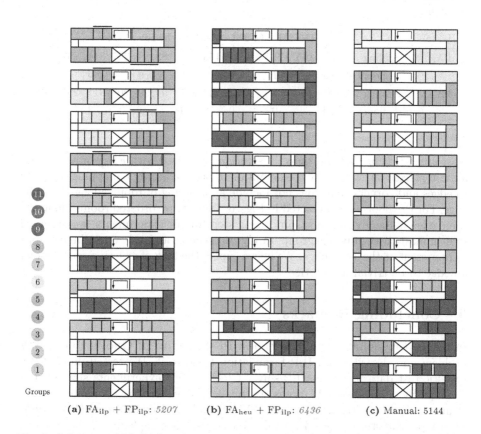

(a) FA_{ilp} + FP_{ilp}: *5207*   (b) FA_{heu} + FP_{ilp}: *6436*   (c) Manual: 5144

Fig. 8. Solutions of M-9M found by first computing a floor assignment with FA_{ilp} and FP_{ilp} as well as a manually crafted solution. Red line segments indicate that the sizes of the rooms next to the segments had to be scaled down slightly. (Color figure online)

by FP_{ilp} after 66 h (Fig. 7c) achieved a lower score (4235) than the objective function value of the manual solution (5144), but appears less coherent. Note, however, that FP_{ilp} placed more groups within one floor (7 vs. 6). On the other hand, in Fig. 7c, the small orange group 2 is spread over four floors (with 1, 1, 2, 1 objects). This yields 14 times the intra-floor distance, that is, the distance between two consecutive floors (which we set to 20 m). Although the larger violet group 8 in Fig. 8c is spread over just two floors, where it occupies four objects on each floor. This yields more, namely 16 times the intra-floor distance. If we want to favor solutions that avoid spreading groups over more than two floors, we simply have to replace our linear floor distance function (see p. 11) by, say, a quadratic one (e.g., $|j - i|^2 \cdot 20$ m for the distance from floor i to floor j).

Table 6. Worst needed scaling factors and the number of floors that needed scaling when solving the (single-floor) FLOOR PLANNING instances created by floor assignment solutions.

Instance		sM-3M	M-3XL	M-9M	M-18S	C-11L	MC-15L
FA_{ilp}	Scaling factor	165/171	–	153/171	82/99	207/311	207/311
	# floors scaled	1	0	6	9	6	7
FA_{heu}	Scaling factor	–	–	144/171	94/99	207/311	207/311
	# floors scaled	0	0	1	1	3	8

Post-processing of ILP Solutions. A solution computed with FP_{ilp} only specifies a room placement, which, by the definition in Sect. 3, is only a mapping of rooms to edges and corners that does not include an ordering of rooms along the same edge. For the floor plans depicted in Figs. 6, 7 and 8, we ordered rooms along the same edge manually in a post-processing step. In our experiments, we observed that in most cases only one group, rarely two groups, and only once three groups were present at the same edge. We hence expect the task of ordering rooms along an edge to be trivial in practice.

9   Concluding Remarks

In this paper, we introduced the FLOOR PLANNING WITH GROUP PROXIMITY problem and described an ILP formulation to solve it. We tested our ILP on realistic test data using the ILP solver CPLEX. Our experiments showed that small problem instances (28 rooms, 4 groups) can be solved to optimality within minutes and with satisfactory results. While medium-sized instances (122 rooms, 11 groups) can still be solved within a few hours, the complexity of the problem makes it unfeasible to tackle larger instances directly.

We further showed that large (multi-floor) instances of the floor planning problem can be tackled by first solving the respective FLOOR ASSIGNMENT WITH GROUP PROXIMITY problem and then solving the resulting (single-floor) instances as before. We have formulated the floor assignment problem, too, as an ILP, but we also suggested a greedy heuristic for it. Our experiments demonstrated that splitting instances via a floor assignment wasn't only much faster, but also improved some of the results for medium-sized instances that FP_{ilp} computed within a fixed runtime limit.

Let us consider the optimal solution found for the smallest instance sM-3M by FP_{ilp} in Fig. 6a. Since the ILP only optimizes according to an objective function, we should not be surprised to find, for example, gaps between rooms or corner rooms that cannot be entered. A planner would directly avoid these and other deficiencies. However, such deficiencies can easily be eradicated in a (manual) post-processing step as shown by the floor plan in Fig. 6b. As mentioned in the introduction, if a user is not satisfied with an optimal solution (say, to an ILP), then this is not a failure of the solver, but a failure of the model and the corresponding objective function. Here, we decided to keep the model as simple as possible to test the potential computational limitations of our approach.

The proposed variants of the floor planning problem and our algorithmic approaches have demonstrated their potential to support planners in the design of architectural floor plans of office buildings. We stress that formulating the floor planning problem as an ILP has the advantage that a planner may add further requirements. For example, one can easily add constraints that force two given rooms to be adjacent, extend the model to less constrained building outlines, or allow rooms to occupy two corners.

References

1. Ahmadi, A., Pishvaee, M.S., Akbari Jokar, M.R.: A survey on multi-floor facility layout problems. Comput. Ind. Eng. **107**, 158–170 (2017). https://doi.org/10.1016/j.cie.2017.03.015
2. Allen, T.J., Fustfeld, A.R.: Research laboratory architecture and the structuring of communications. R&D Manage. **5**(2), 153–164 (1975). https://doi.org/10.1111/j.1467-9310.1975.tb01230.x
3. Barth, L., et al.: Semantic word cloud representations: hardness and approximation algorithms. In: Pardo, A., Viola, A. (eds.) Latin American Symposium on Theoretical Informatics (LATIN 2014), vol. 8392, pp. 514–525. Springer (2014). https://doi.org/10.1007/978-3-642-54423-1_45
4. Bekos, M.A., et al.: Improved approximation algorithms for box contact representations. Algorithmica **77**(3), 902–920 (2016). https://doi.org/10.1007/s00453-016-0121-3
5. Bergman, D., Cardonha, C., Mehrani, S.: Binary decision diagrams for bin packing with minimum color fragmentation. In: Rousseau, L., Stergiou, K. (eds.) Integration of Constraint Programming, Artificial Intelligence, and Operations Research (CPAIOR 2019), vol. 11494, pp. 57–66 (2019). https://doi.org/10.1007/978-3-030-19212-9_4
6. Bortfeldt, A.: A reduction approach for solving the rectangle packing area minimization problem. Eur. J. Oper. Res. **224**(3), 486–496 (2013). https://doi.org/10.1016/j.ejor.2012.08.006

7. Charman, P.: A constraint-based approach for the generation of floor plans. In: International Conference on Tools with Artificial Intelligence (TAI 1994), pp. 555–561 (1994). https://doi.org/10.1109/TAI.1994.346443

8. Del Río-Cidoncha, M., Iglesias, J., Martínez-Palacios, J.: A comparison of floorplan design strategies in architecture and engineering. Autom. Constr. **16**(5), 559–568 (2007). https://doi.org/10.1016/j.autcon.2006.12.008

9. Drira, A., Pierreval, H., Hajri-Gabouj, S.: Facility layout problem: a literature analysis. IFAC Proc. Vol. **39**(3), 389–400 (2006). https://doi.org/10.3182/20060517-3-FR-2903.00208

10. Duarte, J.P.: Towards the mass customization of housing: the grammar of Siza's houses at Malagueira. Environ. Plann. B: Plann. Des. **32**(3), 347–380 (2005). https://doi.org/10.1068/b31124

11. Eppstein, D., Mumford, E., Speckmann, B., Verbeek, K.: Area-universal and constrained rectangular layouts. SIAM J. Comput. **41**(3), 537–564 (2012). https://doi.org/10.1137/110834032

12. Hamacher, A., Kjølsrud, E.: Growing floor plans and buildings - generation and optimisation through algorithms and synthetic evolution. In: Estévez, A.T. (ed.) International Conference of Biodigital Architecture and Genetics (BIODIG2014) (2014). https://researchgate.net/publication/311455480

13. Huang, E., Korf, R.E.: Optimal rectangle packing: an absolute placement approach. J. Artif. Intell. Res. **46**, 47–87 (2013). https://doi.org/10.1613/jair.3735

14. Karmarkar, N.: A new polynomial time algorithm for linear programming. Combinatorica **4**(4), 373–395 (1984). https://doi.org/10.1007/BF02579150

15. Knecht, K., König, R.: Generating floor plan layouts with k-d trees and evolutionary algorithms. In: Generative Art Conference (GA'10). pp. 238–253. Milan (2010), http://www.generativeart.com/on/cic/GA2010/2010_18.pdf

16. Kovács, L.: Knowledge based floor plan design by space partitioning: a logic programming approach. Artif. Intell. Eng. **6**(4), 162–185 (1991). https://doi.org/10.1016/0954-1810(91)90022-G

17. Lobos, D., Donath, D.: The problem of space layout in architecture: a survey and reflections. arquiteturarevista **6**(2), 136–161 (2010). https://doi.org/10.4013/arq.2010.62.05

18. Marson, F., Musse, S.R.: Automatic real-time generation of floor plans based on squarified Treemaps algorithm. Int. J. Comput. Games Technol. 2010, 624817 (2010). https://doi.org/10.1155/2010/624817

19. Meller, R.D., Bozer, Y.A.: Alternative approaches to solve the multi-floor facility layout problem. J. Manuf. Syst. **16**(3), 192–203 (1997). https://doi.org/10.1016/S0278-6125(97)88887-5

20. Meller, R.D., Gau, K.Y.: The facility layout problem: recent and emerging trends and perspectives. J. Manuf. Syst. **15**(5), 351–366 (1996). https://doi.org/10.1016/0278-6125(96)84198-7

21. Michalek, J., Choudhary, R., Papalambros, P.: Architectural layout design optimization. Eng. Optim. **34**(5), 461–484 (2002). https://doi.org/10.1080/03052150214016

22. Shekhawat, K., Duarte, J.P.: Rectilinear floor plans. In: Çağdaş, G., Özkar, M., Gül, L.F., Gürer, E. (eds.) CAAD Futures 2017, pp. 395–411 (2017). https://doi.org/10.1007/978-981-10-5197-5_22

23. Shekhawat, K., Pinki, Duarte, J.P.: A graph theoretical approach for creating building floor plans. In: Lee, J.H. (ed.) CAAD Futures 2019, pp. 3–14. Springer (2019). https://doi.org/10.1007/978-981-13-8410-3_1

24. Shi, F., Soman, R.K., Han, J., Whyte, J.K.: Addressing adjacency constraints in rectangular floor plans using Monte-Carlo Tree Search. Autom. Constr. **115**, 103187 (2020). https://doi.org/10.1016/j.autcon.2020.103187
25. Sierksma, G., Zwols, Y.: Linear and Integer Optimization: Theory and Practice. CRC Press, New York, 3rd edn. (2015). https://doi.org/10.1201/b18378
26. Simon, J.: Evolving floorplans (2017). https://www.joelsimon.net/evo_floorplans.html
27. Upasani, N., Shekhawat, K., Sachdeva, G.: Automated generation of dimensioned rectangular floorplans. Autom. Constr. **113**, 103149 (2020). https://doi.org/10.1016/j.autcon.2020.103149
28. Veloso, P., Celani, G., Scheeren, R.: From the generation of layouts to the production of construction documents: an application in the customization of apartment plans. Autom. Constr. **96**, 224–235 (2018). j.autcon.2018.09.013

Architectural Automations and Augmentations: Design

Type Investigation in the Form of High-Rise Building Using Deep Neural Network

Jinmo Rhee[(⊠)] and Pedro Veloso

Carnegie Mellon University, Pittsburgh, PA 15206, USA
jinmor@cmu.edu, pveloso@andrew.cmu.edu

Abstract. In this paper we propose a deep learning (DL) method to investigate existing types of high-rise buildings and to generate new ones. We collected data comprehending diverse forms of high-rise building from major cities in the world to train a generative DL model (IntroVAE) to capture morphological features. After clustering the features, we can distinguish types of high-rise buildings and use that information to generate novel high-rise building forms. This research demonstrates that generative DL models can uncover the latent types of architectural form in large datasets and can expand the typological interpretation of complex architectural forms. Besides, we demonstrate the potential of the proposed DL method for building massing design by developing a proposal of a high-rise building form based on three techniques: exploration, synthesis, and interpolation.

Keywords: Deep learning · High-rise building · Typology · Form · Neural network

1 Introduction

High-rise building is a representative architectural type of contemporary city and has its own morphological features, which can be noticeably distinguished from other architectural types such as church, residential vernacular, market, etc. These features are derived from design principles and styles and are key to the formal and typological identity of a high-rise building. Typically, high-rise buildings have been categorized by their major characteristics such as their main towers shape [1], types of structural systems [2], the position of their circulation core, their height [3], their program, or the combination of some of these [4].

In this research, we contribute to the existing taxonomies of building types by establishing formal analysis of a large database of high-rise building. By using computational methods to analyze and cluster a database encoding the form of high-rise buildings, we expect to uncover latent building types and provide opportunities to expand architectural creativity by re-interpreting and transforming their forms.

Morphological investigation of high-rise building based on the analysis of geometric features is not straightforward, because high-rise building form requires numerous and complex variables [5] representing various features and aesthetic elements derived from sources such as context and architectural styles. There is no standardized way to

© Springer Nature Singapore Pte Ltd. 2022
D. Gerber et al. (Eds.): CAAD Futures 2021, CCIS 1465, pp. 175–189, 2022.
https://doi.org/10.1007/978-981-19-1280-1_11

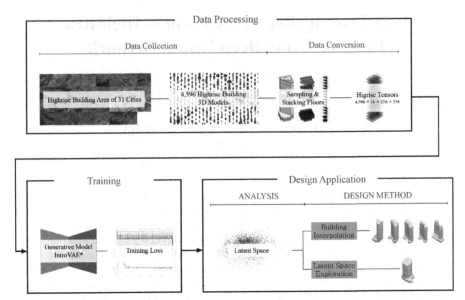

Fig. 1. Overall process of type investigation in the form of high-rise building and its design using deep neural networks.

represent these variables. For example, with the adoption of parametric and geometric modeling techniques by contemporary practitioners, the design of high-rise buildings has adopted curvy and variational shapes that are noticeably different than the rigid shapes of conventional design. Those difference makes more difficult to establish a representation for comprehensive formal analysis and typological studies of high-rise buildings.

With a system that can represent complex morphological features, we can thoroughly examine the form of varied high-rise buildings, address new formal types by re-clustering the features, and potentially explore the new formal identity in the design of high-rise buildings. Therefore, this research aims to implement generative deep learning (DL) model for grasping geometric features, discovering types of high-rise buildings, and supporting new design practices for the massing design of high-rise buildings.

This research is constituted by three steps: data collection and pre-processing, training, and design application (Fig. 1).

In the data collection and pre-processing step, we collected three-dimensional data of high-rise buildings, converted each building into a series of two-dimensional images by horizontal slicing, and stored them in a deep tensor.

We used this tensor database of high-rise building forms to train a generative DL model called autoencoder, which contains two parts: an encoder and a decoder. This encoder compresses the tensor representation of each building into a lower-dimensional vector. The decoder learns how to reconstruct the original tensor based on this vector. By reducing the tensor representation to a lower-dimensional representation (embedding) the model also enables the discovery of the dominant morphological features in the dataset. These embeddings can be used to categorize forms of high-rise buildings into several representative types.

Besides, the decoder can not only reconstruct existing buildings but also generate new ones that are in the parameter space of the embedding. Based on the manipulation of the embedding of a trained model, we investigate three different techniques to generate a high-rise building form: exploration, synthesis, and interpolation. We introduce the potential of these techniques to support architectural creativity by developing the schematic design of two high-rise buildings.

2   Machine Learning-Integrated Architectural Design

The recent innovations and increased accessibility of DL has supported the integration of artificial intelligence technology to architectural design process. Besides, it also shed a new light on the importance of architectural data storing, collection and processing for decision-making in architecture. This section reviews research projects that used generative models from DL and image and voxel-based representations for the generation of architectural forms.

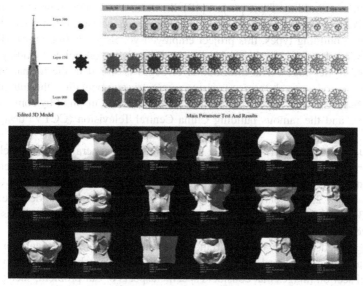

Fig. 2. (Top) Generated stylized plans with gradually changed style weights [6] and (Bottom) Capitals automatically design with machine learning [7].

The Spire of AI [6] project created a new form of architecture using Style Transfer [8] techniques from DL. The authors contoured building forms and obtained the images of horizontal sections of the forms. Then, they trained a Neural Style Transfer model with the section images. After training, users can apply a specific style to the sections and convert the pixel values of the style-transferred sections into voxels, which is a three-dimensional representation of pixels. By vertically stacking the voxel values, they can reconstruct the shape of a building with the style (Fig. 2). The idea of contouring the sections was key to represent the three-dimensional information of the form for image-based DL. Similar idea can be found in Fresh Eyes [9]. However, in this project, DL is only applied to transferring a style to other images, not the process of learning the morphological feature of buildings or synthesizing the overall form of buildings.

The automation of capital design with machine learning is introduced in the research of Artificial Intelligence Aided Architectural Design [7]. The authors trained artificial neural networks based on the detailed configuration of the Roman Corinthian order capitals. The input data format was composed of matrices that includes the sample coordinate values, surface normal vector and volume center plane deviation. The output data are displacement values. By reconstructing a three-dimensional model of an order based on this predicted displacement values, they can generate three-dimensional variations of the new capital forms based on the given input parameters, both purposeful and random (Fig. 2). This research shows the potential of integrating DL to architectural design activities. Both repeatable and predictable design activities can be easily replaced by machine learning tools in the first place, by teaching the system decision-making based on the work performed by the architects.

Voxel-based representation is another way to construct form dataset. Geometric Interpolation of Building Types project [10] illustrates a method to represent a building within a fixed number of voxels and their vectorized connections. For investigating and interpolating building types, this project employed voxels as scaffoldings of building representation. By defining the connection between the voxel points, building can be represented by its edges. The information of edges is stored in the format of tensor which deep neural network can grasp. Using parametric augmentation, the authors create a dataset based on two "types": an abstract castle structure inspired by John Hejduk's architecture and the famous building China Central Television (CCTV) designed by OMA.

David Newton's high-rise building synthesis using a generative DL model [11] is another project to use voxel-based representation of architectural form. He collected "500 building massing models located in downtown New York City" [11] and convert them into "$1 \times 256 \times 256$ voxels" [11]. Using one of the three-dimensional deep generative models, 3D-IWGAN, he trained the formal features of the buildings and was able to synthesize a new building form.

These precedents show the importance of defining a specific representation to capture features of architectural form for generation. Our approach uses a novel variation of stacked section images that considers specific aspects of our problem, such as building scale and parts. Similarly to [10], our research also emphasizes the importance of capturing typological features in the latent space. However, instead of synthesizing the dataset based on two building types, we use our stacked representation to explore a vast dataset of high-rise buildings from different cities.

3   Dataset

In this section we discuss our latest method for encoding of building forms into tensors. We collected three-dimensional data of high-rise buildings, converted them into two-dimensional images by slicing their three-dimensional forms, and finally a deep tensor dataset was created by stacking these sliced images.

3.1 Data Collection

The research started with the collection of three-dimensional form data of existing high-rise buildings from manually selected downtowns of 31 cities around the world and for each city focused on specific areas that have a concentration of high-rise buildings: New York City, Chicago, Atlanta, Los Angeles, Miami, Philadelphia, Pittsburgh, Boston, Seattle, San Francisco, San Diego, Houston, Dallas, Baltimore, Detroit, Indianapolis, Denver, Vancouver, Toronto, London, Paris, Riyadh, Dubai, Abu Dhabi, Hongkong, Shanghai, Taipei, Bangkok, Singapore, Honolulu, Sydney.

To facilitate a large collection of data, we automated the process of scraping high-rise building three-dimensional models from OSM (Open Street Map).

The general height threshold for high-rise building is of 25 m [12]. However, this definition implies that almost every building with more floors than 5 stories is high-rise building. If we naively accepted this threshold, it might be difficult to grasp the distinctive morphological characteristics of a high-rise building. Considering the properties of the collected dataset and intention to manifest the formal characteristics of a high-rise building in this research, we set the threshold into 70 m and picked a proper value which can provide a total of 4,956 high-rise buildings formatted as three-dimensional OBJ models. We used Rhinoceros and Grasshopper for handling and modeling three-dimensional data.

3.2 Data Processing

In order to train the image-based generative model, we developed a technique to represent each three-dimensional building as a set of two-dimensional images. The technique involves slicing and sampling 16 floors of each building and represent them as figure-ground diagrams. To do this, we sliced each building horizontally using the 3 m standard floor to floor height adopted in OSM, and put all slices into three groups based on the range of their relative heights (i.e. 0–33%, 33–66%, 60–100%). Then, we selected the first 6 floors of the first group, the 5 floors in the middle of the second group, the first two floors of the third group, and the last 3 floors of the third group. The sampling of the first six floors of the first group reflects that most buildings have the "podium" typology where their overall forms tend to have larger bases. The sampling of five floors in the middle of the second group reflects the prevailing form typically found in the midsection of high-rise buildings. The sampling of the first two floors of the third group reflects the tendency for high-rise buildings to taper towards the top. The sampling of the last three floors of the third group reflects the tendency for high-rise buildings to have a spire at the top.

We have determined the sampling of 16 floors for encoding each building as a tensor in this research. The sampling number can be larger to represent the original form of three-dimensional high-rise building model more accurately.

Each slice of the high-rise building is a diagrammatic representation of a floor plan boundary encoded as a tensor shape of shape $1 \times 256 \times 256$. By stacking 16 of these slices, we can establish a tensor to represent a high-rise building's morphological features and train a DL model with convolutional layers. The total tensor size for the entire dataset of 4,596 buildings is $4596 \times 16 \times 256 \times 256$ (Fig. 3).

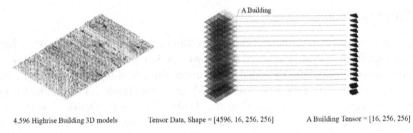

4,596 Highrise Building 3D models   Tensor Data, Shape = [4596, 16, 256, 256]   A Building Tensor = [16, 256, 256]

Fig. 3. Data conversion of 3D model to tensor through image format.

4 Training

4.1 Model

One of the most popular generative DL models is Generative Adversarial Network (GAN). According to [13]:

It consists of two networks: the generator network $Gen(z)$ maps latent z to data space while the discriminator network assigns probability $y = Dis(x) \in [0,1]$ that x is an actual training sample and probability $1 - y$ that x is generated by our model through $x = Gen(z)$ with $z \sim p(z)$. The GAN objective is to find the binary classifier that gives the best possible discrimination between true and generated data and simultaneously encouraging Gen to fit the true data distribution. We thus aim to maximize/minimize the binary cross entropy:

$$\mathcal{L}_{GAN} = \log(Dis(x)) + \log(1 - Dis(Gen(z))) \tag{1}$$

with respect to *Dis/Gen* with x being a training sample and $z \sim p(z)$.

Another of the popular generative DL models is Variational Autoencoder (VAE). According to [13]:

It consists of two networks that encode a data sample x to a latent representation z and decode the latent representation back to data space, respectively:

$$z \sim Enc(x) = q(z|x), \quad x \sim Dec(z) = p(x|z) \tag{2}$$

The VAE regularizes the encoder by imposing a prior over the latent distribution $p(z)$. Typically, $z \sim N(0, I)$ is chosen. The VAE loss is minus the sum of the expected log likelihood (the reconstruction error) and a prior regularization term:

$$\mathcal{L}_{VAE} = -E_{q(z|x)}\left[\log\frac{p(x|z)p(Z)}{q(z|x)}\right] = \mathcal{L}_{llike}^{pixel} + \mathcal{L}_{prior} \tag{3}$$

With

$$\mathcal{L}_{\text{llike}}^{\text{pixel}} = -E_{q(z|x)}\big[\log p(x|z)\big] \tag{4}$$

$$\mathcal{L}_{\text{prior}} = D_{KL}(q(z|x)\|p(z)) \tag{5}$$

where DKL is the Kullback-Leibler divergence.

In terms of the model selection between GAN and VAE for this experiment, our main challenge was in the tradeoff between blurry images and trained latent space. Nowadays, due to the tremendous development of GAN, GAN can generate sharp synthesized images but have less freedom to explore the latent space of the model; with new input data, the latent vector cannot be preserved, and the model must be recalculated to fit the new data. In addition to characteristics of latent space, GANs are hard to train because of its unstable training process and mode collapse [14]. On the other hand, VAE usually generates blurry images, compared to the image quality generated by GAN, but it allows for more freedom to explore the latent space. Since the latent space fits to the entire given dataset, exploring it does not require new fitting calculation. Due to these tradeoffs, we selected IntroVAE as a hybrid model that conciliates the consistent latent space with high quality images from GAN (Fig. 4) [15].

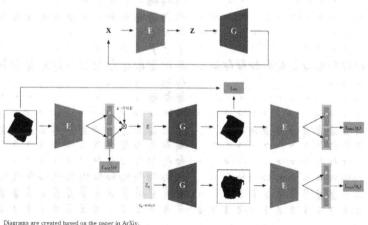

Diagrams are created based on the paper in ArXiv.
Huaibo Huang et al., "IntroVAE: Introspective Variational Autoencoders for Photographic Image Synthesis

Fig. 4. Architecture of IntroVAE, re-drawn illustration fit to the dataset.

This model requires two parts in training to generate an image: one part to discriminate the generated samples from the training data, and another part to mislead the discriminator. Specifically, this model has the approximate inference model of VAEs (encoder) as the discriminator of GANs and the generator model of VAEs (decoder) as the generator of GANs. In addition to performing adversarial learning like GANs, the inference and generator models are trained jointly for the given training data to preserve the advantages of VAEs [15].

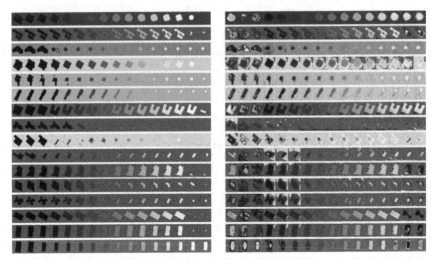

Fig. 5. Results comparison between ground truth (left) and predicted (right) images from dataset with three channels.

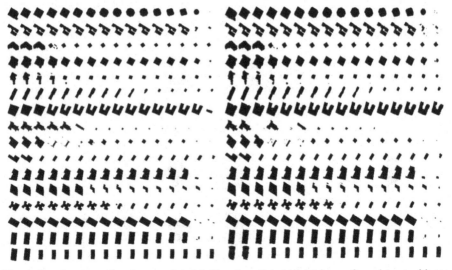

Fig. 6. Results comparison between label (left) and predicted (right) image from dataset with one channel.

4.2 Training

On the onset of the training process, we tested a dataset that included actual height of buildings represented by background colors. This dataset has 3 channels (i.e. RGB) instead of 1. Our initial was that color information would be helpful for easing learning by providing more distinguished features of each building and floor. After training the model based on this dataset for 500 epochs, based on comparison between the original

(a) and decoded (b) images of test dataset (Fig. 5), we found too much randomness, noise, and often incorrect colors on the images. Color information was difficult to learn, because floor plan shape and height value did not have a strong correlation.

To overcome the challenge with colors, we switched to a grayscale representation of the buildings and trained the model for 500 epochs. Based on comparison between the original (a) and decoded (b) images of test dataset (Fig. 6), the decoded images were still a little bit noisy and had minor errors, but they had higher accuracy and sharp edges. For the final learning, the hybrid model was trained on a computer with the following specifications: 'Intel(R) Core (TM) i7-8700k @ 3.70 GHz', 64 GB memory, and two GTX-1080ti graphic processing units. It took almost 200 h to train the data for 1300 epochs. Learning rate of encoder and decoder were $2e-4$, lambda for L1 loss was 100, hidden dimension was 10, beta1 and beta2 were 0.5 and 0.999 respectively, and batch size was 128.

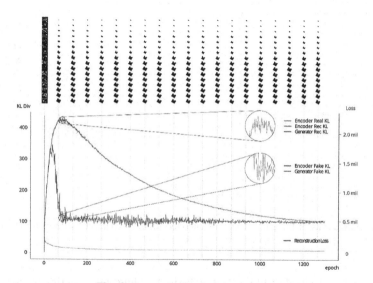

Fig. 7. Losses during training process.

Figure 7 illustrates the quality of the reconstructed sample building diagrams from the test dataset with regard to the losses of the reconstruction and the Kullback–Leibler (KL) – divergences [16]. After almost 20 epochs, the losses started to converge to a stable stage in which their values fluctuate slightly around a balance line [15]. KL divergence between the latent distribution of the decoder and a normal distribution started to converge after almost 100 epochs. After this point, the KL-divergences continuously decreases with little fluctuation of their values. KL divergences of encoder and decoder's face image also decreases and converge after almost 60 epochs. Compared to these divergences from 0 to 60 epochs, the width of fluctuation of these divergences are drastically smaller. Encoder and decoder's reconstruction and real image KL-divergences converged to fake image's KL-divergences.

4.3 Typical Form of High-Rise Building Using Latent Space

Encoded vectors of each building tensor keep the morphological features of the building with the reduced data dimension: every building can be represented with 10 floats in the latent space by the reduced features. However, 10-dimensional space is still hard to visualize. In order to visualize latent space which has 10 dimensions in to 3-dimensional space, we employ t-SNE (t-distributed stochastic neighbor embedding) [17] to reduce dimensionality. The hyper-parameters are: perplexity 35, learning rate 100, iterations 1500. All building in the dataset can be placed and represented in 3-dimensional space as a data point. A data point in the latent space is a high-rise building. The location of the point represents its morphological characteristics, and the distance among the points represents the degree of morphological similarity. The shorter the distance, the more similar form of high-rise buildings.

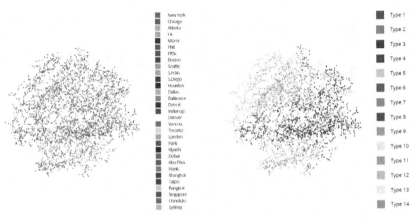

Fig. 8. Visualizations of the latent space of encoded high-rise dataset by cities (left) and types (right).

We discovered a total of 14 types of high-rise buildings by clustering the data points of the buildings in the latent space with Density-Based Spatial Clustering of Applications with Noise (DBSCAN) [18]. Figure 8 demonstrates two different visualizations of the latent space of encoded high-rise dataset: by cities and types.

By tracking the nearest data point of each cluster's center, the typical form of each type can be investigated (Fig. 9). Type 1 has a parallel configuration of a tower and podium. Type 2 is the simplest box shape tower. Type 3 is a twin tower sharing a podium. Type 4 is a narrow rectangular tower with many irregularities. Type 5 is a cake-shaped tower without a podium. Type 6 is a tower with subtraction. Type 7 is a circular-shaped tower. Type 8 has also parallel configuration of a tower and podium like type 1, but the podium is large and tall. Type 9 is a simple slim tower. Type 10 is similar to type 5 with a podium. Type 11 is similar to type 4 with a podium. Type 12 is like type 8 with a podium. Type 13 is like type 4 with lower irregularities. Type 14 is a simple tween tower without a podium (Table 1).

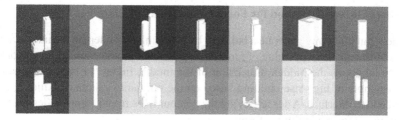

Fig. 9. Typical form of each type of high-rise building.

Table 1. Feature comparison of each type.

		1	2	3	4	5	6	7	8	9	10	11	12	13	14
Profile Shape	Rect.														
	Circle														
Vertical Shape	Regular														
	Irregular														
Tower Ratio	Normal														
	Narrow														
	Skinny														
Number of Tower	1														
	2														
Podium	0														
	1														

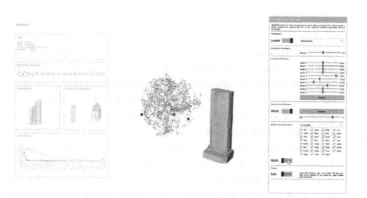

Fig. 10. Interface of Deeprise for high-rise building form analysis and generation.

5 Design Application

We developed this model as a design plug-in prototype of Grasshopper in Rhinoceros, a popular three-dimensional modeling tool in architecture and product design. The prototype, called "Deeprise", has its own interface (Fig. 10) and deploys the trained model with the learned high-rise building form for design and runs it in back-end to provide three different approaches for high-rise building form generation.

5.1 Three Different Method for Form Generation

Designers can randomly generate a building by changing the sliders in the interface. The slider will change the seed value of random function and produce a random vector with the same length as the hidden dimension of the model. Figure 11 shows a schematic design example of high-rise building based on the randomly generated building forms from Deeprise interface. After retrieving the contour lines from Deeprise, designers can use lofting to create a buildings form.

Fig. 11. Design example using random exploration from Deeprise and its design process.

The second method to generate a high-rise building form is synthesis. Designers can synthesize a vector in the latent space by changing slider and assigning a value to each dimension of the latent space. This method allows designers to control the form of a high-rise building with higher precision. After they roughly explores the latent space to discover a proper form of a high-rise building, they can digitally sculpt the form in detail by changing sliders a bit (Fig. 12).

The last method for generation of a high-rise building form is interpolation. Instead of editing a form of high-rise building by manually changing values in a vector, this method can generate a series of interpolated forms between two buildings. Through the SLERP (Spherical Linear Interpolation) function, it generates hybrid forms of two buildings with different combination ratios of them (Fig. 13). This method is significant for architectural design, because designers can mix two different types of buildings by changing the parameters of the interpolation without having to manage the larger parameter space of the model directly. Figure 14 demonstrates a design scenario where the form consists of 75% of a building and 25% of another.

Fig. 12. Series of synthesized forms by slight changes of Z-vectors.

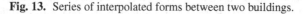

Fig. 13. Series of interpolated forms between two buildings.

Fig. 14. Design example using interpolation.

6   Conclusion

We have categorized high-rise buildings according to their morphological character-istics, discovered typical types of them, and generate new high-rise buildings using a generative DL model. To train the model, we used a dataset with custom representa-tion of morphological information of high-rise buildings. Based on this, we conducted design experiments with the trained model to explore new high-rise building forms for schematic architectural design.

6.1   Discussion and Contribution

This research illustrates a new methodological approach for the analysis of architectural morphology in design. The new design types are drawn by advanced statistical techniques applied to databases with three-dimensional information of high-rise buildings. The method of representation and interpretation of architectural form in this research supports a complex and thorough analysis of the existing buildings. With this method, designers can expand their knowledge of architectural forms by uncovering the latent form and types from the real world. Furthermore, this research demonstrates the potential that the generative DL model can be used not only for directly creating the design results but also for analyzing design objects. Rather than independently generating high-rise building forms with a trained model, this research addressed the integration of formal analysis and generation through the latent space. In this sense, DL could be developed into an efficient tool to help designers to analyze their design problems and generate solutions that are both creative and grounded on real-world data.

Lastly, the expanded exploration coverage of high-dimensional design solution space through DL can provide vast opportunities to discover new architectural forms. Instead of interpreting morphological features of several high-rise buildings and abstracting their formal principles, the generation methods in this paper demonstrates the potential to synthesize a new high-rise form by exploring the design solution space where complex formal principles exist.

6.2 Challenges and Future Study

There are three aspects to be considered for future steps: diversity of building types, using data that extrapolate geometry, and learning directly from a three-dimensional representation of buildings.

This research only used high-rise buildings to capture the morphological features of architectural form. If more architectural types, such as churches, markets, airports, etc., are available, the potential of architectural design exploration with DL can be expanded. For this goal, not only different DL models but also different ways to represent architectural form into learnable data format should be explored.

Besides, we intend to extrapolate geometric and physical characteristics of buildings in order to explore other criteria and relationships that affect the built form. By integrating geometric and other social, economic, cultural, and environmental data, form can be more broadly interpreted as a phenomenon of human activities. As a result, Deeprise will be one step closer to analyze morphological principles and features related to our society.

Technically, the most challenging part of this research is converting three-dimensional form data into stacks of two-dimensional images for learning and then reconstructing three-dimensional forms. These conversion and reconstruction process incur to a certain amount of loss of the original morphological features. Specifically, since the original height information of each floor boundary image was approximated by relative values, the reconstructed form was segmented by extruded geometries in the modeling software.

In the recent years, deep learning research has been pushing the boundaries of representation beyond the structured domain of images. These advancements enable the design of deep learning workflows that do not require conversions between three and two-dimensional data. Some of the examples include geometric representations such as voxels [19], meshes [20], and point clouds [21], which are available in DL libraries such as Pytorch3D [22]. Learning with these representations showed good performance in well-structured geometric domains with small scale variance, such as in models of bodies and faces. Applying these techniques to the domain of building morphology will require systematic exploration and experimentation.

References

1. Szolomicki, J., Golasz-Szolomicka, H.: Technological advances and trends in modern high-rise buildings. Buildings **9**, 193 (2019). https://doi.org/10.3390/buildings9090193
2. Kovacevic, I., Dzidic, S.: Modern structural concepts for high-rise buildings. In: САВРЕМЕНА ТЕОРИЈА И ПРАКСА У ГРАДИТЕЉСТВУ (2018).https://doi.org/10.7251/STP 1813549K
3. CTBUH, T.C. on T.B. and U.H.: Height Criteria (2020). https://www.ctbuh.org/resource/height
4. Moussavi, F., Ciancarella, M., Scelsa, J.A., Crettier, M., Kilalea, K. (eds.): The Function of Style. Actar, Barcelona (2014)
5. Donath, D., Lobos, D.: Massing study support. A new tool for early stages of architectural design, p. 8 (2008)
6. Ren, Y., Zheng, H.: The spire of AI - voxel-based 3D neural style transfer. In: Holzer, D., Nakapan, W., Globa, A., Koh, I. (eds.) RE: Anthropocene, Design in the Age of Humans - Proceedings of the 25th CAADRIA Conference - Volume 2, Chulalongkorn University, Bangkok, Thailand, 5–6 August 2020, pp. 619–628. CUMINCAD (2020)
7. Cudzik, J., Radziszewski, K.: Artificial intelligence aided architectural design. In: AI for Design and Built Environment, pp. 77–84 (2018)
8. Gatys, L., Ecker, A., Bethge, M.: A neural algorithm of artistic style. arXiv (2015). https://doi.org/10.1167/16.12.326
9. Steinfeld, K., Park, K., Menges, A., Walker, S.: Fresh eyes. In: Lee, J.-H. (ed.) CAAD Futures 2019. CCIS, vol. 1028, pp. 32–46. Springer, Singapore (2019). https://doi.org/10.1007/978-981-13-8410-3_3
10. De Miguel Rodríguez, J., Villafañe, M., Piškorec, L., Sancho Caparrini, F.: Generation of geometric interpolations of building types with deep variational autoencoders. Des. Sci.**6**, E34 (2020).https://doi.org/10.1017/dsj.2020.31
11. Newton, D.: Generative deep learning in architectural design. Technol.|Architect. + Des. **3**(2), 176–189 (2019). https://doi.org/10.1080/24751448.2019.1640536
12. Craighead, G.: High-Rise Security and Fire Life Safety, 3rd edn. Butterworth-Heinemann (2009)
13. Larsen, A.B.L., Sønderby, S.K., Larochelle, H., Winther, O.: Autoencoding beyond pixels using a learned similarity metric. arXiv:1512.09300 [cs, stat] (2016)
14. Thanh-Tung, H., Tran, T.: On catastrophic forgetting and mode collapse in generative adversarial networks. arXiv:1807.04015 [cs, stat] (2020)
15. Huang, H., Li, Z., He, R., Sun, Z., Tan, T.: IntroVAE: introspective variational autoencoders for photographic image synthesis. arXiv:1807.06358 [cs, stat]. (2018)
16. Kullback, S., Leibler, R.A.: On information and sufficiency. Ann. Math. Stat. **22**, 79–86 (1951)
17. van der Maaten, L., Hinton, G.: Viualizing data using t-SNE. J. Mach. Learn. Res. **9**, 2579–2605 (2008)
18. Ester, M., Kriegel, H.-P., Sander, J., Xu, X.: A density-based algorithm for discovering clusters in large spatial databases with noise (1996)
19. Wu, J., Zhang, C., Xue, T., Freeman, W.T., Tenenbaum, J.B.: Learning a probabilistic latent space of object shapes via 3D generative-adversarial modeling. In: Proceedings of the 30th International Conference on Neural Information Processing Systems, NIPS 2016, pp. 82–90. Curran Associates Inc., Red Hook (2016)
20. Cheng, S., Bronstein, M., Zhou, Y., Kotsia, I., Pantic, M., Zafeiriou, S.: MeshGAN: non-linear 3D morphable models of faces. arXiv:1903.10384 [cs] (2019)
21. Achlioptas, P., Diamanti, O., Mitliagkas, I., Guibas, L.: Learning representations and generative models for 3D point clouds. arXiv:1707.02392 [cs] (2017)
22. Facebook: Pytorch3D. https://github.com/facebookresearch/pytorch3d

Structural Design Recommendations in the Early Design Phase Using Machine Learning

Spyridon Ampanavos[1]([✉]) [iD], Mehdi Nourbakhsh[2], and Chin-Yi Cheng[2]

[1] Harvard Graduate School of Design, Cambridge, MA 02138, USA
sampanavos@gsd.harvard.edu
[2] Autodesk Research, San Francisco, CA 94105, USA
{mehdi.nourbakhsh,chin-yi.cheng}@autodesk.com

Abstract. Structural engineering knowledge can be of significant importance to the architectural design team during the early design phase. However, architects and engineers do not typically work together during the conceptual phase; in fact, structural engineers are often called late into the process. As a result, updates in the design are more difficult and time-consuming to complete. At the same time, there is a lost opportunity for better design exploration guided by structural feedback. In general, the earlier in the design process the iteration happens, the greater the benefits in cost efficiency and informed design exploration, which can lead to higher quality creative results.

In order to facilitate an informed exploration in the early design stage, we suggest the automation of fundamental structural engineering tasks and introduce ApproxiFramer, a Machine Learning-based system for the automatic generation of structural layouts from building plan sketches in real-time. The system aims to assist architects by presenting them with feasible structural solutions during the conceptual phase so that they proceed with their design with adequate knowledge of its structural implications.

In this paper, we describe the system and evaluate the performance of a proof-of-concept implementation in the domain of orthogonal, metal, rigid structures. We trained a Convolutional Neural Net to iteratively generate structural design solutions for sketch-level building plans using a synthetic dataset and achieved an average error of 2.2% in the predicted positions of the columns.

Keywords: Machine learning · Structure approximation · Convolutional neural net · Design assistance

1 Introduction

Structure is a fundamental element of a building design. When the structural design is developed in parallel and in coordination with the architectural design, it can inform an architect's decisions and lead to a harmonious integration of the two. However, when

© Springer Nature Singapore Pte Ltd. 2022
D. Gerber et al. (Eds.): CAAD Futures 2021, CCIS 1465, pp. 190–202, 2022.
https://doi.org/10.1007/978-981-19-1280-1_12

structure is not taken into account during the early phase of design, reconciling architectural and structural design can be a cause of delays, conflicts between architects and engineers, and undesirable design compromises.

A study investigating the collaboration between architects and structural engineers conducted in New Zealand in 2009 found that among the primary points of friction are the limited understanding of structural engineering from the side of architects and the late involvement of structural engineers in the project [1]. On the other hand, it has been repeatedly argued that the cost of design changes increases the later they are introduced in the process [2, 3]. The term 'cost' is not limited to monetary expenses but can be generalized to the ability of a change to impact the design [4].

Parametric modeling and BIM software have been used by practitioners and researchers to address such collaboration conflicts [4], and specialized software has been used in research and educational settings to facilitate and promote a better understanding between architects and structural engineers [5]. While such solutions have significantly benefited the field, they do not specifically address the conceptual stage of the design.

The conceptual stage of design is commonly described as a divergent process. It is characterized by quick iteration, and often happens outside of a CAD environment, in the form of sketching. In order to achieve a smoother integration of architectural and structural design, structural feedback should be easily available during the conceptual stage. Such feedback cannot and does not need to be precise, as the design itself in this phase lacks precision. In contrast, approximate and directional feedback can be useful for improving a design towards a better solution with respect to its structure.

In this paper, we introduce ApproxiFramer, an automated system with the ability to generate structural design recommendations during the conceptual phase of architectural design. The goal of such recommendations, indicating potential/optimal structural solutions, is to inform the architects' design decisions and ultimately reduce conflicts with the structural engineers when they get involved in the project at a later stage.

A tool targeting the conceptual design phase has to respond to two main challenges. First, the feedback needs to be generated in real-time. Second, the tool should be able to directly handle conceptual sketches without requiring the user to translate them into different software. Recent advances in the field of Machine Learning (ML) have demonstrated an increasing ability to handle irregular types of input data, such as images or sketches. In addition, ML methods have been previously used to successfully accelerate structural design tasks [6–9]. ApproxiFramer employs a machine learning model to tackle both the speed and the integration challenges.

In this paper, we develop and evaluate ApproxiFramer by focusing on a specific structural domain, rigid metal structures. We trained a neural net on a synthetic dataset consisting of sketch-level single floor building plans and their corresponding structural layouts. The neural net generated structural layouts in real-time while achieving an average percentage error of 2.21% in the positions of the structural elements of the test set, confirming the potential of the method for early phase design assistance.

We make two contributions in the area of early-phase design decision support. First, we introduce a method for generating approximate structural solutions for architectural sketches in real-time. Second, we report on the results of an experiment and demonstrate that a machine learning-based system can successfully learn to generalize a consistent set of structural principles.

2   Related Work

Some previous work seeking to assist architects in designing buildings that better conform to various performance criteria has employed various forms of optimization [10]. Such works that focus on the early design phase typically combine procedural modeling and simulation software, with the parameters of the generative model being tuned through an optimization algorithm [11, 12]. Shea et al. elaborate on how parametric modeling and engineering performance feedback can be used to improve architectural designs [13]. Optimization is not necessarily the end goal of these processes but rather a tool to automatically construct solutions that can guide the architect towards design improvements [14, 15]. Other work has focused on integrating designers' preferences through interactive optimization [16, 17]. More recently, Hamidavi et al. proposed a system that uses multiple types of structural optimization and BIM modeling to improve the collaboration of architects and structural engineers [18]. However, setting up a good procedural model for an optimization process is non-trivial, and an optimization framework to guide this process has been suggested as well [19].

In practice, optimization and the performance simulations that it relies on are often too time-consuming to be employed in the early design phase. The use of surrogate models has been suggested as a way to accelerate simulations [20]. Tseranidis et al. provide an overview of multiple ML algorithms for the approximation of structural engineering calculations [21]. While most surrogate models are trained to only work with specific parametric models and structural topologies, some research has addressed generalizable models that work with multiple topologies of 3d trusses [9].

Other work has used machine learning to directly approximate optimal solutions. Support Vector Machines have been trained to optimally solve individual modules of a space frame [6], Bayesian nets have been used for bi-directional inference with the goal of identifying the most promising areas of a design space with respect to structural performance [22], and neural nets have been used to predict optimal parameters describing the bracing of a metal frame [8].

In contrast to previous research that has targeted design using parametric models of the geometry, we are proposing a method for structural design approximation that directly uses sketch-level plans. The goal is to provide real-time guidance in the early design exploration during the actual sketching before an idea is formalized into a CAD drawing.

3   Method

3.1   Approach

ApproxiFramer aims to inform the early phase design exploration in a sketch-based environment through the real-time generation of structural designs. Figure 1 describes how ApproxiFramer can be integrated into such an environment. A user-generated sketch first passes through a pre-processing step that converts a noisy and imprecise input to a clean drawing so that it can be used with our predictive system. This kind of processing is common in commercial design graphics software, so this research considers it given, and further technical elaboration is out of scope.

Consequently, we propose the use of a neural net to solve the problem of real-time structural layout predictions from building plans. Inspired by previous work that suggests the decomposition of structural problems into sub-problems that are easier and more generalizable [8, 23], we do not attempt to estimate the complete structure at once. Instead, we use an iterative approach, only locally solving the problem and predicting a partial structure, gradually extending the solution until no more extensions are necessary. We expect that the neural net will have more chances of identifying patterns when focusing on a small area of the given building at each step, even if every observed building is unique. In order to evaluate and further develop the proposed method, we conducted an experiment where the scope of the problem has been limited, as described in the next section.

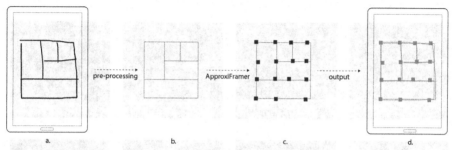

Fig. 1. Sketch-based interaction and structural predictions. The user designs a sketch of a plan (a.), the system converts the noisy sketch to a clean drawing (b.) and passes it to ApproxiFramer that predicts the placement of the structural elements (c.). The structural solution is superimposed on the user's initial sketch (d.).

3.2 Problem Scope

In general, the design of a structure is informed by a series of specifications and constraints. The type of structural system, materials, available structural members, regulation, and others will all affect the solution of the design, so that the same building design may lead to very different structural designs based on these parameters. The current experiment operates in a constrained space where these parameters are assumed to have fixed values.

In this paper, we focus on rigid metal frames, always connected at right angles. No bracing is typically required for such structures. Selection of the appropriate cross-sections and sizing of the structural elements are outside of the scope of this study and are typically of minor importance during the conceptual design phase. As such, the structures that we design can be easily abstracted to a diagrammatic level. A set of coordinates that indicate the locations of the columns is then a sufficient description of such a frame.

The developed system generates structural layouts for orthogonal, sketch-level, single-floor plans. These plans include exterior and interior walls of the building, both represented by single straight lines that are either horizontal or vertical.

3.3 System Architecture

The input of the system is in image format, providing significant flexibility to the user in the selection of design software or medium. In the core of the system lies a convolutional neural net (CNN) that we trained to take an image of a sketch, representing a plan of a building layout, and predict the position of a group of columns. In each iteration, the newly predicted columns are added to the solution, and a new image is rendered that contains both the initial sketch and the columns that have been placed so far. This newly rendered image is then used as the input of the next iteration. Algorithm 1 describes this iterative process.

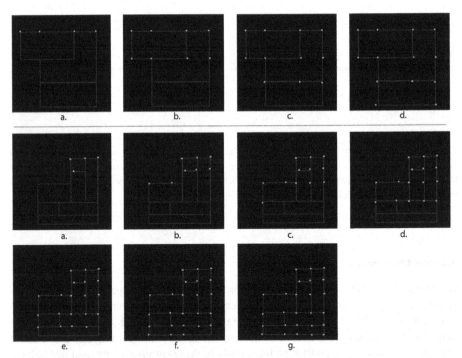

Fig. 2. Representative examples of iterative predictions for a smaller structure in four steps (top) and a larger structure in seven steps (bottom).

Each iteration solves a local sub-problem, scanning the building from left to right and from top to bottom, and adding a fixed number[1] of columns to the solution. The

---

[1] The last iteration will add any number of columns between 0 and that fixed number, as needed.

columns are assumed in a specific order as well: left to right and top to bottom. The model was trained to output a zero vector when there are no more columns to be placed.

Algorithm 1. Predicting positions of all columns.

```
Image ← initialSketch
Columns ← []
repeat
    column ← predictNextColumn(image)
    columns.append(column)
    image ← addColumnToImage(image, column)
until allColumnsHaveBeenPlaced(image, columns)
return image, columns
```

The number of columns for each iteration was defined as $n = 4$, based on initial results after considering alternatives between 1 and 8. Figure 2 demonstrates two examples of structures being predicted in 4 and 7 iterations.

The CNN takes as input an image of 128×128 pixels with four channels. Two channels contain the building layout and the already placed columns, and two contain the pixel coordinates, as suggested in [24]. The image passes through three convolutional layers with kernel sizes 7, 3, 3 and strides 2, 2, 2 and a ResNet block [25], followed by two fully connected layers, an LSTM layer [26], and two output layers which are also fully connected. The first output layer (4×2) contains the coordinates of the four predicted columns. The second output layer (4×3) contains the type classification for each of the predicted columns. The possible types are free-standing, column on corner, or column on wall. All layers use ReLu activation functions, except for the ResNet and the output layers. The ResNet uses linear activations and is followed by batch normalization and leaky ReLu. The coordinates output layer uses sigmoid activation since the coordinates are normalized in the range $[-1, 1]$, and the type output layer uses softmax to convert the output to class probabilities. The number of filters is shown in Fig. 3, which depicts the structure of the neural net in detail.

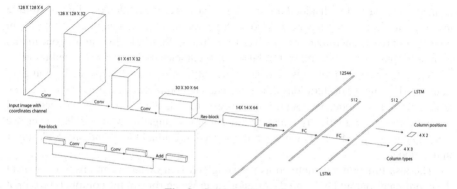

Fig. 3. CNN Architecture.

3.4 Dataset

An appropriate dataset can be sourced from historical or synthetic data. In general, we expect that given a number of buildings and their corresponding structures that follow a specific set of principles, we can train a CNN to abstract these principles and iteratively generate similar structures for more buildings of the same type. In this work, we generated a synthetic dataset of buildings on an orthogonal grid. The structural layouts were generated using heuristics. While this is not the ideal scenario to demonstrate the power of our system, our focus here is to demonstrate the ability of the system to approximate a set of structural designs, which is expected to generalize to other, more sophisticated datasets as well.

We created 35 building layouts, each building including both exterior and interior walls. This initial set of buildings was augmented through 90-degree rotations, scaling, and translations. For each building, we designed a structural layout using the same heuristics: fitting a grid of columns with a predefined maximum span. The resulting dataset contains 10,000 pairs of buildings and structural layouts. Out of these, we used 9,000 for training and validation, and 1,000 for testing. In order to use the training data with the system's iterative approach, we generated the set of all possible configurations of incremental structural designs for each of the buildings. In the incremental structural designs, we determine the next partial solution - i.e., the next group of columns to be placed - by ordering the columns by x and y. After this process, we ended up with 137,644 training samples and 12,514 testing samples.

3.5 Training

In order to obtain a complete structural solution for a design layout, we need to run the model in an iterative way, in each step adding to the observed image the predicted columns of the previous step. However, there are a few challenges in practice. Each time that a column is predicted, the location contains a small error (i.e., the alignment may be one or more pixels off). When this column is added to the input image of the next step, it contributes to a larger error in the next prediction. Eventually, the error accumulates until the model is unable to predict the next column locations in a sensible way.

We used two methods to overcome the problem of the accumulated error. First, we created a new dataset with added noise in the locations of the rendered columns. Adding noise is a technique that has been used with neural nets for data augmentation [27] in order to improve generalization and avoid overfitting. Similarly, by training our model on noisy inputs, we aim to make it robust to inaccuracies during iterative prediction.

Second, we worked towards increasing the output size of each step and, by doing so, reducing the number of iterations that are needed to complete a structural layout. Using an earlier, simpler model, we found that simply increasing the size of the output decreased the performance dramatically when no other major changes were made. However, we were able to get good results when we introduced a residual block and an LSTM layer in the model.

The loss function was defined as the weighted sum of the mean absolute error of the coordinates output layer and the categorical cross-entropy of the column type output layer, with weights 1.0 and 0.2. The network was trained using stochastic gradient descent for 900 epochs.

4 Results

On a single run, the CNN model is outputting predictions for four columns. The output includes the column coordinates and the column type (between free-standing, column on corner, or column on wall) (Fig. 3). While we found that training using a weighted loss on a combination of the column coordinates and the column types improved the model performance compared to training on column coordinates prediction alone, the column type information is not used during inference, and therefore it is also excluded from the following results.

4.1 Single Predictions on Perfect Observations – CNN Evaluation

First, we evaluate the model performance on single predictions (i.e., four columns). We use as input all possible partially completed structures from the test set, with a four columns step, and following the ordering by x and y coordinates. The partial completion is done based on the ground truth so that the model is predicting based on a perfect observation.

The model successfully identified when to stop adding new columns 100% of the time. During training, we used a zero vector to indicate the stopping point of the predictions. During inference, we modified the threshold to be a vector where at least one of the x or y coordinates has a value less than 2.

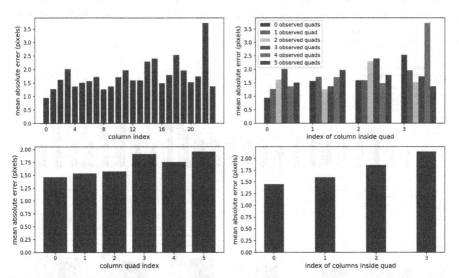

Fig. 4. Left Top: Mean error for first 24 columns. Left Bottom: Mean error for first six prediction steps, where one prediction step outputs four columns (or one "column quad"). Predictions on clean observations. Right Top: Mean absolute error of first six quad predictions, grouped by order inside quad – the first group is first columns of 6 quad predictions, the second group is second columns, etc. Right Bottom: Mean absolute error of same order columns. Predictions on clean observations.

Our CNN achieved an average error of 2.21% in the predicted column positions. This error corresponds to a mean distance of 1.51 pixels between the predicted column locations and the ground truth for our dataset images of 128×128 pixels. Figure 4 Left shows the mean distance between predictions and ground truth in relation to the ordering of the columns. We do not observe a significant change in performance as the size of the observed, partially completed structures increases. However, we notice that within each four columns (or column quads) coming from a single prediction, earlier columns tend to have smaller error. This is better captured in Fig. 4 Right, where columns have been put in four groups based on their order of appearance within a single prediction. The mean absolute error increases from 1.46 pixels for the first columns of each prediction to 2.14 pixels for the fourth columns. This behavior is attributed to the use of the LSTM layer, which introduces a recurrent architecture in the model. Each column of a single prediction depends on the previously estimated columns of the same prediction, and as a result, the error accumulates.

4.2 Iterative Predictions – System Evaluation

Next, we evaluate the system performance on the goal task, which is to estimate all columns for each building. This is accomplished by using our CNN in an iterative way, where each step relies on the output of the previous prediction.

The system predicted the correct number of columns 95.3% of the time. Out of the 1000 buildings of the test set, 47 were solved with fewer or with more columns than the ground truth. These buildings have been removed from the report of the rest of the metrics described below.

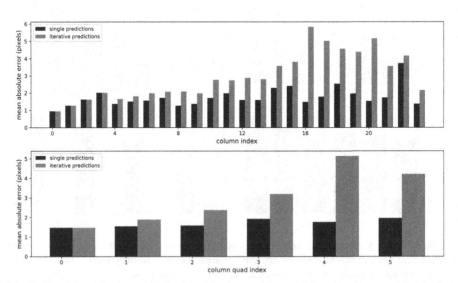

Fig. 5. Top: Mean absolute error for first 24 columns. Bottom: Mean absolute error for the first six prediction quads.

The system achieved an average error of 2.21% in the predicted column positions. The mean distance of the predictions from the ground truth among all iterations was 1.84 pixels. Figure 5 shows in green the mean distance of the predictions from the ground truth in relation to the ordering of the columns. We observe that the error increases for inputs with more columns in the already completed structure. In the previous Subsect. 4.1, we found that the size of the completed structure only has a minimal effect on the CNN performance. Therefore, we attribute this error increase to the accumulated error from the previously predicted columns that are used in the input of the new predictions.

5 Discussion

5.1 ML as an Approximation Means for Early Phase Structural Assistance

Our model performs very well on single predictions, maintaining a very low average error (2.21%) and producing results that are visually coherent. The system also maintains a low average error on iterative predictions (2.21%). In practice, it predicts all column positions well on a large subset of the test set and only fails to output reasonable results while iterating on some building designs. This happens as the model remains susceptible to noisy observations of previous predictions. Currently, the performance tends to decrease both with repeated iterations (Fig. 4 Left and Fig. 5), as well as with later outputs of a single run (Fig. 4 Right). Further investigation is needed on the potential of the two approaches and the optimal combination of them.

The results suggest that the proposed method, which relies on machine learning techniques, constitutes a promising approach for the automatic suggestion of structural designs for early phase architectural sketches or drawings. Once a trained model is loaded, our system only needs a few milliseconds to generate such a structure. Therefore, we believe that a system like ApproxiFramer can provide valuable design assistance during the conceptual design phase.

Furthermore, ApproxiFramer could be combined with other ML methods that propose optimal cross sections of columns and beams [28], based on structural skeletons similar to the ones that our system generates. It can also be used as a complementary tool to parametrization and optimization methods such as the one introduced in SketchOpt [29], providing early, quick estimates before a structural optimization is run.

5.2 Iterative Approach vs. End-to-End Model

The ApproxiFramer system relies on the iterative use of a neural net to predict a complete structure. Early experimentation results, as well as the increasing error between the first and last predictions of a single model inference (Fig. 4 Right Bottom), suggest that the current model architecture is not appropriate for the end-to-end prediction of complete structures.

Apart from performance considerations, we believe that the iterative approach has other advantages, too, over a whole-structure prediction. The model's ability to complete partial structures could be potentially leveraged to interactively guide the design of structure as well as space, following the design paradigm of interactive optimization

[17]. Combined with a different sub-problem parsing strategy in the future, e.g., one where subsequent iterations have increased level-of-detail, this would allow the designer to lead the system towards a specific direction, for example, by modifying the outputs of the initial steps of the structure prediction.

5.3 Generalizability

In this paper, we demonstrated the feasibility of the suggested method in the domain of rigid metal frames connected at right angles. Functioning inside this domain, we were able to simplify a structural design to a set of columns, assuming that beams can be added in a post-processing step using simple heuristics. Even though we considered single-floor plans, the method is easily generalizable to low-rise buildings with a typical plan repeated in all floors above the ground. We expect that our method is also generalizable to different structural systems. However, appropriate modifications will have to be made to accommodate the potential use of multiple types of structural elements in more complex domains. For instance, we have already demonstrated the prediction of labels associated with each column, and it is not difficult to imagine how such labels could be used as classes of multiple types of structural elements.

The dataset used in this work contains orthogonal designs with heuristically generated structures; however, we expect the proposed method to be generalizable to different datasets and human-generated or computationally optimized structures.

5.4 Limitations

The results indicate that the column positioning tends to be noisy, something that may be easier noticeable for columns that should be placed on wall intersections. This is not necessarily an issue in the specific context of early phase sketching since the user's sketches are expected to be similarly rough, and precision is not the goal at this stage. Nevertheless, a post-processing step might be able to fine-tune the positions at a local level.

Last, larger structures are currently more difficult to solve, mainly because of the accumulated error. A larger dataset and different data augmentation techniques are expected to improve this performance.

6 Conclusion

We introduced a method for early phase design assistance with respect to structure, with the goal of promoting more informed design decisions early on and better preparing architects for later stage collaboration with structural engineers. We described a system that can predict structural layouts of single-story building designs from diagrammatic sketches and trained a CNN that performs this task in an iterative way. While we achieved satisfactory performance on our test set, our model only consists a first step to applying this approach to the real world. Further improvements are required in terms of robustness and the ability to take into account structural constraints and parameters.

Future work may explore how to interactively address architectural aspects of the structure by adding the designer in the loop between prediction iterations. Alternative ways of parsing the overall problem into smaller sub-problems may also be investigated in relation to different structural systems. The addition of a fine-tuning step can be investigated as a way to reduce the impact of small inaccuracies in the positions of previously placed columns. Finally, the use of a synthetic dataset generated through optimization or a dataset from historical data - i.e., from a structural engineering practice - will be a significant step towards deploying such a system in the real world.

Acknowledgements. We would like to express our gratitude to Mohammad Keshavarzi for his help with the synthetic data preparation process.

References

1. Charleson, A.W., Pirie, S.: An investigation of structural engineer-architect collaboration. SESOC **22**, 97–104 (2009)
2. Collaboration, Integrated Information and the Project Lifecycle in Building Design, Construction and Operation. Construction Users Roundtable (2004)
3. Paulson, B.C., Jr.: Designing to reduce construction costs. J. Constr. Div. **102**, 587–592 (1976)
4. Davis, D.: Modelled on software engineering: flexible parametric models in the practice of architecture (2013). https://researchbank.rmit.edu.au/view/rmit:161769
5. Charleson, A.W., Wood, P.: Enhancing collaboration between architects and structural engineers using preliminary design software. Presented at the 2014 NZSSE Conference (2014)
6. Hanna, S.: Inductive machine learning of optimal modular structures: estimating solutions using support vector machines. AI EDAM **21**, 351–366 (2007). https://doi.org/10.1017/S08 90060407000327
7. Zheng, H., Moosavi, V., Akbarzadeh, M.: Machine learning assisted evaluations in structural design and construction. Autom. Constr. **119**, 103346 (2020). https://doi.org/10.1016/j.aut con.2020.103346
8. Aksöz, Z., Preisinger, C.: An Interactive structural optimization of space frame structures using machine learning. In: Gengnagel, C., Baverel, O., Burry, J., Ramsgaard Thomsen, M., Weinzierl, S. (eds.) DMSB 2019, pp. 18–31. Springer, Cham (2020). https://doi.org/10.1007/ 978-3-030-29829-6_2
9. Nourbakhsh, M., Irizarry, J., Haymaker, J.: Generalizable surrogate model features to approximate stress in 3D trusses. Eng. Appl. Artif. Intell. **71**, 15–27 (2018). https://doi.org/10.1016/ j.engappai.2018.01.006
10. Evins, R.: A review of computational optimisation methods applied to sustainable building design. Renew. Sustain. Energy Rev. **22**, 230–245 (2013). https://doi.org/10.1016/j.rser.2013. 02.004
11. Keough, I., Benjamin, D.: Multi-objective optimization in architectural design. In: Proceedings of the 2010 Spring Simulation Multiconference, pp. 1–8. Society for Computer Simulation International, San Diego, CA, USA (2010). https://doi.org/10.1145/1878537.187 8736
12. Lin, S.-H.E., Gerber, D.J.: Designing-in performance: a framework for evolutionary energy performance feedback in early stage design. Autom. Constr. **38**, 59–73 (2014). https://doi. org/10.1016/j.autcon.2013.10.007
13. Shea, K., Aish, R., Gourtovaia, M.: Towards integrated performance-driven generative design tools. Autom. Constr. **14**, 253–264 (2005). https://doi.org/10.1016/j.autcon.2004.07.002

14. Caldas, L.: An evolution-based generative design system: using adaptation to shape architectural form (2001). http://dspace.mit.edu/handle/1721.1/8188

15. Bradner, E., Iorio, F., Davis, M.: Parameters tell the design story: ideation and abstraction in design optimization. In: 2014 Proceedings of the Symposium on Simulation for Architecture and Urban Design, p. 26. Society for Computer Simulation International, Tampa, FL, USA (2014)

16. Turrin, M., von Buelow, P., Stouffs, R.: Design explorations of performance driven geometry in architectural design using parametric modeling and genetic algorithms. Adv. Eng. Inform. **25**, 656–675 (2011). https://doi.org/10.1016/j.aei.2011.07.009

17. Mueller, C.T., Ochsendorf, J.A.: Combining structural performance and designer preferences in evolutionary design space exploration. Autom. Constr. **52**, 70–82 (2015). https://doi.org/10.1016/j.autcon.2015.02.011

18. Hamidavi, T., Abrishami, S., Hosseini, M.R.: Towards intelligent structural design of buildings: a BIM-based solution. J. Build. Eng. **32**, 101685 (2020). https://doi.org/10.1016/j.jobe.2020.101685

19. Yang, D., Ren, S., Turrin, M., Sariyildiz, S., Sun, Y.: Multi-disciplinary and multi-objective optimization problem re-formulation in computational design exploration: a case of conceptual sports building design. Autom. Constr. **92**, 242–269 (2018). https://doi.org/10.1016/j.autcon.2018.03.023

20. Wang, G.G., Shan, S.: Review of metamodeling techniques in support of engineering design optimization. J. Mech. Des. **129**, 370–380 (2007). https://doi.org/10.1115/1.2429697

21. Tseranidis, S., Brown, N.C., Mueller, C.T.: Data-driven approximation algorithms for rapid performance evaluation and optimization of civil structures. Autom. Constr. **72**, 279–293 (2016). https://doi.org/10.1016/j.autcon.2016.02.002

22. Conti, Z.X., Kaijima, S.: Enabling inference in performance-driven design exploration. In: De Rycke, K., et al. (eds.) Humanizing Digital Reality, pp. 177–188. Springer, Singapore (2018). https://doi.org/10.1007/978-981-10-6611-5_16

23. Hajela, P., Berke, L.: Neural network based decomposition in optimal structural synthesis. Comput. Syst. Eng. **2**, 473–481 (1991). https://doi.org/10.1016/0956-0521(91)90050-F

24. Liu, R., et al.: An intriguing failing of convolutional neural networks and the CoordConv solution. arXiv:1807.03247 [cs, stat] (2018)

25. He, K., Zhang, X., Ren, S., Sun, J.: Deep residual learning for image recognition. Presented at the Proceedings of the IEEE Conference on Computer Vision and Pattern Recognition (2016)

26. Hochreiter, S., Schmidhuber, J.: Long short-term memory. Neural Comput. **9**, 1735–1780 (1997). https://doi.org/10.1162/neco.1997.9.8.1735

27. Goodfellow, I., Bengio, Y., Courville, A.: Deep Learning. The MIT Press, Cambridge (2016)

28. Chang, K.-H., Cheng, C.-Y.: Learning to simulate and design for structural engineering. In: International Conference on Machine Learning. pp. 1426–1436. PMLR (2020)

29. Keshavarzi, M., Hotson, C., Cheng, C.-Y., Nourbakhsh, M., Bergin, M., Rahmani Asl, M.: SketchOpt: sketch-based parametric model retrieval for generative design. In: Extended Abstracts of the 2021 CHI Conference on Human Factors in Computing Systems. pp. 1–6. Association for Computing Machinery, New York, NY, USA (2021). https://doi.org/10.1145/3411763.3451620

Quantifying Differences Between Architects' and Non-architects' Visual Perception of Originality of Tower Typology Using Deep Learning

Joy Mondal[(✉)] [iD]

WEsearch lab, New Delhi 110019, India
joy@wesearchlab.com

Abstract. The paper presents a computational methodology to quantify the differences in visual perception of originality of the rotating tower typology between architects and non-architects. A parametric definition of the Absolute Tower Building D with twelve variables is used to generate 250 design variants. Subsequently, sixty architects and sixty non-architects were asked to rate the design variants, in comparison to the original design, on a Likert scale of 'Plagiarised' to 'Original'. With the crowd-sourced evaluation data, two neural networks - one each for architects and non-architects - were trained to predict the originality score of 15,000 design variants. The results indicate that architects are more lenient at seeing design variants as original. The average originality score by architects is 27.74% higher than the average originality score by non-architects. Compared to a non-architect, an architect is 1.93 times likelier to see a design variant as original. In 92.01% of the cases, architects' originality score is higher than non-architects'. The methodology can be used to quantify and predict any subjective opinion.

Keywords: Originality · Tower typology · Visual perception · Crowd-sourced · Subjective evaluation · Deep learning · Neural network

1 Introduction

Architecture is a unique discipline where art and engineering meet subjective demands. Architects offer design solution to their clients (predominantly non-architects), but these two groups may not always have the same aesthetic sensibilities. Multiple studies [1–3] have shown that architects and non-architects have different preferences. Jeffrey and Reynolds [4] studied the differences in aesthetic "code" of architects and non-architects, and argued that buildings constructed according to the "code" of architects is less likely to receive popular acclaim. Another study [5] focused on the decision-making in purchasing residential properties. It found that non-architects ranked a property perceived as "family home" higher than a property perceived as "light and outward facing". This finding indicates that subjective factors tend to be more relevant to non-architects than

© Springer Nature Singapore Pte Ltd. 2022
D. Gerber et al. (Eds.): CAAD Futures 2021, CCIS 1465, pp. 203–221, 2022.
https://doi.org/10.1007/978-981-19-1280-1_13

objective design goals. Moreover, Brown and Gifford [6] concluded that typically, architects cannot predict the non-architect's aesthetic evaluation of architecture. Therefore, a potential conflict exists between architects and non-architects, who may have different expectations from a building. This research investigates the differences between architects and non-architects with regard to the visual perception of originality of design of the tower typology.

Fig. 1. Absolute tower d on the right. Clicked by Sarbjit Bahga, reproduced under CC license.

Throughout the history of architecture, the tower typology has been a symbol of power and wealth, the identity of city skylines, and iconoclastic landmarks for human navigation [7]. In many ways; by combining vertical mobility, material innovation, mechanical heating, speed of construction, wind and earthquake resistance, evacuation planning, and service automation [8]; towers (or skyscrapers) exhibit the pinnacle of architectural and engineering design [9]. During the design of a tower, architects are progressively moving away from the extruded box towards non-orthogonal designs [10]. Simple shapes such as rectangle and ellipse are often transformed and varied in the z-axis to conceive complex geometries. Vollers [10] explains *that: "Twisted geometries with repetition of elements are applied not so much for economic gain as for semiotic connotation".*

This research uses the design of Absolute Tower D designed by MAD Architects (see Fig. 1) - a twisting tower with elliptical floor plan [11] - as the original design against which architects and non-architects are asked to evaluate the originality of design variants. The following sections elaborate the background, the research methodology, the use of deep learning for quantification, the results, and the future scope of work.

2   Background

2.1   Originality in Design

The definition of originality in the context of design is rather subjective and open-ended. It is often conflated with innovation, novelty and creativity. Being innovative, novel and/or creative independently may not necessarily mean that a design, a work of art, a theory, or a discovery is original. The phenomenon of simultaneous invention explains that most scientific advancements are made independently and more or less simultaneously by multiple scientists [12], exemplified as early as in 1774 by the simultaneous discovery of oxygen by Scheele and Priestley. In the context of art, Lamb and Easton [13] have argued that science and art are not dissimilar in this regard. The way papers of simultaneous discoveries are same in terms of the core idea, but not same word-for-word; likewise, two painters may independently paint about the same core theme, but their paintings may not be identical stroke-for-stroke. Therefore, originality cannot be independently absolute. It can only be reviewed in comparison to reference(s). Since originality is relative, being innovative, novel and/or creative cannot be linked causally to originality. Instead, they are better understood as features of originality.

The emphasis on originality and individuality as a way of life has been propagated by popular ad campaigns in the twentieth and twenty-first centuries; exemplified by Apple's 1997 ad campaign slogan "Think different". Reinartz and Saffert [14] studied 437 ad campaigns and concluded that the combination of originality and elaboration is the most effective way (96% more than median) to inspire people to view a product favourably and buy it. It is followed by the combination of originality and artistic value (89% more than median). It would be safe to conclude that the perception of originality - be it visual, audio, tactile, or spatial – subliminally plays a significant role in one's appreciation of a product or an act of creativity. However, a universal definition of originality in the context of design is difficult to establish because of the fact that originality in design can be discerned in several ways – in the process, function, and form. Originality of a process may be defined as a byproduct of creativity that makes an idea evolve into a system and then into an artefact [15]. Originality may be sought in the function of a design which typically manifests itself through transformation of scientific or technical research into a product [16]. Originality may also be sought in the form of a design. Often times form is explored with structural performance [17] and/or energy performance [18] in mind. Originality may also accessed as an antonym of plagiarism, i.e., from the perspective of copyright laws. However, existence of copyright laws, their structure (state vs federal), and the extent of the law's ambit varies from one country to the other. The processes to detect and the legal implications of violating originality (or copyright) of design are beyond the scope of this research.

Non-architects are typically not privy to the process of architectural design. Their sense of originality in architecture is primarily derived from visual stimuli [19]. The aim of this research is to compare the visual perception of originality between architects and non-architects. In other words, this research aims to quantify the originality of forms. Evaluating the originality of a given form without an explicit reference would require the evaluator to subliminally conjure all the forms ever seen, and then compare the given form with the conjured forms that act as reference. Therefore, understanding originality of form, similar to understanding the originality of scientific discoveries, is an exercise of (visual) comparison with one's experiences as the reference. For the purpose of this research, instead of relying on visual comparison with the sub-conscious, forms of design variants will be compared to the form of the original (reference) design of the Absolute Tower D.

It is to be noted that non-architects associated with other design fields may understand architecture beyond visual stimuli. It may be argued that "non-architect designers", as a group, sits between "architects" and "non-architects non-designers", with respect to holistic understanding of the process of architectural design. In the context of this paper, the phrase "non-architect(s)" excludes the sub-group of people that are associated with any design field – both as context and in selection of participants for evaluation of originality.

2.2 Machine Learning in Architecture

Artificial intelligence, and in particular machine learning, has become a popular topic in all computational processes across industries. Since the 2010s it has been incorporated in various researches in architectural design as well. Machine intelligence can be utilised to support creativity [20], automate housing layout generation [21], automate implicit design iteration through discretisation [22], transfer 3D style of a geometry to another geometry [23], appropriate performance simulation [24], and calculate urban space perception [25].

Traditional programming requires the programmer to explicitly define rules to (subsequently) generate output. Deep learning is a departure from such a system. It is a subset of machine learning that uses deep neural network with multiple hidden layers to statistically appropriate the entire solution space (see Fig. 2). It does so by training on discrete sample dataset with known input variables (independent variables or features), and known output (dependent variable(s) or label(s)). On completion of training, i.e., after statistically mapping the relationship between the independent and dependent variables (see Fig. 2b), the neural network is capable of predicting output for any new set of input variables. Consequently, with the use of deep learning, the number of data samples that need to be explicitly evaluated by survey participants reduces significantly (from thousands to hundreds). The evaluation of additional data samples can subsequently be predicted with a neural network that is trained with the explicit evaluations (see Fig. 2c).

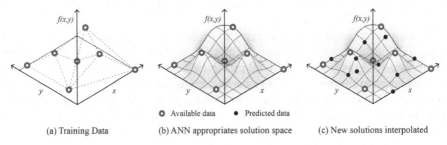

○ Available data   ● Predicted data

(a) Training Data   (b) ANN appropriates solution space   (c) New solutions interpolated

Fig. 2. Training and prediction by neural network

3   Research Methodology

This research uses a computational methodology that can be used to quantify (see Fig. 3) and predict any kind of subjective evaluation of design. The methodology combines crowd-sourced design evaluation with the statistical power of deep learning. The methodology has four key steps mentioned as following –

1. *Defining design variants*: The first step in the process is to collect or generate design variants that can be evaluated. For example, if urban streetscapes are to be evaluated for beauty, images of urban streetscapes need to be collected for evaluation. If a parametric definition is being used for design, design variants need to be generated by varying the parameters (independent variables).
2. *Crowd-sourced evaluation*: The second step in the process is to get the design variants evaluated by relevant group(s). The evaluation data is discrete in nature and does not truly represent the solution space (see Fig. 2a). Therefore, it is not directly used for comparative study.
3. *Neural network training*: The third step in the process is to train a deep neural network with the evaluation data (output or dependent variable) and the design parameters (input or independent variable) that define the design variants. Through training, the neural network learns to appropriate the solution space (see Fig. 2b).
4. *Predicting subjective evaluation*: On the completion of training of the neural network with sufficient accuracy, the fourth step is to predict subjective evaluations of a larger new set of design variants. Since neural networks learn the solution space during training, they can predict the evaluation of new variants by virtue of interpolation (see Fig. 2c). Finally, the predicted values are used for comparative analysis.

The following sub-sections discuss the four steps in detail with respect to this research, along with discussing the preliminary analysis of the evaluation data and the limitations in the process of quantification.

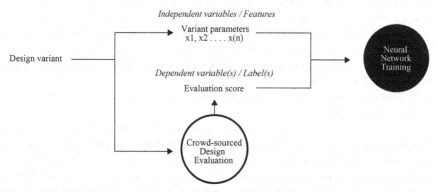

Fig. 3. Computational methodology for training a neural network to quantify and predict crowd-sourced subjective evaluation.

3.1   Step 1: Defining Design Variants

Design variants of Absolute Tower D were defined and generated by a parametric Grasshopper definition of the original design. The original design [11] can be parametrically represented by only five parameters – the two radii of the elliptical floor plan, total number of floors, floor to floor height (or total height) and total rotation (or floor to floor rotation). In order to have design variants of similar visual weightage, total number of floors and floor to floor height are not varied to generate design variants. In addition to the three other parameters of the original design, nine additional parameters are added to the parametric definition. These additional parameters change the shape of the floor plan from ellipse to a four-legged star to a square and to a rhombus, control the nature of the rotation of floor plates (linear or bezier), change the state of the balconies (present or absent), and scale the floor plates towards the top and the bottom of the tower. The floor plan is represented by a NURBS curve instead of an ellipse, and the corner point weights of the control polygon and the position of the mid-points of the control polygon are varied to morph the floor plan into the four shapes. The parameters (see Table 1) are elaborated as following -

- *(x1) Radius 1 of the floor plan (plan_r1)*: Controls the length of the bounding box of the floor plan. When the floor plan is elliptical, it represents one of the radii of the ellipse.
- *(x2) Radius 2 of floor plan (plan_r2)*: Controls the width of the bounding box of the floor plan. When the floor plan is elliptical, it represents one of the radii of the ellipse.
- *(x3) Floor plan corner point weight (crpt_weight)*: Controls the weightage of the corner points of the control polygon of the floor plan. When all the other parameters are kept constant at the values of the original design, *crpt_weight* = 0.0 yields a rhombus floor plan, *crpt_weight* = 0.5 yields the original design, and *crpt_weight* = 1.0 yields a rectangular floor plan (see the first row in Fig. 4).
- *(x4) Floor plan mid-point movement (midpt_move)*: Controls the displacement of the mid-points of the control polygon of the floor plan towards the centre of the floor plan. When all the other parameters are kept constant at the values of the original design, *midpt_move* = 0.0 yields the original design, and *midpt_move* = 1.0 yields a blunt four-legged star floor plan (see the second row in Fig. 4). When all parameters except *crpt_weight* and *midpt_move* are kept constant at the values of the original design, *crpt_weight* = 1.0 and *midpt_move* = 0.5 yield a sharp four-legged star floor plan (see the third row in Fig. 4).
- *(x5) Total rotation (tot_rot)*: Controls the total angle of rotation between the bottom and the top floor plates.
- *(x6–x9) Distribution of the rotation values (rotstart_x, rotstart_y, rotend_x, rotend_y)*: The four parameters control the nature of the distribution of the rotation values of floor plates. The distribution is calculated by a bezier S-curve. Therefore, the four parameters are the two anchor points (*rotstart_y* and *rotend_y*) and the two handles (*rotstart_x* and *rotend_x*) of the S-curve. When all of the four parameters have a value of zero, the S-curve takes the shape of a straight line, thereby making the distribution linear in nature.

- *(x10) Presence or absence of balcony (bal_state)*: Controls the presence or absence of balcony projections using a discrete boolean value. When *bal_state* is zero, the balconies are replaced by glazed facade.
- *(x11–x12) Scaling values of the floor plates (scale_top, scale_bottom)*: Controls the amount of scaling in the top three quarters (*scale_top*) and the bottom quarter of the tower (*scale_bottom*). The scaling of the floor plates start from the top of the bottom quarter towards both the directions.

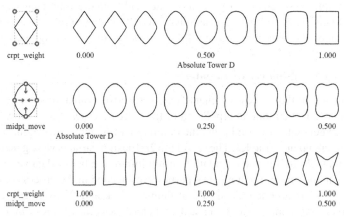

Fig. 4. Variation in floor plan with *crpt_weight* and *midpt_move*

Table 1. Domains of the parameters (independent variables) of the design variants.

Parameters	Minimum	Maximum	Tower D	Number of possible values
plan_r1	10.00	15.00	13.70	500
plan_r2	15.00	25.00	18.50	1000
crpt_weight	0.000	1.000	0.500	1000
midpt_move	0.000	0.500	0.000	500
tot_rot	0	720	208	720
rotstart_x	0.000	0.750	0.535	750
rotstart_y	0.000	0.750	0.000	750
rotend_x	0.000	0.750	0.395	750
rotend_y	0.000	0.750	0.000	750
bal_state	0	1	1	2
scale_top	0.010	1.000	1.000	990
scale_bottom	0.500	1.000	1.000	500

In multivariate studies such as this research, the minimum number of data samples needed to train a neural network can be determined using few industry accepted rules of thumb. Sekaran and Bougie [26] prescribe the 10x rule which states that the number of data samples should be at least ten times the number of independent variables. To test the correlations of independent variables, Green [27] recommends a minimum sample size of 50 + 8k, where k is the number of independent variables. To conclusively make predictions, Green [27] recommends a minimum sample size of 104 + k, where k is the number of independent variables. Given that the number of independent variables is twelve in this research, the rules of thumb recommend a minimum of 120, 146 and 116 data samples respectively. To accommodate additional testing data to measure the accuracy of neural network, this research has used 250 data samples (design variations) for crowd-sourced evaluation. Figure 5 shows 84 out of the 250 design variants.

3.2   Step 2: Crowd-Sourced Evaluation

The second step in the process of quantifying subjective evaluations is to evaluate the design variants. To facilitate interpolation of the extreme cases, fifty of the 250 design variants were generated by combining the minimum, maximum and original design values of the independent variables. The rest of the 200 design variants were generated using random values. Similar to the calculation of minimum number of data variants needed for evaluation, the minimum number of participants needed in the process of evaluation also needs to be ascertained. As a rule of thumb, Clark and Watson [28] recommend ten participants per item on the rating scale, whereas DeVellis [29] recommends fifteen participants per item on the rating scale. This research has used a Likert scale of four items for crowd-sourced evaluation. Thus, following the latter rule of thumb, sixty participants were needed for evaluation. Since this research compares the evaluations of two groups, i.e., architects and non-architects, the 250 design variants are evaluated by sixty architects as well as sixty non-architects. The lowest age amongst the architects was 23. Consequently, all the selected non-architect participants were older than 22. As mentioned in Sect. 2.2 "Originality in Design", the non-architect participants exclude people associated with other design fields. The quantification of the visual perception of originality of the excluded sub-group is not part of this research.

Since the evaluations are comparative in nature, to acquaint the participants to the extremities of the design variants, all the design variants were shown to each of them before the process of evaluation. The architects and the non-architects were asked to rate the design variants against the following question –

"How would you rate the visual relationship of the displayed designs with respect to the reference design?"

A Likert scale was used to collect the evaluations. The design of a Likert scale has two variables – the number of categories in the scale and the description of the categories in the scale. Given that 250 design variants were to be evaluated, six or above categories could lead to decision fatigue [30]. A four-category scale was selected over a five-category scale to reduce the risk of participants avoiding the process of evaluation by selecting the 'Neutral' category. Agree-disagree descriptions yield lower quality data as they suffer from acquiescence response bias [31]. Therefore, qualitative labels of 'Plagiarised', 'Similar', 'Different', and 'Original' were used as categories.

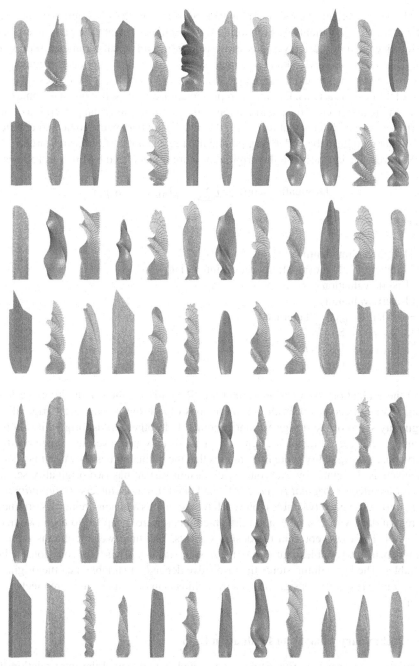

Fig. 5. 84 of the 250 design variants used for crowd-sourced evaluation; top left is the original design, bottom left is a rectangular extrusion variant without balcony.

The design variants were shown alongside the reference design (original design). The order of the design variants was randomised for each participant. The participants were not informed that the design variants are derived from a universal parametric representation. This information was skipped to understand the visual perception of originality of the design variants as an explicit artefact, without any connection to the process of generation of the design variants. Parametric representation may be self-evident to some of the participating architects. The quantification of the difference in visual perception of originality between architects that recognise the underlying parametric logic and architects that do not recognise the underlying parametric logic is not part of this research.

For every design variant i, the originality score is calculated by Eq. 1 as follows -

$$\text{Originality score}_{(i)} = \left(\sum_{n=1}^{r} L_{(i)(n)} \times W \right) / r \tag{1}$$

where,

i = Design variant identifier
r = Number of Likert evaluations per design variant
L = Likert evaluation
L = Likert evaluation
W = 0.00, if $L_{(i)(n)}$ is 'Plagiarised'
0.33, if $L_{(i)(n)}$ is 'Similar'
0.67, if $L_{(i)(n)}$ is 'Different'
1.00, if $L_{(i)(n)}$ is 'Orginal'

If all the evaluations of a design variant are 'Plagiarised', the originality score of the design variant becomes 0. If all the evaluations of a design variant are 'Original', the originality score of the design variant becomes 1. In other words, originality score of 0 implies that the particular design variant is visually perceived as 'Plagiarised' by everyone, and originality score of 1 implies that the particular design variant is visually perceived as 'Original' by everyone. Each design variant has two originality scores – one for architects (*Originality score (ar)*) and one for non-architects (*Originality score (non_ar)*). As a result, two tables (one each for architect's and non-architect's originality scores) of data with 250 rows and thirteen columns were compiled to train two neural networks. The rows represent the design variants. The first twelve columns represent the independent variables that define the design variants. These are common for both the tables. The last column stores the respective dependent variable, i.e., the originality scores of the respective design variants by architects in one and by non-architects in the other.

3.3 Preliminary Analysis of Evaluation Data

Correlation matrices (see Fig. 6) were generated to understand the intra-relationships between all the variables (independent and dependent). Depending on the correlation values, independent variables are excluded from deep learning. Additionally, the correlation values indicate which independent variables play a major role in the visual perception of originality.

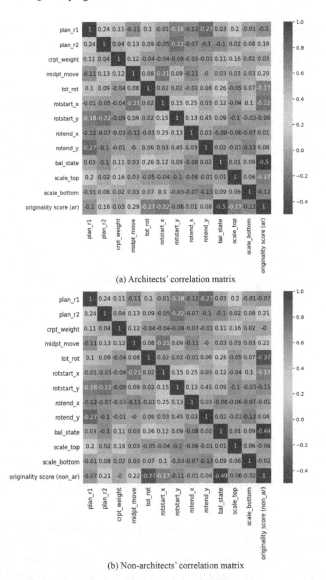

(a) Architects' correlation matrix

(b) Non-architects' correlation matrix

Fig. 6. Intra-correlation matrices of all variables in crowd-sourced data

Exclusion of Independent Variables. Independent variables with high intra-correlation (r > 0.75) are typically excluded from training neural network because they tend to supply the same information to the neural network. Consequently, by including both the independent variables, one adds noise instead of incremental information. As the colour-coded matrices reveal, none of the independent variables have high intra-correlation (r > 0.75). Therefore, all the independent variables are to be included in the training. Additionally, none of the independent variables have high correlation (r > 0.75) to either of the dependent variables. This is a typical feature of data sets that are compiled from subjective observations. The lack of strong correlation makes it almost impossible for traditional statistical methods to appropriate such a data set with a high degree of accuracy (~75% or above).

Significant Independent Variables. The analysis of the correlations of the independent and the dependent variables (see Table 2) reveals that seven out of the twelve independent variables are inversely correlated to the dependent variables. Floor plan mid-point movement (*midpt_move*) exhibits highest correlations with the originality scores of architects (r = 0.29) and non-architects (r = 0.22). As shown in Fig. 7 (see Fig. 4 for 2D plans), this independent variable changes the shape of the plan from an ellipse to a four-legged star.

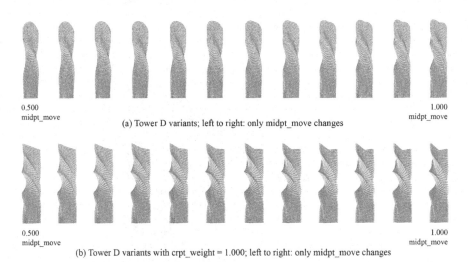

0.500
midpt_move   (a) Tower D variants; left to right: only midpt_move changes   1.000
                                                                                                midpt_move

0.500
midpt_move   1.000
                                                                                                midpt_move

(b) Tower D variants with crpt_weight = 1.000; left to right: only midpt_move changes

Fig. 7. Visual effect of varying *midpt_move* on the 3D of tower D

Radius 2 of floor plan (*plan_r2*) exhibits the second highest correlations with the originality scores of architects (r = 0.16) and non-architects (r = 0.21). As shown in Fig. 8, this independent variable controls the horizontality of design variants, thereby reducing slenderness with increase in value. The nature of the two independent variables with highest correlations with dependent variables (*midpt_move* and *plan_r2*) indicates that when visually comparing towers (or design variants) of the same height, both architects and non-architects tend to subliminally concentrate more on visual features that horizontally control the overall silhouette of the towers.

Presence or absence of balcony (*bal_state*) has the highest inverse correlations with the originality scores of architects (r = −0.50) and non-architects (r = −0.49). Additionally, it exhibits the lowest absolute difference in correlations (|Δr| = 0.01) between architects and non-architects. This suggests that when visually comparing towers for originality (or plagiarism), both architects and non-architects tend to ignore features that affect local surface articulation (see Fig. 9).

Table 2. Correlations of independent and dependent variables in crowd-sourced data.

Independent variables	Originality score (ar)	Originality score (non_ar)	Absolute difference
plan_r1	−0.10	−0.07	0.03
plan_r2	0.16	0.21	0.05
crpt_weight	0.03	0.00	0.04
midpt_move	0.29	0.22	0.07
tot_rot	−0.17	−0.37	0.20
rotstart_x	−0.22	−0.17	0.05
rotstart_y	−0.06	−0.11	0.05
rotend_x	0.01	−0.01	0.02
rotend_y	0.08	0.04	0.04
bal_state	−0.50	−0.49	0.01
scale_top	−0.17	−0.06	0.11
scale_bottom	−0.12	−0.02	0.10

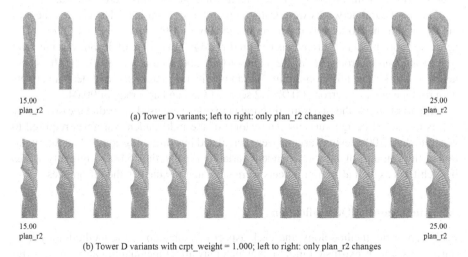

15.00
plan_r2

25.00
plan_r2

(a) Tower D variants; left to right: only plan_r2 changes

15.00
plan_r2

25.00
plan_r2

(b) Tower D variants with crpt_weight = 1.000; left to right: only plan_r2 changes

Fig. 8. Visual effect of varying *plan_r2* on the 3D of tower D

(a) Tower D variant (bal_state = 1)   (b) Tower D variant (bal_state = 0)

Fig. 9. Visual effect of varying *bal_state* on the 3D of tower D

Total rotation (*tot_rot*) exhibits the highest absolute difference in correlations ($|\Delta r|$ = 0.20) between the originality scores of architects ($r = -0.17$) and non-architects ($r = -0.37$). This would appear to indicate that architects tend to be more visually sensitive towards rotation of floor plates. Conversely, non-architects tend to see rotated towers as a visually homogeneous group without much regard to finer differences in the amount of rotation.

3.4 Steps 3 and 4: Neural Network Training and Predicting Originality Score

The third step in the process of quantifying subjective evaluations using deep learning is to train a neural network with the evaluation data. On the completion of training of the neural network with sufficient accuracy, the fourth step is to predict subjective evaluations of a larger set of design variants.

Google Colab was used to write, edit and execute the code for steps 3 and 4. Scikit-learn machine learning library was used for greater flexibility and control over the neural network models. Choosing the hyper-parameters of a neural network (e.g., the number of hidden layers, the number of neurons in each layer, activation function, loss function, batch size, and number of epochs) is a complex process, that affects the network's efficiency. Scikit-learn's 'GridSearchCV' class was used to iteratively train the two neural networks with varied hyper-parameters, until best results were attained.

The two neural networks were trained on 200 design variants. The effectiveness of the two neural networks was tested on the remaining fifty design variants. The select neural network model for architects can predict originality score of design variants with root mean square error of 0.08, R2 score of 0.81 and accuracy of 91.35%. The select neural network model for non-architects can predict originality score of design variants with root mean square error of 0.07, R2 score of 0.83 and accuracy of 90.04%.

As part of step 4, the two trained neural networks were used to predict the originality scores of 15,000 design variants. The values of the independent variables required to generate the 15,000 design variants were calculated by combining equidistant interpolation of the domains of each independent variable. Finally, the 15,000 originality scores by architects as well as non-architects were tabulated to quantify the differences.

3.5 Limitations of Quantification

Each step of the methodology and each aspect of each step of the methodology have intrinsic as well as extrinsic limitations which may affect the quantification of the originality scores. Given the variance in absolute differences between the correlations of independent and dependent variables between architects and non-architects (see Table 2), changes in the selection of parameters to be varied and the range in which they are to be varied to generate the design variants will yield different originality scores. The form used in step 2 of the process, i.e., collection of crowd-sourced evaluation has intrinsic limitations. Changing the perspective from which design variants are seen in the form, and changing the medium of seeing the design variants (rendered, diagrammatic, pictures of model, animated, vs mounted on VR set etc.) will affect the visual perception of objects or buildings. Collection of such comparative data sparks questions about the extrinsic influences, e.g., is the visual perception of architecture by non-architects who are either designers or are trained in artistic fields more correlated to architects instead of non-architects not trained in creative fields? Additionally, the data used to train the neural networks is reflective of the socio-cultural bias of the volunteers. This feature of the nature of crowd-sourced data can be utilised to analyse and predict design preferences specific to groups of people (see section "Conclusion").

4   Result and Discussion

The percentage distribution of the originality scores of 15,000 design variants by architects as well as non-architects are shown in Fig. 10 through ten bins with bin-width of 0.10. Table 3 shows the numerical summary of the predicted originality scores. To understand the nature of the respective originality scores, Fig. 10 and Table 3 are to be read in conjunction. The nature of the percentage distribution bins indicates two differentiators between architects and non-architects. Firstly, the tallest bins of architects are closer to 'Original' (1.00) than the tallest bins of non-architects. It implies that architects tend to be more lenient than non-architects at seeing design variants as original. This observation is corroborated by the higher mean (0.76) and median (0.83) originality scores by architects compared to the mean (0.59) and median (0.62) originality scores by non-architects. In fact, the mean and median originality scores by architects are between the 'Different' and 'Original' categories in the Likert scale, whereas for non-architects they are between the 'Similar' and 'Different' categories.

The second differentiator is that the tallest bins of architects are taller than the tallest bins of non-architects. It implies that architects tend to have a higher consensus than non-architects at reading the visual perception of originality. This observation is corroborated by the lower coefficient of variation (0.22) of the originality scores by architects compared to the coefficient of variation (0.39) of the originality scores by non-architects. In fact, in the case of non-architects, the coefficient of variation is higher than the step value (0.33) of the Likert categories. Additionally, the numerical analysis of the originality scores (see Table 3) reveals that an architect is 17.14 times likely to

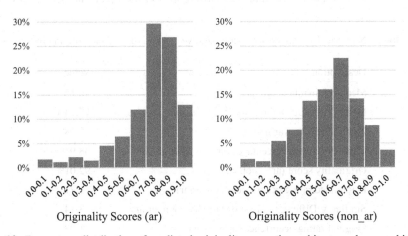

Fig. 10. Percentage distribution of predicted originality scores by architects and non-architects

see a design variant as 'Original' instead of 'Plagiarised', whereas, for a non-architect, the likelihood drops down to 5.26 times.

Table 3. Numerical summary of the predicted originality scores.

Description	Originality score (ar)	Originality score (non_ar)
Mean	0.76	0.59
Median	0.83	0.62
Standard deviation	0.17	0.23
Coefficient of variation	0.22	0.39
Original/Plagiarised ratio	17.14	5.26

Subsequently, the comparative relation between the originality scores by architects and non-architects was ascertained (see Table 4). The average originality score by architects (0.76) is 0.16 higher than the average originality score by non-architects (0.59). In other words, the average originality score by architects is 27.74% higher than the average originality score by non-architects. The mean difference of corresponding originality scores by architects and non-architects is 0.18. In other words, on an average, each originality score by architects is 38.94% higher than the corresponding originality score by non-architects. In as many as 92.01% of the 15,000 design variants, architects' originality score is higher than non-architects'. When the data distribution of the originality scores by architects and non-architects is analysed with respect to the Likert scale categories, the finer differences become clear. Compared to an architect, a non-architect is 1.71 times likelier to see a design variant as 'Plagiarised', and 3.14 times likelier to see a design variant as 'Similar' or 'Different'. On the other end of the spectrum,

Table 4. Numerical summary of the comparative relation of predicted originality scores.

Description	Value
Mean originality score (ar-non_ar)	0.16
Mean originality score percentage ((ar-non_ar)/non_ar)	27.74%
Mean difference (ar-non_ar)	0.18
Mean difference Percentage ((ar-non_ar)/non_ar)	38.94%
Originality score (ar > non_ar)	92.01%
Plagiarised rating occurrence (non_ar/ar)	1.71
Similar & Different ratings occurrence (non_ar/ar)	3.14
Original rating occurrence (ar/non_ar)	1.93

non-architects display the same propensity for being less lenient at labelling a design variant as original. Compared to a non-architect, an architect is 1.93 times likelier to see a design variant as 'Original'.

All of these observations reiterate that architects tend to be more lenient than non-architects at seeing design variants as original. It may be argued that architects are trained to observe the nuances of visual difference between artefacts. Consequently, what may seem like same artefacts to non-architects will look more varied (~ comparatively original) to architects. The other takeaway is that architects have a higher consensus than non-architects at evaluating the originality of design variants. This phenomenon may be attributed to similar rigour of academics and training of architects compared to a more diverse academic and professional background of non-architects. It is to be noted that the reasons speculated behind the varied observations are rather anecdotal in nature. Anecdotal correlations are often not causal. A psychological and/or neuro-response study is needed to explain the subliminal reasons for the observed variance.

The outcome of this study has implications on the decision-making of cityscape preservation and on the architect-client interaction. The decision to preserve or facelift properties and precincts in cities are usually taken by architects, planners, art historians and politicians. The quantitative difference in visual perception of design between architects and non-architects indicates that without direct representation of residents, such a decision making body may miss out on properties and/or precincts that may have sentimental value for the residents. Secondly, the methodology used in this paper can be used to better understand the aesthetic sensibilities of a client. The client may be asked to rate few design options generated by the parametric definition of a design concept. The options may be rated for a variety of keyword-driven subjective opinions, e.g., originality, beauty, appropriateness, etc. Subsequent to training a neural network with the collected data, design may be optimised in tune with the client's subjective opinion(s) along with the regular objective design goals of energy consumption reduction and daylighting.

5 Conclusion

Subjective evaluation of the visual perception of originality of the rotating tower typology by architects and non-architects is predicted by training two deep neural networks with crowd-sourced data of 250 data samples. Use of neural network allows the appropriation of the entire solution space by using limited number of training data samples. Predictions of 15,000 design variants are tabulated to quantify the differences of originality scores between architects and non-architects. It is concluded that the average originality score by architects is 27.74% higher than the average originality score by non-architects. Compared to a non-architect, an architect is 1.93 times likelier to see a design variant as original. In fact, in 92.01% of the cases, architects' originality score is higher than non-architects'. Within themselves, architects tend to have a higher consensus than non-architects at reading the visual perception of originality. Analysis of the correlations of the independent variables revealed that both architects and non-architects tend to subliminally concentrate more on visual features that horizontally control the overall silhouette of towers (such as the shape of floor plan). Additionally, architects tend to be more visually sensitive towards rotation of floor plates, whereas non-architects tend to see rotated towers as a visually homogeneous group without much regard to the amount of rotation. Interestingly, both architects and non-architects tend to ignore local articulation of surfaces when comparing the overall shapes of design variants.

The methodology of training a neural network on crowd-sourced data marks a departure from the top down evaluative guidelines published by experts to a more inclusive bottom up evaluation by end users. The methodology can be used to quantify subjective evaluation of any kind. Beauty or safety perception of urban streetscapes can be predicted as a part of urban design aid by training a neural network with crowd-sourced ratings of photographs of urban streetscapes. The independent variables for such an exercise can be calculated by applying image segmentation [32] on the photographs to extract the areas and mutual positions of roads, signage, greenery, sky, vehicles, etc. Subsequently, the trained network can be used to calculate the fitness function of an evolutionary algorithm to optimise design proposals. The socio-cultural bias embedded in crowd-sourced evaluation can be utilised to analyse and predict design preferences specific to groups of people. For example, in the case of predicting beauty or safety perception of urban streetscapes, evaluation data may be categorised by the cities of residence of the participants. Consequently, the same design proposal will have different predicted scores not only depending on its location, but also on the basis of how it is perceived by different age groups, race, gender, etc.

The future scope of this research is twofold. Firstly, the subjective evaluations by architects and non-architects are to be repeated with the extruded glass box design variant as the reference design. This exercise will establish the effect (if any) of changing the reference design on the visual perception of originality. Secondly, the methodology discussed in this paper is to be applied to different building typologies. A summation of results of multiple typologies will comprehensively quantify the differences of visual perception of originality of design between architects and non-architects in global terms.

References

1. Gifford, R., Hine, D.W., Muller-Clemm, W., Reynolds, D.J., Jr., Shaw, K.T.: Decoding modern architecture: a lens model approach for understanding the aesthetic differences of architects and laypersons. Environ. Behav. 32(2), 163–187 (2000)
2. Llinares, C., Montanana, A., Navarro, E.: Differences in architects and nonarchitects' perception of urban design: an application of kansei engineering techniques. Urban Stud. Res. 1, 1–13 (2011)
3. Ghomeshi, M., Jusan, M.M.: Investigating different aesthetic preferences between architects and non-architects in residential facade designs. Indoor Built Environ. 22(6), 952–964 (2013)
4. Jeffrey, D., Reynolds, G.: Planners, architects, the public, and aesthetics factor analysis of preferences for infill developments. J. Archit. Planning Res. 16(4), 271–288 (1999)
5. Montanana, A., Llinares, C., Navarro, E.: Architects and non-architects: differences in perception of property design. J. Housing Built Environ. 28(2), 273–291 (2013)
6. Brown, G., Gifford, R.: Architects predict lay evaluations of large contemporary buildings: whose conceptual properties? J. Environ. Psychol. 21(1), 93–99 (2001)
7. Moldovan, I., Moldovan, S.V., Nicoleta-Maria, I.: Iconic architecture: skyscrapers. In: Proceedings of 5th International Conference Civil Engineering - Science and Practice, pp. 1461–1468. Zabljak (2014)

8. Ray, P., Roy, S.: Skyscrapers: origin, history, evolution and future. J. Today's Ideas Tomorrow's Tech. **6**(1), 9–20 (2018)
9. Peet, G.: The origin of the skyscraper. CTBUH J. **1**, 18–23 (2011)
10. Vollers, K.: The CAD-Tool 2.0 morphological scheme of non-orthogonal high-rises. CTBUH J. **3**, 38–49 (2009)
11. Lagendijk, B., Pignetti, A., Vacilotto, S.: Case Study: Absolute World Towers, Mississauga. CTBUH J. **4**, 12–17 (2012)
12. Ogburn, W., Thomas, D.: Are inventions inevitable? A note on social evolution. Polit. Sci. Quart. **37**(1), 83–98 (1922)
13. Lamb, D., Easton, S.M.: Multiple Discovery: The Pattern of Scientific Progress. Avebury, United Kingdom (1984)
14. Reinartz, W., Saffert, P.: Creativity in Advertising: When It Works and When It Doesn't. Harvard Business Review, Brighton (2013)
15. Satir, S.: Innovation and originality in design. IJIRES **2**(5), 372–376 (2015)
16. Shibayama, S., Wang, J.: Measuring originality in science. Scientometrics **122**(1), 409–427 (2019)
17. Adriaenssens, S., Block, P., Veenendaal, D., Williams, C.: Shell Structures for Architecture: Form Finding and Optimization. Routledge, London (2014)
18. Tian, Z.C., Chen, W.Q., Tang, P., Wang, J.G., Shi, X.: Building energy optimization tools and their applicability in architectural conceptual design stage. Energy Procedia **78**, 2572–2577 (2015)
19. Sanatani, R.P.: A machine-learning driven design assistance framework for the affective analysis of spatial enclosures. In: Proceedings of CAADRIA 2020, pp. 741–750. Bangkok (2020)
20. Bruno, E.: Commentary/Integrating AI and deep learning within design practice processes: XKool technology. Ardeth. Innov. Happens **05**, 220–226 (2019)
21. Chaillou, S.: AI+ Architecture: Towards a New Approach. Master's Thesis, Harvard University (2019)
22. Koh, I.: Discrete sampling: there is no object or field … Just statistical digital patterns. In: Retsin, G. (ed.) Architectural Design, vol. 89, no. 2, pp. 102–109. Wiley, Hoboken (2019)
23. Ren, Y., Zheng, H.: The spire of AI - voxel-based 3D neural style transfer. In: Proceedings of CAADRIA 2020, pp. 619–628. Bangkok (2020)
24. Yousif, S., Bolojan, D.: Deep-performance - incorporating deep learning for automating building performance simulation in generative systems. In: Proceedings of CAADRIA 2021, pp. 151–160. Hong Kong (2021)
25. Verma, D., Jana, A., Ramamritham, K.: Quantifying urban surroundings using deep learning techniques: a new proposal. Urban Sci. **2**(3), 78 (2018)
26. Sekaran, U., Bougie, R.: Research Methods for Business. Wiley, Hoboken (2016)
27. Green, S.B.: How many subjects does it take to do a regression analysis. Multivariate Behav. Res. **26**(3), 499–510 (1991)
28. Clark, L.A., Watson, D.: Constructing validity: basic issues in objective scale development. Psychol. Assess. **7**(3), 309–319 (1995)
29. DeVellis, R.F.: Scale Development: Theory and Applications. Sage Publications, Newbury Park (2003)
30. Pignatiello, G.A., Martin, R.J., Hickman, R.L., Jr.: Decision fatigue: a conceptual analysis. J. Health Psychol. **25**(1), 123–135 (2018)
31. Revilla, M.A., Saris, W.E., Krosnick, J.A.: Choosing the number of categories in agree-disagree scales. Sociol. Methods Res. **43**(1), 73–97 (2014)
32. Mousavirad, S.J., Ebrahimpour-Komleh, H.: Image segmentation as an important step in image-based digital technologies in smart cities: a new nature-based approach. In: Ismail, L., Zhang, L. (eds.) Information Innovation Technology in Smart Cities, pp. 75–89. Springer, Singapore (2017). https://doi.org/10.1007/978-981-10-1741-4_6

Building Archetype Characterization Using K-Means Clustering in Urban Building Energy Models

Orçun Koral İşeri[✉] and İpek Gürsel Dino

Department of Architecture, Middle East Technical University, Ankara, Turkey
{koral.iseri,ipekg}@metu.edu.tr

Abstract. Population growth in cities negatively affects global climate problems regarding environmental impact and energy demand of building stock. Thus, buildings should be examined for energy efficiency by reaching acceptable internal thermal comfort levels to take precautions against climate disasters. Although building energy simulations (BES) are widely used to examine retrofitting processes, the computational cost of urban-scale simulations is high. The use of machine learning techniques can decrease the cost of the process for the applicability of quantitative simulation-based analyses with high accuracy. This study presents the implementation of the *k-means* clustering algorithm in an Urban Building Energy Modeling (UBEM) framework to reduce the total computational cost of the simulation process. Within the scope of the work, two comparative analyses are performed to test the feasibility of the *k-means* clustering algorithm for UBEM. First, the performance of the *k-means* clustering algorithm was tested by using the observations on the training data set with design parameters and performance objectives. The second analysis tests the prediction accuracy under different selection rates (5% and 10%) from the clusters partitioned by the *k-means* clustering algorithm. The predicted and simulation-based calculated results of the selected observations were comparatively analyzed. Analyses show that the *k-means* clustering algorithm can effectively build performance prediction with *archetype characterization* for UBEM.

Keywords: Urban building energy modeling · Archetype characterization · K-means clustering

1 Introduction

More than 50% of the total world population lives in urban areas. However, intensive urbanization has severe consequences on climate change regarding the high energy use and environmental impact [13]. As a result, cities are under transformation to decrease the environmental impact due to climate change. Local governments have already started to reduce greenhouse gas emissions (GHGs) goals by enforcing necessary regulations. For instance, the City of New York committed to decreasing its GHGs by 80% until 2050 [8]. However, the transformation should begin with the quantitative analysis of the

© Springer Nature Singapore Pte Ltd. 2022
D. Gerber et al. (Eds.): CAAD Futures 2021, CCIS 1465, pp. 222–236, 2022.
https://doi.org/10.1007/978-981-19-1280-1_14

current state and possible intervention actions to improve performance and reduce the environmental footprint in the built environment [6].

City managements started to form datasets for the built environment related to examining the effect of climate change regulations [18]. These datasets are digital representations of the characteristics of the urban building stock, and they can support identifying and analyzing opportunities and corrective actions for sustainable transformation. However, the limitations to data collection make it challenging to analyze retrofit scenarios [14]. For instance, the data sharing process is limited in Turkey [20]. Thus, alternatives are needed for access to urban datasets.

Urban-scale retrofit of buildings is among the climate change adaptation and mitigation strategies. In various studies, high-resolution analyses were applied on urban building stock with different scales and objectives [27], e.g., human-building interaction, micro-climate observation, and building *archetype characterization*. Because building retrofit scenarios should be evaluated from different perspectives for realistic evaluation application of retrofitting process [11, 22], however, a multi-objective approach can be challenging for the consideration of all retrofit alternatives. The evaluation process can be complex, mainly due to the computational cost and a high number of parameters. Therefore, there is a need to examine the urban multi-objective retrofit scenario evaluation process by developing new computational approaches.

The urban-scale retrofit process is evaluated with urban building energy modeling (UBEM) using different approaches such as bottom-up building energy simulations and top-down data-driven algorithms [19, 32]. Data-driven algorithms are preferred in the UBEM process due to their ease of application and evaluation capacity. In particular, one of the critical data-driven approaches in UBEM is *archetype characterization*, which is realized by grouping the building stock according to similar physical and thermal properties [30]. For instance, the grouping criteria can be energy demand values of building units as performance objectives or construction dates as parameters. Although such approaches are preferred in the literature, clustering over a single parameter may be insufficient in evaluating many building stocks in cities. For this reason, the number of evaluation criteria should be increased to understand building stock's properties as explanatory indicators and facilitate the neighborhood analysis with efficient clustering [1, 28].

The data framework in the UBEM compose of different datasets., thus, the analysis process can be laborious for reaching valuable results. Various studies have preferred machine learning (ML) algorithms because of their ability to manage large and heterogeneous data sets [32, 33]. Archetype identification, energy demand prediction, and occupancy pattern detection are purposes for the usage area of ML [16, 17, 19]. Among these methods, clustering algorithms are effective for building *archetype characterization,* which is a suitable approach for pre-processing heterogeneous data [10]. The algorithm can provide acceleration for urban building energy modeling analysis by determining the *archetype characterization* for the building stock in the neighborhood scale.

The clustering algorithm's performance is essential. It can also be modified according to the selection of feature types of training data because the selection of the features is related to the clustering algorithm' performance. There are examples in the literature that propose statistical sensitivity analysis to evaluate the building features' impact on the building performance criteria [26, 38]. The analysis ranks critical parameters. Sensitivity analysis is commonly applied in building energy modeling to quantify the impact of design parameters on the performance objectives [25]. On the other hand, the calculation cost can be reduced by fixing the features that do not affect the model outputs with the analysis.

This study proposes an approach to predict building performance objectives, which can accelerate the neighborhood-scale building energy simulation process with high accuracy. *K-means* clustering algorithm is used for the partitioning of the residential building stock based on their (a) physical/thermal properties and (b) performance objectives. Building simulations were conducted for the whole neighborhood model to calculate the latter dataset. The main reason for choosing the clustering technique is to predict energy use and indoor thermal comfort on the neighborhood scale rapidly by selecting from the partitioned clusters using different selection rates (5% and 10%). This method can provide advantages for analyzing the current condition of the building stock and the quantitative performance evaluation of retrofit alternatives. Since the *k-means* clustering algorithm is sensitive to data distribution (particularly to outliers), two comparative analyses were performed to understand the performance of the clustering algorithm for *archetype characterization*. In the first analysis, two clustering models were separately trained using physical/thermal properties and performance objectives. The first analysis indicates that the selected design parameters can be used to characterize archetypes using clustering, which can be used for the performance objective prediction. In the second analysis, the performance objectives were used as input features of the training data for the *k-means* clustering algorithm. Random selection was applied from the clusters formed previously, then two different selection rates (5% and 10%) were applied from these representative clusters. These selections were simulated and were compared with full model simulation results for the prediction accuracy of the clustering algorithm with the selections from the partitioned clusters. Consequently, the results of the analyses indicated that the random selections from these clusters successfully represent the performance of the studied neighborhood. Thus, clustering algorithm preference before the neighborhood simulation could contribute to the acceleration of the neighborhood-scale building energy simulations.

2 Materials

The *archetype characterization* with the *k-means* algorithm was tested in multiple neighborhoods to measure the success of the process. *Bahçelievler*, *Yukarı Bahçelievler*, and *Emek* neighborhoods in Ankara's *Çankaya* district were included as the study area. Ankara generally has a cold and arid climate, so the ASHRAE climate zone is included in the 4B classification (*CDD10°C ≤ 250, HDD18°C (Heating Degree Days) ≤ 3000*) [3]. Heating energy demand (Q_H) has a high proportion of total energy demand in the building stock of a region. Therefore, cooling energy demand was not calculated for the simulation process.

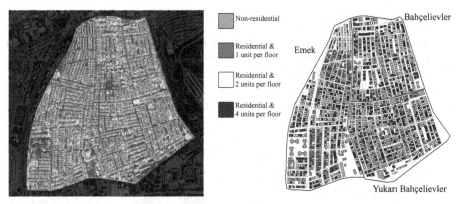

Fig. 1. Selected Neighborhoods in Ankara (left) and 2614 Buildings Based on Building Functions and number of floors.

In Fig. 1, the boundary of the three neighborhoods is shown on the left, and a color-coded representation for the building function with the number of floors is shown on the right. 93% of the study area buildings are residential units, and the remaining buildings are commercial buildings in which are generally located on the ground floors of the buildings. Since occupancy information is not among the data provided by official institutions, it has been obtained from national and city statistical reports by adapting it to the region [35]. During the field visits, the total number of floors and window-wall ratio values of the buildings were collected and entered into the physical properties datasets. Consequently, the building's energy simulations generated the residents' daily energy usage patterns with building energy and comfort standards [4, 5, 34].

3 Methodology

This section presents the proposed methodology of UBEM for the building stock in the selected neighborhoods. The process includes geometrical operations, building simulation (i.e., energy and comfort performance results generation), clustering with machine learning with hyper-parameter tuning, and two-step comparative analysis (i.e., clustering for *archetype characterization*) (Fig. 2).

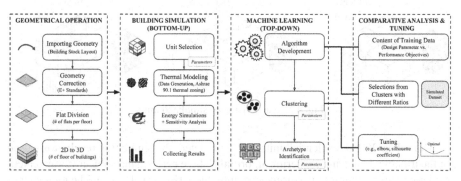

Fig. 2. Flowchart of the proposed method

3.1 Thermal Modeling

The workflow starts by importing the building physical (e.g., layout, # of floors) data to the algorithm. The building footprint curves were simplified into four-edged convex polylines to decrease the building energy simulations' computing cost based on the energy modeling standards defined in [21]. The 2-dimensional building footprints transform to the 3-dimensional thermal zones with the knowledge of the number of floors (Fig. 3).

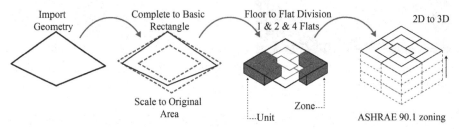

Fig. 3. Geometry correction for thermal modeling

The spatial layout of residential units affects buildings' thermal balance; therefore, it is essential to model in detail. However, the layout data of all buildings in the neighborhood cannot be available in the study regions. For that reason, authors have developed three different types of layouts for units that consist of one, two, and four thermal zones (Fig. 1). The simulation units were divided into different zone types, e.g., bedroom (*B*), living room (*L*), service (*S*). According to the building function, living rooms and bedrooms are the default for all the units. Service areas in which include hallways and bathrooms, do not have external windows (Fig. 4). Each unit has different thermal loads and occupancy schedules compatible with its usage. The distribution of layouts for the building stock was developed by random distribution in parallel with the data obtained from national statistics to the study area.

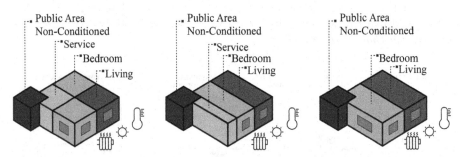

Fig. 4. Zone division of simulated units

More than 25000 residential units were simulated, the simulation results were used for the training data in the clustering algorithm. Three performance objectives are calculated annually by building performance simulations (Fig. 4). These are heating energy demand (Q_H) and lighting energy demand (Q_L), and degrees of overheating (OHD). The heating and lighting demand are two parts of the total energy use calculation that vary according to different design parameter values (Table 1). Since there is no mechanical cooling system, natural ventilation through windows is the only way to cool the zones in residential units. Overheating ratings (OHD) are used to measure thermal disturbance in the summer season. OHD is calculated using a fixed upper-temperature limit for each zone type. The threshold of OHD is 28 °C for the living room and 26 °C for the bedroom [9].

Table 1. Design parameters and performance objectives of the training data

Thermal and physical properties			
Property	Value	Unit	Type
U-value, Wall*	{0.60, 1.88}	W/m^2-K	Pre-defined
U-value, Roof*	{1.88, 3.12}	W/m^2-K	Pre-defined
U-value, Floor*	{0.93, 1.92}	W/m^2-K	Pre-defined
U-value, Window*	{5.1, 2.1}	W/m^2-K	Pre-defined
Heating set point/set back	25.0, 20.0	°C	[6]
Ventilation type	only natural, one-sided	–	Pre-defined
Ventilation limits	21.0, 24.0	°C	Pre-defined
Infiltration	0.0002, 0.0003	m^3/s-m^2	Extra
Window opening ratio	[0.25–0.5]	–	Pre-defined
Occupancy schedule	29 types,1 to 5 people	–	[26]
Window-to-wall-ratio	[0.15–0.30]	–	Extra
Equipment load	{2, 3, 5}	W/m^2	Pre-defined
Lighting density	{5, 7, 10}	W/m^2	Extra
Performance objectives			
Property		Unit	Type
Heating demand (Q_H)		kWh/m^2	Pre-defined
Lighting demand (Q_L)		kWh/m^2	Pre-defined
Overheating degrees (OHD)		°C	Pre-defined

*Before/after 1980

A residential archetype unit can be characterized as related to building physical, thermal, or occupancy properties. A part of the residential unit features is obtained from official institutions for this study [24], e.g., building footprint or per floor unit number. The rest of the building features were generated to the extent specified by national statistics and building energy modeling standards [4, 5], [42], for instance, occupancy properties. However, in several cases, the number of building parameters in the training data may not be sufficient for the clustering algorithm; therefore, the authors added extra parameters to the simulation algorithm.

The extra design parameters are included in the data pool by performing sensitivity analysis according to their impact on energy use performance objectives. Within the scope of this study, the calculation of impact was achieved with sensitivity analysis. Morris's analysis is used as a sensitivity analysis. Morris Sensitivity analysis is a screening local sensitivity analysis with the elementary-effect method based on a finite distribution of input parameters. The analysis works to rank the input factors' relative importance, namely the first-order main effect (Si), by influencing the output parameters [22].

Table 2. Results of Morris sensitivity analysis

Parameter	Range	mu*	Type
U-value, wall	{0.6, 1.2, 1.8, 2.4}	33.386	Pre-defined
Window-to-wall ratio	{0.1, 0.2, 0.3, 0.4}	31.844	Extra
Infiltration rate	{0.0002, 0.0003, 0.0004, 0.0005}	28.452	Extra
Lighting density	{5, 7.5, 10, 12.5}	22.252	Extra

Morris sensitivity analysis was applied to test a pre-defined and three extra design parameters regarding the influence ($mu*$) on a performance objective, which is selected as the heating demand (Table 2). The *U-value Construction* is a pre-defined design parameter, and it was included in the analysis to compare the impact of the three extra design parameters as a proxy [34]. The authors manually defined ranges for these parameters in the building energy simulations (Table 1). Based on the first-order (Si) main effect index results, window-to-wall ratio, infiltration rate, and lighting density for zone parameters were highly influential on Q_H. Consequently, these design parameters were added to the training dataset.

3.2 Occupancy Modeling

Occupancy modeling for residential buildings is one of the critical features for building performance. The subject previously studied residential buildings to monitor occupant actions and cluster activity schedules from performance objectives [7, 12]. Many uncertainties exist with a high degree of influence for energy demand and indoor thermal comfort as occupants interact with building systems (e.g., heating setpoints, natural ventilation) [41]. Nevertheless, most modeling approaches use default occupancy schedules, and they ignore the different occupant profiles and their specific ways of space use and system interaction. The writers of this study have proposed to use a new method for realistic occupancy modeling. The process is a combination of datasets from different resources and different statistical techniques (Fig. 5).

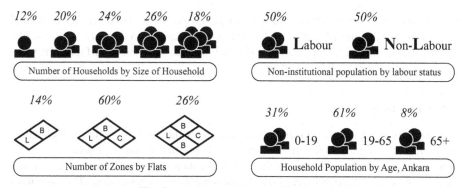

Fig. 5. Occupancy by national statistics

The proposed approach consists of national statistics for occupancy and location-based address registrations for unit information [24]. For the selected neighborhoods, occupancy scenarios are modeled based on three different statistical information and different unit layout modeling, where is mentioned in Sect. 3.1 (Fig. 5). The statistical data were used to map the daily occupant activities for residential units, i.e., household size, labor status, age [36, 37]. The simulation units were divided into different zone types, i.e., bedroom (*B*), living room (*L*), service (*S*). Thirty-one occupancy schedules were matched with the simulation zones based on occupant preferences. Each schedule was assigned randomly to the units, which helped to produce randomly distributed objective performance results in the training dataset.

3.3 Building Energy Simulation

The selected district's digital models were built according to Turkish TS-825, ASHRAE 55, ASHRAE 90.1 standards [4, 5, 34]. Ladybug/Honeybee Visual Coding tools were the simulation tools for the building energy simulations with the EnergyPlus engine [29, 39]. All residential unit simulations were separately simulated. For each simulation, random construction, internal load, and occupancy schedules were assigned (Fig. 6).

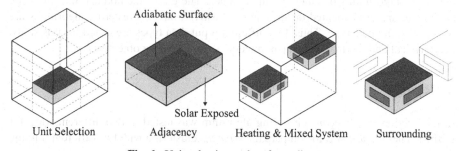

Fig. 6. Unit selection and surface adjacency

Surface types differ for vertical and horizontal positions based on the residential unit position. The internal walls are adjacent, and they are set to adiabatic surfaces. The surrounding geometries were introduced as environmental context surfaces. They are essential due to solar radiation reflections.

3.4 Clustering

Clustering is an unsupervised machine learning algorithm that works for unlabeled data structures—the algorithm search for similarity between the values of parameters. The similarity is a valuable measure for the qualitative data features. However, the distance calculation works better to recognize the numeric data's relationship. The algorithm defines the distances of the instances from each other according to their similarities or dissimilarities [40].

The process starts with the selection of features and feature extraction from training data. Then, the algorithm proceeds with the design of the clustering algorithm suited explicitly to the problem. It evaluates the results to improve the algorithm's performance. Lastly, it completes with the realization and comparison of the results based on statistical formulas [31].

3.5 Partial Clustering

The partitional clustering algorithm defines the center points in the data for non-overlapping clusters [23], e.g., *k-means*, *k-medoids*. The *k-means* approach begins with the random selection of k-different center points for each cluster [2]. The *k-means* algorithm updates the center points by iterative computation. At the same time, the expectation-maximization step repeats until the centroid positions reach a pre-defined convergence value. While the expectation step arranges each point for its nearest center point of the cluster, the maximization step computes all the points for each group and sets the new centroid.

In clustering, there is a trade-off between prediction accuracy and cluster stability [10, 15]. Therefore, the tuning process is essential for the algorithm's performance in terms of accuracy. The process of parameter tuning consists of sequentially altering one of the algorithm's parameters' input values. The elbow method and the silhouette coefficient are implemented during tuning. Lastly, *k-means* clustering algorithms are sensitive to the data type. Thus, two different input data types are tested for this study, i.e., physical and thermal design parameters vs. performance objectives.

4 Results

In this section, the *k-means* clustering algorithm was tested in two different ways for UBEM. The first test was to compare the *k-means* clustering with two different training data types, i.e., design parameters and performance objectives. The second test was realized with a different amount of training data (5% and 10%) from the clusters partitioned by the *k-means* clustering algorithm. The training data sets were taken from the same generated data of the selected built environment for each step.

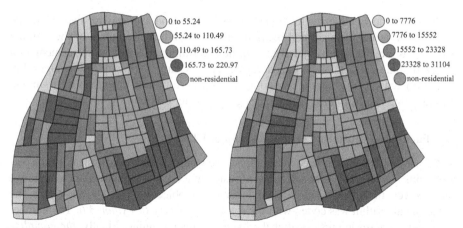

Fig. 7. Spatial distribution of the performance objectives; (left) heating demand (kWh/m²), (right) overheating degrees

Figure 7 shows the spatial distribution of the two performance objectives in the clustering algorithm's training data for the selected region, i.e., Q_H and *OHD*. The colors represent the region-based the generated data that is the average results of all buildings inside the area. The spatial distribution of the two performance objectives has resulted differently.

4.1 Comparative Analysis for the Qualities of Training Data for Clustering Algorithm

In this section, the *k-means* clustering algorithm is trained with two different data types for the training dataset. Firstly, design parameters were introduced in the *k-means* algorithm, e.g., the residential units' physical and thermal properties and occupancy data (Table 1). The number of clusters resulted in seven clusters using elbow and silhouette coefficient tuning techniques. Secondly, performance objectives were introduced, and after the tuning process, the number of clusters was four.

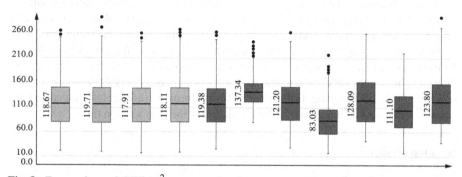

Fig. 8. Energy demand (kWh/m²) averages for cluster outputs of two clustering processes; Performance objective-based clusters (left, red), Design parameter-based clusters (right, blue) (Color figure online)

Figure 8 shows the box plot values of energy demand averages for two clustering processes in terms of mean and distribution. The number of features of the training dataset was more than the objectives, and the algorithm performed more segmentation for the training dataset. While the energy demand averages of the clustering algorithm groups trained with the training data containing the performance objectives were similar, the clustering results trained with the design parameter differed.

4.2 Partitional Clustering for Facilitation of UBEM

In this part, the model was used in a comparative study to validate how the *k-means* perform with a lower number of selections of residential units. The metrics of comparative analysis were the average and standard deviation for objectives Q_H, Q_L, and *OHD*. The clustering algorithm was coded with the sckit-learn library of *Python 3.6*. The number of clusters was seven clusters after the hyper-parameter tuning. Finally, *the complete model* (i.e., brute-force simulations) in Table 3 is the simulation results of all residential units in the selected neighborhoods.

Table 3. The comparative analysis of different selection rates (5% and 10%) from clusters and complete model simulation

Metrics	Five percent sample	Percentage change	Ten percent sample	Percentage change	Complete model
$\bar{x}_{QH}$	124.97 $\mp$ 43.70 SD	%5.87	120.1 $\mp$ 45.00 SD	%1.75	118.04 $\mp$ 44.36 SD
$\bar{x}_{QL}$	14.97 $\mp$ 5.54 SD	%0.00	14.8 $\mp$ 5.60 SD	%1.14	14.97 $\mp$ 5.69 SD
$\bar{x}_{OHD}$	14986 $\mp$ 6882 SD	%2.75	15699 $\mp$ 7096 SD	%1.88	15410 $\mp$ 7150 SD

Five-percent and *ten-percent* selections were the ratios of the random selection data instances from seven clusters. These selected instances were simulated to generate performance objectives. Then, they have compared with the complete model results in terms of the percentage change. The *k-means* algorithm divided the dataset of generated objectives into different clusters. As seen in Table 3, three performance objectives were compared with average, standard deviation, mean percent ratio. *Ten-percent* sample performed more accurately compared to *five-percent* sample for all performance objectives. However, the values were close to less than a 5% confidence interval ratio even for *five-percent* sample of clustering. In conclusion, the performance of the clustering algorithm showed that it could be used for UBEM to decrease the computation time.

5 Discussion

In this study, comparative analyses were carried out to observe the performance of the *k-means* algorithm for *archetype characterization* of the UBEM performance dataset.

The *k-means* clustering models were separately trained with performance objectives (e.g., energy demand and overheating degrees) and design parameters (e.g., physical and thermal properties of building stock). Even though the number of instances was the same in the two training datasets, the training performed with the parameter-based dataset consisted of more features with higher accuracy. The *k-means* has achieved more clusters with the design parameters included training dataset, and each cluster was reliable to differentiate itself from other clusters according to their dissimilarities (Fig. 8). According to these results, it was seen that the design parameters of residential building stock can play significant role for the building performance prediction in the *archetype characterization* process without simulating all residential units in the selected urban regions. Because building energy simulation is an expensive method in terms of computational cost, and the real building performance data may not always be available for building stock analysis.

Secondly, the performance of *k-means* was tested with different selection ratios from the clusters partitioned by the *k-means* clustering, i.e., *complete model* vs. two different sampling ratios (*five-percent* and *ten-percent*). *Ten-percent* selection resulted more accurately to predict to cluster the energy demand and overheating degree values. However, *five-percent* selection ratio also can be used as an alternative instead of simulating all residential units in the selected region. Thus, it has been seen that the partitions formed by the k-means clustering algorithm are successful in representing the performance data of the entire study area. Nevertheless, each clustering process should be tuned to reach an optimal number of clusters with high accuracy. Elbow and silhouette coefficients were applied and tested multiple times as hyperparameter tuning. Because *centroid positioning* of clusters may result differently between trials due to random initiation. Otherwise, this situation may cause false interpretations during the use of the clustering algorithm.

6 Conclusion

UBEM has capacity for analyzing the urban building stock's performance objectives by collecting, managing, and producing large amounts of real or synthetic data. In addition, advanced machine learning algorithms can be suitable for clustering or estimating these performance objectives. This study proposes a methodology to apply the *k-means* clustering algorithm for the UBEM process. Instead of simulating the entire building stock in the neighborhoods, the clustering method was applied for the clusters of similar features of the building stock. However, the qualities of training data are essential in clustering algorithms. Therefore, two different comparative analyzes were realized for the qualities of training data and the prediction performance of clustering algorithm. For the first analysis, two clustering models were trained with the training data consists of design parameters and performance objectives, separately. The clustering algorithm split the design parameters included training dataset into more groups than the performance objective-based training dataset with high accuracy. The comparative analyses results indicated that physical and thermal parameters of residential building stock could be used as training data content in the clustering process for the UBEM *archetype characterization*. In the second analysis, the methodology consists of the comparative analysis for the energy simulation with the different selection ratios from the cluster of buildings partitioned by *k-means* clustering algorithm—energy demand averages were compared between the different number of samples from clusters and complete model. The results showed that the clustering algorithm might be suitable for urban building energy modeling to reduce simulations' computational costs. For further studies, the proposed methodology will be tested in an automated process for different climatic zones without simulating entire settlements.

Acknowledgements. This research is supported by the Scientific and Technological Research Council of Turkey (TUBITAK), Grant No. 120M997.

References

1. Aksoezen, M., et al.: Building age as an indicator for energy consumption. Energy Build. **87**, 74–86 (2015). https://doi.org/10.1016/j.enbuild.2014.10.074
2. Arvai, K.: K-Means Clustering in Python: A Practical Guide – Real Python
3. ASHRAE: ASHRAE climatic design conditions 2009/2013/2017
4. ASHRAE: ASHRAE Standard 55-2004 – Thermal Comfort (2004). https://doi.org/10.1007/s11926-011-0203-9
5. ASHRAE: ASHRAE Standard 90.1-2013 – Energy Standard For Buildings Except Low-rise Residential Buildings (2013)
6. World Bank: Cities and climate change: an urgent agenda. Urban development series knowledge papers. World Bank, Washington DC (2010)
7. Bedir, M.: Occupant behaviour and energy consumption in dwellings: an analysis of behavioral models and actual energy consumption in the Dutch housing stock (2017)
8. Chen, Y., et al.: Automatic generation and simulation of urban building energy models based on city datasets for city-scale building retrofit analysis. Appl. Energy. **205**, 323–335 (2017). https://doi.org/10.1016/j.apenergy.2017.07.128
9. CIBSE: Guide a - Environmental design. The Chartered Institution of Building Services Engineers (2006)
10. Deb, C., Lee, S.E.: Determining key variables influencing energy consumption in office buildings through cluster analysis of pre- and post-retrofit building data. Energy Build. **159**, 228–245 (2018). https://doi.org/10.1016/j.enbuild.2017.11.007
11. El Gindi, S., Abdin, A.R., Hassan, A.: Building integrated Photovoltaic Retrofitting in office buildings. Energy Procedia **115**, 239–252 (2017). https://doi.org/10.1016/j.egypro.2017.05.022
12. Guerra-Santin, O.: Relationship between building technologies, energy performance and occupancy in domestic buildings. In: Keyson, D.V., Guerra-Santin, O., Lockton, D. (eds.) Living Labs, pp. 333–344. Springer, Cham (2017). https://doi.org/10.1007/978-3-319-33527-8_26
13. Hong, T., et al.: CityBES: a web-based platform to support city-scale building energy efficiency (2016)
14. Hong, T., et al.: Ten questions concerning occupant behavior in buildings: the big picture. Build. Environ. **114**, 518–530 (2017). https://doi.org/10.1016/j.buildenv.2016.12.006
15. Hsu, D.: Comparison of integrated clustering methods for accurate and stable prediction of building energy consumption data. Appl. Energy. **160**, 153–163 (2015). https://doi.org/10.1016/j.apenergy.2015.08.126

16. El Kontar, R., Rakha, T.: Profiling occupancy patterns to calibrate urban building energy models (UBEMs) using measured data clustering. Technol. Archit. Des. **2**(2), 206–217 (2018). https://doi.org/10.1080/24751448.2018.1497369
17. Kontokosta, C.E., et al.: A dynamic spatial-temporal model of urban carbon emissions for data-driven climate action by cities (2018)
18. Kontokosta, C.E.: Energy disclosure, market behavior, and the building data ecosystem. Ann. N. Y. Acad. Sci. **1295**(1), 34–43 (2013). https://doi.org/10.1111/nyas.12163
19. Kordas, O., et al.: Data-driven building archetypes for urban building energy modelling. Energy **181**, 360–377 (2019). https://doi.org/10.1016/j.energy.2019.04.197
20. KVKK, K.V.K.K.: Kişisel verilerin Korunması ve işlenmesi Politikası, Ankara (2018)
21. LBNL, L.B.N.L.: Input Output Reference. EnergyPlus (2009)
22. Ma, Z., Cooper, P., Daly, D., Ledo, L.: Existing building retrofits: methodology and state-of-the-art. Energy Build. **55**, 889–902 (2012). https://doi.org/10.1016/j.enbuild.2012.08.018
23. MacQueen, J.: Some methods for classification and analysis of multivariate observations. In: Proceedings of the 5th Berkeley Symposium on Mathematical Statistics and Probability, Volume 1: Statistics, pp. 281–297. University of California Press, Berkeley (1967)
24. NVI, T.C.İ.B.N. ve V.İ.G.M.: Yerleşim Yeri Sorgulama / Adres Sorgulama / Adres Doğrulama - Vatandaş Sorgu İşlemleri
25. Østergård, T., et al.: A stochastic and holistic method to support decision-making in early building design. Proc. Build. Simul. Tian 2013, 1885–1892 (2015)
26. Østergård, T., et al.: Building simulations supporting decision making in early design - a review. Renew. Sustain. Energy Rev. **61**, 187–201 (2016). https://doi.org/10.1016/j.rser.2016.03.045
27. Reinhart, C.F., Davila, C.C.: Urban building energy modeling - a review of a nascent field. Build. Environ. **97**, 196–202 (2016). https://doi.org/10.1016/j.buildenv.2015.12.001
28. Pérez, M.G.R., Laprise, M., Rey, E.: Fostering sustainable urban renewal at the neighborhood scale with a spatial decision support system. Sustain. Cities Soc. **38**, 440–451 (2018). https://doi.org/10.1016/j.scs.2017.12.038
29. Roudsari, M.S., Pak, M.: Ladybug: a parametric environmental plugin for grasshopper to help designers create an environmentally-conscious design. In: Proceedings of BS2013: 13th Conference of International Building Performance Simulation Association, pp. 3128–3135 (2013)
30. Sokol, J., et al.: Validation of a Bayesian-based method for defining residential archetypes in urban building energy models. Energy Build. **134**, 11–24 (2017). https://doi.org/10.1016/j.enbuild.2016.10.050
31. Sola, A., et al.: Simulation tools to build urban-scale energy models: a review. Energies **11**, 12 (2018). https://doi.org/10.3390/en11123269
32. Swan, L.G., Ugursal, V.I.: Modeling of end-use energy consumption in the residential sector: a review of modeling techniques. Renew. Sustain. Energy Rev. **13**(8), 1819–1835 (2009). https://doi.org/10.1016/j.rser.2008.09.033
33. Tardioli, G., Kerrigan, R., Oates, M., O'Donnell, J., Finn, D.P.: Identification of representative buildings and building groups in urban datasets using a novel pre-processing, classification, clustering and predictive modelling approach. Build. Environ. **140**, 90–106 (2018). https://doi.org/10.1016/j.buildenv.2018.05.035
34. TSE: Ts 825: Binalarda Isı Yalıtım Kuralları (2008)
35. TÜİK: TÜRKİYE İSTATİSTİK KURUMU Turkish Statistical Institute (2010)
36. TUIK, T.S.I.: Employment status and participation rate (2020)
37. TUIK, T.S.I.: Indicators related with disability and old age, 2012, 2014, 2016, 2019 (2019)
38. Westermann, P., Evins, R.: Surrogate modelling for sustainable building design – a review. Energy Build. **198**, 170–186 (2019). https://doi.org/10.1016/j.enbuild.2019.05.057

39. Crawley, D.B., Pedersen, C.O., Lawrie, L.K., Winkelmann, F.C.: EnergyPlus: energy simulation program. ASHRAE J. **42**, 49–56 (2000)

40. Xu, D., Tian, Y.: A comprehensive survey of clustering algorithms. Ann. Data Sci. **2**(2), 165–193 (2015). https://doi.org/10.1007/s40745-015-0040-1

41. Yan, D., et al.: Occupant behavior modeling for building performance simulation: current state and future challenges. Energy Build. **107**, 264–278 (2015). https://doi.org/10.1016/j.enbuild.2015.08.032

Interior Design Network of Furnishing and Color Pairing with Object Detection and Color Analysis Based on Deep Learning

Bo Hyeon Park(ID), Kihoon Son(ID), and Kyung Hoon Hyun(✉)(ID)

Department of Interior Architecture Design, Hanyang University, 222 Wangsimni-ro,
Seongdong-gu, Seoul, Korea
{pbk96,gnswn00,hoonhello}@hanyang.ac.kr

Abstract. Furnishing is one of the most important interior design elements when decorating a space. Because every interior design element is colored, it is essential to consider the pairing of furnishing and color during the design process. Despite the importance of the furnishing and color pairing, the decision-making process by which the pairings are made remains a "black-box" of the interior design process. However, the advancement of social networks and online interior-design platforms such as Today's House allows collecting large quantities of actual interior design cases that can be shared publicly. In addition, it has become possible to extract various features and relationships of data through machine learning techniques and network analysis. Thus, this paper proposes a data-driven approach to reveal distinct patterns of furnishing and color pairing through object detection, color extraction, and network analysis. To do that, we collected a large quantity of image data (N = 14,111) from Today's House (ohou.se) online interior-design platform. Then, we extracted furnishing objects and color palettes from the collected images using object detection and color extraction algorithms. Finally, we identified distinctive patterns of furnishing and color pairing through network analysis.

Keywords: Color network · Color-furnishing pairing · Machine learning · Network analysis · Interior style analysis

1 Introduction

Home furnishing is a compound word referring to home interior space and furnishing, which means the process of decorating a space with furniture and accessories. According to Singh & Sharma [1], home furnishing is one of the most important interior design elements when decorating an interior space. Because every interior design element is colored, it is essential to consider the color of the furnishings during the design process [2]. Thus, furnishing and color pairing play essential roles in interior design. However, until recently the design process has been what is known as a "black-box" [3]. Because there has been no universal rule for furnishing and color pairing in interior design, every designer makes design decisions according to different pairing standards and methods.

© Springer Nature Singapore Pte Ltd. 2022
D. Gerber et al. (Eds.): CAAD Futures 2021, CCIS 1465, pp. 237–249, 2022.
https://doi.org/10.1007/978-981-19-1280-1_15

According to Weiss et al. [4], the criteria for defining interior design styles are vague, and there is no systematic classification method. In this context, designers choose interior elements such as furnishings and colors based on past experience [5, 6]. Designers do share information such as popular furnishing pairing [7] and color pairing [8, 9], but the results are often different because they rely mostly on their own personal experiences and accumulated design knowledge. Thus, despite the importance of furnishing and color pairing, the decision process by which the pairings critical to interior design are made, remains as a "black-box" of the interior design process.

However, we are currently living in an environment that allows us to analyze the black box of the design process using big data. Using online interior-design platforms such as Today's House (ohou.se), we can crawl images of the interior spaces of existing homes that offer a variety of interior design trends. In Today's House, people present their home interior design style and share furnishing information used in their interior design with others. Moreover, we have been able to analyze big data about interior design results using the network analysis method to reveal the shared interior design principles of many people. Then we also have been able to process the data to create a structured format for analysis through machine learning-driven object detection and color extraction.

Therefore, the aim of this paper is to reveal the distinct patterns of furnishing and color pairing (hidden for each interior style) by analyzing large-scale data from an online interior-design platform using machine learning techniques and network analysis methods. To achieve this goal, we conducted the following tasks:

1. We collected a large quantity of image data ($N = 14,111$) from Today's House, an online interior-design platform.
2. We extracted furnishing objects and color palettes from the collected image data using object detection and color extraction algorithms.
3. We converted the extracted data into a format for network analysis.
4. We analyzed the principle of furnishing and color pairing through network analysis.

2   Related Works

2.1   Interior Design Element

Design style can be defined as the repetition of a particular design method or the pairings of elements in a design [10]. In interior design, style is related to the design concept and shows design direction. The interior design also has a significant influence on the user's psychological feelings and their behavior [11]. Interior style is composed of furnishings and color pairing. Furnishings are essential elements that create the interior design, affecting the overall interior style [4]. Next, the colors used in the interior create a space's atmosphere [2]. Color is another key element of interior style. In this context, it is crucial to design a harmonious interior style through appropriate color-furnishing pairings [5].

2.2 Color and Object

There are various interior design styles, such as modern and classic, but the criteria for classifying them are vague [4]. Therefore, designers use rule of thumb or interior color design experiences to find appropriate color-furnishing pairings that suit the style [5, 6]. However, designing harmonious furnishing and color pairings is a difficult task. Therefore, many studies have been conducted to analyze the interior style and recommend the best pairings of furnishing and color for an interior style. For example, Weiss et al. [4] proposed a system using deep learning to train interior style with images and recommend furnishings compatible with a style. Chen et al. [6] used a Bayesian network to train furnishing and color pairings from 600 interior images. The system offers new design of color combinations with a similar style. However, both of these papers did not conduct style analysis using pairings with furnishings. On the other hand, Liu et al. [12] conducted a study to analyze the images of 54,000 living rooms on Airbnb (airbnb.com) to investigate the interior style trends according to the culture of individual countries. Liu's research differs from previous studies in that it analyzes the cultural interior design trends by conducting interior element analysis. However, Liu analyzed furnishing and color separately and only used only five types of furnishings. Therefore, to analyze quantitatively ambiguous interior styles and clarify the hidden rules of interior element pairings, it is necessary to analyze the relationships between furnishing and color.

2.3 Network Analysis

Network analysis is a method of quantitatively analyzing the overall process of construction and diffusion or evolution of elements by displaying the relationship between elements as nodes and links. Because network analysis identifies the relationships between the elements, it is possible to identify patterns in the data [13]. For example, Ahn et al. [4] created a bipartite network using shared flavor compounds between food ingredients and formed a food flavor network through bipartite projection. Ahn et al. [4] identified a rule for pairing of food ingredients by calculating prevalence and authenticity in a network. Similarly, Park et al. [13] used 96,000 classical music data to form a network. The latter calculated the degree and centrality of the network and analyzed the growth of their bipartite network through the basic properties of music composers and composer networks. In similar ways, our paper aims to identify relevant furnishing and color pairing rules through analysis of an interior style data network.

3 Methods

3.1 Data Collection

The living room is, by definition, the largest room in the house, the main daily living space, and the space that stands out in a house. In this respect, interior style and furniture design recommendation studies have been conducted mostly about living rooms [12, 15]. Therefore, we used our customized data crawler to collect living room data from that offered by Today's House (ohou.se). This source is an online interior-design

platform that provides for the viewing of interior design trends by sharing real interior design images and information offered by various users. Before sharing interior
designs, the people select their interior design style among the eight predefined styles
of Today's House. Thus, the interior images provided, are also divided into eight styles:
*Natural, Modern, Scandinavian, Vintage, Unique, Classic_Antique, Provence_Romantic,
Korea_Asia.* (Fig. 1). A total of 55,862 interior images belonging to the eight styles were
collected. We rejected images from those collected if they exhibited the following conditions: 1) image was not of a living room; 2) image was taken before or during interior
construction; 3) image did not reveal the entire space due to enlargement of only a specific furnishing; 4) image revealed only four or fewer furnishings. In the end, 14,111
images were used in the analysis.

Fig. 1. These images show four of the eight style images collected through Today's House.

3.2   Data Processing

We conducted the following data processing steps (Fig. 2). First, we detected objects
in an image and saved them separately (Fig. 2-a). The overlapped parts of objects were
removed by cropping (Fig. 2-b). Next, colors were extracted and saved from each object
image (Fig. 2-c). Then, two matrices (object and color data) were generated and stored
from each living room interior image through this process. In each matrix, object names
were set by adding the name of the color with the highest ratio. In the first matrix,
the colors of an object were stored in binary format, and in the second matrix, the
ratio of each color was stored (Fig. 2). We conducted network analysis using these
furnishing and color matrices. The data processing was conducted automatically using
custom-developed software.

Object Detection. To detect objects in interior images, we utilized YOLOv3 [16]. YOLOv3 detects the bounding box of object from an image. Because it predicts multiple bounding boxes and class probability using a single convolutional neural network, it is a state-of-the-art, high-speed, uncomplicated pipeline-object detection method.

For object detection, the representative ten living room furnishings were set to ten object classes. The ten object classes are *art wall, cabinet, ceiling light, chair, curtain, floor lamp, plant, rug, sofa,* and *table*. Thus, we conducted a labeling process to create ground truth data of object detection using 1,899 living room interior images. In 1,899 images, there are 1,899 *art walls*, 3,077 *cabinets*, 1,373 *ceiling lights*, 2,978 *chairs*, 2,098 *curtains*, 527 *floor lamps*, 3,087 *plants*, 1,118 *rugs*, 1,192 *sofas*, and 1,962 *tables* for each class. Through this process, we trained the YOLOv3 model.

YOLOv3 extracts the bounding box information of the detected object. For preventing the objects from often overlapping in the interior image, the following bounding box crop process was conducted on overlapping objects (Fig. 3). First, if object_a is included in another object_b, object_a was saved immediately without cropping (Fig. 3-b). However, if a and b overlap, the overlapped part was removed and saved (Fig. 3-c). If the area removed through the cropping process was over 75%, the object was discarded.

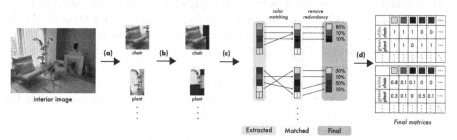

Fig. 2. Data processing methods: (a) object detection; (b) image crop; (c) color detection; (d) matrix generation. (Color figure online)

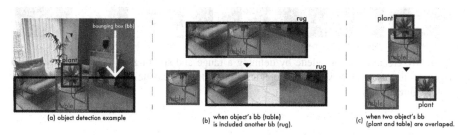

Fig. 3. Bounding box crop process

Color Extraction. We divided the range of colors to show only the representative features of the colors detected in the furnishings using the Korean Standard (KS) of color [17]. KS color is for representing the standard colors used in korea and consists of 12 chromatic colors and 3 achromatic colors: *Red, YellowRed, Brown, Yellow, GreenYellow,*

Green, BlueGreen, Blue, PurpleBlue, Purple, RedPurple, Pink, Black, White, and *Gray*. We made a color palette based on 15 KS colors (Fig. 4-A).

We expressed colors using CIELAB for accurate color detection. CIELAB is expressed using the parameters L*, a*, and b*. L* indicates brightness, a* indicates the degree of red and green, and b* indicates the degree of yellow and blue. Unlike the commonly used RGB and CMYK systems, the CIELAB color space is defined based on human-vision research, so the colors do not change regardless of the display equipment or printing medium used.

For more accurate color matching, colors were subdivided into six types according to the L* value representing the color brightness. In this way, the original 12 colors (*Red, YellowRed, Brown, Yellow, GreenYellow, Green, BlueGreen, Blue, PurpleBlue, Purple, RedPurple,* and *Pink*) were subdivided into 72 colors. The *Black, White,* and *Gray* colors already divided according to the L* values were not subdivided. In the end, we created 75 colors for the analysis (Fig. 4-b).

To extract the color of an interior furnishing image, we utilized a python library, Colorgram, which extracts colors in an image as RGB values and color ratio. We extracted colors from every furnishing image, converted the RGB colors to CIELAB colors, and then matched them to the color palette. To compare CIELAB colors, the color distance (ΔE) value needed for matching was calculated by the CIEDE2000 method [18]. Sharma's CIEDE2000 ΔE calculation method considers lightness, chroma, and hue of color. Using this method, the researchers designated the final matching color as the extracted color and the palette color with the minimum ΔE. Moreover, to discern the influence of the color, we extracted the dominant color without using all the colors detected. Only the colors with the top three ratios, among all the colors detected in the furnishing image, were used as the dominant colors.

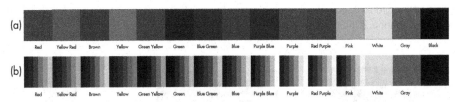

Fig. 4. We created a color palette by defining a range and splitting the color range based on brightness to make color extraction more accurate: (a) Color palette and (b) Subdividing color. (Color figure online)

3.3   Color-Furnishing Network Analysis

To analyze the influence of each color-furnishing pairing in an interior space, we constructed 150 color-furnishing (CF) objects, each of which is a combination of 15 colors (C) and 10 furnishing object classes (F). We calculated the prevalence and authenticity (relative prevalence) using the authenticity algorithm [14] to identify the influence of CF within each style and the relative influence of other styles. We checked the inherent influence of a CF using its prevalence and authenticity. The interior style of the CF is expressed as s, the CF is i, the total number of images per style is N_s, the total number of images with i for each style is n_i^s, and the prevalence of the CF in the image is expressed as P_i^s, where $< P_i^{s'} >_{s' \neq s}$ is the average value of prevalence of different styles. The formula of prevalence and authenticity was taken from Ahn et al. [4] and used as follows:

$$P_i^s = n_i^s / N_s \tag{1}$$

$$A_i^s = P_i^s - < P_i^{s'} >_{s' \neq s} \tag{2}$$

We calculated the prevalence and authenticity of 11,175 CF pairs and 551,300 CF triplets to identify significant pairings for each style. The calculation method is the same as above, and the formulas used are (3) and (4).

$$P_{ij}^s = n_{ij}^s / N_s; P_{ijk}^s = n_{ijk}^s / N_s \tag{3}$$

$$A_{ij}^s = P_{ij}^s - < P_{ij}^{s'} >_{s' \neq s}; A_{ijk}^s = P_{ijk}^s - < P_{ijk}^{s'} >_{s' \neq s} \tag{4}$$

We constructed our CF network using the method of Ahn et al. [14]. To check the CF network through the colors shared by CF, we constructed a bipartite network projection using P_i^s, A_i^s. A bipartite network projection is a network in which nodes are divided into two sets, allowing only connections between nodes of different sets [19]. The CF connects with its color to form a one-mode network (Fig. 5-a). The Fig. 5-b graph is a projection of a one-mode network of CF and color into CF space, and when more than one color is shared, a CF network is formed.

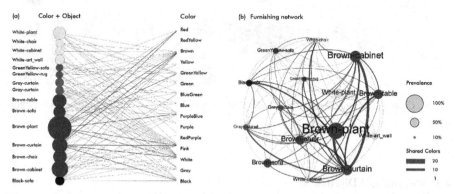

Fig. 5. Network generation process: (a) One-mode projection and (b) CF network. In (a) and (b), three random images of *Provence_Romantic* style were used. The weight of each link is the number of colors shared by the two nodes. The size of the node represents the prevalence of each CF by style. (Color figure online)

4   Implementation and Results

4.1   Object Detection and Color Detection

To evaluate the performance of our newly trained YOLOv3 model, we selected 100 living room interior images that were not used in training. The images used for performance evaluation included ten objects. The average precision of the 10 object values was 0.88, and the average recall value was 0.83. We performed object detection using the trained model and colors extracted from the object images (Fig. 2). These extracted colors were saved using a predefined CIELAB color palette and matching process (Fig. 2-c). Figure 6 illustrates the color-extraction result samples.

4.2   Data Structure

Throughout the 14,111 images, the eight interior styles showed a similar trend in the number of furnishings used (Fig. 7-a). Likewise, the color distribution of the furnishings also showed a similar pattern regardless of the style (Fig. 7-b). *Brown* and achromatic colors (*White, Black, Gray*) were most often used (Fig. 7-c). Compared with the distribution map of all the colors used (Fig. 7-c), the ratios of the rest (i.e., except for *Brown* and achromatic colors) are much lower (Fig. 7-d). We observed that most of the furnishings in the images have a high proportion of *Brown* and achromatic colors and that these affect the overall color sense in the space.

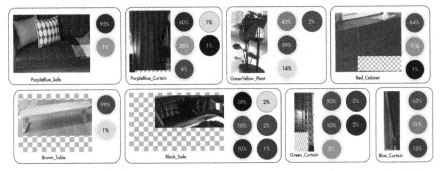

Fig. 6. Color detection example: The highlighted part is the top-3 dominant colors. (Color figure online)

4.3   Results and Discussion

There were significant differences in the CF networks for each style (Fig. 8). For example, in the *Scandinavian* style, White CF is the upper class, whereas in *Korea_Asia* style, *PurpleBlue*, and *Pink* CF tend to dominate. *Brown* CF shows intensive connectivity in most styles, and achromatic CF and *YellowGreen-plant* occupy the upper class. In the Korean living rooms, CF in *Brown* and achromatic colors are mainly used, and *YellowGreen-plants* tend to be commonly used.

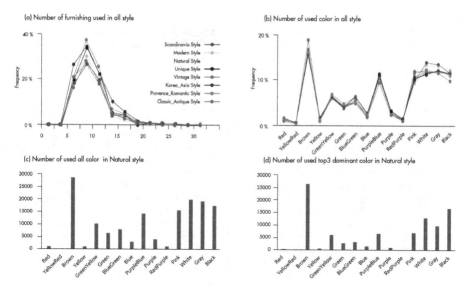

Fig. 7. Distribution of furnishings and colors per interior style: (a) Distribution of the number of furnishings in all styles, (b) Distribution of the number of colors used in all styles, (c) Distribution of all colors used in the *Natural* style, and (d) Distribution of top-3 dominant colors used in the *Natural* style (Color figure online)

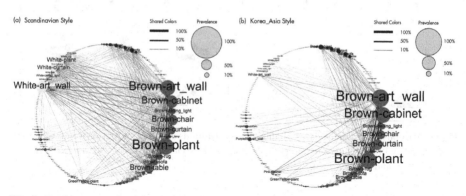

Fig. 8. Backbone graph of the CF networks: We only visualized the edges with the top 10% of the number of shared colors to increase its readability. Please note that we used the full network for the analysis: (a) *Scandinavian* style backbone graph; (b) *Korea_Asia* style backbone graph (Color figure online)

To identify which color pairing makes a difference in style, we measured the authenticity of a single CF, a CF pair, and a CF triplet. In the eight styles, *Brown* CF occupied a large part in common and belonged to all styles of pairing authenticity. In Fig. 9, the six most authentic single CFs, ingredient pairs, and triplets for *Scandinavian* style, *Provence_Romantic* style, *Modern* style, and *Natural* style are organized in a color pyramid. The ranking of the CF in each style shows the difference. The *Natural* style mostly uses *Brown* CF and combines pairs and triplets with many shared colors (Fig. 9-d). On

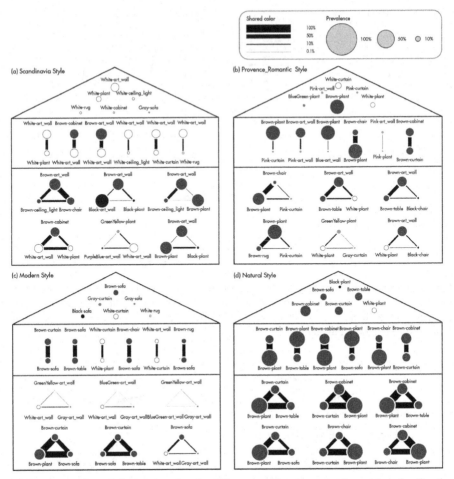

Fig. 9. We constructed eight styles of authenticity pyramids and present four styles of authenticity pyramids with distinctive features: (a) *Scandinavian* style, (b) *Provence_Romantic* style, (c) *Modern* style, and (d) *Natural* style.

the other hand, in the *Scandinavian* style (Fig. 9-a), *White* CF is predominant, and the number of shared colors between pairs and triplets tends to be small compared to other styles. Overall, in the *Scandinavian* style, *White* CF is high, and *Brown* CF is used in most CF pairings. The result supports a widely known description of the *Scandinavian* style. Noh [20] explains that in the *Scandinavian* style, bright colors such as white and neutral, as well as the unique colors of natural wood materials, are often used.

The *Natural* style has many CF pairings with strong connections. Unlike the *Natural* style, the rest of the styles were composed of pairs and triplets with relatively few shared colors. According to Lee [21], *Natural* Style uses natural materials such as wood and soil and uses natural colors such as cream, brown, and green. Our analysis result supports Lee [21], and we can argue that it is essential to use CF pairings of similar colors to design an interior with the *Natural* style.

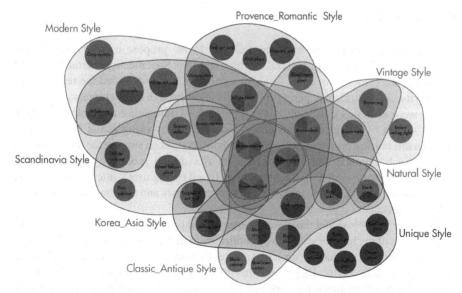

Fig. 10. Co-occurrence diagram: The six most authentic furnishings and furnishing pairs used in specific styles. (Color figure online)

In addition, Lee [22] said *Provence_Romantic* style is intended for a cozy feeling by creating a warm and lovely atmosphere. This results in frequent use of pastel colors and fabrics with pink and floral patterns. The *Modern* style is characterized by a simple and precise atmosphere. Therefore, achromatic colors are often used [23]. Our analysis results are consistent with the description of interior styles from a design expert. First, in *Provence_Romantic* style (Fig. 9-b), the authenticity of *Pink* CF is high, and it is a pairing of *Pink* and *Brown*. *Pink* CF plays the role of point color for the *Provence_Romantic* style. In contrast, we observed that in the *Modern* style, *Brown* and achromatic CFs made up the entire look (Fig. 9-c). In particular, among the achromatic CFs, the *Gray* CF plays a significant role in determining the style regardless of its harmony with other colors. In the end, our analysis based on large-scale data has yielded results that support existing interior style descriptions.

The co-occurrence graph (Fig. 10) can be used to confirm the similarity between interior styles by CF pairing. Figure 10 is a co-occurrence diagram showing the CFs shared by each style. First of all, the CFs of the *Unique* style are most often not shared. The pairings that include *Green-plant*, *PurpleBlue-sofa*, and *Gray-ceiling light*, exist only in the *Unique* style. On the other hand, many CFs of the *Provence_Romantic* style are shared with other styles. Although the latter style uses a pairing of *Blue* and *Pink* CFs not used in other styles, the *Brown* CF is included in most pairings, so it is located close to all styles. In the *Natural* style, all parts are shared with other styles. We observed that the *Natural* style is the least distinctive style.

5   Conclusion

The contributions of our paper are as follows: 1) We propose a novel interior style analysis method based on big data and machine learning. Specifically, a method by which to quantify the principle of color-furnishing pairing. 2) We identified the relationships between the color-furnishing paring and interior style through network analysis methods.

We were able to identify meaningful patterns of furnishing and color pairing in interior design by calculating prevalence and authenticity using actual interior cases. Through this pairing principle, we can create datasets that can support the interior design process. Such data sets can inform actual interior design process by recommending pairing examples or searching for furnishings based on the user's interior design style. In other words, it can expand the possibilities for efficiency and automation in the interior design process.

However, it is clear that the current paper has additional considerations. The representativeness of the interior style can be confirmed in that the interior style data collected in Today's House is an image of an actual resident's home. Therefore, the selected images can represent each interior design style that affects the accuracy of the study. In a future study, we will utilize social network meta data (such as the number of likes and comments on the interior design cases) that we also crawled using the Today's House data source. In addition, we will complement the material detection utilized to increase the robustness of the analysis. Therefore, our study opens up new options for supporting efficient interior design process by providing a systematic understanding of interior styles previously difficult to understand quantitatively, using principles of furnishing pairing.

Acknowledgements. This work was supported by a National Research Foundation of Korea (NRF) grant funded by the Korean government (MSIP: Ministry of Science, ICT and Future Planning) (NRF-2020R1C1C1011974).

References

1. Singh, M., Sharma, M.: Impact of home furnishing awareness programme on the use of fabric in home furnishing. Int. J. Home Sci. **2**(2), 197–200 (2016)
2. Haller, K.: Colour in Interior Design, Colour Design: Theories and Applications, 2nd edn. Woodhead Publishing, Cambridge (2017)
3. Won, P.-H.: The comparison between visual thinking using computer and conventional media in the concept generation stages of design. Autom. Constr. **10**(3), 319–325 (2001)
4. Weiss, T., Yildiz, I., Agarwal, N., Ataer-Cansizoglu, E., Choi, J.W.: Image-driven furniture style for interactive 3D scene modeling. Comput. Graphics Forum **39**(7), 57–68 (2020)
5. Zhu, J., Guo, Y., Ma, H.: A data-driven approach for furniture and indoor scene colorization. IEEE Trans. Visual Comput. Graphics **24**(9), 2473–2486 (2017)
6. Chen, G., Li, G., Nie, Y., Xian, C., Mao, A.: Stylistic indoor colour design via Bayesian network. Comput. Graph. **60**, 34–45 (2016)
7. Quercus Living. Timeless home furnishing and accessories to invest in. https://www.quercusliving.co.uk/knowledge/timeless-home-furnishing-and-accessories-to-invest-in/. Accessed 31 Jan 2021

8. Elle decor magazine. 20 Eye-catching color combinations to elevate your home. https://www. elledecor.com/design-decorate/color/g26629581/best-color-combinations/. Accessed 30 Jan 30 2021

9. Open Gallery: 4 Interior color combinations without failure. https://m.post.naver.com/viewer/ postView.nhn?volmeNo=8523745&meberNo=856760. Accessed 30 Jan 2021

10. Chan, C.-S.: Can style be measured? Des. Stud. **21**(3), 277–291 (2000)

11. Zhang, X.: Discussion on application for interior space design and the application of interior design style. In: the 2016 International Conference on Education, Management and Computing Technology, pp. 2352–5398. Atlantis Press (2010). (2016)

12. Liu, X., et al.: Inside 50,000 living rooms: an assessment of global residential ornamentation using transfer learning. EPJ Data Sci. **8**(4) (2019)

13. Park, D., Bae, A., Schich, M., Park, J.: Topology and evolution of the network of western classical music composers. EPJ Data Sci. **4**(1), 1–15 (2015). https://doi.org/10.1140/epjds/ s13688-015-0039-z

14. Ahn, Y., Ahnert, S., Bagrow, J., Barabási, A.: Flavor network and the principles of food pairing. Sci. Rep. **1**(1), 1–7 (2011)

15. Ogino, A.: A design support system for indoor design with originality suitable for interior style. In: International Conference on Education, Management and Computing Technology, IEEE, Kyoto, Japan, pp. 74–79 (2017)

16. Redmon, J., Farhadi, A.: Yolov3: an incremental improvement. arXiv preprint arXiv:1804. 02767 (2018)

17. Korean Agency for Technology and Standards. https://www.kats.go.kr/content.do?cmsid=83. Accessed 2021/1/31

18. Sharma, G., Wu, W., Dalal, E.: The CIEDE2000 color-difference formula: Implementation notes, supplementary test data, and mathematical observations. Color Research & Application: Endorsed by Inter-Society Color Council, The Colour Group (Great Britain), Canadian Society for Color, Color Science Association of Japan, Dutch Society for the Study of Color, The Swedish Colour Centre Foundation, Colour Society of Australia, Centre Français de la Couleur **30**(1), 21–30. (2005)

19. Zhou, T., Ren, J., Medo, M., Zhang, Y.: Bipartite network projection and personal recommendation. Phys. Rev. E **76**(4), 046–115 (2007)

20. Opinionnews. Scandinavian interior, aesthetics of comfort and modernity. https://www.opi nionnews.co.kr/news/articleView.html?idxno=38206. Accessed 20 Feb 2021

21. Sukbakmagazine. Natural interior. http://www.sukbakmagazine.com/news/articleView.html? idxno=51105. Accessed 20 Feb 2021

22. Sukbakmagazine. Emotional Provence&Romantic concept. http://www.sukbakmagazine. com/news/articleView.html?idxno=50778. Accessed 20 Feb 2021

23. Sukbakmagazine. The concept we choose the most, 'Modern'. http://www.sukbakmagazine. com/news/articleView.html?idxno=50650. Accessed 20 Feb 2021

A Phenotype-Based Representation that Quantifies Aesthetic Variables

Ban Liang Ling[(⊠)] and Bige Tunçer

Singapore University of Technology and Design, Singapore, Singapore
`banliang_ling@mymail.sutd.edu.sg`

Abstract. Design search should provide designers with diverse yet high-quality options. However, current methods to represent parametric options are mainly genotype- or performance-based. Qualitative values such as composition or aggregate shape are often not considered. Thus, the exploration of options is constrained, as the search algorithm is limited to genotype or performance descriptors. The aim of this paper is to improve existing methods to quantify aesthetic variables, thereby leading to more intelligent forms of design search. The paper proposes a phenotype-based representation that captures key features such as geometrical proportions, the framing of negative spaces and shape compositional relationships. A design space that consists of three different parametric models is introduced to demonstrate how the Zernike moments provide a uniform and consistent method for phenotype representation. The paper also proposes ways to measure phenotype similarity between options. Lastly, strategies to navigate, organize, and bring design logic to the design space are also presented.

Keywords: Design search · Parametric design · Zernike moments

1 Introduction

Design search aims to provide architects with a set of high-quality design options. These methods usually occur in a black-box and the derivation of solutions is based on the input information [1, 2, 17]. There exist 3 main categories of information exchanged during design search: a) genotype is a set of numbers that follow associative rules of a parametric model, b) phenotype is the produced design option and c) the performance is measured through various calculations. In design search, a representation of the design space is first generated, before a set of criteria streamline the design space into a few options. Performed within a Computer-Aided Architectural Design (CAAD) environment, designers can use computational tools such as building performance simulation or Multi-Objective Optimization (MOO) to help with the streamlining process [25]. The input information are digital representations of a building, which can be broadly classified as genotype- or performance-based.

D. Gerber et al. (Eds.): CAAD Futures 2021, CCIS 1465, pp. 250–267, 2022.
https://doi.org/10.1007/978-981-19-1280-1_16

1.1 Genotype-Based Representation

Genotype-based representations are widely used in MOO, where the optimization algorithm understands each option as a combination of parametric model inputs. Further computations are then made based on this set of numbers. The qualitative aspect of shapes or forms does not influence the optimization process and designers usually perform qualitative assessment at the end of MOO [17, 27, 28]. Relating a set of numbers to a design option might create many-to-one relationships where the same set of numbers refer to vastly different options. A genotype-based representation also limits a designer from updating the parametric model with new associative rules and appending it to an ongoing MOO process. This is because the new set of numbers will undergo updated shape generation rules, thereby creating a new relationship between genotypes and the generated model.

1.2 Performance-Based Representation

Performance-based representations focus on parts of the building which designers aim to quantify. Examples include space syntax analysis of floorplans [22], solar irradiation incident on the façade [11], or ventilation rate of room volumes [12]. These calculations provide a quantitative method for comparison and the data represents a building's behavior. During design search, where many options are compared, performance-based results can help to select the higher performing ones. However, the metrics can be arbitrary (giving a value to aesthetics [7]) or overly simplified (a trade-off between accuracy and computing time [17]). Both cases could mean that the same metric would give different results during an updated design search. In a design scenario where high-quality options with diverse forms are expected, current genotype- and performance-based representations constrain the design search. This is because the formal expression of parametric models is not considered. The challenge therefore is to identify a phenotype-based representation, which communicates key formal expressions that will influence the design search algorithm.

1.3 Key Challenges

This paper addresses two challenges that limit the influence from a model phenotype during design search (1) Genotype and performance values do not always provide adequate representations of a design option and could therefore limit the explored design space; and, (2) When a parametric model is updated, it usually leads to a new design search process which is independent of the older one. This could lead to abortive computations, as similar options are repeatedly explored. Both challenges are discussed in detail and a phenotype-based solution is proposed.

2 Existing Methods

The main goal of design search is to arrive at a set of diverse and high-quality design options. This section focuses on how a mixture of mathematical and statistical strategies can help to extract high performing solutions and measure diversity.

2.1 Extract High Performing Solutions

The most common method of design search is MOO, which centers around a search for high performance results. MOO has been adapted into tools with various levels of human-computer interaction (HCI). *Biomorpher* is an interactive designer-in-the-loop tool, which considers a user's aesthetic preference. A manual selection matches a designer's choice with a set of genotypes [7]. This is then converted into a fitness value that becomes part of the optimization process. Another tool, *Paragen*, opts to filter the design space based solely on quantitative fitness values, before presenting designers with the pareto optimal options [1]. Other tools expand on *Paragen's* strategy: providing designers with the fitness space and data clusters, thereby supporting navigation of the multi-dimensional solution space [17].

Data clustering assumes that the performance-based values provide a good representation of building phenotypes and each cluster should have similar formal traits. These tools perform design search in a hierarchical manner, whereby quantitative metrics are used to first eliminate the lower performing options. Qualitative analysis only takes place after the filtered options are presented to the designer (Fig. 1).

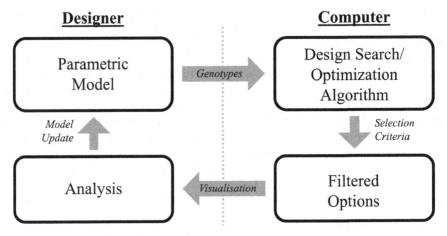

Fig. 1. Key stages in a typical design search process

2.2 Measure Qualitative Values

Yousif et al. attempt to integrate aesthetic variables in two different ways: (1) by including aesthetic parameters as constraints for the parametric model—to always generate 'beautiful' forms [28]; and (2) by analyzing the solution set with a form comparison algorithm—to identify diverse forms [27]. The algorithm works by overlapping two building footprints and recording the minimum shape difference value. Jupp and Gero [14] instead focus on creating qualitative feature based (QFB) graphs that tracks the shape boundary. The QFB graphs generate a unique encoding for each floorplan, making it possible to perform clustering based on shape features [13]. Compared to Yousif's area-based method, the QFB graphs are more complicated and require an existing library of features. Using machine learning techniques, Newton trained a 3D Convolutional neural networks model to measure building forms [20]. His method provides an alternative way of representation as each design option is identified by a set of probabilities. Similar to the QFB graphs, users have to pre-determine a library of qualitative features.

Qualitative values are difficult to compute and one's perception threshold might change once the variety of shapes is updated [20]. For example, a square and a rectangle might be considered different. But this perception will change when a circle is added to the design space. Beyond widely understood features like squares and rectangles, it is also arbitrary to determine qualities like symmetry or protrusion. On the other hand, area-based metrics might oversimplify diversity measures as the global form is prioritized over localized spatial relationships.

Current methods either account for aesthetic variables specific only to the current design, based on *a priori* building features or they are quantified as a type of performance metric. A different approach which allows a computer to appreciate qualitative value will improve design search.

3   Phenotype-Based Representation

A phenotype-based representation method should be consistent and reliable across different scenarios. Although there exist different 3D voxelization methods to quantify phenotypes, these strategies are computationally expensive [24]. Furthermore, at the urban level or early-design stage, phenotypes consist of simple forms which do not necessitate complex voxelization methods. Designers are instead concerned about the types of spaces that are suggested through aggregated shapes [26]. Therefore, this paper proposes an abstraction method where each model will be identified by 10 images (Fig. 2: top, bottom and elevations from the 8 cardinal directions). This means that subsequent computations will be performed on this set of 2D images.

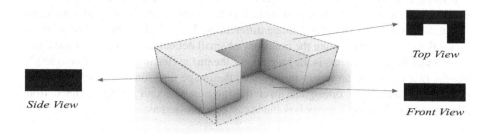

Fig. 2. Extraction of model images

3.1   Zernike Image Moments

Zernike image moments will be derived for each model image. *Zernike moments of an image are based on orthogonal Zernike polynomials defined over the interior of a*

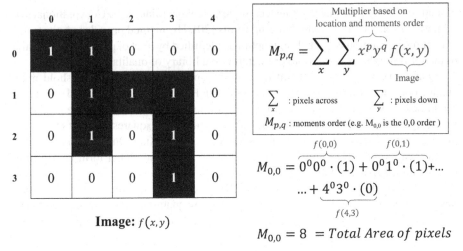

Image: $f(x, y)$

$$M_{0,0} = 8 = Total\ Area\ of\ pixels$$

Fig. 3. General formula for regular moments

unit disc [9, 18]. This method involves 2 main steps: a) calculating the regular image moments and b) mapping to orthogonal Zernike polynomials.

An image f(x, y) has value 1 when the pixel is filled and 0 when the pixel is empty: f(2, 1) = 1. Regular image moment (M_{pq}) is a summation of pixel values, multiplied by the pixel location (x, y), and increased to the power of the moment order (p, q). Figure 3 shows the terms and representations. This method is computationally light, and can be imagined as the counting of pixels, with the pixel location increasing in significance when the order (p, q) increases. Lower order moments include the 0, 0 order (M_{00}= area), the 0, 1 and 1, 0 orders (M_{01}, M_{10} → image centroid).

When applied to building footprints, lower order image moments capture the general shape and rough proportions. Higher order image moments are more sensitive to pixel location as the multipliers gain significance. Therefore, the higher order moments can pick up subtle and localized shape differences. These include minor variations in proportions, small kinks along the boundary or small deflections along the shape edge.

Regular image moments is the mapping of the image function f(x, y) onto $x^p y^q$. Teague [23] expanded on this to instead map the image function onto a set of Zernike polynomials which are orthogonal over the interior of a unit circle. Adapted from Khotanzad and Yaw [16], the polynomials can be expressed as:

$$V_{p,q}(x, y) = V_{p,q}(\rho, \theta) = R_{p,q}(\rho)\exp(jq\theta) \tag{1}$$

where

p Positive integer or zero.
q Positive and negative integers. $p - |q|$ is even, $|q| \leq p$.
ρ Length of vector from origin to *(x, y)* pixel.
θ Angle between vector ρ and x-axis in counterclockwise direction.

Instead of counting pixels across the y- and then x-axis (x, y), Zernike moments start from the center in increasing counterclockwise radial sweeps ($V_{p,q}(\rho, \theta)$). A switch to polar coordinates means that shape rotations do not produce moment fluctuations compared to other mapping functions [10]. The mapping onto Zernike polynomials also ensure that these moments can be rotation, scale, and translation invariant. This means that a square rotated by 10 degrees, scaled smaller proportionally and shifted sideways will still have the same image moments as its original shape.

The python library, *Mahotas*, will be used to calculate the Zernike image moments [4]. A total of 25 independent feature values, which have increasing moment orders, would provide global and local patterns within the image. The key features of each model image, including geometry proportions, framing of negative spaces and the resultant aggregate shapes, will be captured. These are important features that can help designers with space planning.

3.2 Design Space Set-Up

A parametric model with 2 blocks (20 × 8 × 10 m tall) was used as a test case. Each block could rotate and translate along the length and breadth of a square site. Examples can be seen in Fig. 4. A typical model genotype consists of 6 numbers: {block 1- (angle, x & y coordinates), block 2- (angle, x & y coordinates)}. Analysis is focused on the top view, which captures the biggest transformations.

(a)   (b)   (c)   (d)   (e)   (f)   (g)   (h)

Fig. 4. Series of images showing a translation of block 2 along the length of the site (a → e), before a rotation of block 2 from 0 to 135° (e → h).

Figure 5 shows that translation and rotation of the blocks (Fig. 4) have different impacts on the Zernike moments. The first 5 points on each graph tracks the changes in Zernike coefficients as block 2 has a 10 m increment in the y direction. The next 3 points tracks the changes as block 2 rotates with a 45-degree increment. It is evident that some Zernike terms display a larger change during a translation event while others demonstrate larger changes during rotational events. The combination of all 25 terms provide a unique feature vector that describes the composition of blocks for each model option.

3.3 Design Search Based on Qualitative Features

The design space was further expanded to include a total of 3 different parametric models: 2 blocks, 3 blocks and 4 blocks. The aim was to test the sensitivity of Zernike moments in capturing similar aggregate shapes. The 3 parametric models generate options which have the same total area and they follow the same logic where blocks can translate and rotate. However, the model genotype (angle, x & y coordinates) consists of 6, 9 and 12 numbers respectively. Figure 6 shows some generated options from the 3

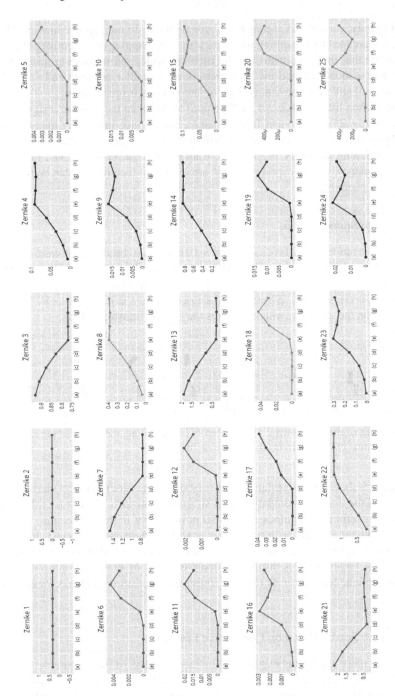

Fig. 5. Each line chart tracks the changes in a Zernike coefficient across a → h.

different parametric models. From the small sample, one can notice that the aggregation of 4 smaller blocks start to resemble some options from the 2 rectangular blocks (blue outline). Some of the composition of 4 blocks also bear similarities to that of options from the 3 blocks model (green outline).

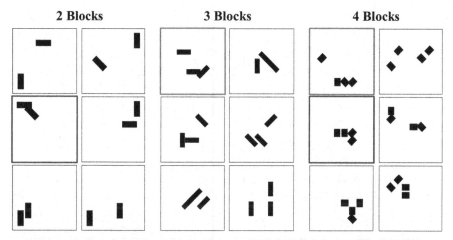

Fig. 6. Generated options from 3 different parametric models (Color figure online)

4   Design Space Visualization

The 3 parametric models combined to generate a design space with 15,672 options. To reflect the relative size of possibilities for each model, there were 3,864 of 2 blocks, 4,608 of 3 blocks and 7,200 of 4 blocks. Each option was recorded as a list of 25 Zernike moments which means that the design space would consist of 25 dimensions.

4.1   Uniform Manifold Approximation and Projection (UMAP)

UMAP was used to reduce the 25 dimensions into 2 key dimensions. This algorithm preserves the relative global positions of each data point, but the resultant 2-dimensional coordinates do not represent the actual distances between options [19]. The UMAP visualization is seen in Fig. 7.

To better understand the UMAP results in Fig. 7, an example which reduces 3 dimensional points into 2 key dimensions is presented (Fig. 8). The points in the left and right plots of Fig. 8 are colored to track their positions after dimension reduction. In 3D, it is obvious that the points are arranged in an open square box. This is also known as the global structure. The local structure refers to the grid-like arrangement of points along each surface. UMAP tries to maintain both global and local structures, sacrificing the actual point-to-point distances when data is projected from higher to low dimensions.

It begins by forming a neighbor graph in high dimension, clumping nearby data into relevant local structures. Next, this neighbor graph can be imagined as the open square

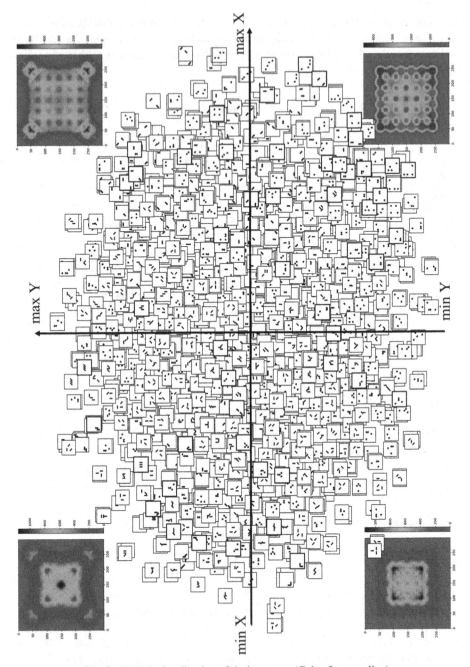

Fig. 7. UMAP visualization of design space (Color figure online)

box, which must be flattened into 2 dimensions that best represents the global and local data structures. In Fig. 8, the UMAP results show that the global 4-sided structure and the points adjacency are retained. On the local scale, the 3D rows of points are also represented by the UMAP linear clumps along each side.

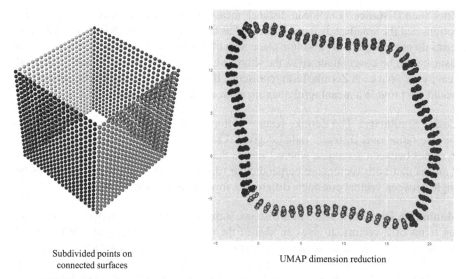

Subdivided points on connected surfaces

UMAP dimension reduction

Fig. 8. Dimension reduction of surface points using UMAP (Color figure online)

This method is unsupervised, which means that the observed organization in Fig. 7 is based purely on the image Zernike moments. As illustrated in Fig. 8, the nearest neighbors and the overall data organization can be communicated to a designer. Producing unsupervised organization is helpful as designers should not be expected to manipulate arbitrary hyperparameters when visualizing the complex design space.

The design space is then split into 4 quadrants, where images in each quadrant are overlapped to count the frequency where a block is present on the site (Fig. 7). Each heatmap reflects the count: red represents an area of high overlapping frequency while blue represents an area of low overlapping frequency. The results show that there are clear trends in block arrangement that emerge. Across the x-axis, arrangements of blocks shift from the center of the site towards the periphery of the site. Across the y-axis, blocks are initially arranged in a square and they open up to be positioned at the corners. The emergence of unsupervised geometric trends show that the Zernike moments provide reliable and unique image descriptions.

A 2-dimension display helps designers to get an intuitive overview of the design space. The mathematical background with which the Zernike moments and UMAP methods are built on, contributes to the emergence of the observed geometrical trends. UMAP also ensures that options with similar shapes will be placed close together.

4.2 Similarity Measure

Since each image can be represented by a Zernike feature vector, a similarity score can be calculated. For a 25-term vector, it is important to account for the changes in amplitude across each positional term.

Euclidean Distance: Euclidean distance measures the point distance between design options and the smallest distance is deemed as the closest match. However, the Zernike terms do not represent the absolute placement of blocks on the site. Since the Euclidean distance can be conceptualized as the shortest connection between 2 points, it depends heavily on what each Zernike term represents. In this case, using the Euclidean distance would not provide a meaningful quantification of design similarity.

Cosine Similarity: The Zernike feature vector should be imagined as a list of values that are ratios over different radial degrees. Cosine similarity measures the difference in angles between 2 vectors, where 1 means that both vectors are parallel (similar) and 0 means that both vectors are perpendicular (different). This would be relevant as the angle between vectors considers differences in value across each Zernike term.

Manhattan Distance: Manhattan distance sums up the difference between corresponding terms of 2 vectors. It gives an idea of the total units a vector must shift to match another vector. This score represents the difference in amplitude for each vector term.

It is important to note that similarity scores are akin to a reduction of information in 25 dimensions into 1 dimension. Therefore, some information is bound to be lost. In this scenario, the Manhattan and cosine distances provide a better similarity measurement as they consider the amplitude and position of each vector term.

4.3 Clustering

UMAP visualization helps by organizing the design space based on shapes, but it is still tedious for a designer to manually sieve through all the options [21]. To further organize the design space, density-based clustering (using cosine similarity) was used to group similar looking design options [3]. Firstly, this reduced the explorable design space from 15,672 to 605 phenotype clusters. Secondly, the group clusters provide an identity for the design options. Essentially, this creates a hierarchy of design logic for the design space. Figure 9 shows some promising cluster results that capture key compositional features. It also demonstrates an ability to identify shape typologies which will form the basis for further levels of data hierarchy.

5 Design Space Navigation

Using cosine similarity and Manhattan distance, the 2 furthest clusters were identified as clusters 79 and 29 (Fig. 10). Cluster 79 is closely packed into an aggregate square shape while cluster 29 is a spread-out L-shape. At increasing distance from cluster 79,

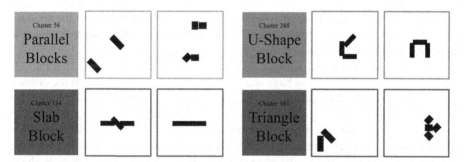

Fig. 9. Group clusters

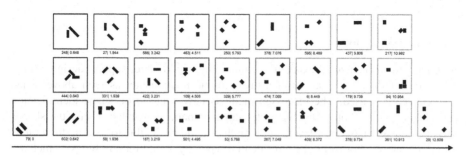

Fig. 10. Clusters identified at increasing distance from cluster 79 towards cluster 29

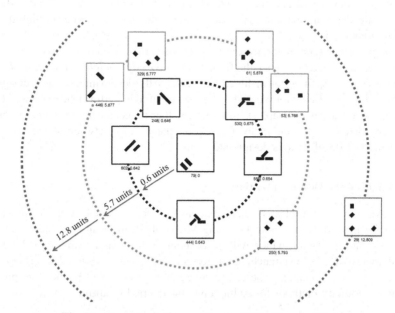

Fig. 11. Conceptual distance diagram (Color figure online)

intermediate clusters demonstrate a widening distance between blocks, as the arrangement slowly morphs to resemble cluster 29. This can be imagined as shifting along the x-axis of the UMAP visualization (Fig. 7).

Measuring similarity with a single distance score can be conceptualized in Fig. 11. Multiple options are deemed to have similar scores due to its proximity to the queried option. However, options found at increasing distances do not always preserve the same block composition logic. Figure 11 shows that at 5.7 units (green), there already exists 3 different composition schemes: a) Clusters 446 and 320 follow a 2-block composition which frames a courtyard space. b) Clusters 61 and 53 arranges the 4 blocks in a linear manner. c) Cluster 250 places the blocks in a U-shape. Despite having notable shape differences, all 3 schemes were classified within the same distance score. This demonstrates divergent design pathways and underlines how a single similarity score limits the design logic.

5.1 Data Hierarchy

A strategy that organizes the data into a hierarchy would better support design search. This would reduce the chances of a many-to-one relationship, as seen above, when a single score is used. The discussed methods have focused on shape identification and further layers should involve architecture domain knowledge. Using the shapes classification as a base, new layers of hierarchy should be flexible to account for various design intentions. For example, a residential typology would require blocks which are spaced further apart while an industrial building would require a certain footprint aspect ratio. Also, there could be situations where a designer is looking for different geometrical strategies to achieve a particular aggregate footprint. Both situations are subsets of design search and the expectation remains the same—to identify a diverse range of high-quality design options.

To help create a data hierarchy, the Function-Behavior-Structure (FBS) framework can be used. Introduced by Gero, the framework *is a design ontology that describes all designed things* [5, 6]. Function in this case can refer to the design intention (i.e. designing a residential block). Behavior would be a set of measurable performance or criteria (i.e. buildings spaced at least 20m apart). Structure will then reflect the performance or criteria and be formalized as a data hierarchy (i.e. measure the distance between adjacent blocks). Table 1 shows FBS of various design intentions.

5.2 Matching the Design Intention

A simple design search tool was created to identify different geometry arrangements for an input footprint (refer to Table 1). Using Microsoft Paint or an equivalent tool, one can sketch out a footprint, which will be an input to the design search tool. Next, the tool will identify the closest matching clusters and provide an additional criterion — the number of blocks. Organizing the results based on the number of blocks is to provide alternative geometry methods for adding up to the queried footprint.

Table 1. Function-Behavior-Structure Ontology Framework for various design intentions

Function	Behavior	Structure
Residential design in tropical climate	– Buildings should be spaced > X m apart for privacy – Angles between buildings should be > Y degrees for privacy – Buildings should face North-South to avoid the sun	1. Extract building outlines a. *Calculate distance between buildings* b. *Calculate smallest angle between buildings* 2. Identify North direction c. *Check building orientation* 3. Eliminate options which fail criteria
Industrial building design	– Buildings should have aspect ratio of X for efficient space planning – Building edges should be 90 degrees for efficient space planning	1. Extract building outlines a. *Calculate aspect ratio* b. *Identify corner angles* 2. Eliminate options which fail criteria
Different geometry arrangements for a footprint	– Options should have various number of blocks – Aggregated shape should resemble input footprint	1. Extract building outlines a. *Count number of blocks* 2. Extract Zernike for input footprint a. *Search clusters for closest cosine similarity* b. *Offer designer selection sorted by cosine similarity*

Figure 12 shows the closest matches to a U-shape sketch. Along the y-axis, the number of blocks for each option is tracked. On the x-axis, the similar options are located nearer to the origin. The results show that with the current design space, options with 2, 3 and 4 blocks arrange themselves in a U-shape and these compositions are displayed to the user. It suggests various alternative ways of achieving a U-shape, some with a continuous slab, others through composition of smaller volumes.

Another search was conducted for a slab block and the results again suggested various methods for generating a similar linear shape (Fig. 13). Designers are exposed to alternative compositions and they can also embark on a different design pathway if they prefer the suggested forms. For example, the initial search was based on a single slab block, with its inherent geometric and space framing constrains. He might be intrigued by the 2-block system, which has a different set of constrains. Based on the geometry, the design intention might be updated, and a new set of measurements could be added as layers of data filter.

The results show that the phenotype-based representation has the potential to suggest designs to users. Zernike moments quantify aesthetic variables and offer data-centric strategies. This means that qualitative value can be considered alongside traditional measurements like solar irradiation or space syntax analysis. The key is to identify

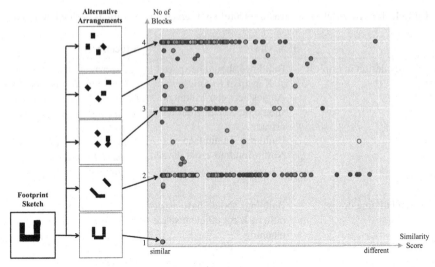

Fig. 12. Closest match graph (no of blocks vs similarity score)

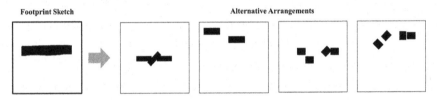

Fig. 13. Search results for a slab block

appropriate behaviors or measurements that best represent the design intention. This will help organize and provide design logic to the design space.

6   Conclusion

This paper proposed a new phenotype-based representation which can be integrated into existing design search methods. Earlier sections also demonstrated how Zernike moments can communicate qualitative considerations to a computer; thereby ensuring reliable organization of the design space. Compared to a genotype- or performance-based representation, the proposed method provides a better solution to quantify phenotype diversity [2].

By generating an intermediate "Zernike image moments" phenotype space, it opens the possibility to combine various parametric models into the same design search process (Fig. 14). Instead of understanding a design option as a set of genotypes, the MOO algorithm can now have a better appreciation of shapes based on the Zernike feature vector. Across generations, the algorithm will also relate each Zernike term to a performance trend. This strategy has the potential to improve design search as the computer can better appreciate aesthetic variables.

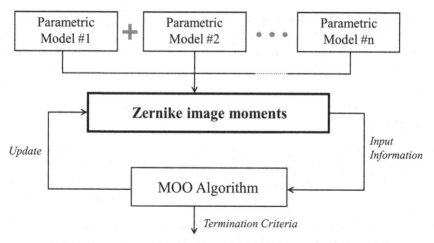

Fig. 14. Proposed MOO process which feeds off unique Zernike image moments

Other potential applications include using the Zernike moments to first identify key form typologies, before performing a targeted performance analysis. In practice where the design phase is much shorter, computation time is valuable, and a targeted search might be more relevant [8]. The proposed method of model abstraction offers quick and reliable shape appreciation. Calculation of Zernike moments for all 15,672 options took less than a minute on a single core 2.6 GHz processor. This means that typology identification can be automated, thereby improving the quality of solutions.

6.1 Human-Computer Interactions (HCI)

This paper aims to study the usage of Zernike moments in quantifying phenotypes. There was a conscious effort to select unsupervised methods like UMAP for dimension reduction and HDBSCAN for clustering. These methods demonstrate stability in high-dimensional datasets and would be reliable for studying the qualitative spread and organization of shapes. The earlier sections show promising results, but there are still challenges with respect to designer HCI. The tool presented in Sect. 5.2 is an early prototype, used to interact with the phenotype space. However, the workflow is still similar to that in Fig. 1, where designers feed off results from black-box algorithms. Further explorations will instead look at direct HCI with the design space (Fig. 7), using Zernike moments as a medium for communication.

6.2 Future Work

The authors identify 2 key areas for further development:

Improve the Organization of the Data Hierarchy. Despite some promising results, more finetuning is required for the data hierarchy. Different design intentions can be studied, and this should provide a larger base of measurement methods. The FBS ontology would continue being a framework for classifying the various measurements. Users should be able to mix and match data filters according to their design intention.

Visualization and Interaction with Design Space. With a 25-term feature vector and potentially a data hierarchy, the design space would consist of many dimensions. Further research will study different visualization techniques and perhaps mixed media modes to improve HCI.

References

1. Beulow, P.: Paragen: performative exploration of generative systems. J. Int. Assoc. Shell Spatial Struct. **53**(4), 271–284 (2012)
2. Brown, N. and Mueller, C.: Quantifying diversity in parametric design: a comparison of possible metrics. In: Artificial Intelligence for Engineering Design Analysis and Manufacturing, vol. 33, no. 1, pp. 1–14 (2018)
3. Capello, R., Moulavi, D., Zimek, A., Sander, J.: Hierarchical density estimates for data clustering, visualization, and outlier detection. In: ACM Transactions on Knowledge Discovery from Data, New York (2015)
4. Coelho, L.P.: Mahotas: open source software for scriptable computer vision. J. Open Res. Software **1**(1), e3 (2013)
5. Gero, J.: The situated function-behaviour-structure framework. In: Design Studies (2004)
6. Gero, J.S., Kannengiesser, U.: The function-behaviour-structure ontology of design. In: Chakrabarti, A., Blessing, L.T.M. (eds.) An Anthology of Theories and Models of Design, pp. 263–283. Springer, London (2014). https://doi.org/10.1007/978-1-4471-6338-1_13
7. Harding, J., Brandt-Olsen, C.: Biomorpher: interactive evolution for parametric design. Int. J. Archit. Comput. **16**, 144–163 (2018)
8. Haymaker, J., et al.: Design space construction: a framework to support collaborative, parametric decision making. J. Inf. Technol. Constr. **23**, 158–178 (2018)
9. Huang, M., Ma, Y.Q., Gong, Q.P.: Image recognition using modified Zernike moments. Sen. Transducers **166**(3), 219–223 (2014)
10. Huang, Z., Leng, J.: Analysis of Hus moment invariants on image scaling and rotation. In: 2nd International Conference on Computer Engineering and Technology (2010)
11. Jakubiec, A., Reinhart, C.F.: DIVA 2.0: integrating daylight and thermal simulations using rhinoceros 3D, Daysim and EnergyPlus In: Proceedings of 12th Conference of International Building Performance Simulation Association, Sydney (2011)
12. Jakubiec, J.A., Doelling, M.C., Heckmann, O., Thambiraj, R., Jathar, V.: Dynamic building environment dashboard: spatial simulation data visualization in sustainable design. J. Technol. I Archit. + Des. 27–40 (2017)
13. Jupp, J., Gero, J.: Qualitative representation and reasoning in design: a hierarchy of shape and spatial languages. Vis. Spat. Reason. Des. **III**, 139–162 (2004)
14. Jupp, J., Gero, J.: A qualitative feature-based characterization of 2D architectural style. J. Am. Soc. Inform. Sci. Technol. **57**(11), 1537–1550 (2006)
15. Jupp, J., Gero J.: Visual style: qualitative and context-dependent categorization. In: Artificial Intelligence for Engineering Design Analysis and Manufacturing (2006)
16. Khotanzad, A., Yaw, H.H.: Invariant image recognition by zernike moments. IEEE Trans. Pattern Anal. Mach. Learn. **12**(5), 489–497 (1990)
17. Ling, B.L., Jakubiec, A.: A three-part visualisation framework to navigate complex multi-objective (>3) building performance optimisation design space. In: Proceedings of BSO 2018: 4th Building Simulation and Optimization Conference, Cambridge UK (2018)
18. Liu, M.F., He, Y.X., Ye, B.: Image Zernike moments shape feature evaluation based on image reconstruction. Geo-spatial Inf. Sci. **10**, 191–195 (2012)

19. McInnes, L., Healy, J., Saul, N., Grossberger, L.: UMAP: uniform manifold approximation and projection. J. Open Source Software **3**(29), 861 (2018)
20. Newton, D.: Multi-objective qualitative optimization (MOQO) in architectural design. Appl. Constr. Optim. **1**(36), 187–196 (2018)
21. Shireen, N., Erhan, H., Woodbury, R., Wang, I.: Making sense of design space. In: Çağdaş, G., Özkar, M., Gül, L.F., Gürer, E. (eds.) CAADFutures 2017. CCIS, vol. 724, pp. 191–211. Springer, Singapore (2017). https://doi.org/10.1007/978-981-10-5197-5_11
22. Tarabieh, K., Nassar, K., Abu-Obeid, N., Malkawi, F.: The statics of space syntax: analysis for stationary observers. Int. J. Archit. Res. **12**(1), 280–306 (2018)
23. Teague, M.: Image analysis via the general theory of moments. J. Opt. Soc. Am. **70**(8), 920–930 (1980)
24. Vantyghem, G., Ooms, T., Corte, W.: VoxelPrint: a grasshopper plug-in for voxel-based numerical simulation of concrete printing. In: Automation in Construction (2021)
25. Vierlinger, R., Hofmann, A: A framework for flexible search & optimization in parametric design (2013)
26. Wang, L.K., Janssen, P., Chen, K.W., Tong, ZY., Ji, G.H.: Subtractive building massing for performance-based architectural design exploration: a case study of daylighting optimization. In: Sustainability (2019)
27. Yousif, S., Tan, W.: Incorporating form diversity into architectural design optimization. In: Proceedings of the 37th Annual Conference of the ACADIA, Cambridge MA (2017)
28. Yousif, S., Clayton, M., Tan, W.: Towards integrating aesthetic variables in architectural design optimization. In: 106th ACSA Annual Meeting Proceedings (2018)

A Multi-formalism Shape Grammar Interpreter

Rudi Stouffs[(⊠)] [iD]

Department of Architecture, National University of Singapore,
4 Architecture Drive, Singapore 117566, SG, Singapore
`stouffs@nus.edu.sg`

Abstract. A shape grammar is a formal rewriting system for producing languages
of shapes. While initially defined to operate over line segments and labeled points,
the theory of shape grammars has numerously been extended to include other spatial and non-spatial entities, including, planar segments and volumes, some types
of curves, weights, colors and descriptions. Over the years, also a number of shape
grammar interpreters have been developed, implementations that support the specification and application of shape rules. However, each of these implementations
has adhered to a single shape grammar formalism, even if the exact formalism
may differ from one implementation to another. This paper reports on the development and application of a shape grammar interpreter supporting multiple shape
grammar formalisms. This is achieved in two ways, first, by supporting a variety
of representational structures as compositions of basic data types, and second,
by providing two alternative matching mechanisms for spatial elements, a non-
parametric and a parametric-associative mechanism. Together, this provides for a
flexible and extensible interpreter for the specification and application of shape
rules, which has been implemented in the Python programming language and is
accessible from Rhino and Grasshopper.

Keywords: Shape grammar · Sortal grammar · Multi-formalism · Shape
grammar interpreter

1 Introduction

A shape grammar is a formal rewriting system for producing languages of shapes [1,
2]. It consists of a set of productions, or shape rules, operating over shapes drawing
from a vocabulary of spatial and, optionally, non-spatial elements, e.g., labels. A shape
grammar may be considered to include an initial shape as the starting point in the
productive (generative) process. The language defined by a shape grammar is the set of
shapes generated by the grammar that do not contain any non-terminal symbols.

Grammar formalisms for design come in a large variety, requiring different representations of the entities being generated, and different interpretative mechanisms for
this generation. Shape grammars also come in a variety of forms, even if less broadly.
Initially, shape grammars emphasized labeled shapes, a combination of line segments
and labeled points [2]. Plane segments, solids [3] and curves [4] have been added later.
Stiny [5] additionally proposes numeric weights as attributes to denote line thicknesses

© Springer Nature Singapore Pte Ltd. 2022
D. Gerber et al. (Eds.): CAAD Futures 2021, CCIS 1465, pp. 268–287, 2022.
https://doi.org/10.1007/978-981-19-1280-1_17

or surface tones. Knight [6, 7] considers a variety of qualitative aspects of design, such as color, as shape attributes. Stiny [8] also proposes to augment a shape grammar with a description function in order to enable the construction of verbal descriptions of designs.

A shape rule combines a specification of recognition and manipulation (search and replace). Expressed in the form $a \rightarrow b$, a specifies the pattern to be recognized, while both a and b participate in the manipulation. Manipulation involves replacing the recognized a by b in the shape under investigation. Generally, recognition is not simply based on a one-to-one mapping of spatial (or non-spatial) elements but, instead, on the existence of a part relationship supporting emergence, that is, the recognition of shapes that may not be anticipated as such. An example of emergence is the third square that arises from two squares overlapping along their common diagonal (Fig. 1).

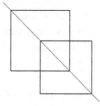

Fig. 1. Two squares overlapping along their common diagonal. A third, smaller square emerges as the result of the overlapping.

Recognition necessarily applies under an allowable transformation, for example, a similarity transformation, and the resulting manipulation must occur under the same transformation for both a and b. That is, a rule $a \rightarrow b$ applies to a shape s under a transformation t if $t(a)$ is a part of s ($t(a) \leq s$), yielding the shape $s - t(a) + t(b)$. What actually constitutes an allowable transformation may differ [9]. Shape grammars commonly consider transformations of similarity, allowing for translation, rotation, reflection and uniform scaling. A few examples constrain the transformations to be isometric or Euclidean (disallowing scaling) (e.g., [10]).

Stiny [11] also suggests a parametric shape grammar, consisting of parametric shape rules operating on (non-parametric) shapes. A parametric shape rule embeds (numerical) parameters that govern the position of some spatial elements (or their boundary elements) within the left-hand-side shape (and corresponding elements in the right-hand-side shape). Stiny does not suggest any implementation, but a number of researchers have implemented a parametric-style shape grammar interpreter adopting a graph-based matching mechanism (e.g., [12–14]). Commonly, line segments, or the infinite lines carrying these segments, and their intersection points, serve as nodes and edges (or vice versa) of the graph. In addition, one or more kinds of associations between line segments (or their intersection points), such as parallelism and perpendicularity, or equal lengths and distances, are considered as invariants for the matching process. Therefore, we adopt the term parametric-associative shape rules.

A shape grammar interpreter is the engine that supports the application of shape rules, including recognition and manipulation. Implementing a shape grammar interpreter requires implementing the part relationship for shapes—with or without attributes—, the operations of sum and difference on shapes, and solving the matching problem, that is, identifying under which transformation a rule may apply to a given shape. Thus, a shape grammar formalism defines the kinds of shapes that are operated upon and the kinds of transformations under which the matching occurs. Beirão [15, pp. 228–236] offers a survey of (implementations of) shape grammar interpreters, and concludes that they have common limitations. Many of them compute only on two-dimensional shapes; and most do not consider a part relationship and therefore do not support emergence. Also, '*very few shape grammar interpreters allow for the implementation of rules operating with symbols*' [15, p. 235], never mind other attributes, or a description function. Since then, just a few 'general', three-dimensional shape grammar interpreters have seen the light, namely, GRAPE, DESIGNA, and SortalGI. GRAPE [16] is a parametric-style shape grammar interpreter supporting parametric-associative shape rules operating on line segments, circles and circular arcs. In addition, GRAPE also supports plane segments, and primitive solids, such as spheres, cylinders and boxes, but these can only be matched as is, that is, no part relationship is implemented over plane segments or solids. GRAMATICA/DESIGNA [17, 18] also implements a parametric-style shape grammar interpreter operating on both line and plane segments, although the part relationship is only fully developed for line segments.

SortalGI [14] is the subject of this paper. The SortalGI shape grammar interpreter supports two alternative matching mechanisms for spatial elements, a non-parametric mechanism matching shapes under similarity transformations (translation, rotation, reflection and uniform scaling) and a parametric-associative mechanism under some topological constraints as well as associations of perpendicularity and parallelism. In addition, sortalGI supports a broad variety of spatial and non-spatial elements, including any compositions thereof. Together, this provides for a flexible and extensible framework for the specification and application of shape grammars.

Here, we should also acknowledge Shape Machine [19] as a recent development of a shape grammar interpreter. Although it is not 'general' in the sense used above—it only supports line segments and arcs, mostly in two dimensions—it offers a different generalization in that it supports multiple kinds of transformations, including isometric, similarity and affine transformations.

In this paper, we argue the importance of a multi-formalism shape grammar interpreter, elaborate on the two ways multiple shape grammar formalisms are supported and implemented in SortalGI, and review the strengths and limitations of the approach using example rules and rule applications.

2 Multiple Formalisms

A shape grammar formalism importantly determines the representational structure used to represent shapes (including non-spatial elements or attributes) and the matching mechanism to match (and apply) a rule onto a shape under investigation. A shape grammar interpreter adopting a single shape grammar formalism necessarily locks the user into shape rules only containing spatial and non-spatial elements supported by the representational structure and only matching shapes under the specific matching mechanism and corresponding allowable transformations.

2.1 Motivation

Most examples of shape grammars in literature are analytical shape grammars that have been derived from the investigation of an existing body of designs in order to formalize the design rules underlying such body. While these are important findings, these grammars generally do not reflect on how a designer might adopt shape rules to facilitate their design development. While a shape grammar generally defines a collection of rules together with an initial shape, from a user's point of view, any collection of rules that serves a particular purpose can be considered a shape grammar, whether or not it requires a particular initial shape. As such, a practical shape grammar interpreter should support both the implementation and exploration of existing shape grammars as well as creative rule development and exploration within a design process. For example, within a design process, it may not be unthinkable to adapt not only the rules but also the formalism as the design progresses or the problem shifts. A design that originally was conceived in plan or section, or both, might be further elaborated in three dimensions, or the designer might want to use both non-parametric as well as parametric-associative shape rules intermittently. Even if the user sticks to one particular formalism, a different design problem may ask for a different kind of shape rules supported by a different formalism.

From a different point of view, multiple formalisms could also be applied in parallel. Stiny [5] shows an example where a series of cubes is described separately in plan—as shapes composed of line segments and labeled points—, in a front elevation—as shapes composed of weighted line segments and labeled points—, and in a side elevation—as shapes composed of shaded plane segments. Obviously, this in and of itself can be considered as a single (meta)formalism but, here too, many variations can be considered. Li [20] considers seven drawings (from plan diagram to plan, section, and elevation) and, additionally, nine descriptions (specifying measures of width, depth, and height, among others), in his specification of a shape grammar for (teaching) the architectural style of the *Yingzao fashi*. Duarte [21] considers separate drawings (sketches, plans, elevation, envelope, etc.) reflecting on different viewpoints, and descriptions reflecting on different features. Drawings may embed line segments, plane segments, volume segments, and various weights and descriptions as attributes. Allowing for such flexibility minimally requires supporting multiple formalisms.

2.2 Sortal Structures

Sortal grammars constitute a class of formalisms for shape grammars. Underlying *sortal* grammars is a modular representational approach, considering shapes as adhering to a compound algebra defined by the sum or direct product on basic and augmented algebras [22]. Each algebra is ordered by a part relation ("$\leq$") and closed under the operations of sum ("$+$") and difference ("$-$"), as well as relevant transformations, enabling shape matching and rule application. Each basic algebra corresponds to a specific data kind, whether spatial or non-spatial; combinations of two data kinds under an attribute relationship are captured by an augmented algebra, e.g., points with attribute labels or, even, labels with attribute points. We also adopt the following terms: a data element is denoted an *individual*, a collection of individuals of the same data kind a *form*, and a collection of forms, each of a different data kind, a *metaform*. Also, any individual may be assigned an attribute form or attribute metaform. Note that attribute forms of different individuals of a same form must necessarily adhere to the same algebra, albeit basic, augmented or compound. For example, if one individual in a form is assigned a label as attribute and another one a color as attribute, all individuals in the form necessarily have an attribute metaform consisting of, at least, a form of labels and a form of colors, although either or both forms may be empty.

Representationally, each basic shape algebra defines a single representational module, and modules can be combined to represent compound algebras under compositional operations of sum, direct product, and attribution. In the terminology of *sortal* grammars, these modules and their compositions are termed *sortal* structures, with *sortal* structures formally composed of other, primitive, *sortal* structures. For example, a representation for points and line segments can be obtained as a composition of the *sortal* structures for points and line segments under the operation of sum, specifying a disjunctive relationship between points and line segments; any point or line segment is either one or the other. A representation for labeled points can be obtained as a composition of the *sortal* structures for points and labels under the operation of attribution, specifying a subordinate relationship, similar to an object-attribute relationship, between points and labels; any labeled point is a point with zero, one or more labels assigned as attribute. Finally, a representation for labeled points and line segments, as a basic representation for many shape grammars, then can be obtained as a two-level composition of, firstly, points and labels under the operation of attribution ('^') and, secondly, the resulting labeled points and line segments under the operation of sum ('+'):

$$line\_segments + points \, ^\wedge \, labels \tag{1}$$

As such, different representational structures underlying different grammar formalisms can be defined by selecting both the primitive components and their compositional relationships.

Each primitive *sortal* structure defines its algebraic behavior, including a part relationship to support the matching mechanism and arithmetic operations of, at least, sum and difference over forms for rule application. As an example, the part relationship for forms of points corresponds to the subset relationship, and the operations of sum and difference over forms of points to the set operations of union and difference, respectively. The same algebraic behavior applies to labels. In general, different *sortal* structures may require different algebraic behaviors, thus, different part relationships, and corresponding operations of sum and difference, to apply [22]. The algebraic behavior of a composite *sortal* structure, on the other hand, can be derived from the behaviors of the component structures taking into account the compositional operator applied.

Implementation-wise, each data kind defines an object class representing individuals of that data kind. Such an *individual* class implements, at a minimum, a constructor and functions to apply a transformation to an instance, to compare two instances (of this class), to check if one instance contains another, to combine two instances into a single instance, if possible, and to determine the complement of one instance with respect to another (Table 1). Each algebraic behavior also defines a class, but independent of the underlying data kind, such that its implementation can be shared by different data kinds, such as in the case of points and labels. Such a *form* class serves to represent a form of individuals of the same kind. It minimally implements a constructor and functions to apply a transformation to an instance and to insert an individual instance (of the appropriate data kind) into a form instance. In addition, the functions part-of, sum and difference operating on two form instances are implemented in terms of the functions contains, combine and complement of the respective *individual* class. Furthermore, a *metaform* class serves to represent a composition of forms into a metaform. Similarly to other *form* classes, it minimally implements a constructor and functions to apply a transformation to an instance, to insert a form instance (of an appropriate data kind) into a metaform instance, as well as the functions part-of, sum and difference operating on two metaform instances. Finally, different *sort* classes allow for the definition of primitive and composite *sortal* structures. Each primitive *sortal* structure specifies the respective *individual* class and (corresponding) *form* class used to represent individuals and forms corresponding to this *sortal* structure.

Table 1. Python functions for the minimal implementation of *individual*, *form* and *metaform* classes. The function bodies are omitted as these are dependent on the actual kind of *individual*, *form* and *metaform*, e.g. labels vs points vs line segments, etc. (see [22] for some pseudo-code elaborations).

Individual	Form	Metaform
def __init__(self, sort, *args, **kwargs)		
def transform(self, transformSet)		
def compareValue(self, other)	def insert(self, ind)	def insert(self, element)
def contains(self, other)	def partOf(self, other)	
def combine(self, other)	def sum(self, other)	
def complement(self, other)	def difference(self, other)	

The SortalGI shape grammar interpreter implements these different classes in the Python language (Table 1). Algorithms can be found, to some extent, in literature: algorithms for the operations of sum and difference and the part relationship for shapes composed of (labeled) points and line segments [23, 24]; algorithms for classifying the boundary of a shape with respect to another shape and for constructing a shape from a given boundary, in support of the operations of sum and difference and the part relationship for shapes composed of plane segments (and solids) [25]; algorithms for shape operations on shapes composed of quadratic Bezier curves [26]; pseudo-code elaborations for various primitive *sortal* structures [22].

As such, multiple *sortal* structures can be defined, each supporting a different (shape) grammar formalism. Each *sortal* structure thus defined specifies an algebraic behavior and supports the part relation underlying shape matching and the operations of sum and difference, underlying rule application, over (meta)forms corresponding to this *sortal* structure.

2.3 Sortal Rules

A shape rule is often understood to imply that both left-hand-side and right-hand-side shapes constitute geometries, possibly including non-geometric attributes, e.g., labels (or descriptions). Then, a description rule would imply that both left-hand-side and right-hand-side of the rule constitute a description. Together, shape rules and description rules may combine into a compound rule; while the different component rules operate in parallel, all component rules must apply together or not at all. In the case of *sortal* grammars, a *sortal* rule operates on (meta)forms corresponding to a *sortal* structure. If the *sortal* structure is composed under the operation of sum, the *sortal* rule can be considered as composed of component rules, one for each component structure and operating on forms of this component structure.

Shape matching requires determining an allowable transformation under which a given shape matches part of a shape under investigation. In the case of rule application, the left-hand-side shape of the shape rule constitutes the given shape. In order to identify potential matches, a matching mechanism must first determine potential transformations, upon which each transformation can be applied to the given shape and the transformed shape can be checked to form part of the shape under investigation. If successful, a valid transformation has been found and the rule (manipulation) can be applied to the shape under investigation, under this transformation. We may either be interested in one or all valid transformations for a given rule and shape under investigation.

Shape (or *sortal*) matching can be considered within a shape algebra and across algebras, that is, for each primitive *sortal* structure separately and for the composite *sortal* structure as a whole. In general, shape matching requires finding a single transformation that applies to all spatial kinds that form part of the given shape and formalism, including, e.g., points, line segments and plane segments. However, in the case of rules applying over multiple drawings (see Sect. 2.1), different transformations may need to apply to drawings embedded in different spaces, even if these drawings specify different views of the same design. For this reason, *sortal* structures distinguish two compositional operators: the operation of sum combines *sortal* structures embedded within the same space—a single transformation must apply to all spatial forms—, while the operation of direct product combines *sortal* structures when not embedded in the same space—different transformations may apply for each component structure [22].

When determining a potential transformation to apply to multiple spatial forms, we can prioritize those forms that can facilitate the search process. In the case of isometric or similarity transformations in three dimensions, mapping three non-colinear points on three other points defines a finite number of potential transformations. First, three non-colinear points uniquely define a plane. If both planes intersect, the line of intersection defines an axis of rotation and there exist two rotation angles (one positive and one negative) to map one plane onto the other (Fig. 2, top left). If parallel, instead, a translation perpendicular to both planes will map one onto the other, while an additional 180 degree rotation may be considered to offer an alternative mapping (Fig. 2, top right). In both cases, a reflection through the plane may offer additional alternatives if the given shape is three-dimensional in nature. Next, the problem is reduced to two dimensions, where twice three points—when considered in a fixed order—uniquely determine an affine transformation. If the point figures are similar, the affine transformation becomes a similarity or, possibly, isometric transformation. If the figure portrays reflective or rotational symmetry, then additional transformations can be found by permuting one set of points. In any case, the number of valid transformations is finite and enumerable.

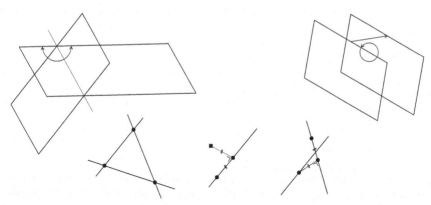

Fig. 2. Determining potential transformations: (top left): two intersecting planes define an axis of rotation and two rotation angles to map one plane onto the other; (top right): two parallel planes define a perpendicular translation to map one plane onto the other, while an additional 180° rotation offers an alternative mapping; (bottom, from left to right): three intersecting lines define three *distinguishable* points (of intersection); one line and a non-collinear point define two additional *distinguishable* points (one foot of the perpendicular and one other at equal distance on the line); two skew lines define three *distinguishable* points (two feet of the common perpendicular and one at an equal distance on one of the lines).

If none or fewer than three points are present in the shape to be matched, then *distinguishable points* can be constructed from distinct carriers of other spatial elements for the purpose of generating a determinate number of valid transformations. These include points of intersection of intersecting lines and the feet of a common perpendicular of two skew lines (Fig. 2, bottom). Note that these *distinguishable lines* may themselves result from the intersection of non-parallel planes. If only two distinguishable points can be determined in this way, a third one can be constructed on one of the lines that has served to determine a distinguishable point at a distance equal to the distance between the two distinguishable points. This ability to construct a third distinguishable point follows from the fact that two distinguishable points already determine a fixed scaling factor. We refer to Krishnamurti and Stouffs [27] for an elaboration of all determinate and indeterminate cases. In an indeterminate case, a base transformation may be able to be determined, e.g., by determining the largest possible scaling factor (based on a line segment's endpoints), or a zero degree of rotation around a determined axis of rotation.

In summary, when only isometric or similarity transformations are allowable, the shape matching algorithm involves the following steps (Fig. 3):

1. Identify three distinguishable points within the left-hand-side shape of the rule; this constitutes the source triplet
2. Identify all distinguishable points in the shape under investigation

3. Enumerate all type-corresponding triplets of distinguishable points from the shape under investigation; these constitute the target triplets
4. Eliminate all target triplets that do not preserve the invariants under the allowable transformations, i.e., angles and length ratios in the case of similarity transformations
5. Determine the transformation matrices that map the source triplet onto each target triplet

Fig. 3. Matching a square (left) onto a figure including three squares (right), under a similarity transformation. Three distinguishable points are identified in the square on the left, all distinguishable points of the same type (intersection points) are identified in the figure under investigation. One triplet of such points is selected (black dots) and a similarity transformation matrix can be determined from the two triplets of points.

Note that when matching over multiple spatial forms, we prioritize points over line segments and plane segments. Labeled points in the left-hand-side shape are especially beneficial, at least when they do not all share the same label(s), as labeled points will only match other labeled points that share at least the same labels.

2.4 Parametric-Associative Matching

We have distinguished two types of shape rules, on the one hand, rules that apply under a similarity (or isometric) transformation and, on the other hand, parametric-associative rules that apply under some topological constraints as well as certain associations. These associations may differ from one approach or implementation to another; topological constraints may also differ to some extent. Here, we will focus on a specific approach adopted for the *sortal* grammar interpreter.

All graph-based matching mechanisms adopted in the context of a parametric-style shape grammar interpreter consider the number of edges of a polygon as an invariant. When matching an irregular triangle of line segments, any triangle of line segments in the shape under investigation can be matched, irrespective of its shape. Similarly, when matching an irregular (convex) quadrilateral, any (convex) quadrilateral can be matched. Rather than considering the number of line segments as the basis for the matching process, we consider the number of distinct infinite line carriers of line segments. A triangle defines three line carriers, all non-parallel, intersecting in the three vertices of the triangle. Thus, considering the infinite lines and their intersection points as, respectively, the nodes and edges of a graph, allows to find any potential triangle as a subgraph composed of three nodes and three edges interconnecting these nodes.

In addition, we consider associations of perpendicularity and parallelism as additional constraints. Thus, in the process of matching a subgraph onto a graph, edges corresponding to perpendicular intersections are marked as such, constraining the matching of a right-angled triangle to only right-angled triangles while, oppositely, an irregular triangle will still match any triangle, including right-angled (Fig. 4).

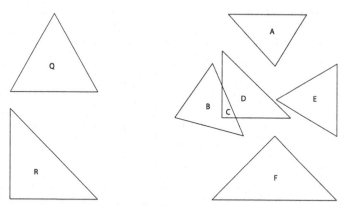

Fig. 4. Considering the figure of six triangles on the right, under a similarity transformation, the equilateral triangle Q only matches the equilateral triangle E, while the right-angled, isosceles triangle R only matches the similarly right-angled, isosceles triangle F. However, under the parametric-associative matching described in this paper, the right-angled triangle R matches every right-angled triangle in the figure, i.e., triangles C, D and F, while triangle Q matches every single triangle in the figure, i.e., A, B, C, D, E and F.

Note that when matching a single triangle or quadrilateral, an affine (in the case of a single triangle) or linear (in the case of a quadrilateral) transformation may be sufficient. While affine or linear transformations generally do not preserve (perpendicular) angles, and in the case of linear transformations, neither parallel lines, here too, constraints of perpendicularity and parallelism can be additionally considered, if present. However, parametric-associative matching easily extends to more complex figures (Fig. 5).

In the case of a quadrilateral, there are four line carriers with up to six intersection points (Fig. 6). In fact, parallel line carriers are considered to intersect at infinity, thereby ensuring that any two nodes in the graph are always linked by an edge. As such, the graph representing any shape of line segments is always complete. Therefore, it is not necessary to adopt a subgraph matching algorithm and, instead, a combinatorial enumeration of potential matches is opportune. Subgraph matching is necessarily nonpolynomial [9]; in comparison, a combinatorial enumeration, searching for k elements within a set of n elements, yields a tight bound of $O(nk)$. Depending on the size of k, this bound is exponential in the worst case (when k approaches n). When searching for a parallelogram, two of the edges in the graph must be marked as parallel, i.e., at infinity. This is also the case when searching for a rectangle, however, in addition, the other four edges must be marked as perpendicular. Distinguishing between convex and concave quadrilaterals is achieved by ordering the intersection points on the infinite line carrier. In the case of a convex quadrilateral, the other two intersection points that are not part of the quadrilateral lie outside of the boundary segments of the quadrilateral. Instead, in the case of a concave quadrilateral, these lie inside a boundary (Fig. 6). Furthermore, in the case of a self-intersecting quadrilateral, one intersection point, other than the four corners, will lie inside not one but two boundary segments, while the other intersection point will lie outside any boundary segment.

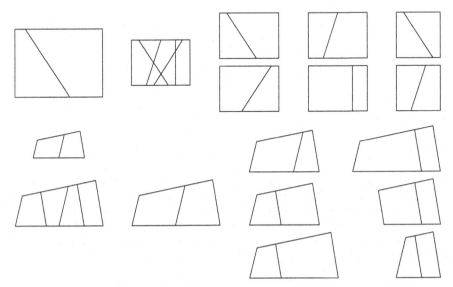

Fig. 5. Two examples of matching, beyond a triangle or quadrilateral, under a similarity transformation versus parametric-associative matching as described in this paper: (top, from left to right) the shape to be matched, the shape under investigation, two matches under a similarity transformation, four additional matches under parametric-associative matching; (bottom, from left to right) the shape to be matched (above—note the two parallel lines) and the shape under investigation (below), the only match under a similarity transformation, six additional matches under parametric-associative matching.

In the case of a convex pentagon, the five infinite lines define ten intersection points, and each line contains exactly four intersection points, including two corner points (Fig. 6). In most cases, the two corner points will lie in between the two other intersection points, but that is not always the case. Thus, considering the ordering of the intersection points too strictly would not allow us to match all convex pentagons with a given convex pentagon. For this reason, we only strictly match the ordering of the intersection points if on a line segment, but not otherwise.

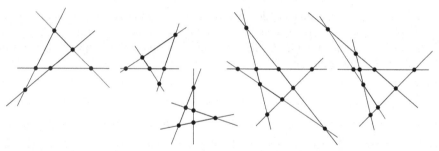

Fig. 6. The graph-based representation underlying parametric-associative matching considers all infinite line carriers as nodes and their intersection points as edges of the graph. Irrespective of the shape of a convex quadrilateral (two left-most figures), its graph is has the same form. Instead, the graph of a concave quadrilateral (bottom center) has a clearly different form. The same cannot be said about convex pentagons. However, ignoring the ordering of the intersection points that lie outside the line segments is sufficient to match among convex pentagons.

While in specific cases it may be difficult to predict the exact results under parametric-associative matching, the matching mechanism broadly follows the following steps (here explained for line segments, note that a similar mechanism applies to plane segments) (Fig. 7):

1. Identify all infinite lines carrying any line segment of the left-hand-side shape of the rule; this constitutes the source tuple
2. Identify all intersection points of lines in the source tuple and note whether the intersection is perpendicular or parallel (point at infinity) and whether the point falls inside, outside or is an endpoint of any line segment on either line; this constitutes the source graph
3. Identify all infinite lines carrying any line segment of the shape under investigation
4. Enumerate all combinations of lines from the shape under investigation that match the cardinality of the source tuple; these constitute the target tuples
5. For each target tuple, identify all intersection points of lines in the tuple and note whether the intersection is perpendicular or parallel (point at infinity) and whether the point falls inside, outside or is an endpoint of any line segment on either line; these constitutes the target graphs
6. Eliminate all target graphs that do not preserve parallelism and perpendicularity as specified by the source graph
7. Eliminate all target graphs that do not correctly match an inside edge for the source graph
8. Eliminate all target graphs where an edge that is an endpoint for the source graph is not matched with an edge that is either an endpoint or an inside point for the target graph
9. For the left-hand-side shape, identify all endpoints of line segments on the respective nodes and note their ordering also with respect to inside edges

10. Do the same for the shape under investigation and eliminate any remaining target graphs that do not correctly match the orderings (i.e., where two edges of the source graph are contained within a single line segment and the corresponding edges in the target graph are not)

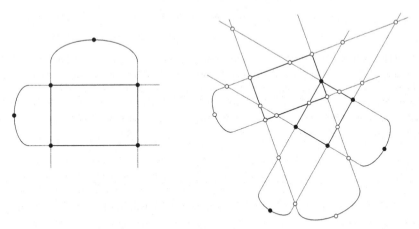

Fig. 7. Matching a rectangle (left) onto a figure including two rectangles (right), under a parametric-associative transformation. The source graph includes four nodes (corresponding the infinite line carriers) and six edges. Four edges correspond to the four vertices and are marked as perpendicular, the other two edges correspond to the two intersection points at infinity and are marked as parallel. One matching target graph is highlighted in the figure (edges denoted as black dots).

2.5 Combining Non-parametric and Parametric-Associative Matching

The two matching mechanisms here described use different geometric representations and *individual* classes, as such, they result in different *sortal* structures. More generally, different formalisms define different *sortal* structures and rules are defined to operate on a particular *sortal* structure, which is the *sortal* structure adopted to represent its left-hand-side and right-hand-side shapes. Obviously, combining different formalisms in a same design process requires the data to be transitioned from one *sortal* structure to another. More particularly, if we wish to use non-parametric and parametric-associative rules interchangeably (see Sect. 3 The SortalGI Sortal Grammar Interpreter), we need a mechanism to translate shapes from one *sortal* structure to another. Ultimately, *sortal* structures are intended to be defined by the user, whether an interface developer or, possibly, a designer, and there is no guarantee that the *sortal* structures defined for both matching mechanisms would be similarly composed. Nevertheless, it remains possible to map two differing *sortal* structures and automatically convert shapes between these structures, at least if two conditions are met.

The first condition is the definition of an equivalence relationship between *individual* classes underlying primitive *sortal* structures that are expected to be mapped. For example, there exist four *individual* classes for points, depending on whether the points are to be represented in 2D or 3D and whether they support a parametric-associative mapping or, instead, a non-parametric mapping under a similarity transformation (Table 2). It is obvious that in order to support the interchangeability of both types of rules, the two individual classes for 3D points (and similar for 2D points) need to be considered equivalent. It is less obvious whether this equivalence can be extended to include both 2D and 3D points. After all, while 2D points can be converted to 3D points, without any loss of data, the opposite does not hold true.

Table 2. Possible equivalence of *individual* classes for automatic conversion between *sortal* structures.

Equivalence class	3D *individual* classes	2D *individual* classes
points	point3D pointP3D	point2D pointP2D
lines	lineSegment3D lineSegmentP3D	lineSegment2D lineSegmentP2D
surfaces	planeSegment3D planeSegmentP3D	planeSegment2D planeSegmentP2D
circles	circle3D circleP3D	circle2D circleP2D
circular arcs	circularArc3D	circularArc2D
ellipses	ellipse3D ellipseP3D	ellipse2D ellipseP2D
elliptical arcs	ellipticalArc3D	ellipticalArc2D
curves	bezierCurve3D bezierCurveP3D	bezierCurve2D bezierCurveP2D

The second condition is to define a subsumption relationship over *sortal* structures. Previously, we have noted that *sortal* structures combine under compositional operations of sum, direct product, and attribution. The operation of sum (and direct product) specifies a disjunctive relationship between the elements of both structures. As such, a subsumption relationship (denoted '$\leq$') can be defined on *sortal* structures as follows:

$$a \leq b \Leftrightarrow a + b = b \qquad (2)$$

where a and b are any two *sortal* structures. It follows that a *sortal* structure composed under the operation of sum subsumes all its constituent structures:

$$a \leq a + b \qquad (3)$$

We can consider an extension of *sortal* subsumption to the operation of attribution (denoted '∧') as follows:

$$a \leq a \wedge b \tag{4}$$

This follows from (2), if we assume that:

$$a + a \wedge b = a \wedge 0 + a \wedge b = a \wedge (0 + b) = a \wedge b \tag{5}$$

where 0 is the identity element for the operation of sum, the operation of sum distributes over the operation of attribution and a and $a \wedge 0$ are considered identical (or equivalent).

Combining the equivalence of primitive *sortal* structures with the subsumption relationship on *sortal* structures, we could write that *sortal* structure a is convertible into b if a is equivalent to a' where a' is subsumed by b (note that a' may be equal to b). However, the actual conversion process may be complicated by the fact that a could be equivalent to both a' and a'' with both a' and a'' subsumed by b. This is possible even without considering 2D and 3D *individual* classes as equivalent. A primitive *sortal* structure is defined, syntactically, by its *individual* class and, semantically, by its name. As such, a *sortal* structure may contain the same *individual* class twice (or more), for example, one representing construction lines with the other one representing other than construction lines.

As such, when combining primitive *sortal* structures corresponding to different types of spatial elements under the operation of sum to support, respectively, non-parametric and parametric-associative rules, some discrepancy between the two composite *sortal* structures may be forgiven. Specifically, as (circular and elliptical) arcs are currently not supported under parametric-associative rules, the conversion from a parametric-associative *sortal* structure to a non-parametric *sortal* structure could be straightforward, on condition that no two primitive *sortal* structures within the same composite structure refer to the same *individual* class. The opposite conversion, from a non-parametric *sortal* structure to a parametric-associative *sortal* structure could result in information loss, if the former allows for circular or elliptical arcs.

Note that these equivalence and subsumption relationships have not been implemented in the SortalGI library yet. This remains as future work. However, in cases where the *sortal* structures are predefined, as is the case for the Grasshopper plug-in elaborated upon below, a conversion mechanism can be explicated rather than implied from equivalence and subsumption relationships.

3 The SortalGI Sortal Grammar Interpreter

The *sortal* grammar interpreter, named *SortalGI*, has been implemented in Python and made accessible within Rhino—as a code library with dedicated Rhino API—and Grasshopper—as a plug-in [14]. The SortalGI library supports both parametric-associative and non-parametric shape rules, operating on *sortal* structures including points, line and plane segments, circular and elliptical arcs, quadratic Bezier curves, labels, weights, colors, enumerative values, and descriptions, in both 2D and 3D. Its functionality is slightly restrained in Rhino due to the need to graphically visualize shapes. Especially, the API and plug-in predefine their *sortal* structures, thereby limiting the support for multiple formalisms but enabling the visualization of shapes in Rhino. In particular, each predefines two *sortal* structures, one for use with parametric-associative shape rules and the other for use with non-parametric shape rules. In the Grasshopper plug-in, ignoring any user-defined description components, the *sortal* structure for non-parametric shape rules is currently defined as:

$$point3D^{\wedge} pLabelD + lineSeg3D^{\wedge} (lLabelD + lColor)$$
$$+ planeSeg3D^{\wedge} (plLabelD + plColor)$$
$$+ circle3D^{\wedge} (cLabelD + cColor) + ellipse3D^{\wedge} (eLabelD + eColor)$$
$$+ arc3D^{\wedge} (aLabelD + aColor) + bezier3D^{\wedge} (bLabelD + bColor) \qquad (6)$$

Here, we present two examples relying upon different formalisms and using both non-parametric and parametric-associative rules, that are implemented using the SortalGI Grasshopper plug-in. The first example is, originally, a three-rule grammar [28]: one rule creates a square from an initial marker, another creates a rotated square inscribed within the original square, and a final rule removes the marker. The role of the marker, a point, is to guide the generation, by moving from one square to the next. Stiny [28] also considers an alternating infill of the squares in black and white, however without explicating the relevant rules. Figure 8 shows a set of four rules generating the inscribed, rotated squares with alternating black and white infill, using points, line segments, plane segments and gray-tone weights as primitive *sortal* structures.

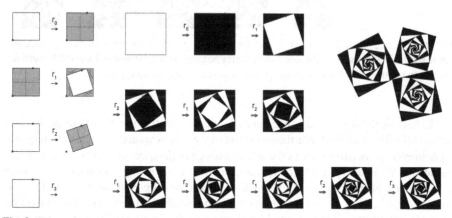

Fig. 8. Four rules generate a series of inscribed, rotated squares with alternating black and white infill. As the rules (left) are drawn in Rhino, they do not disclose their color scheme, unlike the Grasshopper results (middle and right). A single point serves as a marker to guide the generation.

Stiny [28] also shows a variant derivation based on a parametric rule set applied to quadrilaterals. Figure 9 illustrates the use of parametric-associative rules to achieve the same, or similar, including the alternating infill. The only difference is the fact that the inscribed quadrilateral always links the midpoints of the sides of the previous quadrilateral. This is a current limitation of parametric-associative matching.

The second example also adopts both non-parametric and parametric-associative rules, however, within a single derivation. Consider the figure on the right of Fig. 10, adopted for the Serpentine Pavilion in London (by Toyo Ito and Cecil Balmond). While it may seem rather complex, there is a relatively straightforward procedure behind the generation of this pattern. Starting from the outside square, each step takes the last generated square, scales it by a factor of about 0.84 and subsequently rotates it by about 35 degrees (rule r_1). Next, each resulting square has its sides extended until the sides of the original square (rules r_2 and r_3, depending on whether parallel sides extend to adjacent sides of the original square or to the same side). Finally, any partial sides extending beyond the original square are removed (rule r_5). In addition, labels "c" denote the current square and labels "o" the original square. Rule r_0 initializes a square by adding the labels, while rules r_4 and r_5 remove the labels "c" and "o", respectively.

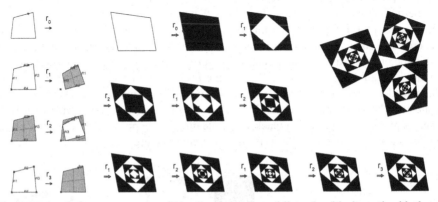

Fig. 9. Four rules generate a series of inscribed, rotated quadrilaterals with alternating black and white infill. Line segments in the rules' shapes may be tagged ('#…') to allow for textual directives to specify parametric relationships between new and existing spatial elements.

Extending the sides of any square until the sides of another square, where the first one occurs at different angles while the second one always appears at the same angle, requires a parametric-associative rule, as the angle between both squares can be considered as a parameter. Without explicating this parameter, the corresponding parametric-associative rule matches the inner square at any angle, as angles are not part of the associations automatically recognized. However, creating a parametric-associative rule to scale and rotate a square with fixed scaling factor and rotation angle is far from straightforward, as neither the scaling factor or the rotation angle is automatically deduced from the rule. Instead, the relationship between both squares would need to be expressed through textual directives specifying distances, angles and/or lengths. Fortunately, a non-parametric rule, implying a similarity transformation will apply the right-hand-side of the rule under the similarity transformation as determined from the matching of the left-hand-side, as such transformation will maintain the scaling factor and rotation angle, irrespective of the actual size and orientation of the square as it is matched. Figure 11 illustrates the derivation of the pattern from the initial square.

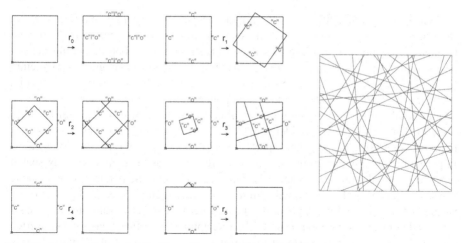

Fig. 10. Rules and pattern of the Serpentine Pavilion in London (by Toyo Ito and Cecil Balmond).

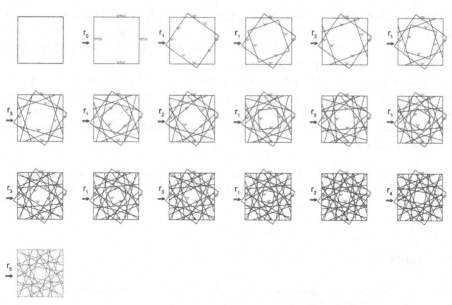

Fig. 11. Derivation of the pattern of the Serpentine Pavilion in London (by Toyo Ito and Cecil Balmond). The rules are shown in Fig. 8. This derivation combines non-parametric and parametric-associative rule applications, as part of the generative process is best captured under a similarity transformation (the iterative rotation and scaling of the square—r_1) while extending the sides of each transformed square until the boundary of the original square can only be achieved using a parametric-associative rule (r_2 and r_3).

When rules adopt a different matching mechanism, they operate on different *sortal* structures, resulting in the need for shapes to be translated from one *sortal* structure to another. As the plug-in prescribes the *sortal* structures for both matching mechanisms, conversion routines can be developed to translate shapes from one *sortal* structure to another. Only the spatial components of the *sortal* structures, and their non-spatial attributes, are prescribed. Descriptional components of the *sortal* structures are custom defined by the user. But as these are not affected by the choice of matching mechanism, the latter components are identical for both *sortal* structures, and no conversion is required. As a result, the plug-in supports the use of both types of rules interchangeably. In particular, the non-parametric *sortal* structure is used as the default and any spatial types, e.g., circular arcs, that cannot be represented within the parametric-associative *sortal* structure are only maintained within the default structure. Specifically, when converting data to allow for the application of a parametric-associative rule, any arcs are set aside and reintegrated into the result of the rule application, after it has been converted back to the non-parametric *sortal* structure. After all, any data not involved in the rule will not be affected or altered.

4   Conclusion

Among 'general' shape grammar interpreters, SortalGI offers the widest flexibility in shape representations and is the only one offering two distinct shape matching mechanisms. As such, it supports a multitude of shape grammar formalisms, including many examples from literature. Remaining limitations with respect to data kinds and matching mechanisms can be addressed by adding primitive *sortal* structures (in the form of *individual* and *form* classes) and matching mechanisms where appropriate. This ignores many idiosyncrasies found in examples of shape grammars in literature that have not been implemented or only partially. The endeavor to demonstrate their implementability remains as future work.

Acknowledgements. I would like to thank Bui Do Phuong Tung for his development work on the SortalGI shape grammar interpreter, and Bianchi Dy for her development work on a previous version of the SortalGI Grasshopper plug-in.

References

1. Stiny, G., Gips, J.: Shape grammars and the generative specification of painting and sculpture. Inf. Process. **71**, 1460–1465 (1972)
2. Stiny, G.: Introduction to shape and shape grammars. Environ. Plann. B. Plann. Des. **7**, 343–351 (1980)
3. Stouffs, R.: The algebra of shapes. Ph.D. thesis. Department of Architecture, Carnegie Mellon University, Pittsburgh, PA (1994)
4. Jowers, Y., Earl, C.: The construction of curved shapes. Environ. Plann. B. Plann. Des. **37**, 42–58 (2010)
5. Stiny, G.: Weights. Environ. Plann. B Plann. Des. **19**(4), 413–430 (1992)

6. Knight, T.W.: Color grammars: designing with lines and colors. Environ. Plann. B. Plann. Des. **16**(4), 417–449 (1989)
7. Knight, T.W.: Color grammars: the representation of form and color in designs. Leonardo **26**(2), 117–124 (1993)
8. Stiny, G.: A note on the description of designs. Environ. Plann. B. Plann. Des. **8**(3), 257–267 (1981)
9. Wortmann, T., Stouffs, R.: Algorithmic complexity of shape grammar implementation. Artif. Intell. Eng. Des. Anal. Manuf. **32**(2), 138–146 (2018)
10. Stiny, G., Mitchell, W.J.: The Palladian grammar. Environ. Plann. B. Plann. Des. **5**(1), 5–18 (1978)
11. Stiny, G.: Ice-ray: a note on the generation of Chinese lattice designs. Environ. Plann. B. Plann. Des. **4**(1), 89–98 (1977)
12. Wortmann, T.: Representing shapes as graphs. Master's thesis. Department of Architecture, Massachusetts Institute of Technology, Cambridge, MA (2013)
13. Grasl, T., Economou, A.: From topologies to shapes: parametric shape grammars implemented by graphs. Environ. Plann. B. Plann. Des. **40**(5), 905–922 (2013)
14. Stouffs, R.: Where associative and rule-based approaches meet: a shape grammar plug-in for Grasshopper. In: Fukuda, T., Huang, W., Janssen, P., Crolla, K., Alhadidi, S. (eds.) Learning, Adapting and Prototyping, vol. 2, pp. 453–462. CAADRIA, Hong Kong (2018)
15. Beirão, J.N.: CItyMaker: designing grammars for urban design. Ph.D. thesis. Faculty of Architecture, Delft University of technology, Delft, The Netherlands (2012)
16. Grasl, T., Economou, A.: From shapes to topologies and back: an introduction to a general parametric shape grammar interpreter. Artif. Intell. Eng. Des. Anal. Manuf. **32**, 208–224 (2018)
17. Correia, R.C.: GRAMATICA: a general 3D shape grammar interpreter targeting the mass customization of housing. In: Achten, H., Pavlíček, J., Hulín, J., Matejovská, D. (eds.) Digital Physicality, vol. 1, pp. 489–496. eCAADe, Brussels (2012)
18. Correia, R.C.: DESIGNA - A Shape Grammar Interpreter. Master's thesis. IST Técnico Lisboa, Universidade de Lisboa, Lisbon, Portugal (2013)
19. Economou, A., Hong, T.-C., Ligler, H., Park, J.: Shape machine: a primer for visual computation. In: Lee, J.-H. (ed.) A New Perspective of Cultural DNA. KRS, pp. 65–92. Springer, Singapore (2021). https://doi.org/10.1007/978-981-15-7707-9_6
20. Li, A.: A shape grammar for teaching the architectural style of the Yingzao fashi. Ph.D. thesis. Department of Architecture, Massachusetts Institute of Technology, Cambridge, MA (2001)
21. Duarte, J.P.: Customizing Mass Housing: A Discursive Grammar for Siza's Malagueira Houses. Ph.D. thesis. Department of Architecture, Massachusetts Institute of Technology, Cambridge, MA (2001)
22. Stouffs, R.: Implementation issues of parallel shape grammars. Artif. Intell. Eng. Des. Anal. Manuf. **32**(2), 162–176 (2018)
23. Krishnamurti, R.: The arithmetic of shapes. Environ. Plann. B. Plann. Des. **7**, 463–484 (1980)
24. Krishnamurti, R.: The construction of shapes. Environ. Plann. B. Plann. Des. **8**, 5–40 (1981)
25. Stouffs, R., Krishnamurti, R.: Algorithms for classifying and constructing the boundary of a shape. J. Des. Res. **5**(1), 54–95 (2006)
26. Jowers, Y., Earl, C.: Implementation of curved shape grammars. Environ. Plann. B. Plann. Des. **38**, 616–635 (2011)
27. Krishnamurti, R., Stouffs, R.: Spatial change: continuity, reversibility, and emergent shapes. Environ. Plann. B. Plann. Des. **24**, 359–384 (1997)
28. Stiny, G.: Computing with form and meaning in architecture. J. Archit. Educ. **39**(1), 7–19 (1985)

Understanding the Span of Design Spaces
And Its Implication for Performance-Based Design Optimization

Likai Wang[(⊠)] [iD]

Nanjing University, Nanjing 210093, Jiangsu Province, China
wang.likai@nju.edu.cn

Abstract. This paper presents a study investigating the impact of design spaces on performance-based design optimization and attempts to demonstrate the relationship between these two factors through the lens of the span of design spaces. The study defines the span of design spaces as the variety of different types of building design that can be embodied by the parametric model; thus, the wider the span, the more likely is the optimization to identify promising types of building design. In order to reveal the relationship between the span of design spaces and performance-based design optimization, the study present a case study that includes design spaces with various spans within a building design optimization problem considering daylighting performance. The result shows that the difference in span can result in significant changes in optimization in relation to fitness and architectural implications.

Keywords: Computational design optimization · Performance-based design · Design spaces · Design exploration · Parametric design

1 Introduction

Applying performance-based design optimization to early-stage architectural design has been considered a promising and efficient approach to assisting architects in design exploration and overcoming data-poor situations. By integrating parametric models, building performance simulation tools, and optimization algorithms, performance-based design optimization can evolve a population of design variants aiming at various performance objectives and identify high-performing solutions. Moreover, by systematically generating and examining a large number of design variants, the optimization process can be also considered a design exploration process that is driven by various building performance goals, which can further offer valuable information and insight to assist architects in their early-stage decision-making and design synthesis [1].

When using performance-based design optimization for architectural design exploration, a critical step is to develop a parametric model for design generation. As the representation of a design scheme, parametric models delineate a design space for the optimization process to search for desirable high-performing solutions. Design spaces encompass all the design variants that can be generated by the parametric model. Meanwhile, the underlying relationship and property of these design variants also characterize

© Springer Nature Singapore Pte Ltd. 2022
D. Gerber et al. (Eds.): CAAD Futures 2021, CCIS 1465, pp. 288–297, 2022.
https://doi.org/10.1007/978-981-19-1280-1_18

the 'shape' of design spaces which can, in turn, have a great impact on the result of the optimization [2]. While certain studies have been carried to investigate the structure of design spaces [3], there is relatively little research on the issue of how design spaces affect performance-based design optimization. Insufficient attention to these two aspects and the relationship between them may compromise the efficacy of performance-based design optimization in architecture, particularly for early-stage design exploration and information extraction.

1.1 The Shape of Design Spaces

In order to understand the relationship between design spaces and performance-based design optimization, the author conceptualized the shape of design spaces and exemplified how design spaces can be iteratively reshaped to facilitate the optimization to identify more favorable or desirable solutions in several preceding studies [2, 4, 5]. These studies highlight that poor-shaped design spaces, for example too small or there are too many invalid design variants, often lead to a failure of performance-based design optimization for identifying legitimate and/or high-performing solutions. These results underline that a better understanding of the characteristics of the design space is a key component of computational design thinking. For designers, conscious awareness of the shape of design spaces can help them to avoid the ill-usage of performance-based design optimization and achieve more desirable results in practice.

The preceding studies investigated how different geometrical modeling approaches [5] and constraint handling strategies [6] can be applied to parametric modeling and refine the design space to facilitate the optimization to deliver better results. These parametric modeling strategies were respectively applied to a specific type of building design, including a high-rise building with sky gardens and a low-rise building with a central courtyard. While the structure of the design space was altered by using these strategies, it is noteworthy that, from the architectural perspective, the boundary of the design space remained unchanged since there were no extra building types included or excluded from the design space by these strategies. Given that the boundary of the design space was unchanged from the architectural perspective, this study defines *depth* as the major characteristics that can cognitively differentiate these design spaces. In this regard, it is also pertinent to inquire into the opposite dimension of design spaces, the *span*, in order to gain a holistic perspective on the shape of design spaces.

For clarifying the differentiation between the depth and span of design spaces, this study refers the depth of the design space to the range of design variation that a parametric model can produce with respect to a specific building type. The deeper the design space, the more accurate and high-performing design variants of the given type of building design can be found by the optimization process. As opposed to that, the span of design spaces can be perceived as the range of design variation that a parametric model can produce with respect to different types of building design. The wider the design space, the more likely is the optimization to identify promising types of building design. In performance-based design optimization, neither the depth nor the span should be ignored. However, considering the *breadth-first-depth-next* strategy for early-stage architectural design [7], it should be more essential to explore a wide range of building types to identify a feasible one than to narrow down the optimization search to a specific type of building design at the outset of the design.

When it comes to the depth and span of design spaces, a similar notion closely related to these two aspects was also discussed by Sheikholeslami a decade ago. Sheikholeslami defined "design alternatives" and "design variations" to distinguish between design variants that can display structural design differentiation and those that cannot [8]. Even so, the issue of design spaces and design variability still receives little attention from the research community as well as computational designers. Thus, following the preceding studies focused on the depth of design spaces, this study delves deeper into the shape of design spaces with the aim of looking at the design space through its 'span' and examine how the span of design space affects performance-based design optimization.

1.2 Paper Overview

For demonstrating the idea of the span of design spaces, the study presented by this paper conducts a series of performance-based design optimization runs fed by design spaces with different spans and compares the optimization results regarding its fitness and architectural implications. To create design spaces with different spans, a generic building massing design generative algorithm is used. The algorithm can be customized to include and exclude certain types of building design in the design space, thus resulting in the change in the span of design spaces [9, 10]. Additionally, the study uses daylighting as a driving factor in the performance-based design optimization runs and investigates how design spaces will affect the way that the optimal building design responds to the goal of greater daylight accessibility for the indoor space.

As the study is focused on the relationship between the design space and performance-based design optimization, issues related to optimization and simulation are out of the scope of this study. Hence, all the optimization and simulation setups adopted in this study are based on prior studies [9, 11], and the detail of optimization and simulation will not be elaborated in this paper. At the same time, although daylighting is used as the driving factor in the conducted optimization run, the main objective of the study is not to discuss how building massing should be optimized for better daylighting. Instead, the study aims to demonstrate how the change of design spaces can affect the result of design optimization as well as the architectural implication revealed by the optimization result, which is also of great significance for the optimization focused on any other performance factors such as solar irradiation or sunlight-hour.

2 Method

This study uses a generic building massing design generative algorithm that is encoded using the subtractive form generation principle and creates building massing design by removing several parts from a predefined box-shaped volume [9, 10]. Thus, the design generation approach can produce building massing design with different architectural features in relation to various building types such as courtyards, stilts, cascading roofs/terraces. Designers can tailor these features using the parameters and constraints within the algorithm, such as the number of voids being made or the constraint on the positions where the voids are allowed to appear in the building massing. By changing these parameters and constraints, different subsets of the possible building massing design variants embodying different building types can be included in or excluded from the design space, which, thereby, differentiates the span of the design space.

2.1 Design Generation

In this study, two parameters and one constraint are applied to control and differentiate the architectural features appearing in the generated design. The two parameters correspond to two types of voids that are created within a box-shaped volume in the generative algorithm (Fig. 1). The first type of voids is aimed to create vertical holes in the initial volume, which creates features such as atriums or courtyards. The second type, on the contrary, generates horizontal voids that create features such as stiles or cascading roofs/terraces. These two types of voids can be controlled by numbers. With different numbers of voids, the generated design embodies different types of building design. For example, when there are zero horizontal voids and more than one vertical void, only building types with atriums and courtyards can be generated.

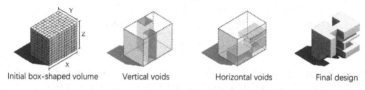

Initial box-shaped volume   Vertical voids   Horizontal voids   Final design

Fig. 1. The elements used in the generative algorithm

For the constraint applied to the design generation, it determines whether the vertical void is allowed to break the boundary of the initial volume. If the constraint is enforced, all vertical voids are kept inside the initial volume in the horizontal direction and create enclosed courtyards or atriums. Contrarily, if the constraint is disabled, the vertical voids are allowed to break the boundary of the initial volume and create a side-opening void. Such voids can result in significant differentiation in architecture because once the boundary is broken, the initial rectangle-plan building volume can be turned into a building with an L-shaped or U-shaped footprint or two separate buildings.

2.2 Creating Design Spaces with Different Spans

With the use of the generative algorithm, design spaces with different spans can be obtained by setting different combinations of parameters and constraints, which allows or disallows certain types of building design to be generated. In this study, eight different setups (*Setup 1* to *Setup 8*) are defined with different unique combinations of the two parameters and the constraint to create the same number of design spaces (Table 1) that are referred to as *Design Space 1* to *Design Space 8*. These setups are decided on the basis of previous studies [9, 10] on the one hand. On the other, it is also attempted to differentiate the number of architectural implications reflected by each of the design spaces. Hence, these setups consider different numbers of the two types of voids under one of the two boundary constraint conditions. While there are other parameters and constraints that can control the generated design, exhausting all possible combinations of parameters and constraints is time-consuming and also not the major concern of this study.

Table 1. Parameter setups

Parameter setups	Vertical voids	Horizontal voids	Boundary constraint
Setup1	3	4	●
Setup2	2	3	●
Setup3	3	4	○
Setup4	2	3	○
Setup5	5	0	●
Setup6	3	0	●
Setup7	5	0	○
Setup8	0	5	N/A

These setups are applied to a middle-rise non-residential building with 20 by 10 column grids in the x- and y- directions and 10 floors in the z-direction. In addition, 85,000 m^2 is set as a target gross area for the generated design. Figure 2 shows groups of random sampling designs generated by using these setups. Although not all of these designs look rational and architecturally interesting, meaningful design differentiation in terms of passive design strategies can be discerned among these designs. Note that, in order to increase design diversity, the number of voids actually defines the maximum number of voids that can appear in the massing, and when two voids are close to each other, they will be merged into one according to a rule encoded in the algorithm [9].

Corresponding to Table 1, the first four setups allow both types of voids to appear in the massing under the two boundary constraint conditions. As a result, these design spaces feature a rich diversity of building massing design variants and encompass design variants displaying various mixtures of architectural features, such as courtyards, atrium, and stilts. A major difference among these four setups is the number of voids, which differ the geometrical complexity of the design variants in these design spaces. Higher numbers of voids defined by Setups 1 and 3 typically produce design variants with more complex configurations or interlocks of different architectural features than those produced by Setups 2 and 4. At the same time, the disabling of the boundary constraint makes Setups 3 and 4 produce building design with different footprints, which allows the optimization to explore building design beyond those with a rectangle footprint as generated by Setups 1 and 2.

The rest four setups allow only one type of voids to appear in the massing, which creates design variants only embodying a smaller number of passive design strategies and their combinations. These setups are aimed to create design spaces with a relatively narrow span as compared with the first four design spaces, and these design spaces can also be perceived as different subspaces of the Design Spaces 1 to 4. As shown in Fig. 2, as the vertical void can create greater design variability from the architectural perspective, more weight is placed on the parameter setups for this type of void. Hence, three setups related to vertical voids are defined with different numbers of voids as well as different boundary constraint conditions. For the horizontal voids, only one setup is defined to create a design space containing the design variants with stilts and/or cascading roofs but without any atrium and interior courtyard.

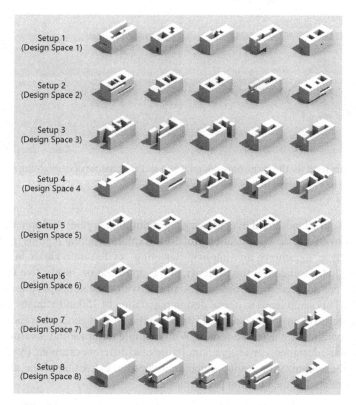

Fig. 2. Random sampling design based on the parameters setups

While all these design variants still bear marked similarity owing to the use of the same generative algorithm, the differentiation among these design variants is expected to result in significant impacts on building performance. It is because these variants represent different families of building massing designs with unique combinations (numbers and positions of voids) of various passive design strategies. The effectiveness of these passive design strategies on building performance improvement can be enhanced or hampered as the result of proper or incompatible combinations of strategies.

2.3   Case-Study Optimization

In the case study, the optimization is set to optimizing the daylighting performance of the building design in a predefined urban environment located in the city of Nanjing, China (Fig. 3). The target building is surrounded by several high-rise and middle-rise buildings, which results in a challenging design condition for achieving favorable daylight accessibility. The optimization is carried out using SSIEA [11], a hybrid evolutionary algorithm; and the daylighting evaluation is conducted using DIVA. The performance indicator is calculated primarily on the basis of Spatial Daylight Autonomy (sDA). At the same time, the indicator is also punished by the difference between the gross area of the generated design and a predefined target value (85,000 m^2) in order to exclude design variants without appropriate functional feasibility.

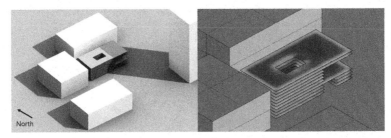

Fig. 3. The site used for the case-study optimization and an example of daylighting simulation (a type of strip window is used in this case study)

SSIEA is an island-based evolutionary algorithm that applies multiple design subpopulations to enable the optimization process to explore different regions in the design space as well as to achieve an "implicit clustering" of the individuals in the design population [11]. To increase the diversity in the optimization result, five design subpopulations are adopted in the optimization runs, with each subpopulation having 30 individuals. This setup leads to a total design population of 150 individuals. For the termination criteria of the optimization, 3150 iterations of design generation and evaluation are used to obtain more accurate results. With multiple design solutions from different subpopulations, designers can analyze similarities and differences among these solutions offered by the optimization result, which facilitates the extraction of the architectural implication related to the building performance.

3   Result

In regards to the result of the case-study optimization, the best solution (elite individual) in each of the five design subpopulations is reinstated as the representative of the optimization result. Figure 4 shows the result from the case-study optimization. As architectural feasibility is not the major concern of this study, the rationality of the generated design will not be discussed.

Comparing with Fig. 2, the differentiation among the design generated by different setups is amplified through the optimization process as the optimization pushes the search to the edge of the design space. In general, a shared optimization tendency is approaching the design with a large spacing by courtyards or atriums to the adjacent buildings on its west and the north sides as displayed in Fig. 3. This tendency reveals an important implication of the design problem that these surrounding buildings significantly undermine the daylight accessibility of the target building, particularly for the area near the north and west facades.

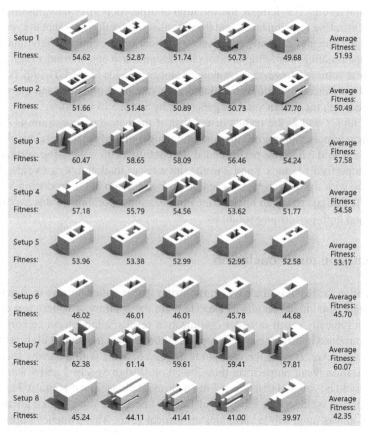

Fig. 4. The results of the optimization (due to the penalty function, the fitness proximately indicates the percentage of indoor area that can receive sufficient annual daylight)

Among these optimization results, the one based on Setup 7 which defines 5 unconstrained vertical voids stands out, followed by Setups 3 and 4 which combine several unconstrained vertical voids and horizontal voids. This indicates that breaking the boundary of the initial volume and creating side-openings can widen the design space by including desirable building design that is positive for daylighting performance.

Within these three setups, despite higher architectural diversity, combining horizontal voids with unconstrained vertical voids as Setups 3 and 4 does not produce more advantageous design variants for daylighting performance comparing with Setup 7. In reverse, the wider span of Design Spaces 3 and 4 (with more possible combinations of courtyards/atriums and stilts/cascading roofs) results in difficulty for the optimization process in identifying the high-performing design similar to the one, derived from the optimization based on Design Space 7, without horizontal voids.

In contrast, Setup 8, defining 5 horizontal voids, and Setup 6, defining 3 constrained vertical voids, make the optimization yield the lowest average fitness among all these optimization results. For Setup 8, the variation of different horizontal voids only creates stilts and cascading roofs/terraces, and the optimization result shows that the variation of these design strategies can only produce a minimum improvement in daylight accessibility for the building. For Setup 6, most elite designs have one internal atrium enclosed by the initial volume. However, the smaller number of voids (3 voids) hampers the optimization from detecting high-performing design variants with larger and more atriums if compared with the design variants generated by Setup 5 (5 voids).

For Setups 1 and 2, the results reveal that, if the boundary constraint is enforced, combining internal atriums with stilts and cascading roofs only makes a relatively limited improvement in daylighting performance. In contrast, with only using vertical voids, Setup 5 outperforms Setups 1 and 2 even with a fewer total number of voids. Thus, in consistence with the difference between Setup 7 and Setup 3/4, introducing horizontal voids to the design space can mislead the optimization algorithm for discovering more accurate and high-performing solutions.

4   Discussion and Conclusion

To summarize the result of the case study, design spaces with a narrow span commonly leads to under-optimized results as expected. On the contrary, a wider span of design spaces can typically result in better optimization results, although, in certain cases, it can also introduce interference for the optimization process to identify high-performing solutions. However, even in the case study, Design Space 7 with a narrow span makes the optimization result outperform those based on Design Spaces 3 and 4 with a wider span, it should be stressed that Setup 7 has a close relationship with Setups 3 and 4, and Design Space 7 is, indeed, a subspace of Design Space 3 or 4 but with a greater depth because it allows more subtle variations on vertical voids. This tendency implies that it is possible to extrapolate from the design space 3 or 4 to Design Space 7 if the architect can understand the implication reflected by the solution found in Design Space 3 or 4 and further focus the optimization on this implication by narrowing down the span of the initial design space.

The above analysis highlights the importance of defining a design space with a wider span at the outset of design for exploration because it is a key premise that the architect can discover promising design subspace and continue to search for a more optimized solution by narrowing down the span of the design space while increasing the depth. In contrast, if the design process starts with a narrow design space, such as Design Spaces 6 and 8, the optimization is unlikely to provide information to help architects obtain a comprehensive understanding of the design space and the design problem and, thereby, it is also less likely to deduce the direction for constructing a design space capable of producing the desirable high-performing solution.

To conclude, this study investigates the impact of design spaces' span on performance-based design optimization and proves that the span of design spaces can play a decisive role in early-stage design exploration. The differences in the architectural implications revealed by the optimization result can navigate the subsequent design process toward entirely different directions. As a result, it determines whether the architect can continuously discover more optimized designs by reflecting on the optimization result or is misled to the design direction toward under-optimized solutions. In regards to future research, more systematic and quantitative investigation on the span and depth of design spaces is needed. In addition, it is also pertinent to establish a cognitive framework that facilitates architects to understand and, thereby, better manipulate the design space to obtain more desirable optimization results.

References

1. Wang, L., Chen, K.W., Janssen, P., Ji, G.: Enabling optimisation-based exploration for building massing design: a coding-free evolutionary building massing design toolkit in rhino-grasshopper. In: RE: Anthropocene, Design in the Age of Humans - Proceedings of the 25th CAADRIA Conference, pp. 255–264 (2020)
2. Wang, L., Janssen, P., Ji, G.: Reshaping design search spaces for efficient computational design optimization in architecture. In: Proceedings of the 10th International Conference on Computational Creativity, ICCC 2019 (2019)
3. Woodbury, R.F., Burrow, A.L.: Whither design space? AIE EDAM Artif. Intell. Eng. Des. Anal. Manuf. **20**, 63–82 (2006). https://doi.org/10.10170S0890060406060057
4. Wang, L., Janssen, P., Ji, G.: Efficiency versus effectiveness: a study on constraint handling for architectural evolutionary design. In: Learning, Prototyping and Adapting - Proceedings of the 23rd CAADRIA Conference, pp. 163–172 (2018)
5. Wang, L., Janssen, P., Ji, G.: Progressive modelling for parametric design optimization. In: Haeusler, M.A., Schnabel, T.F. (eds.) Intelligent and Informed - Proceedings of the 24th CAADRIA Conference, pp. 383–392 (2019)
6. Wang, L., Janssen, P., Ji, G.: Utility of evolutionary design in architectural form finding: an investigation into constraint handling strategies. In: Gero, J.S. (ed.) DCC 2018, pp. 177–194. Springer, Cham (2019). https://doi.org/10.1007/978-3-030-05363-5_10
7. Akin, Ö.: Variants in design cognition. In: Design Knowing & Learning Cognition in Design Education, pp. 105–124. Elsevier (2001)
8. Sheikholeslami, M.: Design space exploration. In: Woodbury, R. (ed.) Elements of Parametric Design, pp. 275–287. Routledge, Abingdon (2010)
9. Wang, L., Janssen, P., Chen, K.W., Tong, Z., Ji, G.: Subtractive building massing for performance-based architectural design exploration: a case study of daylighting optimization. Sustain. **11**, 6965 (2019). https://doi.org/10.3390/su11246965
10. Wang, L., Chen, K.W., Janssen, P., Ji, G.: Algorithmic generation of architectural massing models for building design optimisation: parametric modelling using subtractive and additive form generation principles. In: RE: Anthropocene, Design in the Age of Humans - Proceedings of the 25th CAADRIA Conference, pp. 385–394 (2020)
11. Wang, L., Janssen, P., Ji, G.: SSIEA: a hybrid evolutionary algorithm for supporting conceptual architectural design. Artif. Intell. Eng. Des. Anal. Manuf. **34**, 458–476 (2020). https://doi.org/10.1017/S0890060420000281

Architectural Automations and Augmentations: Fabrication

Adaptive Toolpath: Enhanced Design and Process Control for Robotic 3DCP

Luca Breseghello and Roberto Naboni[✉]

Section for Civil and Architectural Engineering, CREATE - University of Southern Denmark, Odense, Denmark
ron@sdu.dk

Abstract. The recent advances in 3D Concrete Printing (3DCP) greatly impact architectural design, highlighting the disconnection between digital modelling and manufacturing processes. Conventional digital design tools present limitations in the description of a volumetric object, which is constrained to the definition of idealized external boundaries but neglect material, textural, and machinic information. However, 3DCP is based on material extrusion following a programmed toolpath, which requires custom modelling and methods to anticipate the manufacturing results. The paper presents the development of an interactive tool for the preview of 3D printing toolpaths within Grasshopper. Through an experimental campaign and analysis of material results, we integrate geometric, physical and design feedback within the design process. The accurate control of manufacturing variables such as printing speed and dimensions of the layers, together with the simulation and visualization of the results, makes them design parameters, opening to new formal and structural articulations in the design process. The developed instruments are tested on full scale printed prototypes, where their precision is demonstrated. Integrated into a 3DCP-specific design framework, the overall approach contributes to closing the existing gap between the digital environment and fabrication procedures in the construction industry.

Keywords: Toolpath planning · Live-physics simulation · 3D concrete printing · Robotic fabrication

1 Introduction

3D Concrete Printing (3DCP) is a rapidly expanding field both in research and in construction practice. The newly developed technology offers disruptive potential for concrete architecture, promising geometric freedom, reduced material consumption, automation, and a shorter construction chain. Due to its recent introduction, most of its features are yet to be fully exploited, mainly due to material limitations [1], and to a process that is still experimental [2]. In 3DCP, both the manufacturing resolution and the geometrical features of printed artefacts are subject to limitations. This is due to material and technological constraints: (i) the reduced mechanical properties of concrete at its semifluid state make it challenging to print overhanging geometries; (ii) the requirement for a continuous toolpath due to the lack of technological advances in most extrusion

© Springer Nature Singapore Pte Ltd. 2022
D. Gerber et al. (Eds.): CAAD Futures 2021, CCIS 1465, pp. 301–316, 2022.
https://doi.org/10.1007/978-981-19-1280-1_19

systems; (iii) the high dependency of the printed results on the specific material mix and environmental characteristics. Because of these restraints, the design for 3DCP has been severely limited to simplistic outcomes, which are poorly delivered on the premises of highly controllable manufacturing. While most of the current research is dedicated to fundamental material developments and technological advancements [3], little literature exists to date regarding modelling and simulation tools that allow the accounting of the material complexities during the design phase.

This research is aimed at developing a computational tool for interactive 3D modelling, simulation, visualisation and robot-code generation for robotic 3D Concrete Printing (3DCP) [4]. This tool is developed for the parametric environment of Grasshopper for Rhinoceros, and allows to (i) predict accurately the geometric characteristic of a printed concrete filament (Fig. 1); (ii) simulate the printing results of complex material behaviour, taking into account the effect of gravity and the effect of its viscosity; (iii) visualize complex printed formations and generate 3D models for further numerical analysis.

Fig. 1. Robotic 3D Printing process of test specimens for the analysis of printing parameters.

2   Related Research

Every layered AM process is characterized by the discretization of a digital 3D model into layers to generate a machining tool path. This operation is usually performed by extracting horizontal slices from a model at predefined height intervals. Standard 3D printing is performed by using external software to the design environment, namely slicer, which provides the fundamental printing information with a specific machine configuration. Well-established Fused Deposition Modelling (FDM) rely on slicing software, e.g. Cura [5], Simplify3D [6], Slic3r [7], which provides control on a large set of parameters for an automated setup and tuning of the slicing and printing phases, including the visualization of the manufacturing process. The numerous options available allow, for example, to approach the manufacturing of very different objects, use diverse materials, and tailor the printing setup to fit specific needs.

In 3DCP the effects of material and scale affect the printing results in different ways, and their inter-relationships are often the cause of unforeseen failures due to limited dimensional accuracy or mechanical performance [8, 9]. Consequently, 3DCP requires custom design strategies for specific manufacturing and material processes [10, 11]. Recent research has investigated developing software for controlling the printing process with different approaches. The first group of works aim at planning the printing process through a user-friendly slicer similar to FDM software: the Dutch-based company Vertico developed an online slicer, Slicer XL [12], which slices in horizontal layers an uploaded file in.stl format providing a visualization of the object, the motion coordinates and optionally a GCode or RAPID code; RAPCam Concrete [13], developed by RAP Technologies, works interactively with Rhino to analyze, simulate and visualize a printing toolpath, providing collision-detection, overhangs and stability verifications. The work carried out by Comminal et al. [14] exploits the Computational Fluid Dynamics (CFD) software Flow3D and validates the developed model through an experiment that compares the section of the simulation and performed tests. Another group of works looked into the integration of FE Analysis as a verification method for early-age behaviour of the printing process: Wolfs et al. [15] developed a numerical model to analyze the mechanical behaviour of early age 3D printed concrete applying the Mohr-Coulomb theory with time-dependent development of material properties and using the commercial Finite Element (FE) software Abaqus for simulation; in similar perspective, Ooms et al. [16]. Vantyghem et al. [17] developed a parametric tool in the form of a plugin -CobraPrint- for Grasshopper in Rhino that creates a FE mesh to be exported and analyzed in Abaqus for the simulation of the structural behaviour of 3D printed fresh concrete. A buckling simulation of the early age behaviour of the concrete during the printing process has been developed in Karamba within Grasshopper by Vos et al. [18]. However, most of these tools require high computing power, integration of software external to the design environment, or otherwise lack an accurate calculation and visualization of the material and layer behaviour able to provide an agile visualization of the print toolpath for controlled high-resolution printed objects.

3 Methods

An accurate representation of the 3DCP process and results requires a series of data to be gathered, analysed, and integrated. The dimension of the section of the printed filament is determined by a series of design and fabrication parameters. The paper presents a first experiment that investigates the relation between printing speed, layer height and width. The gathered data is then interpolated and implemented into a computational framework that provides a geometric representation of the printed material. Finally, a physics-based simulation is implemented within the same design environment, providing an understanding of the material behaviour and its impact on the design outcome.

3.1 Fabrication Setup

The experiments were performed using the 3DCP robotic fabrication facility of CREATE Lab at the University of Southern Denmark, consisting of an ABB 6650S industrial robot, a control unit, and an extrusion system composed of a conveying pump and a printing head with a circular nozzle diameter of 25 mm (Fig. 2). The employed cementitious material is based on a commercial product for shotcrete with aggregates with a particle size of 2 mm. Admixtures are added to control the curing time and a small dosage of polypropylene fibres. The system is based on a batch-mixing process, where the accelerating admixture is added to the mixer. This process provides a constant material composition but defines a characteristic printing timeframe where the material has mechanical resistance for buildability and the capacity to flow through the pumping system. All the experiments are run adding to the premixed powder 16.5% of water, 1.5% of an accelerator with low content of calcium-chloride and 0.01% of fibres.

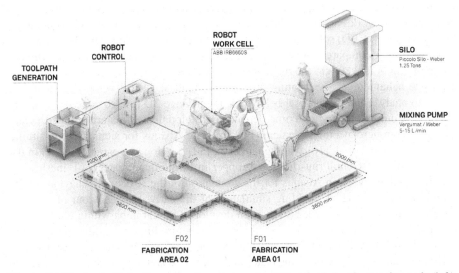

Fig. 2. 3DCP robotic fabrication setup: the robot work cell is provided with a motion code (left) and with a material feedstock (right); the setup has two fabrication areas F01 and F02.

3.2 Filament Calibration Experiment

The capacity to control the printing parameters with specific material characteristics is crucial for a successful design and fabrication with 3DCP. Therefore, a filament calibration experiment was run to investigate the resultant width W of a 3D printed layer and to calibrate a digital model for given ratios of printing speed PS and layer height H. These are the parameters defining the section and the volume of a printed element, assuming a fixed extrusion nozzle dimension, constant pump pressure p and material density, hence a constant material flow.

The experiment is based on a 500 × 500 mm custom-designed specimen, devised as a continuous print path with 100 mm spacing and five layers in height. The print path is run in an alternated fashion, i.e. the endpoint of one layer becomes the beginning of the following one with a vertical movement in height. The specific specimen design provides multiple measuring points where the speed is constant, the possibility to evaluate the influence of the movement accelerations and decelerations in longer and shorter straight segments as well as the material behaviour in the corners. For each specimen, six measurement points M_x are in the middle of each long segment, where the motion speed is constant. In addition, supplementary measures are taken to address the influence of acceleration on the robot speed. These are the points S_x, in the middle of the four short segments, and C_x, taken 30 mm from the corners on the long segments (Fig. 3).

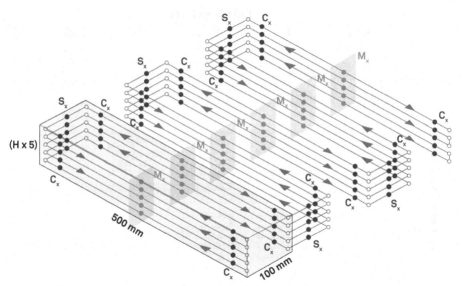

Fig. 3. Custom-design test toolpath for the evaluation of the numerical relationship between printing speed, layer height and width; each specimen is measured at the midpoint of the long segments (M_x), at the extremities of the long segments (C_x), and in the midpoint of the short segments (S_x) to account for the influence of acceleration on the printing speed.

The tested ratios between printing speed and height were chosen with a preliminary printing test. A slow movement combined with a small layer height provides a wider section, which, if excessive, creates irregularities and uncontrolled behaviour of the deposited material. On the other hand, a fast movement and high layer reduce the filament section, with the risk of non-constant extrusion and weak adhesion between the layers. The test is performed on a matrix of printing speeds between 150 mm/s to 600 mm/s with 150 mm/s steps and layer heights from 5 mm to 25 mm with 5 mm steps, which results in 20 printed specimens.

Calibration Analysis. The print of the specimens showed (Fig. 4): *Specimen 5–150* and *5–300* produced an excess of material and an inconstant section; *Specimen 15–150* and *25–150* presented a partial collapse; similarly, specimens 15–300, 20–300, 15–450 produced a full plastic collapse during printing, due to the small printed layer width; *Specimen 10–600* had an inconstant section to the excessive printing speed. A set of specimens were not printed successfully, as the amount of material deposited was insufficient to produce a consistent layer. From a preliminary visual inspection, the print quality is lower when the speed and layer height are low. This is because an excessive amount of material is deposited and accumulated in an uncontrolled fashion; on the other hand, layers smaller than the extruder diameter present a discontinuous material distribution. The influence of the acceleration on the motion speed is visible on the specimens printed at higher speed rates. Measurements of the printed specimens are taken with an electronic calliper with tolerance ±0.01 mm and verified through 2D scanned sections.

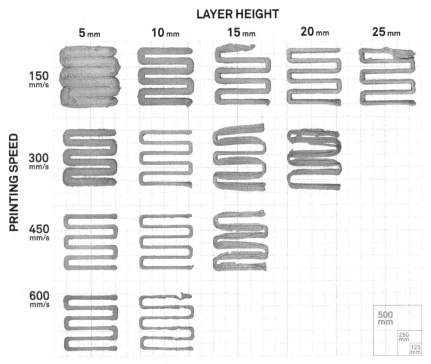

Fig. 4. Matrix of printed specimens with variable proportions of printing speed and layer height; the resultant printing width and the printing quality are evaluated.

Six sections are cut for every specimen at the middle points M_x of each long segment. These are then scanned, and the outer profile is obtained through an automated image processing routine. The sections of each specimen are then measured, overlapped, and analysed to verify the consistency between them and to formulate a geometric construction of a prototypical layer shape and dimension (Fig. 5).

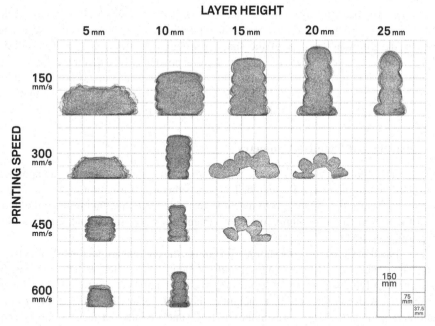

Fig. 5. Matrix of overlapped sections M_x of specimens with variable proportions of printing speed and layer height; the section dimension, their consistency, and the shape and characteristics of the layers are analysed.

Six vertices v are defined and located for each layer: four corners and the two outermost points along with the external profiles. An average for each of these points is used to determine the proportion of the average layer given a defined layer height and printing speed (Fig. 6). As the relation to the printing plane influences the dimensions of the first layer, and the last layer is not pushed from a layer above, these are disregarded from the analysis.

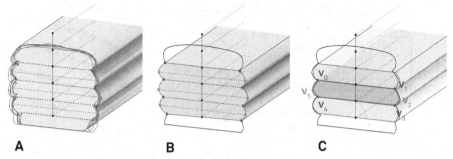

Fig. 6. Layer section analysis through (a) significant vertices v detection; (b) average specimen layers; (c) average layer.

As observed from the layer outline analysis, the specimens with high speed-to-height ratios present an irregular profile, both in the width of the layers in each cut section and between the different sections of the same specimen. The *Specimens 5–150* and *5–300* have an irregular profile and very little correspondence between the different cut sections, which prevent from calculating the shape of the reference layer; on the other hand, *Specimens 10–600, 15–300, 15–450* and *20–300* present discontinuous material extrusion and collapse that prevented from taking all the planned section cuts.

3.3   Modelling of Filament Geometry

When a 3D digital model is sliced into a toolpath for the layered extrusion, an approximation of the original shape is created. To address this transformation, we developed a framework from input 3D design to machine-readable code that, taking from the findings of the filament calibration experiment, integrates visualisation and simulation of the printing toolpath.

Developed in C# within Grasshopper for Rhino, the computational model takes from a toolpath generated slicing a generic input geometry, i.e. surface, BReps, meshes. In the form of polylines generated at points P_x, the toolpath slices are found at a vertical distance H_x, which defines the layer height of the print. As described in 2.2, given constant fabrication parameters, a resultant width W_x and a contact surface S_x are defined. These three parameters define a series of sections s_x, built perpendicular to the toolpath in each point P_x, by the six vertices v_x following the findings of the filament calibration experiment described in 2.2. A low poly mesh is created through the sections and subsequently into a Subdivision Surface (SubD) object. This high precision spline provides a smooth curvilinear object representation from a polygonal mesh input (Fig. 7).

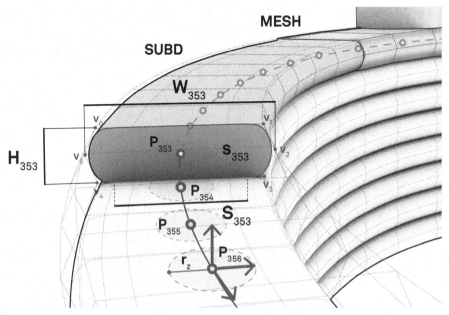

Fig. 7. Visualization of the geometric description of the layer toolpath section: generating vertices *V*, height *H*, width *W*, contact surface *S*, motion planes *P*, robot motion zone r_z.

Parallelly to the visualisation, the layer generation process outputs a set of 3D planes that will serve as motion positions for the linear movement of the robot. The distance between these points is adaptive to the local curvature of the path to minimise the number of points and maintain the closest fidelity to the input shape, and it is approximated. The generated and analysed toolpath is then translated into a RAPID module file with Robots in Grasshopper using the defined motion planes and printing parameters [19]. These are the printing speed, print zones of radius r_z, i.e. a variable that determines the distance at which the robot computes the movement to the next point, and further additional movements for the pre- and post-printing routine. Moreover, a series of quantitative data about the printing process, i.e. total printing time, layer printing time, material quantities, are defined.

The presented framework for visualization and code-generation is tested against a real-case scenario of a circular element of 400 mm circumference with regular geometry. The layers are defined by a 12 mm height and by a variable speed: along with each circular layer, the base speed of 450 mm/s was changed to 200 mm/s in seven predefined areas to increase the layer width.

3.4 Physics-Based Simulation for Complex Toolpath

The large extrusion rate and the density of the material during 3D printing with concrete create emergent behaviours that can be highly unpredictable. The visualization of the material behaviour during the deposition process is highly beneficial to predict failure. It is exploited in the design process to control the toolpath or play with emergent material behaviours. A particle-based simulation integrated into the computational framework is developed taking advantage of the live-physics engine of Kangaroo in Grasshopper and here used to visualize and preview such behaviour and inform the design process (Fig. 8). The simulation is run onto the set of points and lines forming the toolpath, considering the dimensional parameters of the layer. The process considers the set of physical forces generated by the behaviour of concrete when printed. It translates them into a set of motion vectors and constraints in the digital environment. The gravity is directly reflected through a constant negative vertical force g that applies onto each point P_x of the toolpath; the interaction between the segments is calculated through an interaction force i between the points; to avoid excessive deviation of the points, a length constraint l is applied; to reduce the number of points needed in the simulation while keeping the semi-fluid behaviour of the cementitious material, a bending force b applied to consecutive lines creates a resistance mechanism reducing the occurrence of sharp angles in the polyline-based simulated path. Moreover, a loose axial constraining force h applied onto the points that block movements on the horizontal plane is used to avoid consecutive layers to slip and to simulate the friction occurring between them.

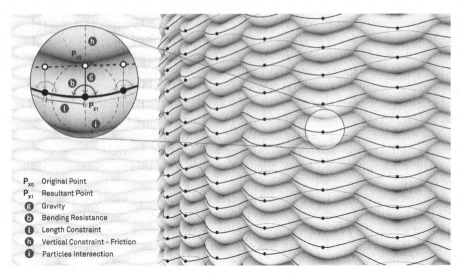

Fig. 8. Print visualization of a physics-based simulation of a patterned toolpath: a combination of forces shapes the previewed toolpath rendering the emergent behaviour of the material.

The presented simulation method is tested against a fabricated hollow circular column designed with a diameter of 400 mm and a total height of 2300 mm. Printed with a constant speed of 300 mm/s, a layer height of 10 mm, and a resulting width, the column and its pattern highlight the relevance of the emergent material behaviour during the printing process. Conceived to enhance its structural stability as well as to provide an unprecedented surface finishing, the toolpath has 230 horizontal layers characterized by a pattern with an amplitude of 150 mm at the base linearly approaching zero and a linear path at the full height of the column.

4   Results and Discussion

In this work, we introduced a modelling framework where the layered shape of the printed object and the machining toolpaths are controlled simultaneously. Taking advantage of such an approach, we generated a larger body of design opportunities through enhanced control over the appearance of the layered of the printed object and the emergent material behaviours during the printing process.

4.1   Layer Dimension Data

Results from the analysis of the resultant width from variable proportions of layer height and printing speed are plotted (Fig. 9). *Specimen 5–150* presents uneven boundaries and an average width of about 89 mm. On the other hand, *Specimen 15–450* has an average measured width of 19.8 mm, smaller than the extruder dimension. This is reflected in a discontinuous extrusion which left voids along the print path. The mean deviation for each set of M_x measures is on average 2.42 mm, increasing to 4.2 mm for wider sections where the material deposition is less controlled; *Specimen 10–300* and *Specimen 15–150* presents a deviation of less than 1 mm. The supplementary measures S and C highlight an overall section increase of about 2 mm on average, due to the acceleration and deceleration of the robot motion around the corners. For all the tested speeds, a non-linear proportion between layer height and layer width is observed. As the layer height increases, the difference in the resultant width decreases. *Specimens 25–150* and *20–150* presents an average width in the points M_x of 31.41 mm and 34.93 mm, with only 3.52 mm of difference between the two; on the other hand, the *Specimens 5–300* and *5–150* presents a difference of 26.64 mm on average.

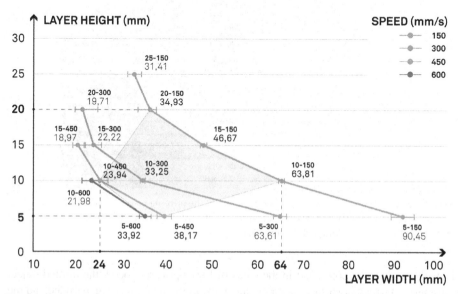

Fig. 9. Graph representation of the average resulting width from a matrix of printing speed and layer height; each width value was measured in the midpoints M_x of the long segments in the specimens.

4.2 Proof of Concept: Speed Variation

The geometric filament description developed in Sect. 3.3. was tested against the real-case scenario of a circular hollow column, to which variations in the sections are implemented through the variation of the printing speed. Tuned through the filament calibration results, the visualization framework previewed a variation in thickness from 53.7 mm in the thicker sections with a printing speed of 200 mm/s and about 24.1 mm. The speed was set at 450 mm/s. The printed element showed an average variation between 56.3 mm and 25.2 mm in the two extremes, i.e. areas with the lowest and highest print speed, with a resultant average deviation of 3.6 mm and 1.2 mm from the digital model (Fig. 10).

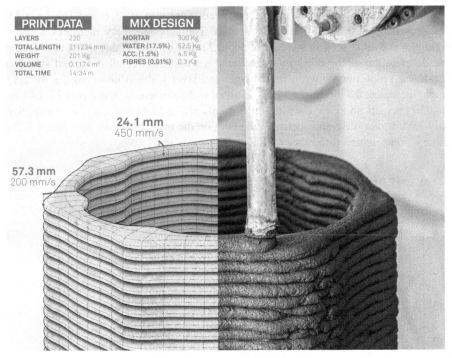

PRINT DATA		MIX DESIGN	
LAYERS	230	MORTAR	300 Kg
TOTAL LENGTH	211234 mm	WATER (17.5%)	52.5 Kg
WEIGHT	201 Kg	ACC. (1.5%)	4.5 Kg
VOLUME	0.1174 m³	FIBRES (0.01%)	0.3 Kg
TOTAL TIME	14:34 m		

24.1 mm
450 mm/s

57.3 mm
200 mm/s

Fig. 10. Comparison of the geometric visualization of the printing process (left) and the proof-of-concept print of a circular hollow element with variable printing speed. The computational model provides data about the print and the mix design.

However, it is to be noted that in the areas of lower printing speed, the printed object presented an irregular surface finishing due to the high rate of material extruded and the consequent pressure exerted by the printing nozzle on it.

4.3   Proof of Concept: Woven Column

The architectural-scale hollow column is characterised by a zigzag pattern with a linearly variable amplitude that gets to a circumference at its full height. The pattern emerging from the effect of the gravity on material that is not supported from the layer below reflects the designed gradient of the toolpath: higher amplitude corresponds to a more considerable amount of material collapsing and sagging onto the layer below. Particularly in the first half of the printed element, the effect causes an interlocking mechanism of consecutive layers. Similarly, the simulation shows a gradual weaving effect (Fig. 12). This programmed design feature increased the buildability of the column during the printing process and generated a surface pattern hardly achievable with conventional concrete casting fabrication methods.

The computational model is built upon an empirical reconstruction of the physical forces observed on physical artefacts. This implies a degree of uncertainty in the model's flexibility, which will require further investigation through observation and numerical analysis of various emerging behaviours. However, the model demonstrates a relevant design tool that provides a relatively fast and reliable preview of the toolpath and the emerging effects due to material behaviour that characterize the printed artefact (Fig. 11). This allows for more rapid design iterations and the integration and exploitation of the emergent effects of the extruded concrete as design features.

Fig. 11. Concrete 3D Printed hollow column presenting a woven toolpath with a gravity-induced deformation.

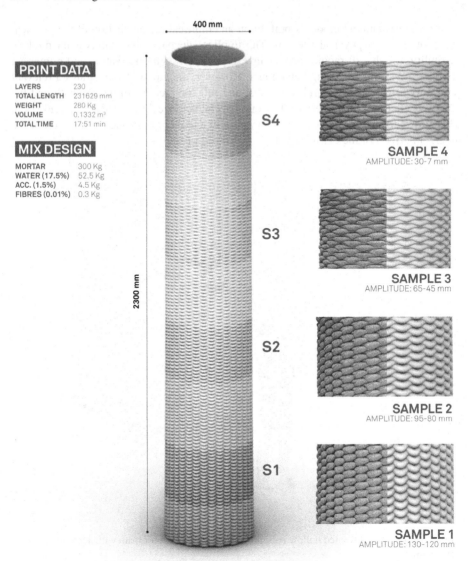

Fig. 12. Comparative visualization of the proof-of-concept column characterized by a variable-amplitude weaving pattern.

5   Conclusions

The presented experimental work addressed and exploited the design and 3D modelling paradigm shift imposed by the manufacturing process of extrusion-based 3DP. The technique imposes a new modelling approach that supports new design affordances when compared to standard concrete casting. 3DCP offers the possibility to control the material layout in every part of an object and use highly specific material patterns informed by structural or environmental analysis. The developed computational model can integrate a series of analysis on the design and fabrication process, i.e. print overhangs, centre of mass, and curvature, developed in a parallel work by the authors [20]. The proposed approach introduces realistic means of previewing and assessing complex printing patterns. This is particularly useful whenever a non-standard printing toolpath is used in the printing of architectural objects for ornamental or structural reasons. This enhanced level of design control enables various design possibilities, as it can provide higher control on the superficial and sectional features of a 3D printed object. Such patterns can be applied to characterize the aesthetics of facade elements specifically (Fig. 13). Future works will focus on analytical methods to assess the early-age physical behaviour of printed concrete and tuning the physics-based simulation parameters. Moreover, the research will be steered to developing a real-time feedback-loop workflow to analyse the printing process, compare it to the digital model, and correct the toolpath to compensate for deviations. In parallel, the model will be tested against a series of architectural-scale prototypes, developing design strategies that take advantage of its flexibility and emergent behaviours.

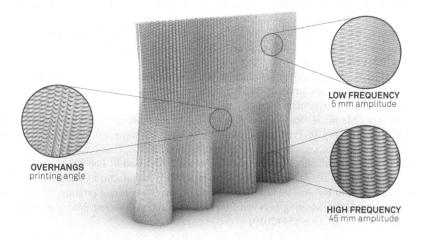

Fig. 13. Architectural outlook: a complex wall geometry and its toolpath are designed; the structure features a locally varying weaving pattern which is physically simulated to detect overhangs and visualize the emergent outlook of the planned toolpath.

Acknowledgements. This work was carried out at the CREATE Lab, a research infrastructure for digital fabrication at the University of Southern Denmark, in cooperation with industrial partner Hyperion Robotics.

The project has been developed with the help of Philip James Douglas, Mads Sørensen, Simon Andreasen, and students from the CREATE Summer School 2020. The authors wish to thank Weber Saint-Gobain for in-kind support of concrete material, and Danish Fibres for the in-kind support of polypropylene fibers.

References

1. Roussel, N.: Rheological requirements for printable concretes. Cem. Concr. Res. **112**(May), 76–85 (2018)
2. Suiker, A.S., Wolfs, R.J., Lucas, S.M., Salet, T.A.: Elastic buckling and plastic collapse during 3D concrete printing. Cement Concrete Res. **135**, 106016 (2020)
3. Craveiro, F., Nazarian, S., Bartolo, H., Bartolo, P.J., Pinto Duarte, J.: An automated system for 3D printing functionally graded concrete-based materials. Additive Manuf. **33** (2020)
4. Breseghello, L., Sanin, S. & Naboni, R.: Toolpath simulation, design and manipulation in robotic 3D concrete printing. In: Projections - Proceedings of the 26th International Conference of the Association for Computer-Aided Architectural Design Research in Asia, CAADRIA 2021. In: Globa, A., van Ameijde, J., Fingrut, A., Kim, N., Lo, T.T.S. (eds.) Hong Kong: The Association for Computer-Aided Architectural Design Research in Asia (CAADRIA), vol. 1. pp. 623–632 (2021)
5. Cura. https://ultimaker.com/software/ultimaker-cura. Accessed 07 Feb 2021
6. Simplify3D. https://www.simplify3d.com/. Accessed 07 Feb 2021
7. Slic3r. https://slic3r.org/. Accessed 05 Feb 2021
8. Wolfs, R.J.M., Bos, F.P., Salet, T.A.M.: Early age mechanical behaviour of 3D printed concrete: numerical modelling and experimental testing. Cem. Concr. Res. **106**, 103–116 (2018)
9. Kruger, J., Cho, S., Zeranka, S., Viljoen, C., van Zijl, G.: 3D concrete printer parameter optimisation for high rate digital construction avoiding plastic collapse. Compos. Part B: Eng. **183**, 107660 (2020)
10. Ahmed, Z.Y., Bos, F.P., Wolfs, R.J.M., Salet, T.A.M.: Design considerations due to scale effects in 3D concrete printing. In: 8th International Conference of the Arab Society for Computer Aided Architectural Design, pp. 115–124 (2016)
11. Carneau, P., Mesnil, R., Roussel, N., Baverel, O.: An exploration of 3d printing design space inspired by masonry. In: Proceedings of the IASS Annual Symposium, November 1–9 (2019)
12. slicerXL. https://www.slicerxl.com. Accessed 01 Feb 2021
13. RAPCAM Concrete. https://www.raptech.io/rapcam. Accessed 01 Feb 2021
14. Comminal, R., et al.: Advancing precision in additive manufacturing modelling of material deposition in big area additive manufacturing and 3D concrete printing. In: Joint Special Interest Group Meeting between Euspen and ASPE Advancing Precision in Additive Manufacturing, pp. 151–154 (2019)
15. Wolfs, R., Bos, F.P., Salet, T.A.: Early age mechanical behaviour of 3D printed concrete: numerical modelling and experimental testing. Cement Concr. Res. **106**, 103–116 (2018)
16. Ooms, T., Vantyghem, G., Van Coile, R., De Corte, W.: A parametric modelling strategy for the numerical simulation of 3D concrete printing with complex geometries. Additive Manuf. 124187 (2020)
17. Vantyghem, G., Ooms, T., De Corte, W.: VoxelPrint: a Grasshopper plug-in for voxel-based numerical simulation of concrete printing. Automat. Constr. **122**, 103469 (2021)
18. Vos, J., Wu, S., Preisinger, C., Tam, M., Xiong Neng, N.: Buckling Simulation for 3D Printing in Fresh Concrete (2020). https://www.karamba3d.com/examples/moderate/buckling-simulation-for-3d-printing-in-fresh-concrete
19. Soler, V., Huyghe, V.: Robots (2016). https://github.com/visose/Robot
20. Naboni, R., Breseghello, L.: High-resolution additive formwork for building-scale concrete panels. In: Bos, F.P., Lucas, S.S., Wolfs, R.J.M., Salet, T.A.M. (eds.) DC 2020. RB, vol. 28, pp. 936–945. Springer, Cham (2020). https://doi.org/10.1007/978-3-030-49916-7_91

PRINT in PRINT: A Nested Robotic Fabrication Strategy for 3D Printing Dissolvable Formwork of a Stackable Column

Mehdi Farahbakhsh[1]([✉]), Alireza Borhani[2], Negar Kalantar[2], and Zofia Rybkowski[1]

[1] College of Architecture, Texas A&M University, College Station, TX 77840, USA
mfb@tamu.edu
[2] California College of the Arts, Oakland, CA 94618, USA

Abstract. In this paper, the fundamentals of a 3D nested construction method for 3D-printing stackable tower-like structures are explained, taking into consideration the transportation, storage, assembly, and even disassembly of building components. The proposed method is called "PRINT in PRINT." This paper also documents the authors' experience of and findings from designing and printing a column erected out of a series of 3D printed components in a short stack. Employing the design principles of 3D printing in a nested fashion, the authors showcase the main parameters involved in dividing the column's global geometry into stackable components. By converting formal, technical, and material restrictions of a robotic-assisted 3D printing process into geometric constraints, the paper describes how the column components are divided, namely that one component shapes the adjacent one.

Keywords: Robotic fabrication · Clay 3D printing · Concrete 3D printing · Nesting · Stackable geometry

1 Introduction

In recent decades, architects have witnessed and facilitated the emergence of new tools for digital fabrication to bridge the gap between design and fabrication. While the relationship between conception and production has evolved significantly, there is still a considerable gap between what we can draw and what we can build. At this point, the process of fabricating a complex architectural form is relatively expensive and time-consuming and often wasteful. Digital fabrication in architecture is evolving from building artifacts towards the construction of full-scale buildings. The size of available tools, however, remains a challenge to the integration of digital technologies within construction processes. In recent years, the implementation of larger tools into mainstream construction has made significant contributions. Although employing large tools has great relevance and value for some projects, it can prevent architects from benefiting from the affordability, ease of use, and accuracy of digital fabrication technologies.

© Springer Nature Singapore Pte Ltd. 2022
D. Gerber et al. (Eds.): CAAD Futures 2021, CCIS 1465, pp. 317–328, 2022.
https://doi.org/10.1007/978-981-19-1280-1_20

While robotic fabrication technologies have enabled the building industry to produce nonstandard architectural forms, the robot's reach and freedom of movement in the space have remained challenging, affecting the size of the final product. Solutions like cable robots [1], gantry systems [2], telescopic booms [3], 6-axis robotic arms equipped with external axis [4] have been developed to address the issue. The size of the final product, however, is still limited to the size of the employed robot. For instance, ICON, a 3D printing company in the US, is using a 3-axis gantry solution. Their latest 3D printer can print objects up to 15.5 feet tall [5]. To print an object larger than the actual robot, it is possible to add linear rails that permit the robot to move forward or upward. For instance, in an experimental project called Casa Covida, Emerging Objects, a California-based architectural practice, used a 3-axis lightweight printer to construct structures larger than itself while adding an external fourth axis to lift the printer in the Z direction [6].

In this paper, the main objective is to leverage robots to enhance both the sustainability and efficiency of large-scale construction and maintain a consistent relationship between architectural expressions and making. Therefore, we propose a nested 3D printing strategy as a novel method in large-scale robotic fabrication to address the limited height of geometries per the current approaches in the literature [7].

2   Methods

Nesting is not a new concept in the manufacturing industry. It is used to minimize waste and maximize efficiency. The nesting strategy, however, has often been referred to as cutting or shaping patterns on a flat surface [8]. Here, we proposed a nested fabrication method called "PRINT in PRINT" that emerged from the constraints of existing paste-extrusion 3D printers. Our nested printing strategy introduced a technique for robotic additive manufacturing of compound nonstandard architectural forms that are taller than the robot's reachable area. These compound geometries consist of several smaller parts that nest within each other. By using the proposed method, the global geometry of a tower-like structure is divided into stackable components. When congruent surfaces are attained between the stacked pieces, the lower component shapes the upper one to preserve the adjacent bodies' tangential continuity. In other words, the inner side of a lower component in a stack is coincident with the outer side of the upper one. The design criteria are explained in more detail in the following design section.

Previous research projects used thin plastic shells as formwork to cast concrete [9, 10]. While the printed plastic dries fast and is expected to make it easy to cast concrete right after the print process, it is still wasteful and difficult to remove the plastic shell. Using a more sustainable material like clay as temporary formwork has some advantages over plastic. In particular, clay dissolves in water and can be easily removed and recycled.

Clay is a fragile material with mechanical properties that make it challenging to use it as a final product in the additive manufacturing of architectural forms. However, it does possess some interesting features that could be beneficial for temporary elements in additive manufacturing processes.

Fresh concrete placed adjacent to clay absorbs clay's moisture over time, and consequently, the clay parts dry faster than air-drying. The fast-drying process results in a cracked and shattered surface on the clay that can be easily removed by hand, pressure-washed, or dissolved in water (Fig. 4c). This is similar to water-soluble materials used for temporary supports in 3D printing with plastic-based materials [11]. The PRINT in PRINT strategy uses clay as a temporary material to 3D print formworks for concrete elements.

Our method consists of three approaches: Design Strategies, Fabrication Processes, and Assembly of a nested column as follows:

2.1 Design Strategies

By using our method, two types of geometries can be easily nested as a 3D composition, including closed and open forms. Here, a closed form refers to a geometry with an enclosed area and continues mass when an empty space is left in between. Such a geometry has a closed curve or polyline in its horizontal section. An open form is a geometry with a non-closed area with no defined space between its mass. Every closed form can be divided into several open forms. The focus of this paper is on closed forms; open forms are the subject of authors' future work. In this paper, we study a conical column with a twisted enclosure. This lofted geometry has an empty core.

The general design step is to intersect a geometry with a series of planes. In this paper, we use surfaces that are parallel to the initial construction plane. The number of the planes (n) and the distance between them (H) determine the number of nested components and the height of each component respectively (Fig. 1a). PRINT in PRINT strategy requires a few design criteria and can create any geometry that follows all below criteria:

- Each slice of the compound geometry should be slightly smaller than the one underneath, so they can nest within each other. In other words, the horizontal section of nested components at any height in the stack should not intersect the adjacent one (Fig. 1b).
- The global geometry can include zero, negative, positive, and double curvatures. If the curvature changes in the Z, the geometry should be sliced and fabricated in separate stacks. In other words, all nested components in a stack must have consistent surface curvatures. The reason for this is that nested components should be taken out of their stack to be assembled later. A component with both positive and negative curvatures along its height can be created in this method but cannot be taken out of the stack (Fig. 1c). In their stack, nested components should be designed in a way that we can easily slide them out. Besides sliding the component linearly, it is possible to take out the nested pieces while rotating and moving them simultaneously.
- An important aspect of the proposed method is to design every component to be matched with its adjacent pieces, creating the desired form once assembled.

   Therefore, a given free-form geometry should be sliced when the top face of a lower component (S1) has the same shape as the upper module's bottom face (S2) (Fig. 2). In this case, the seamless assembly of the final product is guaranteed.

The workflow in this paper includes a parametric algorithm for designing and slicing the geometry in Rhino and Grasshopper. HAL, a plugin for Grasshopper, was used to program the robot and create the toolpath.

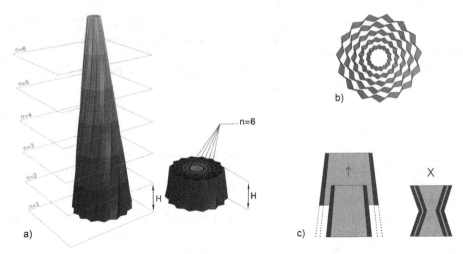

Fig. 1. a) A series of horizontal planes slice the main geometry and break it into several components (n). The fixed distance between planes (H) defines the height of the nested mass. b) Components in a stack should not intersect. c) The curvature of the surface in each stack should be consistent, otherwise, the components cannot be taken out of the stack.

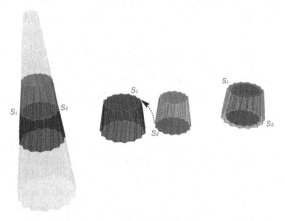

Fig. 2. The lower component's top face (S_1) and the upper component's bottom face (S_2) are identical.

2.2 Fabrication Processes

The nested components were created in two steps. A fast-setting concrete was cast between a series of shells that were robotically 3D printed with clay. The logic behind the design and fabrication of the nested components is that each piece of the compound geometry includes a thin outer shell, which, when nested within other pieces, creates the inner shell of the next component, and those shells together define the boundaries of concrete cores (Fig. 3). The outer shells were later dissolved in water so the concrete cores could be extracted for the assembly step. The concrete's thickness depends on both the design of the main geometry and the number of nested components (n). As the number of nested components increases, the resulting concrete cores are thinner, and the nested elements are shorter (H is smaller). The prototype presented in this paper had six components (n) with a height of 23 cm (H) in the nested stack. The thickness of the concrete in most of the components was 3 cm and the smallest one in the center of the mass was fabricated with a 1.5 cm thickness of the concrete to host the 5 cm steel structure.

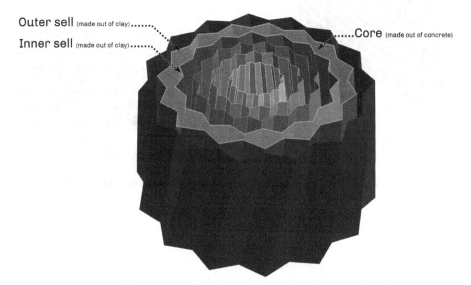

Outer sell (made out of clay) ....Core (made out of concrete)
Inner sell (made out of clay) ....

Fig. 3. The clay shells create a temporary formwork for the concrete cores.

The authors observed failure in building geometries with long straight segments in their horizontal section (Fig. 4a, b). Fabricating thin shells with long and tall straight segments resulted in either collapse or deformation of the printed object. One solution is to substitute straight segments with smaller zig-zag or chevron patterns in the design. Similarly, Burger et. al [12] reported a delamination issue with 3D printing a shell with plastic filaments. While clay remains soft for a while, pouring concrete right away will destroy the soft shells. One solution is pouring concrete in multiple steps with limited height during or after the print process.

IRB 1200 industrial robot, one of the smallest ABB robots, was used in this paper to 3D print a twisted column as proof of concept. The maximum height of the robot while holding an extruder perpendicular to the print bed is 70 cm. The column was built with 40 cm in diameter and 140 cm in height (Fig. 5).

In this experiment, after 48 h, the concrete cores completely absorbed the clay shells' moisture in the lab environment, which had decent ventilation. Most of the outer clay surface was removed manually while nested components were left in a bucket of water for 24 h to dissolve the clay shells. While only 70% of the clay dissolved in water, the result was promising, and the concrete parts were extracted with gentle force and ultimately pressure-washed to remove the remaining clay (Fig. 6).

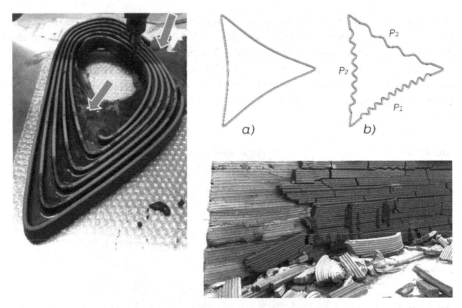

Fig. 4. a, b) Zig-zag patterns are applied to increase the stability of the thin clay shells. c) The dried clay can be easily detached.

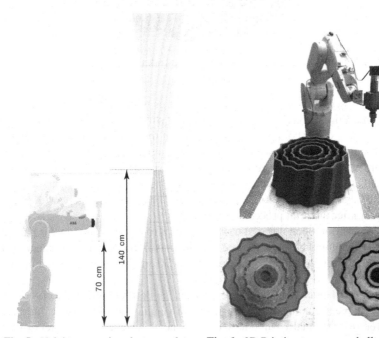

Fig. 5. Height comparison between the robotic arm and the designed column

Fig. 6. 3D Printing temporary shells, casting concrete, and removing the clay shells

2.3 Assembly

Since each concrete piece's bottom surface (S2) was identical to the top surface of the module underneath (S1), the assembly process was straightforward. The nested components were sequentially lifted, giving shape to the main geometry. Three scenarios can be imagined designing the connection between stacked components: First, the printed components can be connected to a network of cables to create a post-tension structural system. Second, concrete is cast in the hollow core of the compound geometry and reinforced with steel bars to shape a single homogenous component [13]. Finally, a separate steel structure can be employed to bear the structural loads and hold the compound geometry pieces together. The third approach was used in this paper, wherein an independent steel structure was designed, and the concrete components were fixed in place using bolts and steel spacers.

The flat surface under the nested components during the printing and casting process resulted in a clean flat surface at the bottom of concrete pieces. There were some minor imperfections on the top surfaces of the concrete components that required a thin (2 mm) gap between each piece to make it visually pleasing during the assembly process. Steel spacers with a 10 mm offset from the edge of the concrete were used to create the gap and fix the parts in place (Fig. 7).

The assembly process was relatively easy and could be accomplished by a single individual on the Texas A&M University campus (Fig. 8).

Fig. 7. The quality of the concrete elements and the detail of the independent steel structure

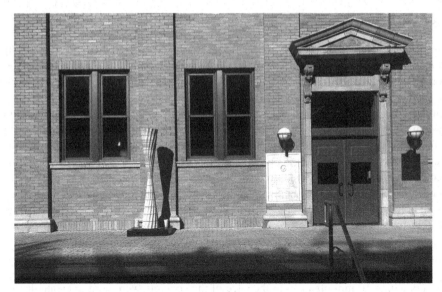

Fig. 8. Two twisted columns with a total height of 280 cm were robotically fabricated.

3   Discussion

In this project, the goal was to establish a feedback loop between geometry, material, simulation, and fabrication. The design criteria explained in Sect. 2.1 are the results of technical considerations in the proposed nesting method. Those rules push toward designing a tapered geometry with either negative or positive curvature or both, so the sliced components can easily nest within each other and be taken out of their stack later.

We were aware of the challenge of matching the concrete cast pieces together in the column when each of them came out of the nested mass and addressed the aesthetic concerns by adding steel spacers in the assembly process.

While HAL plugin was used to generate the robot toolpath, the degree of conformity to the desired geometry was afforded through the proposed algorithmic method, which translates the impacts of manipulation of successive layer width, thicknesses, and buildup rates into the robot's operation. Besides the size of our robot and the printing frame, the final printed geometry was influenced by various deposition parameters. Factors included controlling the rheology of the clay paste [14], deformation of a wet material deposited under a load generated by the subsequent layer deposition [15], surface contact between consecutive printing layers [16], and restricted corbeling slope of overhanging sections to control structure buckling, as well as the drying time of the concrete mix [17]. While accomplishing the ease of transportation and cost reduction, the entire twisted column and its stackable components were affected by the robot's size in the XY plane. Mounting the robot on an external axis will push the boundaries even further. The form, final look, and architectural expression of the column emerged from clay and concrete material properties.

Nesting components have three benefits: first, they minimize material consumption and waste, as each piece shares at least one surface with others. Second, this technique can be more affordable, as a robot can create complex geometries that are taller than the maximum reach of the robot in the Z direction. Third, the nested printing approach minimizes the length of the toolpath, where the robot does not extrude any material. In traditional fabrication approaches where the toolpath is not continuous, moving from one path to another adds to the total fabrication time. Nesting components minimize the travel time between separate paths.

Nesting a tall structure in shorter stacks decreases the surface area and consequently facilitates transportation in terms of the required space. In the prototype presented in this paper, the assembled column had a bounding box of $280 \times 40 \times 40$ cm. Nesting the column in four stacks required a bounding box of $92 \times 40 \times 40$ cm. This means the components can be carried out to the installation site with an SUV with a small cargo space rather than using a truck with a minimum 280 cm bed in length or height (Fig. 9).

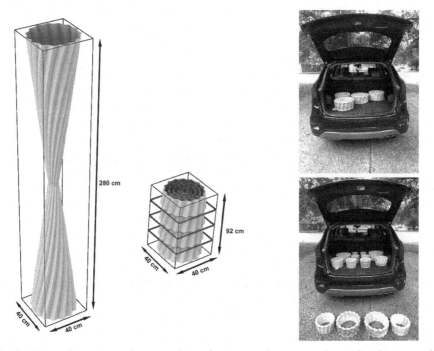

Fig. 9. The nesting strategy decreases the surface area and consequently requires less space for transportation.

As long as the design criteria explained in this paper are followed, a designer can adopt a wide variety of geometries to make stackable tower-like structures in a nested fashion (Fig. 10).

This paper aims to establish a link between the geometry of stackable components and clay 3D printing of a series of centric shells. While serving as the column surface constituents, these shells can be used as formworks of concrete bodies. In this research, the printed clay serves as the formwork for concrete casting with promising results. While clay's moisture adjacent to cast concrete benefits the concrete's curing process, the dried clay shell can be easily removed. The authors will examine the possibility of using clay as a dissolvable support material for real-scale concrete printing, with plans to expand this work to a larger scope. Also, the authors aim to promote the following aspects in the future:

- Use the method to 3D-print different building components on a larger scale,
- Investigate the best technique to reinforce components in the final assembly,
- Address the best way to joint adjacent components while providing the opportunity to disassemble the structure.
- Study the most convenient approach to assemble the component when using fewer tools, scaffolds or lifting mechanisms.
- Simulate the structural behavior of printed components and analyzing their performance.

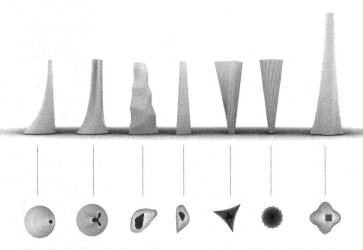

Fig. 10. Various design options are explored and depicted in this picture.

4   Conclusion

In addition to introducing the potential of nested 3D printing, the authors discuss the opportunities and challenges of using a small robot to fabricate an artifact larger than itself. By examining the merits of clay as a medium, hardware, and software details of the robotic system involved, this paper introduces the process of discretization of a twisted column into a limited number of uniquely shaped stackable components, showcasing a close tie between the column overall geometry, its final form, the number of its components, and the method of its fabrication.

The proposed method allows for constructing diverse 3D printed structures with complex surfaces, including zero, negative, positive, and double curvatures. The method can be applied to different types of components such as structural elements, cladding, complex shell structures, and interior surfaces.

On a broader scale, our proposed method can significantly reduce material waste, fabrication time, and production area while using a smaller printer to create artifacts larger than itself. Subsequently, lower investment in tools, facilities and resources will required. Since the method can generate the most compact components, the fabricated pieces occupy minimal storage space, and the pieces can be transported to the site in stacks. This process reduces both transportation and construction costs.

Acknowledgments. This work was partially funded by the X-Grants Program: A President's Excellence Fund initiative at Texas A&M University. Additionally, the authors like to express their gratitude to Dr. Patrick Suermann, head of the Construction Science department, for his unwavering support.

References

1. Izard, J.-B., et al.: Large-scale 3D printing with cable-driven parallel robots. Constr. Robot. 1(1–4), 69–76 (2017). https://doi.org/10.1007/s41693-017-0008-0
2. Khoshnevis, B.: Automated construction by contour crafting, related robotics and information technologies. Autom. Constr. **13**, 5–19 (2004). https://doi.org/10.1016/j.autcon.2003.08.012
3. Keating, S.J., Leland, J.C., Cai, L., Oxman, N.: Toward site-specific and self-sufficient robotic fabrication on architectural scales. Sci. Robot. **2** (2017). https://doi.org/10.1126/scirobotics.aam8986
4. Zhang, X., et al.: Large-scale 3D printing by a team of mobile robots. Autom. Constr. **95**, 98–106 (2018). https://doi.org/10.1016/j.autcon.2018.08.004
5. ICON. https://www.iconbuild.com/vulcan. Accessed 06 Feb 2021
6. Rael, R., San Fratello, V.: Casa Covida. https://www.rael-sanfratello.com/made/casa-covida. Accessed 06 Feb 2021
7. Craveiro, F., Duarte, J.P., Bartolo, H., Bartolo, P.J.: Additive manufacturing as an enabling technology for digital construction: a perspective on Construction 4.0. Autom. Constr. **103**, 251–267 (2019). https://doi.org/10.1016/j.autcon.2019.03.011
8. Nee, A.Y.C., Seow, K.W., Long, S.L.: Designing algorithm for nesting irregular shapes with and without boundary constraints. Ann. CIRP **35**, 107–110 (1986)
9. Naboni, R., Breseghello, L.: Fused deposition modelling formworks for complex concrete constructions. SIGraDi **2018**, 700–707 (2018). https://doi.org/10.5151/sigradi2018-1648
10. Burger, J., Lloret-Fritschi, E., Scotto, F., Demoulin, T., Gebhard, L.: Research collection. 3D Print. Addit. Manuf. **7** (2009). https://doi.org/10.1089/3dp.2019.0197
11. Doyle, S.E., Hunt, E.L.: Dissolvable 3D printed formwork. In: ACADIA, pp. 178–187 (2019)
12. Burger, J., et al.: Eggshell: ultra-thin three-dimensional printed formwork for concrete structures. 3D Print. Addit. Manuf. **7**, 49–59 (2020). https://doi.org/10.1089/3dp.2019.0197
13. Anton, A., Yoo, A., Bedarf, P., Reiter, L., Wangler, T., Dillenburger, B.: Vertical modulations computational design for concrete 3D printed columns. In: Ubiquity and Autonomy - Paper Proceedings of the 39th Annual Conference of the Association for Computer Aided Design in Architecture ACADIA 2019, pp. 596–605 (2019)
14. Mansoori, M., Palmer, W.: Handmade by machine: a study on layered paste deposition methods in 3D printing geometric sculptures (2018)

15. Bard, J., Mankouche, S., Schulte, M.: MORPHFAUX: probing the proto-synthetic nature of plaster through robotic tooling. In: ACADIA 2012 - Synthetic Digital Ecologies : proceedings of the 32nd annual conference of the Association for Computer Aided Design in Architecture, October 2012, pp. 177–186 (2012)
16. Farahbakhsh, M., Kalantar, N., Rybkowski, Z.: Impact of robotic 3D printing process parameters on bond strength. In: 40th Annual Conference of the Association for Computer Aided Design in Architecture ACADIA 2020 (2020)
17. Buswell, R.A., et al.: 3D printing using concrete extrusion: a roadmap for research. Cem. Concr. Res. **112**, 37–49 (2018). https://doi.org/10.1016/j.cemconres.2018.05.006

Path Optimization for Multi-material 3D Printing Using Self-organizing Maps

Diego Pinochet[1,2](✉) [iD] and Alexandros Tsamis[3](✉) [iD]

[1] School of Design, Adolfo Ibañez University, Santiago, Chile
[2] Massachusetts Institute of Technology, Cambridge, USA
dipinoch@mit.edu
[3] Rensselaer Polytechnic Institute, Troy, USA
tsamis@gmail.com

Abstract. Shape generation based on scalar fields opened up the space for new fabrication techniques bridging the digital and the physical through material computation. As an example, the development of voxelized methods for shape generation broadened the exploration of multi-material 3d printing and the use of Functionally Gradient Materials (FGM) through the creation of shapes based on their material properties known as Property representations (P-reps) as opposed to Boundary representations (B-reps) [1]. This paper proposes a novel approach for the fabrication of P-reps by generating optimized 3d printing paths by mapping shape internal stress into material distribution through a single optimized curve oriented to the fabrication of procedural shapes. By the use of a modified version of the traveling salesman problem (TSP), an optimized Spline is generated to map trajectories and material distribution into voxelized shape's slices. As a result, we can obtain an optimized P-Rep G-code generation for multi-material 3d printing and explore the fabrication of P-Rep as FGMs based on material behavior.

Keywords: 3D printing · Path optimization · Multi-material 3d printing · Algorithms · Machine learning · Functional gradient materials

1 Introduction

1.1 Background

New techniques for the generation of shapes in 3d environments demand new methodologies for fabricating them. For example, procedural modeling based on voxelized scalar fields through Machine Learning-although very useful for exploring novel tectonics [2] and user interfaces-demands new methodologies for their fabricability. The generation of shapes on-demand using machine learning challenges the traditional fabrication methods such as additive manufacturing as new shapes are processed in real-time, constantly modifying their topological structure. Also, the generation of new scalar field surface modeling methods opened the door for new modeling and fabrication techniques shapes considering material properties and their distribution in space as the way to compute them digitally and physically fabricate them. An optimization method is

© Springer Nature Singapore Pte Ltd. 2022
D. Gerber et al. (Eds.): CAAD Futures 2021, CCIS 1465, pp. 329–343, 2022.
https://doi.org/10.1007/978-981-19-1280-1_21

proposed to address the fabricability of procedural shapes using 3d printing by mapping the shape's internal stress [3] into material distribution. This project builds upon two projects: SKeXl, a 3d modeler software based on sketches as input [4], and the alchemic printer, a multi-material 3d printer machine [5].

Having a software and hardware platform for the design and fabrication of P-Reps, this paper proposes a novel approach for generating a protocol for multi-material 3d printers. This paper starts by a simple question. Can we map quantitative performance metrics to material gradients into a single path shaped as the infill of a given geometry in an optimized way in terms of computation time and length? In the case of this paper, the optimization of shapes to efficiently distribute materials in shapes, minimizing airtime and avoiding self-intersections, is addressed toward informed material metrics delivered, for example, by the principal stress mesh derived from topological optimization.

Using optimization methods for an efficient path generation and material distribution, a 'P-Rep G-code' is generated to distribute two material concentrations along a single continuous path according to inside an infill layer. The main goal is to relate material performance to material distribution toward the fabricability of voxelized generated structures.

Fig. 1. Tsamis, Alexandros. 2012. https://dspace.mit.edu/handle/1721.1/77777

Whereas Additive Manufacturing tool path optimization research focuses on time reduction, this paper considers tool path optimization to relate shape performance metrics and material properties using the most efficient single path applied to infill areas. While there is a significant body of work using algorithms such as the travel salesman problem (TSP) for path optimization in the world of CNC milling toward reduction of air time [6] (the time the machine moves in z direction to perform rapid movements) or by covering a planar area efficiently by the shortest path possible in the minimum amount of time [7], the optimization of tool paths opens the door for further novel applications in alternative FDM[1] processes. This paper proposes using self-organizing maps (SOM) to produce an

---

[1] Fusion deposition modeling.

optimized path based on Kohonen [8] and Brocki [9] to solve a geometric-fabrication problem of multi-material 3d printing.

2 Related Work

The transition from a conception of the digital shape from a boundary representation (B-rep) toward a property representation (P-rep) has been investigated by the pioneering work of Tsamis [1] and implemented into software packages such as the experimental Vspace[2] and the commercial product from Autodesk Monolith[3]. Property representation of shapes is a fertile field of exploration for design and architecture in which the development of software must find a counterpart in hardware for the proper physical manifestation of the shapes designed.

2.1 Software and Hardware Implementations

Tsamis' seminal work on Property Representations through the use of scalar spaces to represent shape boundaries (B-reps) and material distribution in space opened the door for the exploration for the design of architectural shapes using FGMs (Fig. 1). Tsamis' material approach proposed a design methodology based on material gradients responding to different environmental conditions setting a novel way to implement multi-material 3d printing. By bridging software [10, p. 48] and hardware [10, p, 302] implementations, Tsamis's work established a new design ethos that reconfigured the whole to parts relationships in architecture, seeking a new topology based on continuity instead of discretization. Along the same lines and based on Tsamis research, the work of Michalatos and Payne [1] shows the use of a voxelized-based approach to volume 3d modeling considering material distributions in Monolith [11], a discontinued software bought by Autodesk. Monolith emerged as the paradigm for voxel modeling implementations as a general-purpose modeling software based on material distributions, closely related to the work of other authors such as Richard and Amos [12], Grigoriadis [13], Oxman [14] oriented to physically manufacturing FGMs.

In addition, new techniques for the development of fluid interfaces for design and analysis of shapes that take advantage of Machine Learning (ML) algorithms and the use of 3 dimensional Generative Adversarial networks present challenges for novel fabrication techniques of procedurally generated shapes. As an example, new software implementations such as SKeXL, Sketch to Shape: A Generative design tool using three-dimensional generative adversarial networks by the use of 3d Generative adversarial Networks (Fig. 2). By using 3d GANs, the systems allow the generation of volumetric shapes from 2d sketching input. The system works as an interactive modeling tool that opens a new way to simple 3d modeling taking advantage of machine learning and 3d datasets. The system works as an interactive modeling tool that opens a new way to simple 3d modeling taking advantage of machine learning and 3d datasets. SKeXL represents fertile 3D GANs exploration that works as a novel approach for design generation related

---

[2] https://github.com/tsamis/VSpace.

[3] Michalatos, Panagiotis- Autodesk https://www.food4rhino.com/app/monolith.

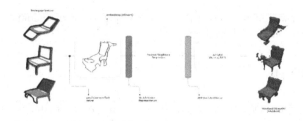

Fig. 2. SKeXL, Sketch to Shape: A Generative design tool using three-dimensional generative adversarial networks by using user sketches to generate procedural shapes. Renaud Danhaive and Diego Pinochet, https://diegopinochet.com/portfolio#42f7b021-2e36-42c0-9527-bf10495f2cd5.

to qualitative criteria. Nevertheless, there is a need for the generation of quantitative metrics toward the fabricability of these shapes. Considering the work developed in the past ten years about voxel shape generations and material computation and distribution using scalar fields, could the work on multi-material 3d printing inform a process to add quantitative metrics for the fabrication of shapes such as the ones produced in novel software packages such as Skexl?

This paper implements a methodology based on the related work shown above to bridge shapes' digital design and possible fabrication on demand by generating fabrication protocols that could be fed into multi-material 3d printers such as the 'Alchemic printer' (Fig. 3).

2.2 Toolpath Optimization for NC Machines

Considerable work in tool path optimization has been produced in the past ten years. Several techniques and algorithms have been used to optimize parameters, for example, air time, the vertical movement for rapid end effector travel in 2D, covered (visited) area, and path length. Lechowicz et al. [15] proposed the use of Greedy algorithms (2opt and Annealing) to optimize 3d printing time. Alternatively, the works of Dreifus et al. [16] or Ganganath et al. [17] use the Chinese Postman Problem [18] and the Traveling Salesman problem using the Cristofides algorithm, respectively, to optimize similar problems like the ones from Lechowicz et al. Whereas the use of the TSP algorithm and variations has shown promising results in optimizing printing time while reducing travel of the end effector, as detailed information has not been provided concerning calculation times.

The traveling salesman problem is a well-known problem in computer science. It consists of finding the shortest route possible that traverses all cities in a given map only once. Considered as an NP-complete hard problem -implying that the difficulty to solve it increases rapidly with the number of nodes to calculate, a general solution that solves the problem is unknown, resulting in an algorithm that tries to find good enough solutions when applied. The TSP and variations of it have been used to solve tool path optimization for 3d printing and milling because of its advantages in finding reasonable solutions that can be applied directly to fabrication problems.

The implementation of an alternative approach to solving the TSP is used in this project through Kohonen's Self-Organizing maps SOM. In 1975, Teuvo Kohonen introduced a new type of Neural network using competitive, unsupervised learning [19]. Kohonen proposed the description of a self-organizing map as a 2D nodes grid, inspired by a Neural Network (NN). By relating the idea of a map to a model, the purpose of Kohonen's technique was to represent a model in a lower number of dimensions while at the same time maintaining the similarity relations of the nodes contained. Kohonen's approach was through the use of Winner Takes All (WTA) and Winner Takes Most (WTM) algorithms. As explained by Brocki [20], by a combination of both algorithms, it is possible to, on the one hand, use WTA to calculate the neuron whose weights are most correlated to a current input as the winner, and on the other, use WTM – that has better convergence- to adapt neuron's synaptic weights in one learning iteration making the neuron's 'neighborhood' also adaptable [21].

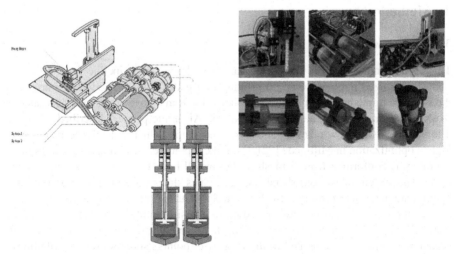

Fig. 3. The Alchemic printer. The system is based in a custom material mixing and extruding system to generate Functional gradient materials from 2 or more components. Diego Pinochet 2019. https://diegopinochet.com/portfolio#2fa9c4d8-59d6-454c-b981-d84f2dba8cde

Kohonen's technique was modified by Brocki introducing a change in Kohonen's algorithm, considering a circular array of nodes instead of a grid. By applying such modification, a dimension reduction is performed, so the neighborhood condition of neurons considers a neuron in front and back of it. This change in Kohonen's technique allows the SOM to behave like an elastic ring while getting closer to the nodes while minimizing length by the neighborhood condition. According to Brocki, this is represented by showing fast-global self-organization at the beginning and local adjustment behavior in the end. Convergence is ensured by applying a learning rate α controlling the exploration -at the beginning- and exploitation -at the end- of the algorithm. A decay in the neighborhood and the learning rate are applied to ensure the proper function of the algorithm. While decaying, the learning rate ensures low displacement of the neurons in the model. Decaying the neighborhood produces moderate exploitation of local minima. The regression expressed by Brocki is based on Kohonen's result in the following equation.

$$\alpha_{t+1} = \gamma_\alpha \cdot \alpha_t, h_{t+1} = \gamma_h \cdot h_t$$

Similar to Q learning function, α is the learning rate in time, γ is the discount, and h the neighbor dispersion. The calculation of the equation is done by traversing the ring starting from an arbitrary point and sorting the nodes by order of appearance of their winner neuron -associated with the corresponding node.

3 Methodology

3.1 Overview

This section introduces the computational methodology and pipeline for the generation of optimized multi-material 3d printing tool paths. This methodology introduces the software implementation to generate toolpaths from a voxelized shape generated in SkeXl to be fabricated in a system such as the Alchemic printer, described in Fig. 4. The proposed workflow takes a solid voxelized shape, which is analyzed using topological optimization in Millipede [22]. From the topological optimization, the resulting geometry is conformed from 2 meshes; the overall boundary and the principal stress mesh. The system takes both shapes and generates slices according to an initial layer height parameter determining areas for material distribution. Both areas are populated with different point distributions according to a desired infill percentage and printing nozzle. An optimization process that solves the traveling salesman problem using self-organizing maps is implemented from the array of points using different distributions - random, square, radial, hexagonal, and triangular. A single spline is generated, traveling most efficiently through all the point-set, minimizing the length of travel and eliminating airtime (reducing print time) and self-intersections (to avoid undesired collisions with the printing nozzle) while maintaining the lowest calculation time possible.

Applying the TSP algorithm for toolpath optimization presents some challenges. It is considered an NP-complete hard problem, implying that the difficulty to solve it increases rapidly with the number of nodes, and we do not know a general solution that solves the problem. This is why through the algorithm, we can find good – enough solutions for fabrication problems that could reduce, in the case of this implementation, printing time and path calculation. Because of this reason, an alternative technique to solve the TSP problem is implemented.

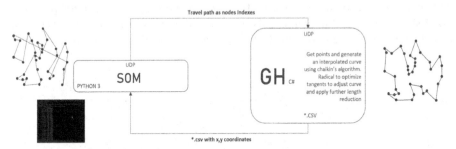

Fig. 4. Pipeline for the optimization. The system sends point's location information from Rhinoceros to a Python 3 component to make the SOM calculations, an ordered array with points' ids is sent back to Grasshopper to generate an optimized path.

3.2 Implementation of Two Versions of TSP Compared to Self-organizing Maps to Optimize Toolpaths

The implementation of an alternative approach to solving the TSP based on Kohonen's Self-Organizing maps SOM described in the previous section (2) is used in this project. Based on Vicente's implementation to implement's Brocki's modification, a Python 3 script is written and connected to Grasshopper (GH) to obtain the optimized path. A CSV file is generated by taking a distributed point set inside a shape slice containing the node's data as IDs and x y coordinates. The python script takes the information and tries to generate an optimized path in the lower amount of iterations possible (10000 per epoch) while avoiding any self-intersection. If any of the parameters in the equation decays over a certain useful threshold, the optimization stops and sends over a socket communication, the ids of the nodes back to GH. After receiving the id of the nodes, the GH definition takes the index pattern and generates an interpolated curve using Chaikin's algorithm [23] implemented in C#. Once a path has been obtained, A final optimization using Constrained Optimization by Linear Approximations (COBYLA) [24] is applied to the tangents of the generated curve to get an extra length optimization of the path (Fig. 5). Finally, a function that calculates Nurbs' points distances to the areas resulted from the topological optimization. This set of distances calculates a material concentration between material A or B to be applied during the printing process. The function to principal stress data to a path results in a gradient path that will be printed as the infill of the printed object (Fig. 6).

The methodology to test the application of SOMs for optimized path generation consider the application of two additional versions of the TSP algorithm. A vanilla version using the Cristofides [25] algorithm and a 2Opt [26] (implemented in Python3) versions of the TSP are used to have comparative optimization metrics through SOMs. Whereas the Cristofides version is considered the most optimal approach to solve the TSP, it doesn't ensure the avoidance of self-intersections for the generated path. A 2-opt TSP solver is used to compare the results of the SOM TSP approach. The 2 Opt calculation is performed using py2Opt [25], a fast Python3 library, to solve the TSP.

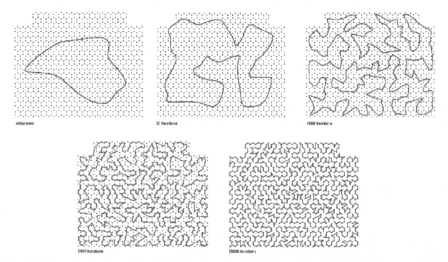

Fig. 5. Path optimization using SOMs implemented in Python 3 and GH. The sequence shows the progression of the calculation using the SOM. In this example, the path calculates the most efficient way to travel all the points in a uniform grid to generate a single path (to reduce air time and calculate material continuity in the extrusion process.)

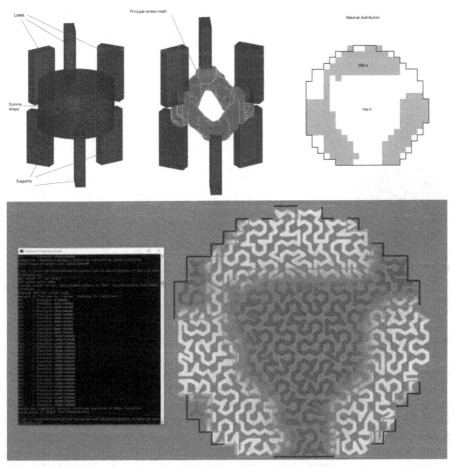

Fig. 6. Applying the SOM optimization to a dummy shape's slice after topological optimization is applied to the initial mesh. The geometry is processed and calculated using topological optimization, then sliced and populated with a point distribution (in this case an ordered grid) and sent to the program for gradient material distribution calculation.

3.3 The Experiment

The three algorithms are tested using the same data. By taking a dummy mesh's slice, different point set distributions -random, circular, square, hexagonal, and triangular- are distributed inside the area defined by the slice polygon (Fig. 7). Each test considers the distribution of different numbers of points (50, 200, and 500) to observe how the algorithms respond in relation to execution time to the increasing number of nodes to solve.

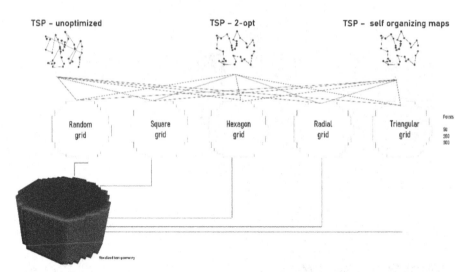

TSP – unoptimized   TSP – 2-opt   TSP – self organizing maps

Random   Square   Hexagon   Radial   Triangular
grid   grid   grid   grid   grid

Fig. 7. Diagram of the experiment. Five types of point distributions are tested using three TSP algorithms (TSP Cristofides, 2-OPT, and SOM) to compare the effectiveness of using self-organizing maps to optimize single path generation.

4   Results

On a random distribution of 50 points, we can observe that while the execution time of the simple TSP performs fast, the resulting path is 12% longer than the results of the SOM, and 9% longer than the 2Opt while creating 8 intersections. The 2 Opt performs 75% faster than the SOM while producing a path 3% longer. By Increasing the number of points within a random distribution to 200 and 500, we observe interesting results that justify the use of SOMs for this project. With 200 points, while the simple implementation of the TSP performs faster than the other two, the path calculation is longer, and the number of intersections increases. The path generated by the simple TSP is 11% longer than the 2opt version and 13% longer than the SOM while generating 44 self-intersections. In the case of the 2Opt TSP, the generated path's length is almost 3% longer than the SOM TSP but with an execution time that is 87% longer than the SOM. At 500 points, the test demonstrates a particular result between the simple TSP and the other two algorithms. The generated path is performed in 3.61 seconds (more 1300 times faster than the 2Opt and 88% faster than the SOM TSP) but with 126 self-intersections (Fig. 8).

Using either a square, hexagonal or triangular pattern, the SOM performs the best in path length and execution time while not presenting any self-intersection. For example, using a square grid distribution and 50 nodes confirms the tendency of the simple

TSP and the 2OPT TSP to perform faster than the SOM TSP (99.77% and 99.93%, respectively), although generating longer paths[4]. The tendency is confirmed at 200 and 500 points where the SOM maintains stable execution times (between 20–28 s), shorter

---

[4] In the experiments, we see that with the exception of the random distribution, the simple TSP performs poorly in length and self intersections number.

a/g	time	length	npoints	epochs	self int	dist
tsp cristofides	0.25	925.39	50	1	8	random
2opt tsp	3	843.21	50	3	0	random
som tsp	20	818.63	50	9	0	random

a/g	time	length	npoints	epochs	self int	dist
tsp cristofides	0.31	1822	200	1	44	random
2opt tsp	186	1621.61	200	3	0	random
som tsp	25.1	1602	200	24	0	random

a/g	time	length	npoints	epochs	self int	dist
tsp cristofides	3.61	2097.66	500	1	126	random
2opt tsp	4816	2558.53	500	4	0	random
som tsp	29	2548.5	500	27	0	random

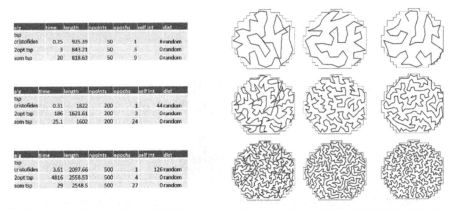

Fig. 8. Results for a random grid (time in seconds, length in mm). The TSP Cristofides algorithm (first column of shapes) calculates the path outperforming by a large margin the other two algorithms (1300 times faster) but generating many self-intersections.

paths, and no self-intersections. At a higher number of points, the 2Opt finds a good path solution with no intersections but with execution times that go from 25 to 45 min, making it unfeasible to use in a system that needs a fast response to generate paths on-demand (Fig. 9, 10, 11 and 12).

Over 1000 points, the SOM starts performing worse in execution times (over 29 s) but solving a non-intersecting path efficiently. This behavior might be produced because of the interaction with GH and how long it takes a software like Rhinoceros 3D to compute in real-time more than 1000 points without using multithreading or GPU computing implementations.

a/g	time	length	npoints	epochs	self int	dist
tsp cristofides	0.25	3088.08	50	1	7	square
2opt tsp	1.23	788.04	50	3	0	square
som tsp	18	741.21	50	15	0	square

a/g	time	length	npoints	epochs	self int	dist
tsp cristofides	0.439	13063	200	1	550	square
2opt tsp	129.6	1608.19	200	4	0	square
som tsp	23.1	1575.31	200	23	0	square

a/g	time	length	npoints	epochs	self int	dist
tsp cristofides	2.6	29641.36	500	1	40466	square
2opt tsp	2109	2580.83	500	4	1	square
som tsp	27.45	2535	500	27	0	square

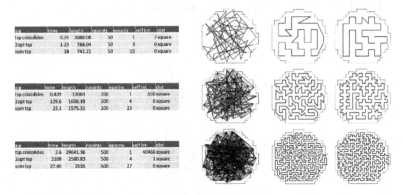

Fig. 9. Results for square point distribution (time in seconds, length in mm)

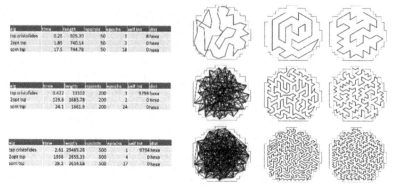

alg	time	length	npoints	epochs	self int	dist
tsp cristofides	0.25	925.39	50	1	8	hexa
2opt tsp	1.85	740.14	50	2	0	hexa
som tsp	17.5	744.78	50	18	0	hexa

alg	time	length	npoints	epochs	self int	dist
tsp cristofides	0.472	13102	200	3	9794	hexa
2opt tsp	129.6	1683.78	200	2	0	hexa
som tsp	24.1	1661.9	200	24	0	hexa

alg	time	length	npoints	epochs	self int	dist
tsp cristofides	2.61	29469.28	500	1	9794	hexa
2opt tsp	1938	2655.13	500	4	0	hexa
som tsp	28.2	2634.58	500	27	0	hexa

Fig. 10. Results for hexagonal point distribution (time in seconds, length in mm)

alg	time	length	npoints	epochs	self int	dist
tsp cristofides	0.25	3190.1	50	1	600	radial
2opt tsp	1.26	668.74	50	2	0	radial
som tsp	20.35	696.9	50	17	0	radial

alg	time	length	npoints	epochs	self int	dist
tsp cristofides	0.37	13288	200	1	10268	radial
2opt tsp	138.6	1251.07	200	4	0	radial
som tsp	23.9	1293.99	200	24	0	radial

alg	time	length	npoints	epochs	self int	dist
tsp cristofides	2.8	30852	500	1	45552	radial
2opt tsp	2003	1948.92	500	4	0	radial
som tsp	30.31	2167.55	500	24	0	radial

Fig. 11. Results for radial point distribution (time in seconds, length in mm)

alg	time	length	npoints	epochs	self int	dist
tsp cristofides	0.25	3281	50	1	668	tri
2opt tsp	1.65	754.99	50	3	0	tri
som tsp	17.69	752.39	50	16	0	tri

alg	time	length	npoints	epochs	self int	dist
tsp cristofides	0.47	12846.1	200	1	9744	tri
2opt tsp	303	1548.17	200	4	0	tri
som tsp	23.23	1528.61	200	24	0	tri

alg	time	length	npoints	epochs	self int	dist
tsp cristofidas	2.65	30646.54	500	1	49690	tri
2opt tsp	2012	2486	500	4	0	tri
som tsp	27.41	2445.31	500	27	0	tri

Fig. 12. Results for triangular distribution (time in seconds, length in mm)

Applying different solvers based on Constrained Optimization by Linear Approximations (COBYLA)[5], Particle Swarm Optimization (PSO)[6], and SPEA-2 and HypE algorithms[7] on the path tangents for further length optimization produces results that range from 3.56% to 5.17%. The use of an extra layer of optimization was not further developed as it increased calculation times over Millipede's topological optimization and path optimization using SOMs. Finally, the material distribution mapped to a polyline is calculated at every epoch of the optimization, not increasing execution times significantly (GH profiler widget calculates this mapping operation in 107 ms. approx. for 500+ points). The resulting output is a customized version of G-code controlling two extruders according to the line gradient mapped values (Fig. 13).

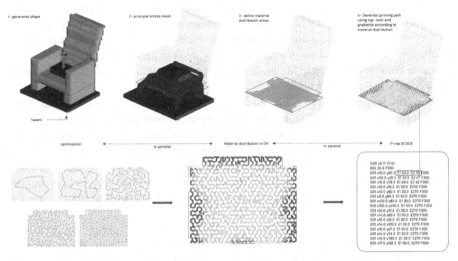

Fig. 13. The SOM algorithm applied for the production of a P-Rep G -Code

5   Conclusion

New techniques for the generation of shapes based on material properties and functions in 3d environments demand new methodologies for fabricating them. In the case of Functional Gradient Materials, parameters such as material continuity between components is crucial. Ensuring a proper gradient deposition between material components using 3D printers opens new paths for exploration for the design, exploration, and performance of FGMs. This work introduces a novel approach to bridge new software developments for volumetric modeling and alternative multi-material 3d printed methods. In this project,

---

[5] Through Radical component from Digital Space Exploration plugin for Grasshopper http://digitalstructures.mit.edu/page/tools#design-space-exploration-tool-suite-for-grasshopper.

[6] Through Silvereye optimization plugin for Grasshopper. https://www.food4rhino.com/en/app/silvereye-pso-based-solver.

[7] Through the use of Octopus for Grasshopper. https://www.food4rhino.com/en/app/octopus.

the motivation was to bridge the work produced in the implementation of the software SkeXl that generates procedural 3d Shapes by using Generative Adversarial Networks and the Alchemic printer, an interactive multi-material 3d printer. It started from a simple premise, Can we map quantitative metrics into material gradients into a single path shaped as the infill of a given geometry in an optimized way in terms of computation and length? In the case of this paper, the optimization of shapes to efficiently distribute materials in shapes, minimizing airtime, and avoiding self-intersections, was addressed toward informed material metrics delivered, for example, by the principal stress mesh derived from topological optimization.

By solving a 3d printing path by mapping material gradients to a single distributed path across an area defined by a boundary can be achieved efficiently using a modified version of the TSP problem. As a result, the comparison of three different approaches of the TSP, demonstrated that when optimizing to minimize path length and print time, self-intersections, airtime, and path-calculation times, self-organizing maps show promising results for their consideration in interactive fabrication systems and in the production of printable Functional Gradient Materials (FGM) and the production of a suitable version of Gcode for P-Reps fabrication.

Nevertheless, more work is needed for the efficient integration of a platform to perform optimized path calculations. As an example, Grasshopper is the more obvious option when generative design and performative analysis is needed in conjunction; Nevertheless, Grasshopper limitations related to its interoperability with external tools such as Python (no Python 3 support, resulting in problems of using GH-Remote and NumPy for example), demand alternative solutions such as the use of extra layers of processing like server communication between different software packages and data files, increasing calculation paths.

Acknowledgements. I would like to thank Dr. Caitlin Mueller and Yijiang Huang for all the support and great lectures at 4.450 Computational Structural design and Optimization at MIT. Their knowledge and dedication motivated me to propose and develop an atypical project that helped me move forward with important areas of my Ph.D. research. Also, I would like to thank the Design and Computation group at MIT and the school of design of Adolfo Ibañez University for their valuable support.

References

1. Tsamis, A.: Software tectonics. Ph.D. thesis, MIT Department of Architecture, Cambridge, MA (2012)
2. By the use of software packages such as Vspace (Tsamis, Alexandros. https://github.com/tsamis/VSpace) or Monolith Michalatos, Panagiotis- Autodesk https://www.food4rhino.com/app/monolith
3. Kaijima, Michalatos: Using millipede for topologycal optimization. http://www.sawapan.eu/
4. Danhaive, R., Pinochet, D.: SKeXL, Sketch to Shape: A Generative design tool using three-dimensional generative adversarial networks by using user sketches to generate procedural shapes. https://diegopinochet.com/portfolio#42f7b021-2e36-42c0-9527-bf10495f2cd5
5. Diego Pinochet (2019). https://diegopinochet.com/portfolio#2fa9c4d8-59d6-454c-b981-d84f2dba8cde

6. Dreifus, G., et al.: 3D Print. Addit. Manuf. 98–104 (2017). https://doi.org/10.1089/3dp.2017.0007

7. Castelino, K., D'Souza, R., Wright, P.K.: Toolpath optimization for minimizing airtime during machining. J. Manuf. Syst. **22**(3), 173–180 (2003)

8. Kohonen, T.: The self-organizing map. Neurocomputing **21**(1), 1–6 (1998)

9. Brocki, L.: Kohonen self-organizing map for the traveling salesperson. In: Traveling Salesperson Problem, Recent Advances in Mechatronics, pp. 116–119 (2010)

10. Michalatos and Payne. Working with Multi-scale Material Distributions. http://papers.cumincad.org/cgi-bin/works/Show?acadia13_043

11. https://static1.squarespace.com/static/54450658e4b015161cd030cd/t/56ae214afd5d08a9013c99c0/1454252370968/Monolith_UserGuide.pdf

12. ACADIA 14: Design Agency. In: Proceedings of the 34th Annual Conference of the Association for Computer Aided Design in Architecture (ACADIA), Los Angeles, 23–25 October 2014, pp. 101–110 (2014). ISBN 9781926724478

13. Grigoriadis: Living systems and micro-utopias: towards continuous designing. In: Proceedings of the 21st International Conference on Computer-Aided Architectural Design Research in Asia (CAADRIA 2016), Melbourne, 30 March–2 April 2016, pp. 589–598 (2016)

14. Oxman, N., Keating, S., Tsai, E.: Innovative Developments in Virtual and Physical Prototyping: In: Bártolo, P.J., et al. (eds.) Proceedings of VRAP: Advanced Research in Virtual and Rapid Prototyping. Taylor & Francis

15. Lechowicz, P., Koszalka, L., Pozniak-Koszalka, I., Kasprzak, A.: Path optimization in 3D printer: algorithms and experimentation system. In: 2016 4th International Symposium on Computational and Business Intelligence (ISCBI), Olten, pp. 137–142 (2016). https://doi.org/10.1109/ISCBI.2016.7743272

16. Dreifus, G., et al.: Path optimization along lattices in additive manufacturing using the Chinese postman problem. 3D Print. Addit. Manuf. **4**, 98–104 (2017). https://doi.org/10.1089/3dp.2017.0007

17. Fok, K., Ganganath, N., Cheng, C., Tse, C.K.: A 3D printing path optimizer based on Christofides algorithm. In: 2016 IEEE International Conference on Consumer Electronics-Taiwan (ICCE-TW), Nantou, pp. 1–2 (2016). https://doi.org/10.1109/ICCE-TW.2016.7520990

18. Dreifus, G., et al.: 3D Print. Addit. Manuf. 98–104 (2017). https://doi.org/10.1089/3dp.2017.0007

19. Kohonen: The self-organizing map. Neurocomputing **21**(1), 1–6 (1998)

20. Brocki, Ł., Korzinek, D.: Kohonen Self-Organizing Map for the Traveling Salesperson Problem (2007). And Brocki, L. (2010).

21. Kohonen: Self-organizing map for the traveling salesperson. In: Traveling Salesperson Problem, Recent Advances in Mechatronics, pp. 116–119

22. Kaijima, S., Michalatos, P.: Intuitive material distributions, Architectural Design (2011). Millipede. An analysis and optimization tool. http://www.sawapan.eu/ based on the work shown in

23. Chaikin, G.M.: An algorithm for high-speed curve generation. Comput. Graph. Image Process. **3**(4), 346–349 (1974)

24. Implemented using RADICAL, a component part of the DSE toolkit. http://digitalstructures.mit.edu/page/tools#design-space-exploration-tool-suite-for-grasshopper

25. https://en.wikipedia.org/wiki/Christofides_algorithm

26. https://towardsdatascience.com/how-to-solve-the-traveling-salesman-problem-a-comparative-analysis-39056a916c9f

CDPR Studio: A Parametric Design Tool for Simulating Cable-Suspended Parallel Robots

Ethan McDonald[1]([⊠]) (ID), Steven Beites[2]([⊠]) (ID), and Marc Arsenault[1]([⊠]) (ID)

[1] Bharti School of Engineering, Laurentian University, Sudbury, ON, Canada
{emcdonald1,marsenault}@laurentian.ca
[2] McEwen School of Architecture, Laurentian University, Sudbury, ON, Canada
sbeites@laurentian.ca

Abstract. The research explores the design and fabrication of a cable-driven parallel robot (CDPR) as an innovative and alternative method for *in-situ* robotic construction. CDPRs consist of a mobile platform attached to a fixed frame by several cables that are controlled through actuating winches. Their increased mobility reveals an immense potential for *in-situ* design-fabrication solutions emphasizing the importance of local materials and local economies. The paper presents a facet of this research focusing on the development of a novel design tool for the control and simulation of cable-suspended parallel robots in order to facilitate architectural design-fabrication explorations while addressing the complexities associated with parallel robots.

Keywords: Robotics · In-situ fabrication · Optimization · Interface · Cable-driven parallel robot · Cable-suspended parallel robot · CDPR · CSPR

1 Introduction

The research explores the development of a mobile robotic construction platform, introducing digital tools and technologies as a catalyst for a creative economy and local building solutions. It begins with the design and fabrication of a cable-driven parallel robot (CDPR) as an innovative and alternative method for *in-situ* robotic construction. CDPRs consist of a mobile platform attached to a fixed frame by several cables that are controlled through actuating winches. They have several advantages over more traditional industrial robotic arms due to their ability to span large distances, their increased mobility, portability and lower associated costs. Employed historically in material/cargo handling applications, high-speed tracking photography and live broadcasting (*e.g.* Sky-Cam), CDPRs have been seldom explored within architectural practice yet have immense potential to transform current methods of construction. As current building processes rely on prefabricated architectural components that are manufactured in global facilities, transported and assembled on-site by highly skilled labour, CDPRs provide an opportunity to circumvent these conditions by facilitating custom, *in-situ* design-fabrication solutions emphasizing the importance of local materials and local economies.

D. Gerber et al. (Eds.): CAAD Futures 2021, CCIS 1465, pp. 344–359, 2022.
https://doi.org/10.1007/978-981-19-1280-1_22

CDPRs are parallel robots, and in contrast to serial robots, they utilize flexible cables as opposed to motor-actuated joints. The flexible cables are driven using several winches, allowing the mobile platform to move in 3D space by varying the cable lengths. Since a CDPR uses cables to move the mobile platform in 3D space, all cables must be maintained in tension within a certain range; otherwise, the mobile platform could become under-constrained. CDPR trajectory planning is challenging because CDPRs have many cables that can potentially collide with objects in the CDPRs workspace.

Accordingly, research is underway focusing on the hardware implementation of the aforementioned platform in addition to the development of an interface to facilitate design explorations. The ability to simulate the CDPR within a design environment prior to physical making is critical to its early adoption within an architectural design framework. Design fabrication tools such as Kuka|prc [1], which allows for the control and simulation of serial robots, have provided designers (with little to no previous experience in robotics) with unlimited design exploration capabilities. The research presented in this paper begins to highlight the methods explored towards the development of a parametric design tool for the control and simulation of CDPRs in order to facilitate architectural fabrication innovation while addressing the complexities associated with parallel robots.

2 Background

2.1 CDPR Components

CDPRs (Fig. 1) use a set of cables to control the position and orientation (also known as the pose) of the mobile platform. This is achieved through the use of winches in order to precisely control the length of the cables. There are two main types of CDPRs; a) suspended and b) fully constrained CDPRs [2]. As the name suggests, suspended CDPRs suspend the mobile platform from cables positioned at the top of the frame. Fully constrained CDPRs, however, require cables that extend from both the top and bottom of the frame.

When evaluating and comparing CDPRs, several vital parameters must be considered; they include the wrench-feasible workspace (henceforth simply referred to as the workspace), cable interferences, and controllable number degrees of freedom (DOF). The workspace refers to the region in 3D space where the mobile platform can be positioned and oriented while maintaining a minimum tension in the required number of cables without exceeding the breaking strength of any cable [2]. Cable interferences occur when one or more cables collide with another cable or an object in its workspace [2]. The ability to analyze and gain visual feedback through the simulation of these parameters are critical factors for ensuring a seamless design-to-fabrication workflow and enhancing the designer's ability to test iterations accurately and efficiently.

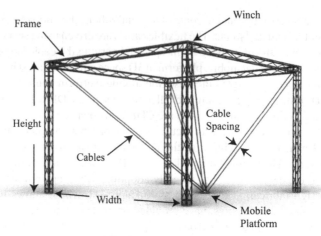

Fig. 1. CDPR architecture.

2.2   CDPRs Used in Architecture

Although not novel to architectural applications, CDPRs have been seldom explored within design/fabrication frameworks. Notable precedents include:

[3] proposed a contour crafting CDPR, improving upon the portability and cost of previously proposed contour crafting solutions. The CDPR geometry minimizes cable collisions by including a movable crossbar that allows the lower cables to translate upwards as required to avoid collisions with the building being constructed. Kinematic and static models were required to simulate and control the CDPR using a control system implemented in MATLAB/Simulink.

[4] introduced the SPIDERobot: a four-cable CDPR capable of pick-and-place operations in 4 DOF. The SPIDERobot is in a suspended configuration which helps to prevent collision with objects in the CDPR's workspace. Implementation of a dynamic position system that is a vision-based control scheme allowed the control system to adaptively correct for error in cable length and avoid collisions. Pick-and-place operations were performed by feeding object coordinates from the Rhinoceros 3D modeling environment into a comma-separated values (CSV) file, then directly into the control system.

[5] attached a 3D printing extruder to the CoGiRo: a large-scale, eight cable CDPR capable of operating in 6 DOF. 3D print trajectories were created using Rhinoceros and a custom script made in Grasshopper that converts optimal trajectories into a G-code file containing the required position and printing speed information. The G-code file was then used to control the CoGiRo by feeding the G-code into a G-code postprocessing script implemented on a programmable logic controller.

Mamou-Mani [6] and Arup [7] have collaborated to create the POLIBOT for the Sir John Soane's Museum. The POLIBOT is an eight-cable CDPR created to demonstrate the potential of CDPRs in architecture through pick-and-place operations within a controlled environment.

As precursors to the potential of CDPRs within architectural praxis, the projects differ significantly from the proposed research in terms of their programming and simulation capabilities. The CDPRs trajectories, presented above, were typically planned using a series of coordinates exported from a computer-aided design (CAD) program such as Rhinoceros in the form of either a CSV file or a G-code file. The CSV/G-code file was then sent directly to the CDPR's control system, and the trajectory is executed on the CDPR. The presented research however focuses on the development of a dynamic interface within the Grasshopper modeling environment to assist designers in generating and analyzing CDPR trajectories for architectural applications, and to simulate them for visual feedback and error detection prior to executing the trajectories. The utility of the proposed interface is demonstrated throughout this paper on an eight-cable suspended CDPR with a parallelogram cable arrangement which will be designed for performing *in-situ* construction tasks (i.e. 3D printing, pick-and-place operations, etc.).

2.3 CDPR Modeling Methods

The mechanical design of CDPRs is simple as they only consist of a small number of mechanical parts. However, the modeling of CDPRs poses some interesting challenges. To facilitate analysis and design activities, engineering professionals often rely heavily on numerical computing software such as MATLAB. MATLAB allows users to analyze data, find numerical solutions to equations, and perform optimizations, but it does not include any CDPR analysis or design tools. In addition to MATLAB, several platforms have emerged for cable-robot analysis and simulation, including kinematics, dynamics, control, and workspace analysis, for example:

WireX. WireX [8] is open-source software for the analysis and design of cable-driven parallel robots. It allows users to select CDPR geometry to analyze or model new configurations. Users can model forward kinematics (FK) and inverse kinematics (IK); compute the structure matrix and stiffness matrix of a given pose; detect collisions; determine workspace, and plan paths from G-code files [8].

CASPR. CASPR [9] is Similar to WireX, but it operates within the MATLAB environment. CASPR is an open-source platform that allows users to model FK and IK, forward and inverse dynamics, CDPR control and determine workspace feasibility. CASPR is useful for modeling and analyzing CDPRs.

However, a need exists for a design tool that combines the modeling and analysis of CDPRs with fabrication logistics through integrated CAD functionality. Accordingly, the objective of the research presented in this paper is to develop a design interface capable of CDPR analysis and simulation in order to facilitate the design and fabrication of architectural assemblies. Within the family of CDPRs, this tool will focus specifically on a cable-suspended parallel robot henceforth referred to as CSPR. Other CDPR systems will not be explored in this paper.

3 Methods

In order to initiate a design-to-production workflow while simultaneously maintaining the simulation capabilities necessary for parallel robots, the research explored the potential integration of CSPR analysis tools with design and fabrication capabilities typically found in CAD environments. McNeel's Rhinoceros was selected for its wide-reaching presence in architectural design communities and for its parametric plugin Grasshopper, a graphical algorithm editor that is tightly integrated with Rhino's 3D modeling tools. The methods described in the following sections will be used to analyze and simulate a CSPR.

3.1 MATLAB/Grasshopper Integration

Early explorations focused on the integration of MATLAB into Grasshopper as a means to facilitate design explorations yet maintain the modeling, analysis and simulation accuracy of CSPRs. IronPython programming language is native to Grasshopper yet incompatible with MATLAB. As a result, a different version of Python was used with Grasshopper to overcome this limitation. GH CPython [10] was selected for its ability to use Anaconda (Python distribution platform) and its various libraries, such as NumPy.

Once Anaconda was integrated into Grasshopper using GH CPython, several scripts were written in MATLAB to check the wrench feasibility, FK, IK and cable tensions, etc. Despite the accuracy of the simulations, the connection between MATLAB and Grasshopper was inadequate for visualizing trajectories due to the extensive time required to run the simulation, as calculations were being performed at every division point across a curve/path. Overall, MATLAB is well suited for modeling and analyzing CSPR designs; however, its integration within Rhinoceros 3D was problematic as it did not facilitate efficient design explorations and/or the rapid simulation of CSPR fabrication processes (Fig. 2).

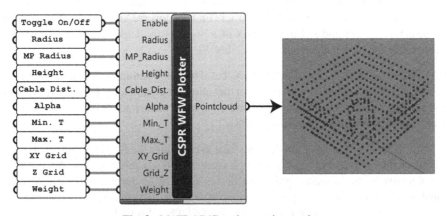

Fig. 2. MATLAB/Grasshopper integration.

3.2 CDPR Studio

The unsuccessful integration of MATLAB/Grasshopper led to the design and development of "CDPR Studio," a suite of Grasshopper components capable of analyzing and simulating CSPRs within the Rhinoceros modeling environment. All components of

CDPR Studio use GH CPython to run the Anaconda Python distribution. The Anaconda Python distribution allows Grasshopper to use the NumPy Python library [11] to perform all required trajectory and wrench feasibility calculations. CDPR Studio can generate trajectories for a variety of fabrication applications, calculating the FK and IK, the kinematic sensitivity of a given pose, cable tensions outside the acceptable range and alerting the designer of any collisions. It also provides users with a method to visualize Cartesian space and joint-space trajectories and out of range cable tensions while calculating the cable lengths and cable length change at every point of the Cartesian space trajectory.

For a pose to be wrench feasible, the external wrench applied to the mobile platform must be balanced using positive cable tensions that are above the minimum cable tension and do not exceed the maximum acceptable cable tension. Each pose's wrench feasibility was checked using techniques identified in *Cable-Driven Parallel Robots: Theory and Applications* [2]. Trajectories were planned based on methods described in *Introduction to Robotics: Mechanics and Control* [12]. The calculations of kinematic sensitivity were performed based on methods outlined in *Kinematic-Sensitivity Indices for Dimensionally Nonhomogeneous Jacobian Matrices* [13].

Figure 3 shows the core components of CDPR Studio, including the wrench feasibility checker, Cartesian space trajectory planner, joint-space trajectory planner, kinematic

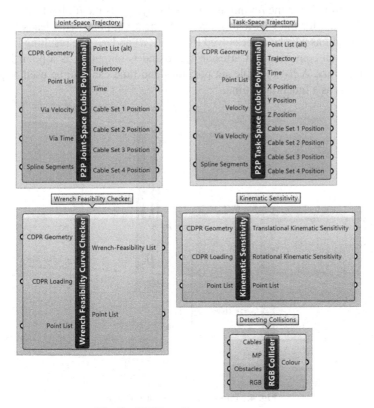

Fig. 3. CDPR studio core components.

sensitivity calculator and the collision detector. Each component accepts specific data types as inputs and outputs new data types required for the analysis and simulation of the desired fabrication applications (*e.g.* pick-and-place, 3D printing, etc.).

Trajectory Planning. CDPR Studio includes two point-to-point trajectory planning methods. The first plans trajectories in joint-space, and the second plans trajectories in Cartesian space.

Joint-Space Trajectory Planner. The joint-space trajectory planner requires CDPR geometry, a point list, via velocity, via time and spline segments to generate the trajectory (Fig. 4). The joint-space trajectory is based on cubic polynomial trajectory interpolation. The ith cable length as a function of time is [12]:

$$\rho_i(t) = a_{0,i} + a_{1,i}t + a_{2,i}t^2 + a_{3,i}t^3 \tag{1}$$

The coefficients in Eq. (1) are calculated by imposing the following bounds:

$$\rho_i(0) = \rho_{0,i}, \ \rho_i(t_f) = \rho_{f,i}, \ \dot{\rho}_i(0) = \dot{\rho}_{0,i}, \ \dot{\rho}_i(t_f) = \dot{\rho}_{f,i} \tag{2}$$

Where $\rho_{0,i}$ is the ith initial cable length (found using IK), $\rho_{f,i}$ is the ith final cable length (found using IK), $\dot{\rho}_{0,i}$ is the ith initial cable velocity, $\dot{\rho}_{f,i}$ is the ith final cable velocity, and t_f is the time for the CDPR to move to the next via point. This leads to a linear system of equations which can be solved for a_0, a_1, a_2 and a_3. The resulting joint-space trajectory is then converted to Cartesian space and is outputted as a point list and a polyline trajectory. A list of cable set positions and time can also be used to plot the cable positions as a function of time.

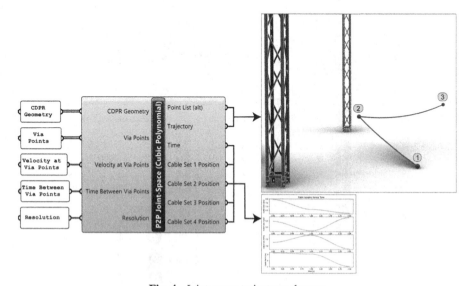

Fig. 4. Joint-space trajectory planner.

Cartesian Space Trajectory Planner. The Cartesian space trajectory planner requires CDPR geometry, a point list, velocity, via velocity and spline segments to generate the trajectory (Fig. 5). The Cartesian space trajectory is based on cubic polynomial trajectory interpolation and uses a combination of methods used in [12] and [14]. The Cartesian space position as a function of time is:

$$\mathbf{p}(t) = \mathbf{p}_0 + \sigma(t)(\mathbf{p}_f - \mathbf{p}_0) \tag{3}$$

where $\mathbf{p}_0$ is the initial point of the trajectory, $\mathbf{p}_f$ is the final point in the trajectory, and:

$$\sigma(t) = b_0 + b_1 t + b_2 t^2 + b_3 t^3 \tag{4}$$

The coefficients in Eq. (4) are calculated by imposing the following bounds:

$$\sigma(0) = 0, \sigma(t_f) = 1, \dot{\sigma}(0) = \frac{\dot{p}_0}{d}, \dot{\sigma}(t_f) = \frac{\dot{p}_f}{d} \tag{5}$$

where $\dot{p}_0$ and $\dot{p}_f$ are the initial and final Cartesian space speeds, respectively, and d is the Euclidean distance between $\mathbf{p}_0$ and $\mathbf{p}_f$. This leads to a linear system of equations which can be solved for b_0, b_1, b_2 and b_3. The resulting trajectory is outputted as a point list and a polyline trajectory. A list of cable set positions, Cartesian space coordinates and time can also be used to plot the cable positions and mobile platform position as a function of time.

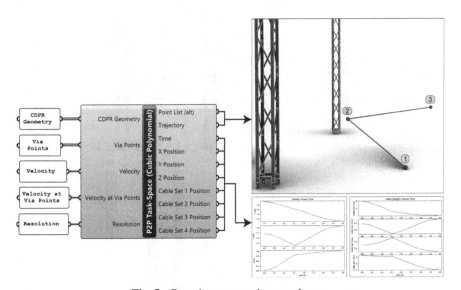

Fig. 5. Cartesian space trajectory planner.

Wrench Feasibility Verification. When all cables of a CDPR are within their acceptable tension range and the external wrench applied to the mobile platform is balanced, it is said to be in a wrench-feasible pose [2]. CDPR Studio checks a pose's wrench feasibility by determining if the pose satisfied the following equations [2]:

$$\mathbf{A}^T\mathbf{f} = \mathbf{w}_p \tag{6}$$

$$\mathbf{f}_{min} \preccurlyeq \mathbf{f} \preccurlyeq \mathbf{f}_{max} \tag{7}$$

where $\mathbf{A}^T$ is the structure matrix, $\mathbf{f}$ is the cable tension vector, $\mathbf{w}_p$ is the external wrench applied to the mobile platform, and $\mathbf{f}_{min}$ and $\mathbf{f}_{max}$ are the minimum and maximum permissible cable tensions respectively [2]. Moreover, "$\preccurlyeq$" implies the component-wise comparison of the vectors. If any of the tensions that balance the external wrench are found to be out of the acceptable range, the wrench feasibility checker alerts the designer that the pose is not wrench feasible. The designer can also determine an entire curve's wrench-feasibility using the wrench-feasibility checker by discretizing the curve into points. They must define the CDPR geometry, loading conditions and convert the curve to a point list for the wrench-feasibility checker to check all points. The wrench-feasibility checker outputs a wrench-feasibility list and a point list that can be used to find points along the curve where cables are out of the acceptable tension range. Figure 6 shows an example of the Grasshopper definition that uses the wrench-feasibility checker to check points along the curve.

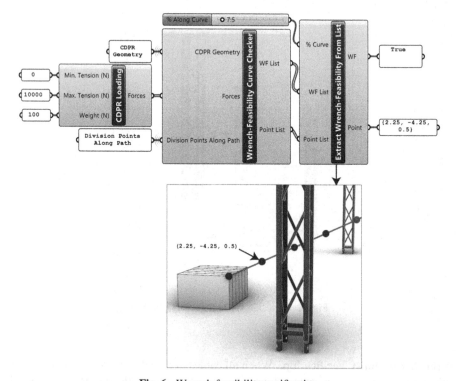

Fig. 6. Wrench feasibility verification.

Kinematic Sensitivity. Kinematic sensitivity (KS) is a measure of how uncertainties in the actuator displacements translate to uncertainties in mobile platform displacements [13]. In essence, it can be used to estimate the positional and rotational accuracy of a mobile platform's pose given the actuator accuracies. The user can determine the kinematic sensitivity along a curve by discretizing the curve into points. They must define the CDPR geometry, loading conditions and convert the curve to a point list. The kinematic sensitivity calculator then outputs a translational kinematic sensitivity list, a rotational kinematic sensitivity list and a point list that can be used to find the KS along the curve (Fig. 7). The translational kinematic sensitivity is calculated using the following equation [13]:

$$\sigma_p = \frac{1}{\sqrt{\min(\lambda_{1,p}, \lambda_{2,p}, \lambda_{3,p})}} \tag{8}$$

where σ_p is the translational kinematic sensitivity, and $\lambda_{j,p}$ is the jth eigenvalue of $\mathbf{K}_p^T \mathbf{P}_r \mathbf{K}_p$ where $\mathbf{K}_p$ is the translational partition of the CDPRs inverse Jacobian matrix, and:

$$\mathbf{P}_r = 1 - \mathbf{K}_r \left(\mathbf{K}_r^T \mathbf{K}_r \right)^{-1} \mathbf{K}_r^T \tag{9}$$

where 1 is the identity matrix, and $\mathbf{K}_r$ is the rotational partition of the CDPRs inverse Jacobian matrix. Similarly, rotational kinematic sensitivity is calculated using the following equation [13].

$$\sigma_r = \frac{1}{\sqrt{\min(\lambda_{1,r}, \lambda_{2,r}, \lambda_{3,r})}} \tag{10}$$

where σ_r is the translational kinematic sensitivity, and $\lambda_{j,r}$ is the jth eigenvalue of $\mathbf{K}_r^T \mathbf{P}_p \mathbf{K}_r$ and:

$$\mathbf{P}_p = 1 - \mathbf{K}_p \left(\mathbf{K}_p^T \mathbf{K}_p \right)^{-1} \mathbf{K}_p^T \tag{11}$$

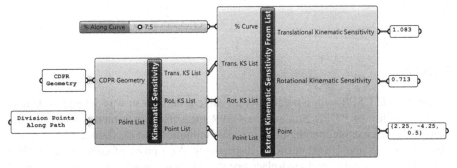

Fig. 7. Kinematic sensitivity calculator.

The estimated overall worst-case position uncertainty can calculated using [15]:

$$\max\left(\|\delta p\|_2\right) = \sqrt{m}\sigma_p\delta\rho \qquad (12)$$

where m is the number of cables, δp is the mobile platform position uncertainty and $\delta\rho$ is the cable uncertainty.

Collision Detection. Collision detection allows designers to observe collisions in their trajectories before executing them on a CDPR. One method for collision detection is outlined in [16] where the authors propose a solution that calculates the distance between each cable and all objects in the CDPR's workspace using an iterative method based on Rosen's minimization. The research presented in this paper however utilizes a current built-in tool within Grasshopper called the One|Many component, allowing for collisions between selected objects to be detected. Collisions are thus detected in CDPR Studio using this RGB collider component. The RGB collider accepts cables, the mobile platform, obstacles and RGB as inputs. If collisions are not detected, the color output is defined by the RGB input value. If collisions are detected, the color is overridden to be red. The RGB collider's output can change the cables' color, allowing designers to visualize collisions quickly. The definition in Fig. 8 is an example of collision detection being implemented to change the CSPR cables' color depending on whether a collision is present. If a collision is detected, the cable color is red. If collisions are not detected, the cable's color is defined by the RGB value going into the RGB collider.

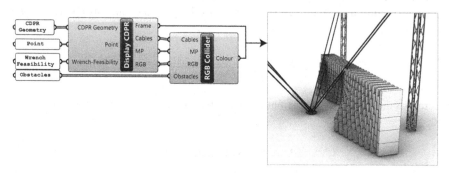

Fig. 8. Collision detection.

4   Results

An eight-cable CSPR with a parallelogram cable arrangement was defined to demonstrate the utility of CDPR Studio. The CSPR measured seven meters wide and five meters tall with a cable spacing of 20 cm. The minimum and maximum cable tensions were set to zero and 10000 newtons, respectively. As a proof of concept, the CSPRs was tasked to simulate a pick-and-place operation for the fabrication of a masonry brick wall. Pick-and-place trajectories were generated using the joint-space and Cartesian space trajectory planners, demonstrating CDPR Studio's trajectory planning features. The mobile platform was then moved to a location outside of its workspace to demonstrate how CDPR Studio detects poses that are not wrench-feasible. Finally, an obstacle was placed in the pick-and-place trajectory to demonstrate how CDPR Studio can detect collisions. Figure 9 shows the CSPR in the Rhinoceros environment, demonstrating the pick-and-place operation. This setup will be used to demonstrate the results of CDPR Studio's features.

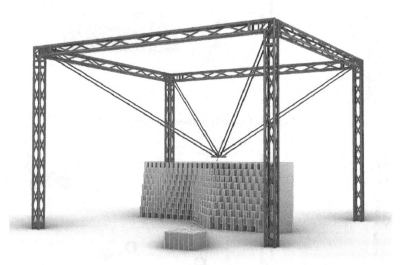

Fig. 9. CSPR – pick-and-place operation.

4.1 Trajectory Planning

Figure 10 shows an example joint-space trajectory of a CSPR picking and placing a block on a platform. CDPR Studio can display the joint-space trajectory in Cartesian space, allowing the designer to see whether the joint-space path is acceptable. Trajectories planned in joint-space are typically curved paths when mapped back into Cartesian-space. The trajectory is shown in Fig. 10, where most of the trajectory appears as a curve.

It is usually desirable to plan trajectories in joint-space for pick-and-place operations; however, the user can also plan the trajectory in Cartesian space if required. Trajectories planned in Cartesian space are inherently linear and are desirable when the trajectory must follow a straight line or approximate a path as accurately as possible (*i.e.* 3D Printing). Cartesian space trajectories are suitable for following curved paths by discretizing the path into various segments. Figure 11 identifies a pick-and-place trajectory planned in Cartesian space. All trajectories appear as straight lines.

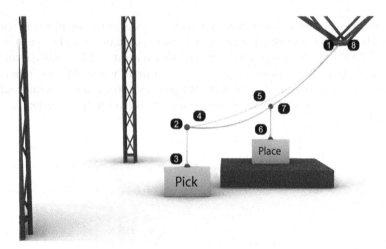

Fig. 10. Example joint-space trajectory.

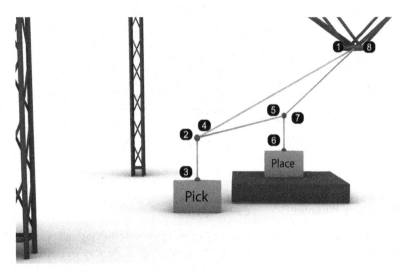

Fig. 11. Example of Cartesian space trajectory.

4.2 Wrench Feasibility

If any of the cables are out of the tension range, the wrench-feasibility checker will display the cables in red. This visual feedback allows the designer to avoid poses that are not wrench-feasible while following a trajectory. Figure 12 demonstrates this alert when the mobile platform is moved out of the workspace, and the cables cannot balance out the mobile platform wrench with positive cable tensions.

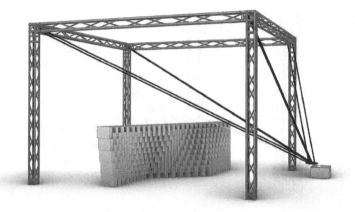

Fig. 12. Example of a CSPR pose failing the wrench-feasibility check.

4.3 Kinematic Sensitivity

A smaller kinematic sensitivity value corresponds with a more accurate location in the workspace. If possible, designers should minimize the kinematic sensitivity along a trajectory to maximize the accuracy of the CDPR. For example, suppose the kinematic sensitivity near the edge of the workspace is two, and the kinematic sensitivity near the center of the workspace is one. In that case, the CDPR should be operated closer to the center of the workspace. Minimizing the kinematic sensitivity is especially important for fabrication processes, including 3D printing and pick-and-place operation. While 3D printing, large kinematic sensitivities can result in poor dimensional accuracy of the printed part. During pick-and-place operations, if the kinematic sensitivity is large, the object can be placed in the incorrect locations. To demonstrate the utility of the kinematic sensitivity, the mobile platform was moved to the position $(x, y, z) = (0, 0.1, 2)$. The translational kinematic sensitivity at that pose was 1.23. If, for example, the actuators had an uncertainty of one millimeter, the worst-case estimated accuracy of the mobile platform position would be 3.48 mm.

4.4 Collision Detection

The RGB collider feature of CDPR Studio helps designers visualize collisions present when following a trajectory. During a collision, the cables are displayed in red, alerting the designer that a collision is present and to adjust their trajectory accordingly. The collision shown in Fig. 13 is a cable-obstacle collision occurring between an obstacle and one or more of the cables. The RGB collider can detect and display collisions between the mobile platform, cables, and as many obstacles as the user defines.

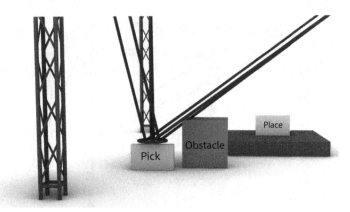

Fig. 13. Pick-and-place cable-obstacle collision.

5   Conclusion and Future Work

The work presented is the first step within a larger body of research that focuses on the design and fabrication of a CSPR as an innovative and alternative method for *in-situ* robotic construction. The development of a mobile robotic construction platform would allow for making at multiple scales, from small objects to furniture to architectural components. In addition to its manufacturing capabilities (*e.g.* subtractive and additive), the CSPR platform has the potential to assist with the assembly process through pick-and-place operations of larger, heavier assemblies.

Several platforms have emerged for cable-robot analysis and simulation that focuses exclusively on CDPR system engineering without the ability to evaluate architectural design variants. This paper presents a novel design tool capable of dynamically and accurately controlling a CSPR in order to maximize manufacturing capabilities and design explorations within an architectural framework. The research resulted in the development of "CDPR Studio," a parametric design tool allowing users, with little to no robotics experience, to model and simulate CSPR within the Rhinoceros design environment. Users can define CSPR parameters such as size, cable spacing and a cable tension range. Trajectories can be generated in joint-space or Cartesian space depending on the user's requirements. Using the wrench feasibility checker and kinematic sensitivity calculator components, important parameters such as wrench feasibility and kinematic sensitivity can be verified at various points along the path. CDPR Studio provides visual feedback of wrench feasibility or collisions, alerting the user if a pose is either not wrench feasible or if a collision is detected.

Future work on CDPR Studio will include new trajectory planners, including planners based on quintic polynomials instead of cubic polynomials. Using quintic polynomials allows users to control the acceleration at the start and endpoints of paths, allowing for smoother motions and transitions. Support for custom CDPR creation, custom tool creation and, online and offline CDPR programming will be further developed. Upon completion, CDPR Studio will be released as open-source software to facilitate the development of custom components that can be added to the software suite. It is possible for CDPR Studio to be extended to include analysis and simulation of general CDPRs (suspended and full constrained) through the addition of appropriate algorithms.

Beyond CDPR Studio, the development of the mobile robotic construction platform is underway. This includes dynamic modeling and analysis in order to identify a suitable CSPR design, the hardware requirements and the tolerances necessary for the intended fabrication applications. The CSPR's workspace, kinematic sensitivity and stiffness will be optimized to operate within the required parameters.

References

1. Braumann, J., Brell-Çokcan, S.: Parametric robot control: integrated CAD/CAM for architectural design. In: ACADIA 2011: Integration Through Computation: Proceedings of the 31st Annual Conference of the Association for Computer Aided Design in Architecture (ACADIA), pp. 242–251. Association for Computer Aided Design in Architecture (2011)
2. Pott, A.: Cable-Driven Parallel Robots: Theory and Application. STAR, vol. 120. Springer, Cham (2018). https://doi.org/10.1007/978-3-319-76138-1
3. Bosscher, P., Williams, R.L., Bryson, L.S., Castro-Lacouture, D.: Cable-suspended robotic contour crafting system Autom. Constr. **17**, 45–55 (2007). https://doi.org/10.1016/j.autcon.2007.02.011
4. Moreira, E., et al.: Cable robot for non-standard architecture and construction: a dynamic positioning system. In: 2015 IEEE International Conference on Industrial Technology (ICIT), pp. 3184–3189 (2015). https://doi.org/10.1109/ICIT.2015.7125568
5. Izard, J.-B.: Large-scale 3D printing with cable-driven parallel robots Constr. Robot. **1**(1–4), 69–76 (2017). https://doi.org/10.1007/s41693-017-0008-0
6. Mamou-Mani, A.: POLIBOT. https://mamou-mani.com/project/the-polibot/. Accessed 02 June 2021
7. Edge, A.: Are cable robots the future of construction? https://www.arup.com/perspectives/building-the-future-are-cable-robots-the-future-of-construction. Accessed 02 June 2021
8. Pott, A.: WireX – an open source initiative scientific software for analysis and design of cable-driven parallel robots. In: IFTOMM World (2019). https://doi.org/10.13140/RG.2.2.25754.70088
9. Lau, D., Eden, J., Tan, Y., Oetomo, D.: CASPR: a comprehensive cable-robot analysis and simulation platform for the research of cable-driven parallel robots. In: 2016 IEEE/RSJ International Conference on Intelligent Robots and Systems (IROS), pp. 3004–3011 (2016). https://doi.org/10.1109/IROS.2016.7759465
10. Abdelrahman, M.: GH CPython (2018)
11. Harris, C.R.: Array programming with NumPy. Nature **585**, 357–362 (2020). https://doi.org/10.1038/s41586-020-2649-2
12. Craig, J.J.: Introduction to Robotics: Mechanics and Control. Pearson, New York, USA (2018)
13. Cardou, P., Bouchard, S., Gosselin, C.: Kinematic-sensitivity indices for dimensionally non-homogeneous Jacobian matrices IEEE Trans. Rob. **26**, 166–173 (2010). https://doi.org/10.1109/TRO.2009.2037252
14. Tremblay, L.-F., Arsenault, M., Zeinali, M.: Development of a trajectory planning algorithm for a 4-DoF rockbreaker based on hydraulic flow rate limits Trans. Can. Soc. Mech. Eng. **44**, 501–510 (2020). https://doi.org/10.1139/tcsme-2019-0173
15. Barnett, E., Gosselin, C.: Large-scale 3D printing with a cable-suspended robot. Addit. Manuf. **7**, 27–44 (2015). https://doi.org/10.1016/j.addma.2015.05.001
16. Lahouar, S., Ottaviano, E., Zeghoul, S., Romdhane, L., Ceccarelli, M.: Collision free path-planning for cable-driven parallel robots. Robot. Auton. Syst. **57**, 1083–1093 (2009). https://doi.org/10.1016/j.robot.2009.07.006

Gradient Acoustic Surfaces

The Design, Simulation, Robotic Fabrication of a Prototype Wall

Kiefer Savage$^{(\boxtimes)}$ (ID), Nicholas Hoban (ID), and Brady Peters (ID)

John H. Daniels Faculty of Architecture, Landscape, and Design, University of Toronto, Toronto, Canada
kiefersavage@gmail.com, {nicholas.hoban,
brady.peters}@daniels.utoronto.ca

Abstract. While acoustics is a critical part of building performance, acoustic surfaces are often not integrated into a room's architectural language. Acoustically-performing surfaces are often considered separately or even applied post-construction on top of existing surfaces. This has numerous detrimental impacts including: material inefficiency, design incompatibility, and reduced visual and acoustic performance. This research proposes an approach that integrates acoustics, architectural design, and digital fabrication within one system. The experiments use methods of parametric design, associative geometry, FDTD and FEM acoustic simulations methods, and combines these with prototyping experiments and the development of a robotic fabrication system. The key finding is the development of the concept of the gradient acoustic surface: which we define as a single architectural tectonic system, that can perform as all three types of acoustic surfaces – absorber, diffuser, or reflector, where the acoustic performance coefficients can gradually change over the surface. Experiments were carried out through the development and production of a full-scale wall sample. Design, simulation, and fabrication workflows were developed and we outline these as well as the fabrication challenges we encountered with the robotic fabrication of a stacked-brick gradient acoustic surface.

Keywords: Architectural acoustics · Acoustic simulation · Robotic fabrication

1 Introduction

Sound is an inevitable part of architecture, and a room's acoustic qualities are one of the important ways people interpret and interact with architectural space [1]. Architecture's aural attributes are, however, one of its most overlooked aspects, and the prevalence of visual design facilitates the design of spaces that Pallasmaa [2] calls "stage sets for the eyes" – spaces devoid of sensorial depth. Sensorial depth should be created from the authenticity of a tectonic logic that addresses architecture's impact on all of the senses. It is known that the articulation of an architectural surface has a direct impact on incident sound waves, and this directly impacts the way space is aurally perceived [3]. The design of architectural surfaces has both visual and aural consequences, and this research prioritizes design for the aural condition. Design techniques are developed that are capable of producing complex surfaces that integrate an architectural tectonic language with a corresponding acoustic effect through the utilization of parametric design.

2   Background

2.1   Acoustic Performance of Architectural Surfaces

When considering architectural acoustics, room shape, surface area, and material characteristics are important. However, it is at the surface where sound waves interact with architecture. When sound waves impact an architectural surface, three different things that can happen: the sound energy can be absorbed, it can be specularly reflected, or it can be diffused – or scattered – into many directions [3]. These three surface conditions need to be considered when designing different acoustic environments.

Absorptive surfaces remove sound energy from a space. Reverberation time, that is, the time it takes for a sound to decay to inaudibility, is one of the most important acoustic characteristics [4], and acoustic absorption is the primary mechanism by which reverberation time can be controlled. Most acoustic absorbers are soft, porous materials such as carpets, heavy curtains, and fiberglass ceiling panels. The porosity of these materials enables sound waves to enter the material between fibers or particles, where the movement of the air is decreased due to friction and changes in direction [5]. Acoustic absorbers are defined by their absorption coefficient, which is found through material testing [6]. The absorption coefficient defines the amount of sound energy that is absorbed by the material compared to the amount of sound energy reflected back into the space.

Reflective surfaces are characterized by dense, smooth materials that preserve and redirect sound energy back into a space [5]. Having too many reflective surfaces is often a negative attribute, as this increases reverberation time and reduces the clarity of speech in a room. The reflection characteristics of surfaces can be most easily modelled using ray-tracing techniques where the angle of incidence equals the angle of reflection; however, one must be careful as surface size also plays a role due to diffraction effects.

Diffusive surfaces are typically characterized by hard materials and irregular geometries; they scatter sound while preserving its energy [3]. Sound scattering generally refers to the amount of directional variation when sound is reflected, while the concept of sound diffusion includes both spatial and temporal sound variation. These types of surfaces are thought to create a "lively" aural environment free from acoustic defects and noticeable reflections. They are typically found in rooms for music such as concert halls and recording studios. By controlling the depth and size of the surface aggregation, a diffusor can be tuned for a specific frequency of sound [3, 7, 8]. Sound diffusing acoustic surfaces are most commonly defined by their scattering coefficient [9], a parameter that can then be used in computer room acoustic simulations.

One of the problems with many commercially available acoustic products is that they are not integrated with the material language of the architectural design; many acoustic panels are designed to be additive features, affixed to walls or ceilings after the architecture is already built. As acoustic performance is often an overlooked aspect of building performance, this reinforces the attitude that acoustics is something to address after a building is built, rather than as an integrated part of the design. This research seeks to develop acoustic surfaces that are an integrated part of an architectural language.

2.2 The Computational Design of Acoustic Surfaces

There are three ways that computation has, and will continue to have, an impact on the design of surfaces for architectural acoustic performance. First, the use of computational tools has enabled the design of architectural surfaces that explore new potentials of highly-detailed complex geometry [10]. Building designers are interested in how the geometry and materiality of these surfaces can positively impact building performance such as acoustics [11]. Secondly, with greater integration of computer simulation techniques into the architectural design environment, building designers can not only explore the formal potentials of these computationally-generated surfaces, but their quantitative performance potentials. Commercial acoustic simulation software is now widely available and round robin tests indicated that for "commercial room simulation programs the results show good agreement" and for all programs "a certain level of performance is found which seems to indicate the state of the art"; however, it should be noted that results also indicate a "limit of performance under the currently applied algorithms" with noted improvements to be made in the areas of low frequency sound, diffraction, and complex impedance [12]. Thirdly, the development of digital fabrication techniques alongside computational design tools has simultaneously enabled the exploration of the tectonics of complex surface geometries alongside geometry and performance. It has been proposed that a linked approach that considers computational design techniques, digital fabrication processes, material properties, and tectonic construction is necessary to explore this new design territory of "digital materiality" [13].

In this research project, we develop new computational design techniques, leverage new research in computer acoustic simulation, and develop new techniques for the digital fabrication of acoustic surfaces. We build on previous research in the creation of customized acoustic environments through the manipulation of complex surfaces [14–16]. We developed a parametric computational model to generate surface geometries that can achieve different aural design objectives. Much previous research in this area has typically focused on the diffusion of sound [7, 17–19], neglecting the two other types of acoustic surface interactions: reflection and absorption. In our research, parametric modelling techniques were used to develop design options and test the extents of surface geometry variability; acoustic simulation techniques were used to test design options for their reflection, absorption, and diffusive qualities; and, robotic fabrication methods were used to prototype the proposed acoustic surface designs to test their constructability and architectural aesthetic qualities.

3 Experiments

3.1 Defining the Gradient Acoustic Surface

In this research project, our central research questions were: how can a single constructive system incorporate reflection, absorption, and diffusion? how can this system vary its form and geometry to create a wide variety of acoustic conditions? and how can new acoustic surfaces be designed to be part of a larger architectural tectonic concept so that acoustic performance, architectural design, and digital fabrication are linked together in one system?

In the process of finding answers to our research questions we developed the concept of the "gradient acoustic surface", which we will define here. As explained previously, the acoustic performance of surfaces is central to the development of a strategy to control the acoustic performance of an architectural space, and acoustic surfaces are classified as absorbers, reflectors, or diffusers – each defined by their own metrics and design techniques. A gradient acoustic surface is defined as a single architectural tectonic system, that can perform as all three types of acoustic surfaces – absorber, diffuser, or reflector – depending on its configuration. The configuration of a gradient acoustic surface varies spatially. A gradient acoustic surface varies from one performative acoustic condition to another, and as a result, the acoustic performance coefficients gradually change over the surface. These gradients of acoustic performance are designed using an associative parametric system that defines the changing relationships between material and geometry, and these changing relationships create different acoustic surface conditions. To be clear, a gradient acoustic surface is not a variable or adaptive acoustic surface, which is a surface that changes its acoustic performance characteristics through dynamic changes in its configuration. A gradient acoustic surface is a static construction, but one that has changing acoustic performance characteristics across its surface that are a result of gradients of geometry and material. We hypothesize that through the use of this new tectonic acoustic surface, architecture can be better acoustically-tuned for its programmatic needs, and acoustic performance can be more closely integrated with architectural design strategies.

3.2 Design of the Gradient Acoustic Surface

We tested the concept of the gradient acoustic surfaces through the design, simulation, and fabrication of one particular instance of a gradient acoustic surface. There are numerous properties of acoustic surfaces that can be varied to create different acoustic effects, and we chose a few of these properties to explore in this experiment. To control reflection, surfaces were oriented in different directions, to control absorption, aperture size was controlled and enabled incoming sound waves to reach an absorptive inner surface, and to control sound scattering, surfaces could be offset in relation to their nearest neighbors. To explore the design potentials of this geometric and acoustic variation an integrated method of computational design, acoustic simulation, and robotic fabrication was developed. Our developed design techniques are described in the following sections: 3.2. Algorithmic Design Logic, 4. Acoustic Computer Simulation of Wall Geometry, and 5. Robotic Prototyping of the Gradient Acoustic Surface.

The design of this gradient acoustic surface was developed through aggregating small, individually-formed, block-like components to form a larger continuous structure, see Figs. 1 and 2. This system was inspired by the previous work of the research group of Gramazio Kohler [20, 21]. This gradient acoustic wall system is a multi-layer system combining blocks with an airspace behind and a second layer of absorptive material. To produce an acoustically reflective surface, the block's end condition was cut to create a smooth surface, following a smooth design surface defining the wall's overall geometry. The form of the wall could then be tuned to create specific reflective effects. In the reflective condition, blocks were densely packed with no air gaps, thus maximizing the reflection of sound back into the space. The logic of the diffusive condition is achieved by randomly offsetting the blocks. Cutting the block's outward-facing end with shallow angles allowed the geometric variability of the wall's surface to be increased. The absorptive condition was created by spacing blocks apart – introducing a porosity to the outer surface – to a maximum of 65% of the block width. As blocks are stacked diagonally on top of each other, this strategy maintains the structural integrity of the stacked surface while allowing sound to enter the sound absorptive wall cavity behind. In the absorptive condition, the end of the block was cut at a sharp angle to reduce reflections and attempts to direct the sound wave into the wall cavity.

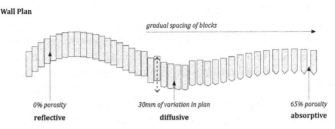

Fig. 1. Design of a gradient acoustic surface, plan.

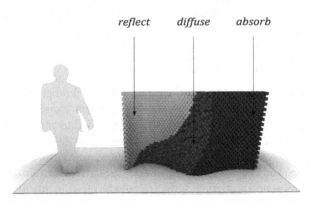

Fig. 2. Design of a gradient acoustic surface, perspective

3.3 Algorithmic Design Logic

The wall's geometric logic was linked to acoustic performance goals through an algorithm developed in the parametric design software Grasshopper, outlined in Fig. 3. The wall's surface geometry is generated by its top and bottom profile curves, which are made in Rhinoceros 3D. Once the curves are set in Grasshopper, they are lofted to produce the wall's base surface geometry. The creation of the input curves in Rhinoceros 3D

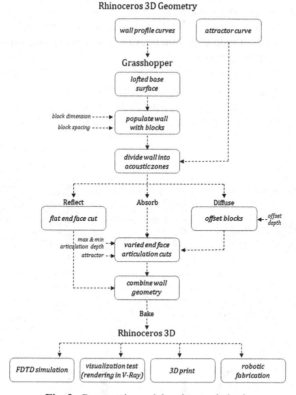

Fig. 3. Parametric model and tectonic logic

allowed for quick adjustment to the wall's size and curvature, making it easy to produce design iterations for simulation, and fabrication tests. The wall's surface was populated with blocks based on the block's dimensions and desired spacing. The spacing could be set at a consistent dimension or be generated at a gradual spacing allowing the wall to produce different reflective or absorptive properties along its length.

To prescribe the type of acoustic surface used along the wall's length, an attractor curve utilizing distances to block faces were used to divide the wall into different acoustic conditions. This division was used in Grasshopper to link the individual block's geometry to varying workflows in the script that created the three different acoustic conditions. This allowed the wall's geometry to adapt parametrically to changes in the attractor curve's shape, allowing the designer to choose the locations of the different acoustic conditions. For the reflective zone, the block's end face was either left flat or faceted to follow the wall's curvature. In the diffusive zone, the block's end was cut into a shallow, randomly angled wedge profile, set to a prescribed depth. The randomization of the angle makes the blocks more conducive to scattering sound reflections. To enable the wall's diffusive segments to be tuned to specific frequencies blocks were offset positively and negatively along their long axis by a defined distance. In the absorptive zone, the block's articulation was optimized to produce acute angles. The cut's depth was controlled to ensure stability between rows and provide enough surface area needed to hold the block with the robotic arm. For walls that blend from diffusive to absorptive conditions, an attractor was used to manipulate the amount of block articulation, allowing the blocks to gradually shift from obtuse to acute angles.

Once the script's parameters were defined and the wall's geometry was set, block geometries were baked from Grasshopper in Rhinoceros 3D. The baked geometry was then used for either acoustic simulations, rendered visualizations, 3D printing, or imputed into a Grasshopper script that formulated the commands needed for robotic fabrication, as described in Sect. 5.

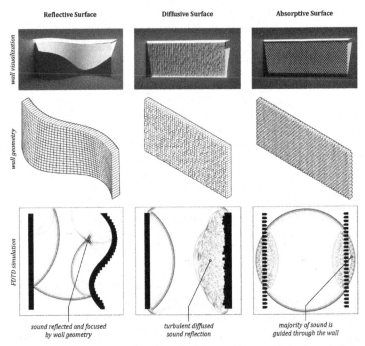

Fig. 4. Three types of architectural acoustic surface: reflective, diffusive, and absorptive. Top images show the interplay of light and geometry, middle images show axonometric drawings, and bottom images show a sound wave simulation of what happens when a single coherent wave is incident upon these surface types.

3.4 Designing for Perception: Light, Sound, and Texture

The wall system's gradient characteristics were designed to vary spatially and materially, impacting the perception and experience of both sound and light. Studies of the impact of geometry on sound were carried out using acoustic computer simulation techniques, further explained in Sect. 4, and studies investigating light were done using computer renderings with the software V-Ray, seen in Fig. 4, and with the creation of small 3D printed and hand-built study models. Surface conditions vary along the wall's face, characterized by changing texture and porosity created by the blocks' face articulations, offset depths, and spacing. This manipulates not only the aural effect of the environment but the visual characteristics of the wall as well. When the wall's surface becomes more complex and articulated, it results in a more haptically textured surface, with visual effects created through the creation of shadow patterns. The ability of the tectonic assembly to make connections between these sensorial attributes creates a sensorily rich environment, thus improving the multi-dimensional experience of the space [22]. This sensorily holistic approach allows designers to manipulate spaces in a multitude of ways that would otherwise be underutilized.

4   Acoustic Computer Simulation of Wall Geometry

There are different computer simulation techniques that can be used to evaluate acoustic performance [23]. The techniques used by the majority of acoustic simulation software are based on ray-tracing algorithms. This technique uses a surface representation of a room as input, abstracts sound as a ray, and traces rays from a sound source to a sound receiver computing the reduction of sound energy from absorption when rays reflect off of walls and computing any re-direction of the sound ray due to scattering effects. However, ray-tracing does not take into account the wave nature of sound and so is challenged to compute diffraction, diffusion, and wave interference effects. Wave-based methods such as Finite Element Modelling (FEM), Boundary Element Modelling (BEM), and the Finite-Difference Time-Domain (FDTD) simulation technique can be used to investigate the impact of sound waves on surface geometries; however, due to long computation times, these methods do not scale to large rooms and high frequencies [3]. Because of the focus on surface geometry in this experiment, FDTD and BEM techniques were chosen over ray-based acoustic simulation for their ability to represent the interaction of sound with the highly complex surface conditions.

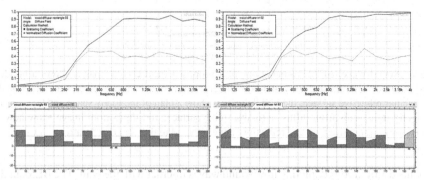

Fig. 5. Scattering coefficient of two randomized block geometries. Results calculated using BEM method in AFMG Reflex software.

In our experiments, 2D sections through the gradient acoustic wall were analyzed using BEM and FDTD techniques. BEM analysis using Reflex software, computed scattering performance, and FDTD analysis, using custom software written in Processing,

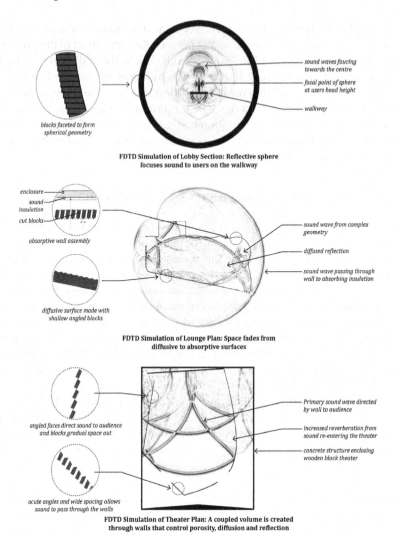

Fig. 6. Sound wave (FDTD) simulations of a proposed theatre. Top images shows a building section through the theatre entry and how the gradient acoustic surface creates an intentionally spectacular reflective surface; the middle images show a plan of the lobby space and how surface geometry transforms from diffusive to absorptive surface; and the bottom image show how all three (absorption, diffusion, and reflection) are combined in the plan of a theatre.

produced animations depicting how sound waves interact with the wall. The BEM analysis of the wall geometry enabled the prediction of the scattering coefficient for different geometric configurations, see Fig. 5. The FDTD simulation works by computing the transmission of energy through a Cartesian grid of nodes [22]. Each node is given a solid or open value enabling the simulation script to solve how sound energy translates through the space at any given point. A speculative design for a theatre was developed to test a variety of acoustic conditions. Figure 6 illustrates some of the sound wave visualizations produced with the FDTD simulations and demonstrates how different acoustic effects can be created with a gradient acoustic surface.

5   Robotic Prototyping of the Gradient Acoustic Surface

One of the key questions investigated during the research of the acoustic surfaces was how to fabricate a highly complex surface capable of self-support structurally as a wall tectonic. To answer this question, a 1:1 prototype was designed to test the constructability and aesthetic qualities of the gradient acoustic surface concept. Building on previous experiments [16, 18, 20, 21, 25], robotic fabrication was utilized for its ability to perform precise and repetitive tasks. Because of the flexibility of robotic digital fabrication processes, the parametric model could be precisely translated from digital model to physical prototype from the micro-scale of the blocks, to the macro-scale of the doubly-curved wall geometry, see Fig. 7.

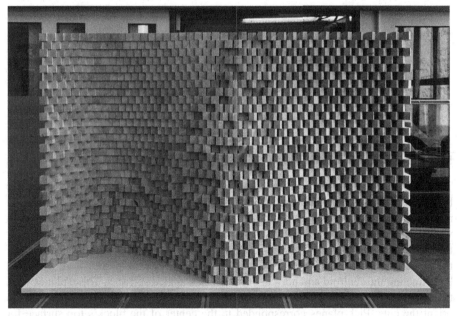

Fig. 7. Gradient Acoustic Surface Prototype. Final version was constructed of 1,174 pine blocks and measured 2412 mm long, 723 mm deep, and 1423 mm tall.

While industrial robots can be used to perform a variety of tasks, they first need to be modified and customized to be used for a specific purpose. An aim of the research was the development of a robotic fabrication process that had minimal tooling changes and a minimum of human intervention; however, it was found that a cooperative human process was needed for initial material loading and the application of adhesive during final placement, see Fig. 8.

Unlike other digital fabrication processes such as 3D printing, laser cutting, and CNC milling, where the process of translating the digital model to fabrication instructions is relatively straightforward, robotic fabrication processes necessitate the development of new digital and physical processes. While this takes more time and presents technical challenges, it also enables innovations in construction manufacturing techniques and the design of architectural tectonics. Using the KUKA PRC plug-in for Grasshopper, the parametric model was modified to generate the necessary data for kinematic simulation and for the production of KUKA robot code for fabrication. In this experiment, we developed two new end-effectors for our KUKA KR150 robot.

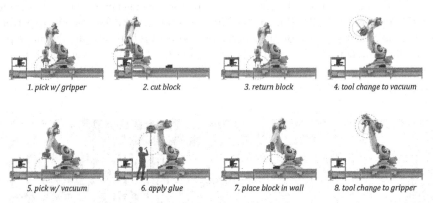

1. pick w/ gripper   2. cut block   3. return block   4. tool change to vacuum

5. pick w/ vacuum   6. apply glue   7. place block in wall   8. tool change to gripper

Fig. 8. Robotic fabrication process.

5.1 Computing the Fabrication Process

To use a robot to build the wall using robotic fabrication, the individual blocks were broken into a series of key planes in Grasshopper. The planes corresponded to specific movements and actions required for each pick, cut, and placement action. The blocks were each broken into four planes, two planes were needed to orient the angled cuts for the block's articulated face with the saw, and two planes for pick operations – one for the pneumatic gripper and one for the vacuum gripper. The cut-planes were generated by extracting the centroid and normal of the block's angled surfaces. These planes were then aligned with a plane on the saw blade's front face, allowing the robotic arm to orient the cuts. Pick-planes corresponded to the center of the block's top surface for the vacuum gripper, and slightly above the center of the block's bottom surface for the pneumatic gripper's location. The pneumatic gripper's location was selected to ensure the block was held securely while being cut, mitigating the amount of deflection caused by the saw.

Once the block was cut the block is temporarily placed while the end effector reorients from the stronger pneumatic gripper to the vacuum gripper to allow for a precise final placement. During the final placement process, the operator works within a cooperative robotic environment to apply adhesive to the block for assembly. Through a series of loops, the necessary translation frames are calculated for robotic path planning to ensure collision-free motion paths. The commands and IO signals were integrated into the robotic control PLC and generated using the KUKA PRC robotic simulation environment to produce the necessary instructions to control the external and end-of-arm tooling functions. We found that special care was needed for the location and orientation of the planes used for the tool-change operation between the vacuum and the pneumatic gripper, as initial attempts at the tool change operation resulted in over-rotation errors in the KUKA robot's A6 axis, requiring the planes to be reoriented to "unwind" the rotation of the robot's axis as it moved through the tool-change process.

All robot fabrication commands were woven together, allowing a single row of the wall to be constructed without interruption. A single row of the wall resulted in 1,599 commands generated in Grasshopper and translated to the KUKA robotic arm through KUKA PRC, see Fig. 9.

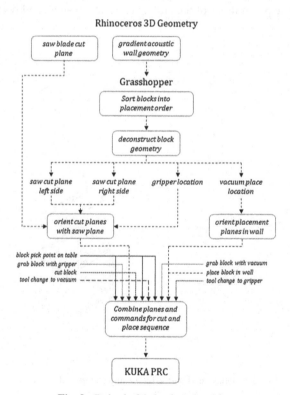

Fig. 9. Robotic fabrication algorithm

5.2 End-Effector Integration

Two custom tools were developed for the fabrication of the wall; an automated saw table to cut the blocks, and a dual-head end-effector that re-oriented between a vacuum and pneumatic gripper. The gripper end-effector enabled the robot to securely hold blocks with the pneumatic gripper while they were being cut, and also provided unobstructed block placement with the vacuum gripper. Unobstructed block placement was essential, as it enabled blocks to be placed immediately beside each other. It was found that the development of a dual-head end-effector was needed due to the significant time needed to do an automatic tool change between two singular end-effectors, which would have increased fabrication time by 10x. The various positions' paths were analyzed in the digital environment for dual end-effector limits and constraints prior to tooling construction. The integration of a sawing station, see Fig. 10, enabled all of the variable single-axis block cuts to be done within the robotic framework. By integrating external and internal robotic tooling a fabrication environment was developed which can encompass the various parametric possibilities of the gradient acoustic wall assembly (Fig. 11).

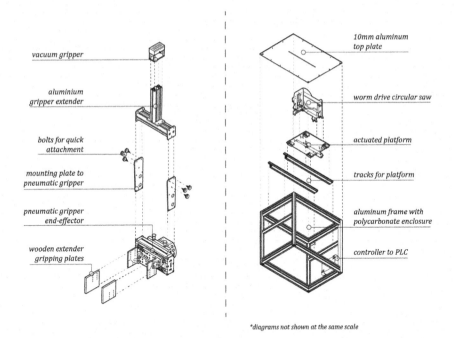

diagrams not shown at the same scale

Fig. 10. Dual-head end-effector and integrated saw table

Fig. 11. From left: pneumatic gripper holding block as actuated saw table performs cutting action; Dual-head end-effector in pneumatic gripper mode – used for cutting; and, Dual-head end-effector in vacuum gripper mode – used for close placement of blocks.

6 Discussion and Future Work

In this research, we have developed the concept of the gradient acoustic surface, and we have undertaken an experiment that sought to develop the necessary workflows and tools to design, simulate, and fabricate a 1:1 prototype. While gradient acoustic surfaces could take many forms, we have chosen to develop and test a single instance. The concept of the gradient acoustic surface we speculate is a useful concept for architectural design as it enables a constructive and performative system that can be integrated with architectural aesthetics, materiality, and a structural concept that can respond to different acoustic performance goals. The system can be modified to create acoustic effects, see Fig. 6, and can be used to create sensory effects, see Fig. 3.

The use of digital fabrication enabled the production of a sample gradient acoustic surface that looked to test the feasibility of construction, producing multi-dimensional variation: block spacing, angled cuts, and surface offsets. It is difficult to imagine how this type of surface would be possible without some sort of digitally-augmented approach. While the development of the fabrication process in terms of software, control, and end-effectors, took considerable time, once this process is established, the production of the gradient acoustic surface was relatively fast. Two end-effectors were developed: a double-ended pneumatic/vacuum gripper, and a saw-table. The double-ended gripper introduced tremendous time savings due to removing the need to carry out tool-change operations; however, it was found that care needed to be taken in defining tool rotations so the robot cabling did not get wound-up. Because of the nature of the stacking process, small dimensional errors in the blocks compounded to create some placement discrepancies, as seen in the blocks in the top-left corner of Fig. 12. Either the dimensional accuracy of the bricks must be improved, or a fabrication process that addresses the slight dimensional errors of the blocks needs to be developed.

To test the acoustic performance of different surfaces, wave-based numerical simulations were used rather than ray-based geometric approaches. Both FDTD and BEM

Fig. 12. Gradient acoustic surface prototype

methods were used to understand the acoustic consequences of different geometric configurations. The FDTD approach was found to be preferable to map sound reflection effects – particularly sound focusing, visualize the spatial and temporal nature of sound scattering, as well as visualize the variable effects of sound absorption and sound redirection produced by the gradient acoustic surfaces. The BEM simulation was useful for calculating and comparing the scattering coefficient.

This experiment does point to the need for future work in several areas. First, while the simulations were informative at demonstrating sound interaction with the various surface geometries; however, due to existing software limitations, simulations were carried out in 2D. Because of the complex 3D nature of the gradient acoustic surface geometry, 3D simulations may also prove useful. The second area of future work would be in the simulation of the spatial variation of acoustic performance. One of the resulting effects of a surface that changes is acoustic performance across its surface is that acoustic performance may not be constant across a space, but spatially variant. The impact and use of this feature would be interesting to explore. The third area of future work presents a challenge, and that is the measurement of performance of the resulting acoustic surface, in-situ, and how these measurements could then be used to enable a more informed approach to designing subsequent surfaces. This is challenging due to the particular nature of the "gradient acoustic surface." As the geometry and material of the surface is ever changing, so too is its acoustic performance. To understand the nature of gradient performance, this would require the measurement of discrete non-gradient conditions at specific intervals, then these results could be interpolated to predict the in-between conditions. This requires constructing and measuring many different wall prototypes in discrete, non-gradient forms, and testing these individually in a controlled setting. In addition to this, as the wall contains both absorption and scattering features, and so these measurements require different samples and different measurement techniques. These testing procedures were beyond the scope (in terms of time, scale, and cost) of this experiment, and this was why we used simulation was used to predict performance.

While much time and effort went into the design of the robotic fabrication process, it could be improved. We speculate that a fully-automated method, with less human interaction, could dramatically decrease the time it takes to fabricate. Further investigations into adhesion methods, specifically ones capable of mitigating the inconsistency of construction materials would be beneficial. A full-sized room-scale installation is planned. Here we can test the pre-fabrication of the system in smaller sections that will be connected onsite. Another area of future research is the design and fabrication of a two-sided gradient acoustic wall. Further investigations into different construction materials and their acoustic and fabrication properties would help define the performative potentials of the gradient acoustic surface.

7 Conclusions

The main contribution of this paper is the concept of the "gradient acoustic surface", which we define as a single architectural tectonic system, that can perform as all three types of acoustic surfaces – absorber, diffuser, or reflector – depending on its configuration, whose geometric and material configuration varies across its surface gradually, and as a result, the acoustic performance coefficients also gradually change over the surface.

This paper describes an experiment in which we sought to understand how a single constructive system could incorporate acoustic reflection, absorption, and diffusion through variation in its geometry, and how these acoustic surfaces could be designed to be an integrated part of a larger architectural tectonic concept so that acoustic performance, architectural design, and digital fabrication are linked together in one system. We found that one of the ways a "gradient acoustic surface" could be realized is through a method of stacking and spacing blocks; however, it is not the conclusion of this experiment to suggest this is the only method, simply a successful approach that we developed for this experiment.

This paper presents an integrated design workflow from the parametric modelling of the basic wall components, through to computer acoustic simulation, and finally the creation of robot instructions for the digital fabrication of a 1:1 prototype. The associative nature of the geometric relations implied by the varying characteristics of the gradient acoustic surface suggests an algorithmic and computational approach, and we found that this was an ideal way to sketch designs, explore the space of options, and carry out performance simulations.

This research project demonstrates how architectural design can be undertaken more holistically, through the integration of the visual and aural attributes of complex surfaces; and how, with the consideration of the novel constructive tectonics made possible through the use of robotic fabrication, architects can generate more sensorily-considered environments.

Acknowledgements. This experiment was funded by the Natural Sciences and Engineering Research Council of Canada's Discovery grants program, and the Canadian Foundation for Innovation.

References

1. Blesser, B., Linda, S.: Spaces Speak, Are You Listening? MIT Press, Cambridge (2007)
2. Pallasmaa, J.: An architecture of the seven senses. In: Questions of Perception, pp. 30–37. a+u Publishing Co. Ltd., Tokyo (1994)
3. Cox, T., D'Antonio, P.: Acoustic Absorbers and Diffusers: Theory, Design and Application. Taylor & Francis, London (2004)
4. Bradley, J.S.: Review of objective room acoustics measures and future needs. Appl. Acoust. **72**, 713–720 (2011)
5. Long, M.: Architectural Acoustics. Elsevier, Burlington (2006)
6. International Standards Organization (ISO): ISO 354:2003 Acoustics - Measurement of sound absorption in a reverberation room. ISO (2003)
7. Peters, B., Tobias, O.: Integrating sound scattering measurements in the design of complex architectural surfaces. In: eCAADe Conference Proceedings 2010, pp. 481–491 (2010)
8. Schroeder, M.R.: Diffuse sound reflection by maximum−length sequences. J. Acoust. Soc. Am. **57**(1), 149 (1975). https://doi.org/10.1121/1.380425
9. International Standards Organization (ISO): ISO 17497-1:2004 Acoustics — Sound-scattering properties of surfaces — Part 1: Measurement of the random-incidence scattering coefficient in a reverberation room. ISO (2004)
10. Moussavi, F.: The Function of Ornament. ACTAR, Barcelona (2008)
11. Peters, B.: Parametric acoustic surfaces. In: ACADIA Conference Proceedings 2009, pp. 174–181 (2009)
12. Bork, I.: Report on the 3rd round robin on room acoustical computer simulation – part II: calculations. Acta Acust. United Acust. **91**, 753–763 (2005)
13. Gramazio, F., Kohler, M.: The Robotic Touch – How Robots Change Architecture. Park Books, Zurich (2014)
14. Peters, B., Tamke, M., Nielsen, S., Andersen, S., Haase, M.: Responsive acoustic surfaces: computing sonic effects. In: eCAADe Conference Proceedings 2011, pp. 819–828 (2011)
15. Williams, N., Burry, J., Davis, D., Peters, B., Pena de Leon, A., Burry, M.: FabPod: designing with temporal flexibility & relationships to mass-customisation. Autom. Constr. **51**, 124–131 (2015)
16. Peters, B., Hoban, N., Yu, J., Xian, Z.: Improving meeting room acoustic performance through customized sound scattering surfaces. In: Proceedings of the International Symposium on Room Acoustics 2019 (2019)
17. Bonwetsch, T., Baertschi, R., Oesterle, S.: Adding performance criteria to digital fabrication: room-acoustical information of diffuse respondent panels. In: ACADIA Conference Proceedings 2008, pp. 364–369 (2008)
18. Vomhof, M., et al.: Robotic fabrication of acoustic brick walls. In: ACADIA Conference Proceedings 2014, pp. 555–564 (2014)
19. Walker, J., Foged, I.: Robotic methods in acoustics: analysis and fabrication processes of sound scattering acoustic panels. In: eCAADE Conference Proceedings 2018, pp. 835–840 (2018)

20. Bonwetsch, T., Kobel, D., Gramazio, F., Kohler, M.: The informed wall: applying additive fabrication techniques on architecture. In: ACADIA Conference Proceedings 2006, pp. 489–495 (2006)
21. Gramazio Kohler Research: gramazio kohler improves a cafeteria's acoustic qualities with computationally designed walls. https://www.designboom.com/architecture/gramazio-kohler-cafe-acoustic-computationally-designed-walls-10-09-2019/. Accessed 09 June 2021
22. Kleine, H.: The Drama of Space, pp. 9–15. Birkhauser, Basel (2008)
23. Siltanen, S., Lokki, T., Savioja, L.: Rays or waves? Understanding the strengths and weaknesses of computational room acoustics modeling techniques. In: Proceedings of the International Symposium on Room Acoustics, ISRA 2010 (2010)
24. Takatoshi, Y., Sakamoto, S., Tachibana, H.: Visualization of sound propagation and scattering in rooms. Acoust. Sci. Technol. **23**, 40–46 (2002)
25. Eversmann, P., Gramazio, F., Kohler, M.: Robotic prefabrication of timber-structures: towards automated large-scale spatial assembly. Constr. Robot. **1**, 49–60 (2002)

Evaluating Team Fluency in Human-Industrial Robot Collaborative Design Tasks

Alicia Nahmad Vazquez[✉] [iD]

School of Architecture, Planning and Landscape, Univesity of Calgary, Calgary, Canada
`alicia.nahmadvazquez@ucalgary.ca`

Abstract. Trust, reliance, and robustness have been identified as key elements for team fluency between teams. They are also crucial elements for successful collaboration between humans and robots (HRC). Robot arms have become integral to numerous digital design and fabrication processes allowing new material forms, more efficient use of materials and novel geometries. It will not be long before close proximity HRC design becomes standard. However, little research has been directed at understanding team fluency development between industrial robots and humans (industrial HRC). Even less to understand the evolution of HRC in creative tasks and factors that influence elements like trust to be established between industrial robot arms and designers. Team fluency is a multidimensional construct, heavily dependent on the context. It is crucial to understand how team fluency develops when designers interact with industrial robots. To this end, in this study, a team fluency measurement scale suitable for industrial HRC in design activities was developed in two stages. In the first stage, HRC literature was reviewed to establish a measurement scale for the different team fluency constructs and identify team fluency-related themes relevant to the design context. A corresponding pool of questionnaire items was generated. In the second stage, an exploratory HRC design exercise was designed and conducted to collect participant's opinions qualitatively and quantitative. Questionnaire items were applied to participants. The results were statistically analyzed to identify the key factors impacting team fluency. A set of curriculum recommendations is made, and a team fluency scale is proposed to measure HRC in design activities.

Keywords: Computational design research · Caad curricula · Computational literacy · Education · Methodologies · Robotic-assisted design · Robotic fabrication · Cooperative systems · Human-robot interaction · Man-machine systems

1 Introduction

Team fluency is defined as the ease of collaboration between the designer and the robot, "the coordinated meshing of joint activities between members of a well-synchronized tea" [1]. It is a perceived and shown lack of friction between the different agents throughout the design task. Fluency in a joint action is a quality of agents performing together and adapting and coordinating with each other in an adaptive way. Fluency is a quality

© Springer Nature Singapore Pte Ltd. 2022
D. Gerber et al. (Eds.): CAAD Futures 2021, CCIS 1465, pp. 378–402, 2022.
https://doi.org/10.1007/978-981-19-1280-1_24

observed in a variety of human behaviours and, recently, in the last ten years, has started to interest researchers in the area of HRI. However, it is worth noting that researchers in the last decade have not come to an agreement on what constructs fluency in HRC; hence it remains a vague and ephemeral concept.

Nonetheless, it can be contended that fluency is a quality that can be recognized in a team and assessed when compared to a non-fluent scenario. A fluent teammate evokes appreciation and confidence [1]. Hoffman (2019, p. 01) describes, "if robotic teammates are to be widely integrated in a variety of workplaces to collaborate with nonexpert humans, their acceptance may depend on the fluent coordination of their actions with that of their human counterpart".

Researchers have tried to relate fluency to efficiency [2, 3]. However, they have found that both are not correlated. Participants would rate their experience as more fluent, even when there was no difference in task completion efficiency [4]. This suggests that fluency is a separate feature of a joint collaborative activity that requires its own individual metrics [2]. This research contends that fluency is a quality that can be positively assessed by analyzing the individual components that facilitate team collaboration. By exploring the features that make fluent teams, the aim is to propose a framework to evaluate fluency in a design human-industrial robot collaboration (HIRC) task and help inform the future design of successful robotic teammates. The main fluency parameters are further subdivided into specific aspects, including subjective and objective metrics. These are then evaluated through questionnaires, field notes, videos and semi-structured interviews; the last three are specifically essential to capture the qualitative notions of fluency in a collaborative task.

1.1 Defining the Team

It is crucial to define the concept of a team, the team division and task assignments before establishing the parameters to be studied on the Human-Robot Team. A team can be defined as a system. Its behaviour and models arise from the interactions between the actors within the team. The dictionary defines teamwork as "work done by several associates with each doing a part but all subordinating personal prominence to the efficiency of the whole" [5]. Discussions on HRC literature address whether the robot should be regarded as a team member or as a tool to be used by humans [6]. Researchers argue that until advances in robot autonomy and intelligence are manifested in unstructured field conditions, the concept of the Human-Robot Team is a matter of phraseology to refer to humans and robots interacting together as teams or to people using robots as a team resource or tool [6, 7].

To form efficient teams, humans need to believe in the concept of a team, that there is a benefit in working together with others and that they can achieve better results as a team than if they work on their own [8]. Team orientation [9] will define individual belief in whether it is worthwhile to stay in the group or leave it. There are only specific circumstances that will encourage people to work as a team. These circumstances include those in which there is a clear, higher goal that cannot be achieved without various individuals' coordinated efforts [10]. When circumstances are not met, the team's performance will be below expected levels, and the team is most likely to encounter problems working together [8].

Social factors defining team performance cannot be translated to apply to human-robot teams. Robot agents and general automation will work towards the task goal without being affected by a belief in the concept of a team. Moreover, robot agents do not have beliefs, a sense of responsibility, conflict resolution skills, nor a need for social acceptance [8]. Hence there is a need to explore new protocols on task assignment, communication flows and how interactions are defined to establish human-robot design teams, understand their behaviours and evaluate their performance. Understanding what and when human designers expect the robot to communicate becomes crucial for designing the team structure, the task, and collaboration. The main thing to consider is that human-robot teams are a new kind of automation that is evolving in its formation, role definition and task assignment. Additionally, as teams develop, humans might not need to retain leadership over all the task aspects. The paradigm of the human as the decision-maker or system supervisor, as set out by Licklider J.C.R., (1960), may need to give way to a future in which the authority in a human-robot team is given to the most appropriate team member, independently of it being human or machine.

1.2 Fluency Metrics in HRC

Hoffman and Breazeal first introduced HRC Fluency metrics in 2013 [2, 3]. As researchers started to use this set of metrics, they have further analyzed, compared and discussed. Fluency metrics has been split into objective (measuring the degree of fluency in the interaction) and subjective metrics (measuring people's perception of fluency in an interaction as related to the qualities of the robot) [1]. However, currently, there are no accepted measures, practises or methods to evaluate industrial fluency in HRC [1]. The metrics refer to non-industrial types of robots. Accepted metrics rarely refer to the cases when the robot is a robotic arm [12]. Additionally, there are no metrics for design or intellectual, creative joint tasks.

Team fluency is crucial for collaborative human-robot teams to exist. Robot's actions need to be perceived as fluent by their human partner for robots to be accepted as collaborators and productive members of a human-robot team. However, fluency in nonrepetitive, non-practised tasks has very different implications. Humans would not be expecting coordinated physical interactions as they would on repetitive tasks [4]. Still, they would be expecting understanding and ease of intellectual communication, including design intent and steps towards its realization. These qualities are observed in various human behaviours but are virtually absent in human-robot collaboration. This paper aims to set a tested toolkit to evaluate fluency in human-robot collaboration within the context of design tasks. The developed metrics could help benchmark advances in creative HRC tasks and lead to better strategies for incorporating robotics in the architectural curriculum as design teammates rather than another fabrication tool.

1.3 Setting Up the Team Fluency Scale

The proposed team fluency scale is a newly developed and validated set of metrics based on a comprehensive review of previous isolated HRC measuring scales for trust [12], fluency [1, 4], and an analysis of the working alliance inventory [13]. None of the reviewed set of metrics was designed for intellectual or creative tasks. This is a fundamental difference, as they are not dealing with unknown outputs (i.e. outputs are usually established like pick and place a ball into a cup) or authorship issues. Both of which are considered on this scale. The metrics related to fluency are not specific to robot arms but have been adapted to them. The metrics related to trust are adjusted from metrics developed for industrial arms in industrial tasks. The proposed set of metrics also includes additional items specific to the design task, direct interaction with path planning and robot programming and the material feedback, non-present in other existing scales.

Team fluency is evaluated as a construct of four main components: **trust, collegiality, robustness, and improvement,** each with its own set of downstream measures (Fig. 1). The four constructs were evaluated through quantitative Likert-scale questionnaires, qualitative semi-structured interviews and field notes during the case study.

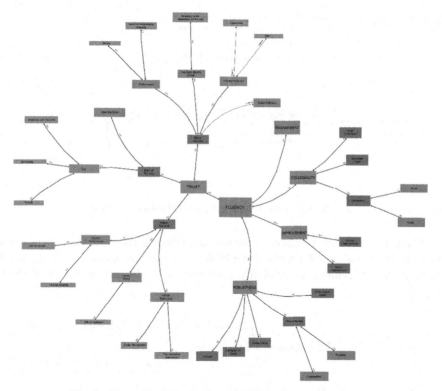

Fig. 1. Hierarchical distribution of the team fluency constructs.

The trust construct has a more significant number of parameters defining it (Fig. 2). Trust is described in the literature as a primary contributing factor to reliance on automation [14, 15]. Trust is composed of trust in the human participant and its behaviour, trust in the robot and trust in external elements (i.e. the setting, materials, room, etc.). Furthermore, without trust in the capabilities and intentions of the team partner, it is safe to assume there won't be team dynamics [16]. The specific abilities in which the human feels they can trust the robot will become the basis of the collaborative task.

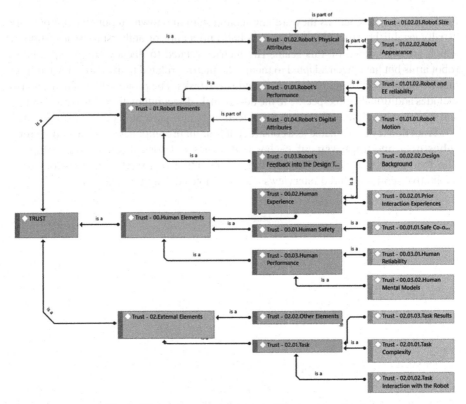

Fig. 2. Elements evaluated as part of the trust construct

Robustness in teams is related to the task and labour division for team members in the task. To assess the robustness of the HIRC, five main parameters were identified: reliance, shared identity, responsibility, attribution of credit and attribution of blame (Fig. 3).

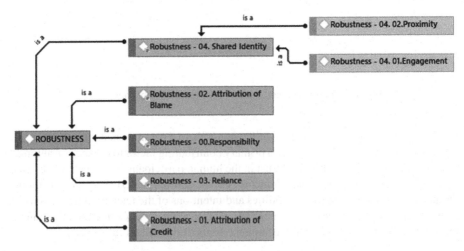

Fig. 3. Elements evaluated as part of the robustness construct

Collegiality is defined as the relationship between colleagues united in a common purpose where both respects each other's abilities to work towards that purpose [5]. It can also be defined as the cooperative interaction amongst colleagues. The abilities of the colleagues do not necessarily need to be the same, but they need to be complementary to achieve the common goal. For the purpose of this research, collegiality evaluates the robot's perceived character traits related to those of a team member, which could qualify it as similar to a human partner. Collegiality consists of the three following indicators (Fig. 4).

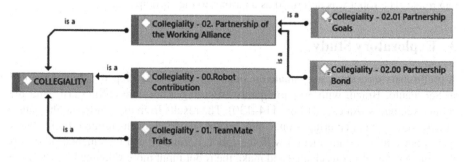

Fig. 4. Elements evaluated as part of the collegiality construct

The last construct of the Fluency scale refers to improvement. Improvement is important in a true collaborative scenario. Collaboration requires partners to iteratively adjust and learn from each other over time. Learning can be considered one way to adapt [17], and as team members learn and adjust, the team is expected to improve. Improvement is divided into human improvement and robot improvement.

2 Measuring Team Fluency

The following steps detail the setup of a design task to collect quantitatively and qualitative data and measure the reliability and internal consistency for the items that construct team fluency. The data is analyzed to identify the relevance of the different questions and their contribution to the team fluency metrics.

2.1 Research Problem, Aims and Objectives

Technical devices and digital fabrication tools allow for new practices and are capable of opening new understandings of matter, new ways of organizing, and new complex and irregular relationships that expand material processes. These forces create new non-linear workflows that can lead to a novel language characteristic of the human-robot collaborative era in architecture. A human-robot collaborative design task was devised to evaluate the team fluency metrics. It integrates robotic, material and human agency in an iterative process where they continuously influence each other. The aim was to enable a man–robot symbiosis. The robot could guide the designer's decision-making according to a complex set of local and global criteria that might have been ignored otherwise (i.e. comparing with original design, performing the structural analysis [18], etc.).

The design scenario and the design task's characteristics are based on a functional team structure [7, 19]. Specific task roles are assigned to the robot relying on its features, such as precision, information processing etc. Design decision-making is assigned to humans. The robot offers feedback, suggestions and information to which the designer can relate. The robot was described as a partner to the participants.

3   Exploratory Study

A preliminary case study to evaluate the fluency scale was developed and implemented in Sao Paulo, Brazil, with 30 participants executing a hot-wire cutting task with the robot (Nahmad Vazquez, 2019 pp. 114–130). The results from that study and the nature of subtractive manufacturing – once cut, cannot be added again- limited the feedback and opportunities for interaction with the robot. Formative and incremental processes that allow for continuous change and make the robot input more valuable [18, 21] were studied before setting up the exploratory study. This section describes the tasks and procedures that took place during the study.

3.1   Method

Design
The human study was set up for the robot to work on a collaborative design task with untrained robot-user designers. The goal was to design a form-found concrete shell structure collaboratively. The rationale that creating a shape together with the robot would open the possibility of establishing a bond between designer and robot. Participants were encouraged to do their own designs rather than performing the collaborative exercise on a design given to them to promote intellectual ownership over the results.

Materials
Participants were told that their job was to design a two-dimensional (2D) pattern and its resultant 3D shell structure after pop-up (using a physics solver to simulate the result) (Fig. 5).

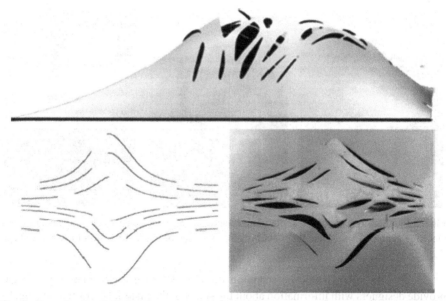

Fig. 5. Top: simulated 3D resultant geometry. Bottom Left: 2D cutting pattern. Bottom Right: 3D print of the results expected form.

A phase-changing material technology was explored within a pop-up process, based on patterns that embed the shape into the material rather than prescribe it, allowing for an experimental approach to digital and physical design as the material exhibits probable but not certain behaviour (Fig. 6). A feedback loop was introduced to build a form-finding, human-robot collaborative design process.

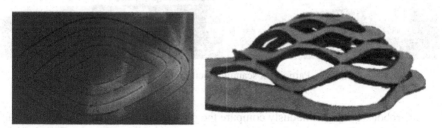

Fig. 6. 2D pattern defined by the participant and cut by the robot and resultant 3D concrete shape after plunging and hydrating.

Task

The task entailed working with the robot during the cutting and forming of the physical concrete shell -which was previously simulated-version. The robot would initially cut the designed pattern on the concrete cloth and then plunge it during the material's forming phase (3 h between hydration and curing as a hard concrete shell) (Fig. 7). The plunging would deform the material into shape.

Fig. 7. Plunging sequence. The robot plunges and massages the concrete cloth at each position. Once the plunging is finished, the robot flips the end effector to use the Kinect to scan the shape.

Finally, the robot would be scanning the model during the formation process to provide designers with information about the status of the material deformation (Fig. 8).

Fig. 8. Participant scanning their model after plunging

The robot would continuously compare the physical model with the digital one and advise where to plunge next, based on the comparison results (Fig. 9). Based on this same framework, the robot would generate new tool paths for the next steps of the deformation. The software was programmed so that robot suggestions would always take the participant closer to achieve their initial desired design (as per the physic solver simulation) [22].

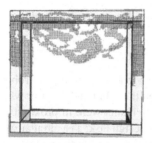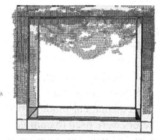

Fig. 9. Comparison between the different plunging iterations. A scan is taken after each before deciding the next step (compare with the original and continue plunging or free-forming it jogging the robot). New robot plunging coordinates are generated from this comparison.

Identical task conditions were provided to all participants. The aim was to design and then iteratively explore possibilities with the robot to form-find a concrete shell. Matching the physical to the digital was seen as a soft objective. It wasn't required from participants to achieve a match but to achieve a design that they felt comfortable with (Fig. 10).

Fig. 10. Participants scanning the shell and deciding on a plunging sequence.

Participants

Twenty-eight participants participated in the case study in WSA at Cardiff University: 12 female and 13 male. The participants' age range was between 18 and 29 years, with an $M = 20$ and $SD = 2.26$. Participants were in their second year of study ($n = 17$), in their third year ($n = 10$) and postgraduate students ($n = 1$). Eleven participants reported no experience with digital fabrication machines, while 17 reported having some experience with digital fabrication machines. The laser cutter was the most popular, with 15 participants having experience with it, followed by the 3D printer. Participants were coded for their age, gender, year of study, previous digital experience, previous digital fabrication machines experience, and design authorship. The number of participants is in line with other research studies in HRI and HCI, which recommend between 15–30 participants before the law of diminishing returns begins to apply and more participants stop equating further insights [23].

Robot

A single-arm industrial robot with a 60 kg payload was used. The robot was placed inside a cage to ensure safe separation between the robot and the user when running in automatic mode. For manual mode tasks, participants were in proximity to the robot. Three end-effectors were used to complete the design task: 1) a circular rotary blade; 2) a Kinect scanner; 3) a wooden sphere mounted at the end of the Kinect.

Data Collection

Data was collected via a Likert-scale questionnaire with 36 questions covering all the team fluency scale constructs. Additionally, semi-structured interviews, field notes and Videos of the design task were also collected. The researcher was always present during the exercise taking notes and providing technical support to the participants.

Data Analysis

The data collected were analyzed in two stages following a qualitative dominant crossover mixed analysis [24]. Quantitative data were analyzed as a function of participants' age, sex and year of study. A quantitative scale was established to help interpret the results from the qualitative interviews and cross-reference the findings. The data analysis process was iterative and reflective, moving back and forth across empirical data and theoretical resources following a 6-step analytical procedure [20, 25].

Atlas.ti, a qualitative analysis software package, was used for the quantitative and qualitative information's coding process. Qualitative information from the interviews, field notes and videos were transcribed verbatim and analyzed using template analysis [26]. The aim was to establish key team fluency themes, emergent themes and the relationship between them.

The video material was specifically coded looking for non-verbal indicators of team fluency, such as the proximity of the participant to the robot (i.e. videos were coded for when participants remained at the edge of the cage versus moving freely around and close to the robot to compare this with their accounts of feeling comfortable or not with the robot and find potential inconsistencies). Attributes like responsibility, trust and attribution of blame were also correlated between participants' accounts and video information (i.e. cases in which the material broke due to the robot's action and resulted in a high attribution of blame, regardless of the participant instructing the robot of the depth of the plunging vs cases were attribution of credit and blame are not related to specific robot actions).

The team fluency construct of *reliance* was only evaluated from the videos and field notes as follows: Reliance evaluates the level of engagement or *reliance* of the participants in the more ambiguous situations described as those in which the participant could choose to solicit input or not from the robot (e.g. after the concrete deformation). The nature of the task creates various moments where the robot has better and more accurate information than the participant over the geometry's status in reference to the digital design. This situation happens specifically after each scan and in-between plunges. The robot, after scanning, would have quantitative information on the deformation. In contrast, the participant would only have visual information of the deformed material but lacked any reference to their original design and the deformed piece. Reliance is related to the number of scans (none to more than 3) and the actions participants take after each scan, and the suggestions given by the robot (ignore, follow, half follow/half juggle).

4 Validating the Fluency Metrics

Once all the data had been analyzed, the test's internal consistency and reliability were measured through a Cronbach's alpha. The closer the Cronbach's alpha coefficient is to 1, the greater the items' internal consistency in the scale. A generally accepted rule is that an alpha above 0.7 indicates acceptable reliability, with 0.8 or higher being considered as appropriate reliability [27, 28]. Values higher than 0.95 are not desirable as they may indicate that the items are entirely redundant. Researchers suggest that when measuring psychological constructs, values below 0.8 can also be acceptable [27]. The team fluency questionnaire has a Cronbach's alpha of 0.83 (Table 1).

Table 1. Reliability and scale statistics fr the team fluency questionnaire

Cronbach's Alpha	SD	Mean	Variance	Sample size	# of items
0.83	18.45	178.94	341.42	28	36

4.1 Validation of the Fluency Metrics

This section presents a review of the measures that have been used to evaluate the different underlying constructs that comprise team fluency. Each construct includes all the downstream measures for each theme. Response to the items was assessed using a five-point Likert scale from 1 (strong disagreement with the evaluated parameter) to 5 (strong agreement with the evaluated parameter). Although all the measures are currently phrased to assess the participant's perception of team fluency, they can be adjusted for an observer scenario [2, 3]. The Cronbach's alpha is reported for each measure to represent its internal consistency:

1. **Trust in the robot elements:** This composite downstream measure evaluates the trust that the robot evokes in the human; it includes the physical and digital aspects of the robot (i.e. the information it is conveying). It consists of the following indicators

• Perceived Robot and End Effector Reliability	
"I felt the robot was going to do what it was supposed to do"	(very unreliably – very reliably)

(*continued*)

(*continued*)

"The scanning, cutting and pushing seemed like they could be"	(highly reliable – highly unreliable)
• Perceived robot motion	
"The way the robot moved made me feel"	(highly uncomfortable – highly comfortable)
"The speed at which the robot performed its tasks made me feel"	(highly uncomfortable – highly comfortable)
"The robot moving in the expected way was"	(strongly concerning – strongly non-concerning)
• Perceived reliability of the information the robot is conveying to the design task	
"I felt the information given by the robot was"	(very useful – very useless)

Cronbach's alpha for this measure was found to be **0.643**. Although this figure is below the accepted cut-off value of 0.7, lower values in the literature are generally accepted when measuring psychological constructs [27].

2. ***Trust in the human elements:*** This composite measure evaluates the trust that the human feels in the robot due to his or her own performance, previous experience and expertise. It consists of the following indicator items:

• Perceived human performance	
"By looking at the deformation of the material, I could decide my next move."	(strongly disagree – strongly agree)
"I believe the robot likes me"	(strongly disagree – strongly agree)
• Perceived human safety	
"I felt while interacting with the robot"	(very unsafe – very safe)
"I was comfortable the robot would not hurt me"	(strongly disagree – strongly agree)
"I trusted that to cooperate with the robot was"	(very unsafe – very safe)

Cronbach's **alpha** for this measure was found to be **0.767**.

3. ***Trust in the external elements:*** This composited downstream measure evaluates the trust that the human perceives due to the characteristics of the task. It consists of the following indicator items:

• Perceived task complexity	
"The complexity of the task made working with the robot"	(very uncomfortable – very comfortable)

(*continued*)

(*continued*)

• Perceived interaction with the robot	
"The task made interaction with the robot."	(very easy – very complex) (reversed scored)

Cronbach's **alpha** for this measure was found to be **0.753**.

4. *Collegiality:* This composite downstream measure evaluates the robot's perceived character traits related to it being a team member and consists of the following indicator items:

• Robot contribution	
"The success of the task was largely due to the things I did."	(reverse-scored)
"Our success was largely due to the things the robot did."	1–7
• Partnership: This downstream measure has been taken from Hoffman's adaptation (Hoffman, 2013a) from the 'Working Alliance Index'. It consists of the '*bond*' and the '*goal*' sub-scales:	
– The *bond sub-scale* consists of the following indicator:	
"The robot was a partner for me in this task."	1–7
– The *goal sub-scale* consists of the following indicators:	
"To what extent did you feel that your performance on this task was out of your hands?"	1–7
"The robot helped me perform better on this task."	1–7
• The complete measure additionally includes the following individual, open question:	
"How will you call the robot? (suggest a name)."	1 Male; 7 Female; 3 Neutral

Cronbach's alpha for this measure was found to be **0.764.**

5. *Improvement:* This composite downstream measure evaluates the perceived improvement understood as the team members' adaptation to each other. Learning can be considered one way to adapt [17]. In this case, the robot is not learning nor adapting, so the measure is taken in relation to the human. It consists of two independent indicators and one individual question:

• Human Improvement	
"My performance in this task improved over time"	1–7
• Robot Improvement	
"The performance of the robot in this task improved over time"	1–7
• The complete measure additionally includes:	
"The robot and I improved our performance on this task over time."	1–7
"The robot helped me perform better on this task."	1–7

Cronbach's alpha for this measure was found to be **0.901**.

6. ***Robustness:*** This composite downstream measure evaluates the functionality of the team, as opposite to its dysfunction. It consists of the following indicator items:

• Responsibility	
"To what extent did you feel it was only your job to perform well on the task?"	1–7 (reverse scored)
"To what extent did you feel ownership of the task?"	1–7
• Attribution of credit	
"I am responsible for most things that we did well on the task"	1–7 (reverse scored)
"The robot should get credit for most of what is accomplished on this task."	1–7
"The success of the task was largely due to the things I did"	1–7 (reverse scored)
"Our success was largely due to the things the robot did."	1–7
• Attribution of blame	
"I hold the robot responsible for any errors that were made on this task."	1–7
"The robot is to blame for most of the problems that we encountered in accomplishing this task"	1–7 (reverse scored)

Cronbach's alpha for this measure was found to be **0.711**.

Eliminating the items that are not contributing to overall reliability is a step towards improving the scale and understanding what factors are more relevant to evaluating team fluency in a human-robot design scenario. A 'correct item-total correlation test that indicates the relation between the overall score of the test and the item's score after excluding the item in question from the total score was performed. If this last correction is not performed, inflation of the item-total correlation may happen [29]. A range between 0.15 and 0.3 is suggested for item removal [30]. The indicators that were not successful for measuring team fluency in an industrial design HRC scenario have been indicated in Table 2.

Table 2. Removed items from the team fluency scale

Item number	Item	Correct item total correlation
6	I knew the scanning information from the robot would be (highly accurate – highly inaccurate)	0.07
7	The information from the scanning was useful for my downstream design decisions (very helpful – very unhelpful)	0.02

(*continued*)

Table 2. (*continued*)

Item number	Item	Correct item total correlation
10	The size of the robot was (highly intimidating – highly encouraging)	0.18
11	The robot cutting and pushing tools seemed (highly unreliable – highly reliable)	0.06
15	If I had more experiences with other robots I would feel about this task (highly concerned – highly unconcerned)	0.10
17	If the task was more complicated and I had to work with the robot I might have felt (highly concerned – highly unconcerned)	0.00
18	I might not have been able to work with the robot had the task been more complex (strongly agree – strongly disagree)	0.17
21	To what extent does the robot have the characteristics you would expect from a human partner doing the same tasks?	0.05
32	The performance of the robot in this task improved over time	0.06

After removing these nine items, the reliability analysis is rerun on the remaining 27 items. The **Cronbach's alpha** of the team fluency scale improves to **0.85,** suggesting that the revised team fluency scale of 27 items has increased its reliability. It would be worth examining them with different task conditions before eliminating them from team fluency analysis in industrial robot – designer collaboration design tasks.

The analysis has ensured the reliability of the quantitative data collected to evaluate team fluency in design HIRC. It has also enabled the detection and removal of the weaker items and increased the scale's reliability.

5   Robot Elements

Based on the revised fluency scale, different robot elements are relevant for designers to trust as collaborative partners. Standard parameters from the literature like the type, size, proximity and behaviour of the robot do not affect designers' trust. The main issues affecting the flow of the design task and their trust in the robotic partner are:

1. **Digital attributes.** Translating design thoughts into robot motions. This as opposed to a partner who understands your goals and helps achieve them. Literature on HRC, including collaboration with industrial robots, rarely speaks about the digital aspects of programming and interacting with the robot; this may be task-specific as designers need to be continuously exploring and testing rather than relying on a pre-programmed robot a defined task.

2. **The lack of transparency and legibility of the robot's behaviour.** A signalling system between robot and human teammates could be implemented to ease collaboration and make their behaviour and motion plans legible to their human partners.

3. **Lack of sensors.** Awareness of its environment and itself will promote trust in the robot from human co-workers.

4. **Design exploration.** This is a divided area; some participants find that the robot requires precise instructions, which limit its possibilities for design explorations as designers would not have this precise information during the early stages. Others consider the playfulness of using the robot, either by manually jogging or quickly iterating different motions, positive for experimental design.

5. **Ceding agency.** Designing with robotic thinking, considering the machine characteristics, allowing more agency to the robot yields better results than trying to control everything. Comments include doing more playful designs, more significant curves that can deform more, etc., to allow the robot and the material more agency.

6. **Reliability of the end effector.** It did not appear to have an impact on trust. This seems a context-specific aspect that has not appeared in previous literature. This is particularly relevant to design HIRC tasks in which end effectors are flexible and varied (i.e. different from single task industrial grippers, etc.).

7. **The robot not being too clever was often mentioned as increasing trust.** Participants felt that it could not harm them or have unexpected behaviours because the robot was not smart.

6 Curriculum Recommendations

6.1 Analysis of Variance

The year of study of the participants became a variable of interest for further consideration. It appeared that the robot elements have different influences on participants from different years of study. After analyzing the data and observing that participants from earlier years were more inclined to play with the robot and less concerned about reliability, this hypothesis was made. They trusted less on the robot's recommendations and preferred using it for plunging by jogging it manually. The final shape, in these cases, is resultant from their intuition jogging the robot and plunging rather than following the robot suggestions. However, more mature participants, from higher years, seemed to be interested in developing a more systematic way of thinking and incorporating parametric design methods into their process. During the design exercise, they considered the input from the robot and used it at each iteration. They also reflected on the robot feedback as valuable information that allowed them to understand the material process quantitatively and how the design was evolving with respect to their initial simulated form. Through an ANOVA, it is investigated whether there was a statistically significant difference in the perceptions of team fluency between the responses obtained from the second year and third-year participants.

6.2 One-Way ANOVA

ANOVA is a parametric test based on six assumptions. The first three are related to the design of the study and include having:

1. One independent variable measured at a continuous level.
2. Two independent variables where each one consists of two or more independent groups.
3. Independence of observations, meaning the groups are independent, and participants in one group are not part of the other.

The second three are related to how the data is fit for the test and involve the following:

1. There are no significant outliers.
2. Data is approximately normally distributed.
3. Variance should be equal (homogeneity of variance).

Therefore, before carrying out the ANOVA test, a boxplot inspection for outliers, a test of normality and a test of homogeneity were carried out to ensure that all the assumptions are met. Data were entered into SPSS to carry out the three tests.

Boxplot Test for Outliers
The boxplot test from SPSS was used. It was found that there were no outliers in the data, as assessed by visual inspection of a boxplot for values greater than 1.5 box-lengths from the edge of the box (Fig. 11).

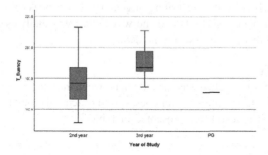

Fig. 11. Assessment of outliers by boxplot visualization

Saphiro-Wilk Test for Normality
The Saphiro-Wilk test for normality was used as it is normally recommended for small sample sizes (<50 participants) [31]. Team fluency was normally distributed for the second year and third-year students, as assessed by Saphiro-Wilk's test (p > 0.05). The Saphiro -Wilk test scores for team fluency for the second year are Sig = 0.92, p > 0.05 and for third year are Sig = 0.83, p > 0.05 (Table 3).

Table 3. Saphiro-Wilk test for year of Study

| Year of study | Saphiro - Wilk | | |
	Statistic	df	Sig
2nd year	0.98	17	0.92
3rd year	0.96	7	0.83

Homogeneity of Variance

A Levene test was used to test the homogeneity of variances. For the team fluency scores obtained between the different years of study, there was homogeneity of variances, as assessed by Levene's test for equality of variances ($p = 0.50$).

Table 4. Homogeneity of variance test across years of study

Levene statistic	df1	df2	Sig
0.45	1	22	0.50

Summary

The collected data for evaluating team fluency meets the three assumptions for parametric analysis. There were no outliers, as assessed by boxplot; data was normally distributed for each group, as assessed by Saphiro-Wilk test ($p > 0.05$); and variances were homogeneous, as assessed by Levene's test of homogeneity of variances ($p = 0.50$). Therefore, parametric statistical analysis tools like one-way or two-way analysis of variance can be used to perform statistical analysis on the data.

One-Way ANOVA

A one-way analysis of variance was carried out between the years of study to explore any difference between design maturity and perception of team fluency in the data obtained from the exercise. The results are presented in Table 4.

Table 5. Descriptive statistics for the data across years of study

Year of study	Mean	SD
2nd year	177.5	16.0
3rd year	191.0	12.0

Participants in the third year experienced a higher level of team fluency ($M = 191.0$, $SD = 12.0$) when compared to participants in their second year ($M = 177.5$, $SD = 16.0$). An ANOVA is statistically significant when not all the group means are equal in the population ($p < 0.05$). Alternatively, when $p > 0.05$, there are no statistically significant differences between the group means. The results of the one-way ANOVA are shown in Table 5. The perception of team fluency was statistically different between participants in different years of study, $F (1, 22) = 4.0$, $P = 0.06$ (Table 6).

Table 6. One-way ANOVA output

	Sum of squares	df	Mean square	F	Sig.
Between groups	890.2	1	890.20	4.0	0.6
Within groups	4945.5	22	224.8		
Total	5835.7	23			

6.3 Two-Way ANOVA

After determining that the year of study affects the designer's perception of team fluency, it became interesting to understand if gender also affects this perception. Research in HRC suggests that males and females exhibit different responses in the way they relate to robots during collaborative tasks [1, 32]. A two way ANOVA was set up. However, there was no statistically significant interaction between the year of study and gender for the 'Team Fluency' score, $F (1, 20) = 0.80$, $p = 0.38$. As represented by the year of study, the effect of design maturity on the perception of team fluency ignores gender. Both males and females have an increased perception of team fluency in a HIRC design scenario with higher design maturity levels.

6.4 Discussion

From an observer's point of view, participants in higher years would be more inclined to follow the plunging pattern suggested by the machine. Statistical results from an ANOVA test confirms that year of study, understood as maturity in their design thinking, impact the designer evaluation of team fluency. The third year of undergraduate school can be recommended as an ideal moment to introduce robots in the architectural curriculum. Students have a good understanding of design processes and are curious to accept other agencies into their design whilst looking for rigorous form-finding and spatial allocation methods.

7 Discussion and Practical Implications

The output of this work provides several theoretical and practical implications. These are discussed below:

7.1   Theoretical Contributions of the Team Fluency Scale

Five main themes emerged from the qualitative analysis of the data as the main drivers for human integration and comfort in the robot teammate.

The robot performance was one of the most discussed themes. This includes the robot's reliability, the robot end effectors and the feedback provided by it to the design task. This aligns with previous and more recent literature [15, 33]. In their meta-analysis of human-robot factors, Hancock et al. (2011) classify the reliability of the robot performance as having the highest impact on trust. The work of Charalambous and van de Brule [33–35], confirms and highlights how the robot performance on the task influences human trust. An unreliable robot will eventually decrease human trust and acceptance of the robot. What is important to consider, and that hasn't appeared in previous literature to the best of our knowledge, is that the robot system includes the end effector's reliability as part of it. Designers don't make a specific differentiation between both. This is of particular relevance to design HIRC. Robot arms have been perfected in their design through the years, and industrial end effectors such as grippers, welding and painting guns etc. However, designers are constantly making new, experimental, untested end effectors, and when their reliability decreases, so will the human trust in the robotic partner.

Physical attributes received little attention from the participants, with most of them describing having a big robot as encouraging and empowering. Making the robot make what they want gives them a sense of control over the design results. This is the opposite of what the researcher expected and to previous literature in which smaller-sized robots increase the human's trust. The robot appearance did not seem to be a factor or contribute to how designers felt about the robot. The literature provides contradicting results, with some research suggests that robots should not be too human in appearance, while others suggest that a more human-like appearance is more engaging to people [36–38]. In both cases - anthropomorphic and tecnomorphic robots-the robot appearance should match the robot abilities. This avoids generating unrealistic expectations on the human user, which will harm the relationship later when they are not met [36, 39]. A possible explanation for this measure is that designers perceive the industrial robot as a tool designed to complete a task; hence its appearance is not essential.

Lack of behaviour legibility was a cause of concern for most participants who desired the robot to indicate its subsequent actions. As humans take cues from other humans and know-how to collaborate with them, similar protocols for human-robot collaboration are expected. This is not a problem solely limited to industrial robots, as the exact requests for legibility of the robot's actions and intentions have been found for self-driving cars [40]. Features to provide the human partner with an understanding of the robot intentions or indications about the robot next moves are also desirable for designers.

Robot feedback was embraced by the participants and constantly described as one of the best things from interacting with the robot. Participants from the third year and higher engaged further with the scanning process, doing it from different angles, and continuously comparing the physical and digital models. Even if manually jogging the robot, they will consult the comparison with the screen with the scan and the digital model. The younger participants, 2nd year, were more inclined to follow their intuition and disregard the machine information or consider it only as a curiosity. A level of maturity seems to be needed to understand the feedback and give some agency to others (humans or non - humans) and accept external comments over their design process, which might differ from their own intuition.

Although the task was performed over three 2-h sessions, the improvement of the human-robot team was positive across participants. The robot was not improving either learning through the design exercise. However, participants rated the team improvement as high with comments like "using the robot made me feel highly empowered when it was doing what I asked him". Participants also credit the robot to adapt and empower them to do more as the task progressed over time. "First it was a little scary for me, and then it became really exciting and rewarding" (CS2-003), "it increased my reach, possibilities and vision, and my enthusiasm" (CS2-019). One went so far as to claim that "by the end of the session, we were good friends, the robot was understanding me, now I feel that I love him", another participant commented, "He is adorable. Oh, I love talking to him".

The robot motion was a positive factor in the participants' perception of it. The motions are very controlled but emotionally were described as "the robot becoming alive". There is a valley between human rational and instinctive reactions to machines. Participants know the robot - especially in this case with an industrial robot- is nothing more than a programmed automaton [41]. However, describing it as a live creature or how it goes from dead to live when it starts to move and becomes part of their process is recurrent. It is important to note that participants describe as alive an industrial robot arm that huge metal parts joined together. They are not describing a humanoid or a softer robot. A possible factor influencing humans attributes of liveliness to the robot is the human "like-me" perception of the robot [32] and the tendency to anthropomorphize even simple interactions by assigning them intention [39].

7.2 Practical Implications of the Scale

The output of this work has significant practical implications. To our knowledge, this is the first empirically developed psychometric scale for measuring team fluency in industrial HRC design tasks. Additionally, this scale can be a powerful tool for curriculum designers, researchers, and practitioners to introduce industrial HRC in architectural and design endeavours. It provides guidance on how the team's configuration and the task affect the designers' perception of team fluency. For instance, the scale identified four key aspects fostering team fluency in industrial HRC design tasks: feedback, robot motion, human performance and task complexity. These three areas appear to be the major determinants for fluency in an industrial HRC design task. Humans will eventually adapt to the artificial motions of the robots regardless of how we shape these interactions [42]. However, as robots become more common in our design practices, we need to do more than minimize cost and maximize durability. The fluency scale aims to understand how the interaction with the robot changes humans' sensory and behavioural statistics and experiences. It aims to aid in developing better sequences.

Teams of humans and robots designing together is new territory. A new type of robot and design workflow where humans do not retain leadership through the entire process must be developed. Relinquishing some of the control has to be accepted as part of a multi-agent design process. Humans might set the rules, but they should be prepared to accept a scenario where control and authority at specific points are given to the most appropriate team member, irrespective of whether it is a human or a machine. A futuristic robotic teammate will most likely fail, regardless of its technological advancements, if the humans misunderstand what it is doing. Scoping how designers relate to robots and what characteristics would make a successful human-robot relation becomes crucial for understanding and designing robots and workflows that designers will engage with in appropriate and successful ways.

8 Future Work

This study's results can provide the basis for further work in team fluency and its constructs (trust, responsibility, robustness, improvement). It marks the first attempt to understand the development of team fluency in industrial HRC for design tasks, university students and recent graduates. The majority of the participants were design students without robotic experience nor enrolled in a specialized robotics course. Therefore, it is essential to validate the results with individuals with an in-depth understanding of industrial robots in digital design processes. Another consideration is that this study was carried in laboratory conditions. Future work would be geared to investigate how the team fluency scale applies in real-world scenarios and whereas the results and trends from the study apply. Another aspect of future work would be to investigate how the specific components of the scale can be individually affected and their implications on overall team fluency. For instance, the feedback was one of the more decisive factors to team fluency; how would a machine learning-enabled robotic feedback determine team fluency. This could provide specific regions to optimize for each construct (robustness, trust, reliance and improvement). Additionally, teams with more than one robot and one human should be tested.

References

1. Hoffman, G.: Evaluating fluency in human–robot collaboration. IEEE Trans. Hum.-Mach. Syst. 1–10 (2019). https://doi.org/10.1109/THMS.2019.2904558
2. Hoffman, G.: Evaluating fluency in human-robot collaboration. HRI Work. Hum. Robot Collab. **2013** (2013)
3. Hoffman, G.: Evaluating fluency in human-robot collaboration. Robot. Sci. Syst. Work. Hum. Robot Collab. **381**, 1–8 (2013)
4. Hoffman, G., Breazeal, C.: Cost-based anticipatory action selection for human-robot fluency. IEEE Trans. Robot. **23**, 952–961 (2007). https://doi.org/10.1109/TRO.2007.907483
5. Merriam-Webster: Merriam - Webster dictionary. http://www.merriam-webster.com/dictio nary/teamwork
6. Groom, V., Nass, C.: Can robots be teammates? Benchmarks in human-robot teams. Interact. Stud. **8**, 483–500 (2007). https://doi.org/10.1075/gest.8.3.02str
7. Gao, F., Cummings, M.L., Solovey, E.: Designing for robust and effective teamwork in human-agent teams. In: Mittu, R., Sofge, D., Wagner, A., Lawless, W.F. (eds.) Robust Intelligence and Trust in Autonomous Systems, pp. 167–190. Springer, Boston (2016). https://doi.org/10.1007/978-1-4899-7668-0_9

8. Joe, J.C., O'Hara, J., Hugo, J.V., Oxstrand, J.H.: Function allocation for humans and automation in the context of team dynamics. Procedia Manuf. **3**, 1225–1232 (2015). https://doi.org/10.1016/j.promfg.2015.07.204

9. Dickinson, T.L., McIntyre, R.M.: A conceptual framework for teamwork measurement. In: Team Performance Assessment and Measurement: Theory, Methods, and Applications, pp. 19–43. Lawrence Erlbaum Associates Publishers, Mahwah (1997)

10. Larson, C.E., LaFasto, F.M.J.: Teamwork: What Must Go Right, What Can Go Wrong. Sage Publications, Newbury Park (1989)

11. Licklider J.C.R.: Man-computer symbiosis. IRE Trans. Hum. Factors Electron. **HFE-1**, 4–11 (1960)

12. Charalambous, G.: The development of a human factors tool for the successful implementation of industrial human-robot collaboration (2014)

13. Horvath, A.O., Greenberg, L.S.: Development and validation of the working alliance inventory. J. Couns. Psychol. **36**, 223–233 (1989)

14. Lee, J., Moray, N.: Trust, control strategies and allocation of function in human-machine systems (1992). https://doi.org/10.1080/00140139208967392

15. Lee, J.D., See, K.A., City, I.: Trust in automation: designing for appropriate reliance. Hum. Factors Ergon. Soc. **46**, 50–80 (2004)

16. Kruijff, G.-J., Janıcek, M.: Using doctrines for human-robot collaboration to guide ethical behavior. In: AAAI Fall Symposium: Robot-Human Teamwork in Dynamic Adverse Environment, pp. 26–33 (2011)

17. Terveen, L.G.: Overview of human-computer collaboration. Knowl.-Based Syst. **8**, 67–81 (1995). https://doi.org/10.1016/0950-7051(95)98369-H

18. Johns, R.L.: Augmented materiality modelling with material indeterminacy. In: Fabricate: Making Digital Architecture, pp. 216–223 (2014)

19. Macmillan, J., Entin, E.E., Serfaty, D.: Communication Overhead: The Hidden Cost of Team Cognition, Washington, DC (2004). https://doi.org/10.1080/03637759309376288

20. Nahmad Vazquez, A.: Robotic Assisted Design: A study of key human factors influencing team fluency in human-robot collaborative design processes (2019)

21. Nicholas, P., et al.: Adaptive robotic fabrication for conditions of material inconsistency. In: Fabricate 2017, pp. 114–121 (2017)

22. Nahmad Vazquez, A., Jabi, W.: Robotic assisted design workflows: a study of key human factors influencing team fluency in human-robot collaborative design processes. Archit. Sci. Rev. 1–15 (2019). https://doi.org/10.1080/00038628.2019.1660611

23. Wright, P., McCarthy, J.: The politics and aesthetics of participatory HCI. Interactions **22**, 26–31 (2015). https://doi.org/10.1145/2828428

24. Onwuegbuzie, A.J., Leech, N.L., Collins, K.M.T.: Qualitative analysis techniques for the review of the literature. Qual. Rep. **17**, 1–28 (2012)

25. Braun, V., Clarke, V.: Using thematic analysis in psychology. Qual. Res. Psychol. **3**, 77–101 (2006)

26. King, N.: Template analysis. In: Symon, G., Cassell, C. (eds.) Qualitative Methods and Analysis in Organizational Research, pp. 118–134. SAGE Publications, Michigan (1998)

27. Kline, P.: The Handbook of Psychological Testing. Routledge, Milton Park (2000)

28. Kulić, D., Croft, E.: Physiological and subjective responses to articulated robot motion. Robotica **25**, 13–27 (2007). https://doi.org/10.1017/S0263574706002955

29. Kline, T.: Psychological Testing: A Practical Approach to Design and Evaluation (2005). https://methods.sagepub.com/book/psychological-testing. https://doi.org/10.4135/9781483385693

30. Loewenthal, K.M.: An Introduction to Psychological Tests and Scales. UCL Press Limited, London (1996)

31. Maxwell, S.E., Delaney, H.D.: Designing Experiments and Analyzing Data: A Model Comparison Perspective. Psychology Press, New York (2004)

32. Hoffman, G., Breazeal, C.: Effects of anticipatory perceptual simulation on practiced human-robot tasks. Auton. Robots. **28**, 403–423 (2010). https://doi.org/10.1007/s10514-009-9166-3

33. van den Brule, R., Dotsch, R., Bijlstra, G., Wigboldus, D.H.J., Haselager, P.: Do robot performance and behavioral style affect human trust? Int. J. Soc. Robot. **6**(4), 519–531 (2014). https://doi.org/10.1007/s12369-014-0231-5

34. Hancock, P.A., Billings, D.R., Schaefer, K.E., Chen, J.Y.C., De Visser, E.J., Parasuraman, R.: A meta-analysis of factors affecting trust in human-robot interaction. Hum. Factors. **53**, 517–527 (2011). https://doi.org/10.1177/0018720811417254

35. Charalambous, G., Fletcher, S., Webb, P.: The development of a scale to evaluate trust in industrial human-robot collaboration. Int. J. Soc. Robot. **8**(2), 193–209 (2015). https://doi.org/10.1007/s12369-015-0333-8

36. Bartneck, C., Kulic, D., Croft, E.: Measuring instruments for the anthropomorhism, animacy, likeability, perceived intelligence, and perceived safety of robots. Int. J. Soc. Robot. **1**, 71–81 (2009). https://doi.org/10.1007/s12369-008-0001-3

37. Broadbent, E., Stafford, R., MacDonald, B.: Acceptance of healthcare robots for the older population: review and future directions. Int. J. Soc. Robot. **1**, 319–330 (2009). https://doi.org/10.1007/s12369-009-0030-6

38. Rau, P.L.P., Li, Y., Li, D.: A cross-cultural study: effect of robot appearance and task. Int. J. Soc. Robot. **2**, 175–186 (2010). https://doi.org/10.1007/s12369-010-0056-9

39. Saerbeck, M., Bartneck, C.: Perception of affect elicited by robot motion. In: Proceeding 5th ACM/IEEE International Conference on Human-Robot Interaction - HRI 2010, pp. 53–60 (2010). https://doi.org/10.1145/1734454.1734473

40. Dragan, A.D., Lee, K.C.T., Srinivasa, S.S.: Legibility and predictability of robot motion. In: 2013 8th ACM/IEEE International Conference on Human-Robot Interaction (HRI), pp. 301–308 (2013). https://doi.org/10.1109/HRI.2013.6483603

41. Gannon, M.: Human-Centered Interfaces for Autonomous Fabrication Machines (2018)

42. LaViers, A.: Make robot motions natural. Nature **565**, 422–424 (2019)

Architectural Automations
and Augmentations: Environment

A Study on Urban Morphology and Outdoor Wind Environment of Riverside Historical Industrial Estate

Linxue Li[(✉)] , Chidi Wang , and Guichao Li

Tongji University, Shanghai, China
academy@atelier1plus.com, {chidiwang,lgc}@tongji.edu.cn

Abstract. The urban morphology has important implications for urban wind environment. In urban planning and design, the optimization of wind environment by urban form controlling has been receiving increasing attention. Taking Shanghai Yangshupu Industrial Estate as a real case, the study analyzes urban morphology and ventilation effect, introducing wind environmental indicators: wind velocity ratio (F_v), factor of wind dispersion (F_d), factor of wind recession (F_r), ratio of wind comfort area (F_c) and morphological indicators: ratio of public space (R_p), ratio of public space distribution (R_{pd}), average frontal façade area (R_{fa}), average sky view factor (R_{svf}). With three-dimensional modeling and CFD simulation, the result shows that these two series of indicators have significant correlation. The present findings are supposed to provide a strategy for optimizing urban morphology and wind environmental at pedestrian level in the early stage of urban design.

Keywords: Urban morphology · Wind environment · Correlation analysis

1 Introduction

Outdoor wind environment, which is one of the key elements of urban climate, plays an important role in mitigating urban heat island effect, alleviating air pollution and improving outdoor comfort [1, 2]. Especially at the pedestrian level, the ventilation performance is very important to human health and outdoor activities. In recent years, with the rapid development of cities, the continuous increase in building density and morphology transformation have profoundly influenced the ground wind environment [3]. The relationship between the urban morphology and the local wind condition has increasingly become the focus of attention.

As one of the key factors controlled in urban design and practice, urban morphology is difficult to be greatly adjusted in the later stage of design [4]. Therefore, in order to optimize the wind environment, it is of great significance to establish a correlated design strategy between urban morphology and the general wind environment in the early stage of urban design. Previous studies have revealed significant correlations among wind velocity at pedestrian level and urban morphology of plan area density [5], building

© Springer Nature Singapore Pte Ltd. 2022
D. Gerber et al. (Eds.): CAAD Futures 2021, CCIS 1465, pp. 405–418, 2022.
https://doi.org/10.1007/978-981-19-1280-1_25

height [6], façade area ratio [7] and building aspect ratio [8]. The majority of such studies are carried out based on in-depth analysis of built urban streets or specific case of building configuration [9]. In view of urban design, operable morphology indicators and their interactional rules with ventilation still need further examination.

This study takes the Yangshupu Industrial Estate as a real case. Since 1870s, the Yangshupu Industrial Estate has witnessed the industrial transformation and urban development of Shanghai, undergoing from industrial boom to factories vacancy. The remaining sectional urban context of wharf, factory, urban road and residence makes the city completely cut off from the river. At present, Yangshupu Industrial Estate has become a typical case for urban renewal on the basis of architectural conservation. Figure 1 shows the various urban morphologies of 14 plots. The riverside location of the site makes wind influence on plot environment more significant. Meanwhile, with the transformation of Yangshupu Industrial Estate, the wind environment calls for further differentiated design according to future image of city function, which gives this research practical value (Fig. 2).

Fig. 1. Mapping of plot division and the present morphology based on Google Earth

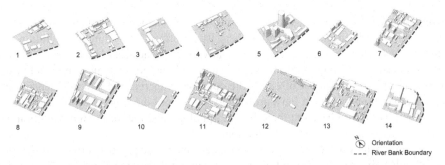

Fig. 2. Simplified three-dimensional modeling and number index of 14 plots

Taking 14 plots of Yangshupu Industrial Estate in Shanghai as real cases, this study quantitatively analyzes the ventilation effects of different urban morphologies under typical summer climate condition. A research framework for integrated analysis of urban morphology and wind environment is applied, including data collection and information systems, core indicators and evaluation system construction, performance simulation, and design strategies derived from correlation analysis (Fig. 3). The aim of this study is to examine the correlations among general indicators of urban morphology and outdoor wind conditions on pedestrian level.

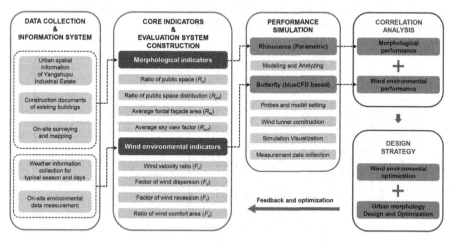

Fig. 3. Framework of morphology and wind environment research

2   Methodology

2.1   Modeling and Setting

Full-scale measurements, reduced-scale wind-tunnel experiments and numerical simulation with Computational Fluid Dynamics (CFD) are the three methods commonly used in urban wind flow assessment [10]. Full-scale measurements, limited by uncontrollable boundary and dynamic meteorological conditions, is rather expensive and time-consuming [11]. Besides, this approach is considered difficult to be applied in early stages of design and evolutionary studies [12]. However, on-site measurements embrace the best data authenticity. Reduced-scale wind-tunnel measurements allow full control over the initial and boundary conditions with a high level of accuracy [13]. Due to the limitation of physical model modification and measurement points setting, this method could be expensive. Numerical simulations (CFD) provide significant advantages over other methods. They allow full control of initial and boundary conditions, as well as easy adjustment of model parameters and configurations [14]. As simulation efficiency and result reliability improved, this method shows better operability and integration with design [15]. Therefore, in the early stages of design, CFD simulations can be used to effectively evaluate the general wind environment performance for design optimization.

In the past 50 years, CFD simulations has evolved into an increasingly used assessment method in urban physics and computational wind engineering, showing some strong advantages compared to the rest of the available tools. Though limited by the computational power for simulating large-scale urban morphology and complex architectural layout, this approach has been proved to have good sensitivity and accuracy for simplified geometry [16, 17]. Therefore, in view of the complex local wind condition in the riverside area and the high efficiency requirements of design and application, the simulation model was idealized and simplified. Referring to parameter settings of previous research [18], this study takes geometric characteristic and general wind environment assessment at pedestrian level as research objects.

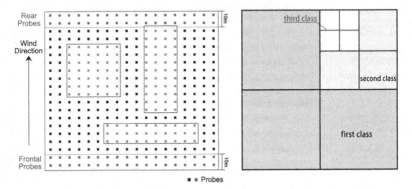

Fig. 4. An example of plot probe value setting and grid refinement

Morphology details (materials, greenery, structures, etc.) and complex climate changes are not considered in the simulation. The model data is initialized in a typical summer condition of Shanghai from June to August according to the Chinese Standard Weather Data (CSWD). The initial wind velocity in calculation adopts the southeast wind of 3.7 m/s, and the gradient wind height is 10 m. The data of wind velocity is collected at 1.5 m pedestrian level based on the grid division of 2 m units with probes evenly arranged on the base plane with 5 m units, keeping away from the building coverage area (Fig. 4). The area 10 m from the front-line and back-line boundary in the windward direction is selected as a data collection area for frontal and rear wind velocity calculation.

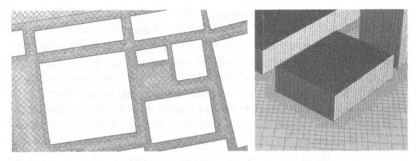

Fig. 5. An example of grid setting in wind simulation

In this study, a wind tunnel covering the plot and all the buildings is created as calculation simulation area. The grid type is unstructured grid. In order to ensure the grid coverage of the smallest space in the plot, 2 m as the smallest unit is used to divide the wind tunnel simulation area, and the grids around the buildings are three-level densified (Fig. 5). Taking the first plot as an example, the plot area is 87058 m², and the number of grids is more than 2.7 million (excluding the number of grids densified around the buildings). The simulation selects the Reynolds-averaged Simulation Turbulence Model (RAS) as the turbulence prop. Each plot calculation is iterated 200 times with each calculation time step setting to 150 to cover the entire plot area, which ensure the convergence and reliability of the calculation.

2.2 Factors and Objectives

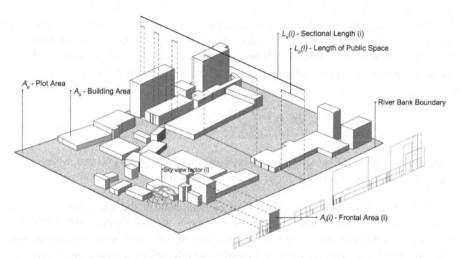

Fig. 6. Schematic diagram of morphological indicators

Presently, because wind velocity is closely related to the overall comfort and outdoor activities [19, 20], and is easy to be measured, perceived and evaluated, it has become an important factor to evaluate the urban wind environment [21]. In the past, different parameters based on wind velocity calculation have been proposed and used to evaluate the ventilation performance of urban areas, as wind velocity ratio (F_v) [22], Factor of Wind Dispersion (F_d) [19], Factor of Wind Recession (F_r) [23], Ratio of Wind Comfort Area (F_c). Similarly, four widely used urban morphology parameters are introduced into this study (Fig. 6).

Wind Environmental Indicators

Wind Velocity Ratio (F_v): Considering the complexity of urban wind environment, the wind velocity at any two points in the same plot will vary greatly. Moreover, the approaching wind is always ever-changing. The wind velocity ratio (F_v) is thus proposed to indicate how much of the wind availability of a location and to eliminate the fluctuated influence of wind by normalization approach [24].

$$Fv = \frac{1}{n}\sum_{i=1}^{n}\frac{V(i)}{V0} \tag{1}$$

$V(i)$ (m/s)is the wind velocity of measurement point i. V_0 is the wind velocity at the reference point (fixed station where it is not affected by buildings). n is the count of measurement points.

Factor of Wind Dispersion (F_d): Due to the complexity of urban airflow, it is difficult to achieve acceptable wind conditions in all areas within the urban region, the wind velocity standard deviation is used to characterize the spatial dispersion degree (Eq. (2)) [25]. Higher value of F_d indicates a higher probability of strong or weak wind conditions in local areas, which may lead to wind discomfort, danger, or pollution.

$$Fd = \sqrt{\frac{1}{n}\sum_{i=1}^{n}(V(i) - Vav)^2} \tag{2}$$

$V(i)$ (m/s) is the wind velocity of measurement point i. V_{av} (m/s) is the average wind velocity of the plot. n is the count of measurement points.

Factor of Wind Recession (F_r): The effect of buildings on the wind environment of pedestrian height in near-field areas has been extensively investigated using CFD models. The F_r is thus defined as the ratio of wind velocity after and before passing through the plot in order to describe the variation of wind affected by the building groups [26] (Eq. (3)). The smaller value of F_r represents the better infiltration effect of the plot that it has less shielding influence on the wind.

$$Fr = Vr/Vf \tag{3}$$

V_f (m/s) and V_r (m/s) respectively represent the frontal and rear average wind velocity.

Ratio of Wind Comfort Area (F_c): Due to the wind influence on the dispersion of heat, humidity and pollution in urban environment, not only strong wind, but also weak wind environment will affect the comfort of outdoor space activities [27]. In the NEN 8100 (2006) wind comfort criteria, the mean wind velocity of 5 m/s is set as the threshold of top limitation for all levels of pedestrian activities with comfort. Having no completely unified low wind velocity standard of discomfort, the threshold value of 0.2 for wind velocity ratio is now widely used to deal with the ever-changing urban wind environment

[28]. Therefore, the F_c is defined to indicate the ratio of wind comfort areas at the height of 1.5 m (Eq. (4)).

$$Fc = Ac/Ap \qquad (4)$$

A_c (m^2) is the area of wind velocity higher than 1 m/s but lower than 5 m/s. A_p (m^2) is the plot area.

Morphological Indicators

Ratio of Public Space (R_p): The plot area (A_p) shows the basic scale of the analyzed plot, and the area occupied by buildings (A_b) is measured in modeling. The ratio of public space (R_p) is given simply by division (Eq. (5)).

$$R_p = (A_p - A_b)/A_p \qquad (5)$$

A_p (m^2) is the plot area. A_b (m^2) is the building area.

Ratio of Public Space Distribution (R_{pd}): With the same density of public space, the actual spatial distribution can still have great variety. In this study, as the wind in summer generally blows from the river to the city, coordinates are established in the direction of river bank. The R_{pd} is defined as the standard deviation of the proportion value of public space in each perpendicular section of the plot (Eq. (6, 7)). The larger value of R_{pd} represents the more uneven distribution of public space in the plot.

$$Rpd(i) = Lp(i)/Ls(i) \qquad (6)$$

$$Rpd = \sqrt{\frac{1}{ns}\sum_{i=1}^{ns}(Rpd(i) - \overline{Rpd})^2} \qquad (7)$$

Annotated in Fig. 4, $L_p(i)$ (m) is the length of public space in section i. $L_s(i)$ (m) is the total length of section i. $R_{pd}(i)$ indicates the public space ratio at one specific section i of the plot. Given the $\overline{Rpd}$ as a mean value of all the $R_{pd}(i)$, the indicator R_{pd} is defined by the standard deviation calculation.

Average Frontal Façade Area (R_{fa}): The wind environment at pedestrian level has shown close relationship to the height variation and vertical dimensions of buildings in urban morphology [29]. In view of the real condition of wind blowing from the river to the city, given the summing vertical projected area of buildings towards the river (A_f), the R_{fa} is defined as the ratio of A_f to the plot area A_p (Eq. (8)). The value of R_{fa} indicates the frontal area density on wind direction and shows general blocking ability to wind on the vertical height [30].

$$Rfa = \frac{1}{Ap}\sum_{i=1}^{N}Af(i) \qquad (8)$$

$A_f(i)$ (m^2) is the vertical projected area of building towards the river. N is the count of buildings. A_{ps} (m^2) is the plot area.

Average Sky View Factor (R_{svf}): Sky view factor represents the fraction of the overlying hemisphere occupied by the sky, which describes a three-dimensional spatial characteristic of the plot [31]. This value reflects the surrounding shelters of the sky and global radiation to the location with no universal standard for the threshold value [32, 33]. The R_{svf} is given by the mean value of all the measurement points in the plot (Eq. (9)), as a dimensionless measure parameter between 0 and 1, that 0 represents completely closed and 1 for fully open.

$$Rsvf = \frac{1}{n} \sum_{i=1}^{n} SVF(i) \tag{9}$$

$SVF(i)$ is the sky view factor of measurement point i. n is the count of measurement points.

3 Simulation and Analysis

3.1 Wind Simulation Analysis of 14 Plots

Table 1. Wind environmental indicators of 14 plots

No.	1	2	3	4	5	6	7	8	9	10	11	12	13	14
F_v	0.42	0.35	0.53	0.55	0.51	0.35	0.37	0.34	0.33	0.50	0.39	0.50	0.32	0.32
F_d	0.63	0.66	0.14	0.09	1.15	0.69	0.79	0.81	0.82	0.63	0.84	0.56	0.69	0.82
V_f	1.88	1.72	1.80	1.90	1.79	1.85	2.55	1.41	1.47	1.76	1.66	1.66	1.41	1.64
V_r	1.27	0.71	2.07	2.12	1.22	1.55	0.84	1.25	0.86	2.05	1.75	1.48	1.02	0.78
F_r	0.67	0.41	1.15	1.12	0.68	0.83	0.33	0.88	0.58	1.16	1.05	0.89	0.73	0.48
F_c	0.84	0.73	1.00	1.00	0.83	0.72	0.73	0.63	0.64	0.90	0.75	0.92	0.67	0.67

F_v – Wind velocity ratio	F_d – Factor of wind distribution	V_f – Frontal average wind velocity (m/s)
V_r – Rear average wind velocity (m/s)	F_r – Factor of wind recession	F_c – Ratio of wind comfort area

Figure 7 shows the visualization of wind simulation at pedestrian level for the 14 case studies. Following the wind environmental indicators listed in Table 1, the result shows that Plot 4 has the highest wind velocity ratio ($F_v(4) = 0.55$), while Plot 13 and 14 have the same lowest one ($F_v(13) = 0.32$, $F_v(14) = 0.32$). It is examined that with the similar value of wind velocity ($F_v(3) = 0.53$, $F_v(5) = 0.51$), the wind velocity dispersion could be significantly different ($F_d(3) = 0.14$, $F_d(5) = 1.15$), which calls for specific urban planning according to actual wind comfort requirement. The indicator F_r of Plot 3, 4, 10 and 11 are all larger than 1.00, which means that the rear average wind velocity is even higher than the frontal side after the wind flows through these plots. However, Plot 7 shows the greatest hindrance to the wind that the velocity of outflow has only one third of inflow ($F_r(7) = 0.33$), in which the influence on adjacent areas could not be ignored. As the result generally shows, the plot with larger value of F_v are more likely to have higher proportion of wind comfort area. Comparatively, Plot 8 has the lowest value of wind comfort area ($F_c(8) = 0.63$), which means that the wind environmental optimization should be taken into consideration in urban renovation process of this plot.

	WIND SPEED m/s
	5.00<
	4.50
	4.00
	3.50
	3.00
	2.50
	2.00
	1.50
	1.00
	0.50
	<0.00

Fig. 7. Visualization of wind simulation results of 14 plots

3.2 Morphological Analysis of 14 Plots

Table 2. Morphological indicators of 14 plots

No.	1	2	3	4	5	6	7	8	9	10	11	12	13	14
A_p	87.06	107.45	96.75	123.40	101.75	60.82	112.97	90.88	141.19	112.44	209.48	204.38	138.52	92.37
A_b	19.47	36.57	22.79	15.11	33.82	20.58	39.94	32.36	56.31	8.65	70.28	13.92	50.42	52.16
R_p	0.78	0.66	0.76	0.88	0.67	0.66	0.65	0.64	0.60	0.92	0.66	0.93	0.64	0.44
R_{pd}	0.36	0.47	0.48	0.28	0.19	0.47	0.46	0.50	0.49	0.20	0.43	0.32	0.48	0.52
N	24	36	33	31	17	29	55	73	26	5	110	57	65	13
H	12.71	12.08	9.85	8.55	31.94	11.55	14.00	9.93	18.46	13.00	11.36	7.98	11.54	18.46
R_{fa}	0.13	0.17	0.09	0.06	0.30	0.15	0.22	0.20	0.20	0.03	0.20	0.04	0.22	0.22
R_{svf}	0.86	0.65	0.84	0.82	0.47	0.65	0.56	0.60	0.47	0.94	0.63	0.92	0.55	0.38

A_p – Plot area ($\times 10^3 \text{m}^2$)			A_b – Building area (m^2)			R_p – Ratio of public space		
R_{pd} – Ratio of public space distribution			N – Number of buildings			H – Average building height (m)		
R_{fa} – Average frontal façade area (m^2/m^2)			R_{svf} – Average sky view factor					

According to the urban morphology indicators calculated in the model and listed in Table 2, different plots can be quantitatively compared. In terms of planar form, Plot 10 and 12 have R_p of more than 0.90, reflecting the fact that the plot has been mostly demolished and transformed into vacant space presently. Plot 2 and 5 have a similar value of R_p ($R_p(2) = 0.66$, $R_p(5) = 0.67$), but the indicator R_{pd} is largely different ($R_{pd}(2) = 0.47$, $R_{pd}(5) = 0.19$). It can be found that the buildings in Plot 2 are mostly arranged in a centralized way, leaving a large and complete public space. Thus, the indicator R_{pd} accurately reveals this spatial feature. From the perspective of height, Plot 5 has been totally reconstructed into commercial complex and office buildings, showing a highest value of R_{fa} ($R_{fa}(5) = 0.30$) in 14 plots. With the similar plot area and building density compared to Plot 5, due to the conservation of historical industrial buildings, Plot 2 shows a relatively low value of the mean building heights ($R_{fa}(2) = 0.17$). In three-dimensional analysis, Plot 14 has the lowest value of R_{svf} ($R_{svf}(14) = 0.38$) while Plot 10 is of the highest ($R_{svf}(10) = 0.94$), indicating the difference of spatial openness in these plots. Overall, the four indicators can be used to comprehensively evaluate urban morphology of a specific plot related to the horizontal and vertical dimensions of buildings as well as their three dimensions.

3.3 Correlation Analysis

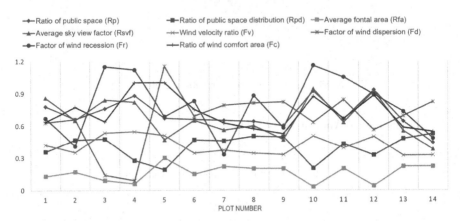

Fig. 8. Line chart of core indicators on morphology and wind environment

Combining the morphological indicators and wind environmental indicators of 14 plots, the Pearson correlation test was used for statistical analysis. (Table 3) The Fig. 9 shows that there are 14 pairs of indicators having relatively strong correlations, which are labeled with colors to differentiate negative or positive effect. Additionally, the indicator with the most significant correlation is outlined in the figure. While some conclusions are obvious, others are new and conducive to further understanding of the coupling relationship between three-dimensional urban morphology and resultant wind environment. Nevertheless, as more attention could be paid to the correlation weight of each indicator set and their variation trends, the study provides a generalized research method and a specific insight into interactive performance-oriented design approach under the complex urban environment (Fig. 8).

F_v - F_v appears to be strongly correlated with four referred morphological indicators. As a result, increasing R_p (r = 0.778**) and decreasing R_{pd} (r = −0.771**) are more likely to enhance the average wind velocity. It indicates that higher openness and more evenly spatial distribution are of the most significance of the overall wind environment according to four indicators.

F_d - F_d has a strong correlation with R_{fa} (r = 0.779**) and R_{svf} (r = −0.663**), with absolute value higher than R_p (r = −0.556*). Subsequently, R_{pd} (r = 0.030) shows no related impact on wind velocity dispersion. Therefore, when wind velocity dispersion becomes a design reference index, three-dimensional spatial characteristics should be taken into consideration.

F_r - Among the morphological indicators examined, R_p (r $= 0.643*$) and R_{svf} (r $= 0.639*$) shows almost equally positive influence on F_r, which means the larger R_p and R_{svf}, the less attenuation of the wind flowing through the plot. Adversely, R_{fa} (r $= -0.621*$) presents strongly negative correlation with F_r, revealing the blocking effect of building façade.

F_c - F_c is strongly related to all the morphological indicators. It illustrates that, under the simulation condition used in the paper, the ratio of wind comfort area on pedestrian level would significantly increase due to the higher proportion of public space (r $= 0.804**$), the more uniform public space distribution (r $= -0.642*$), the smaller frontal area per unit (r $= 0.702**$), and the higher level of sky visibility (r $= 0.779**$).

Table 3. Pearson correlation coefficient for the indicators examined

	F_v	F_d	F_r	F_c
R_p	0.778**	−0.556*	0.643*	0.804**
R_{pd}	−0.771**	0.030	−0.401	−0.642*
R_{fa}	−0.562*	0.779**	−0.621*	−0.702**
R_{svf}	0.672**	−0.663**	0.639*	0.779**
R_p – Ratio of public space	R_{pd} – Ratio of public space distribution		R_{fa} – Average frontal façade area (m^2/m^2)	
R_{svf} – Average sky view factor	F_v – Average wind velocity (m/s)		F_d – Factor of wind distribution	
F_r – Factor of wind recession	F_c – Ratio of wind comfort area			

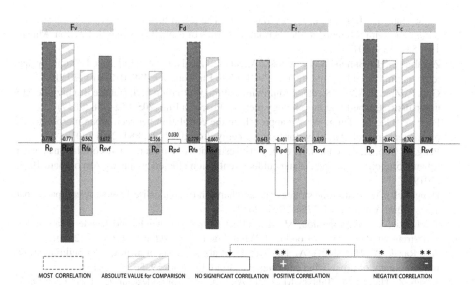

Fig. 9. Diagram of correlation analysis indicators

4 Conclusion and Discussion

Taking Shanghai Yangshupu Industrial Estate as a real case, the study puts forward a series of urban morphology extraction methods and quantitative indicators for wind evaluation. As a result, wind velocity ratio (F_v), factor of wind dispersion (F_d), factor of wind recession (F_r), ratio of wind comfort area (F_c) can comprehensively describe the wind environment of the site. The four morphological indicators of ratio of public space (R_p), ratio of public space distribution (R_{pd}), average frontal façade area (R_{fa}), average sky view factor (R_{svf}) generally present the non-uniformity of urban space and reflect three-dimensional features. They have been proved having significant correlations with the wind environment at the pedestrian level.

The concern to the urban wind environment reflects the ecological concept of design. The relevant conclusions of the study are of certain significance for the preliminary understanding of built environment according to architectural form, development density and intensity in the early stage of urban planning. Therefore, it is possible to generally optimize and predict the urban morphology and wind environment. However, considering more detailed modeling and comprehensive climate conditions, other comparable scientific indicators, and experiments with on-site measurements, this is an initial study that needs to be followed up with more scientific analysis so that the findings can be generalized to other situations.

References

1. Mohajerani, A., Bakaric, J., Jeffrey-Bailey, T.: The urban heat island effect, its causes, and mitigation, with reference to the thermal properties of asphalt concrete. J. Environ. Manag. **197**, 522–538 (2017)
2. Zakaria, N.H., Salleh, S.A., Asmat, A., et al.: Analysis of wind speed, humidity and temperature: variability and trend in 2017. Earth Environ. Sci. **489**, 012–013 (2020)
3. Guo, F., Zhu, P., Wang, S., et al.: Improving natural ventilation performance in a high-density urban district: a building morphology method. Procedia Eng. **205**, 952–958 (2017)
4. Peng, Y., Gao, Z., Ding, W.: An approach on the correlation between urban morphological parameters and ventilation performance. Energy Procedia **142**, 2884–2891 (2017)
5. Peng, Y., Gao, Z., Buccolieri, R., et al.: An investigation of the quantitative correlation between urban morphology parameters and outdoor ventilation efficiency indices. Atmosphere **10**, 33 (2019)
6. Hang, J., Li, Y.: Ventilation strategy and air change rates in idealized high-rise compact urban areas. Build. Environ. **45**, 2754–2767 (2012)
7. Tsichritzis, L., Nikolopoulou, M.: The effect of building height and façade area ratio on pedestrian wind comfort of London. J. Wind Eng. Ind. Aerodyn. **191**, 63–75 (2019)
8. Abd Razak, A., Hagishima, A., Ikegaya, N., et al.: Analysis of airflow over building arrays for assessment of urban wind environment. Build. Environ. **59**, 56–65 (2013)
9. Kaseb, Z., Hafezi, M., Tahbaz, M., et al.: A framework for pedestrian-level wind conditions improvement in urban areas: CFD simulation and optimization. Build. Environ. **184**, 107191 (2020)
10. Ramponi, R., Bert, B., et al.: CFD simulation of outdoor ventilation of generic urban configurations with different urban densities and equal and unequal street widths. Build. Environ. **92**, 152–166 (2015)

11. Schatzmann, M., Leitl, B.: Issues with validation of urban flow and dispersion CFD models. J. Wind Eng. Ind. Aerodyn. **99**, 169–186 (2011)

12. Blocken, B., Stathopoulos, T., Carmeliet, J., et al.: Application of computational fluid dynamics in building performance simulation for the outdoor environment: an overview. J. Build. Perform. Simul. **4**, 157–184 (2011)

13. Antoniou, N., Montazeri, H., Wigo, H., et al.: CFD and wind-tunnel analysis of outdoor ventilation in a real compact heterogeneous urban area: evaluation using air delay. Build. Environ. **126**, 355–372 (2017)

14. van Druenen, T., van Hooff, T., Montazeri, H., et al.: CFD evaluation of building geometry modifications to reduce pedestrian-level wind speed. Build. Environ. **163**, 106293 (2019)

15. Antoniou, N., Montazeri, H., Neophytou, M., et al.: CFD simulation of urban microclimate: validation using high-resolution field measurements. Sci. Total Environ. **695**, 133743 (2019)

16. Hang, J., Sandberg, M., Li, Y.: Effect of urban morphology on wind condition in idealized city models. Atmos Environ. **43**, 869–878 (2009)

17. Blocken, B.: 50 years of computational wind engineering: past, present and future. J. Wind Eng. Ind. Aerodyn. **2014**(129), 69–102 (2014)

18. Pingzhi, F., Jun, S., Qiang, W., et al.: Numerical study on wind environment among tall buildings in Shanghai Lujiazui Zone. J. Build. Struct. **34**(9), 104–111 (2013)

19. Hea, B.J., Dinga, L., Prasad, D.: Relationships among local-scale urban morphology, urban ventilation, urban heat island and outdoor thermal comfort under sea breeze influence. Sustain. Cities Soc. **60**, 102289 (2020)

20. Du, Y., Mak, C.M.: Improving pedestrian level low wind velocity environment in high-density cities: A general framework and case study. Sustain. Cities Soc. **42**, 314–324 (2018)

21. Wang, J.W., Yang, H.J., Kim, J.J.: Wind speed estimation in urban areas based on the relationships between background wind speeds and morphological parameters. J. Wind Eng. Ind. Aerodyn. **205**, 10424 (2020)

22. Chen, L., Hang, J., Sandberg, M., et al.: The influence of building packing densities on flow adjustment and city breathability in urban-like geometries. Procedia Eng. **198**, 758–769 (2017)

23. Tsang, C.W., Kwok, K.C.S., Hitchcock, P.A.: Wind tunnel study of pedestrian level wind environment around tall buildings Effects of building dimensions, separation and podium. Build. Environ. **49**, 167–181 (2012)

24. Ng, E.: Policies and technical guidelines for urban planning of high-density citieseair ventilation assessment (AVA) of Hong Kong. Build. Environ **44**(7), 1478–1488 (2009)

25. Britter, R.E., Hanna, S.R.: Flow and dispersion in urban areas. Ann. Rev. Fluid Mech. **35**, 469–496 (2003)

26. Stathopoulos, T., Wu, H., Bédard, C.: Wind environment around buildings: a knowledge-based approach. J. Wind Eng. Ind. Aerodyn. **41–44**, 2377–2388 (1992)

27. Du, Y., Mak, C.M., Kwok, K., et al.: New criteria for assessing low wind environment at pedestrian level in Hong Kong. Build. Environ. **123**, 23–36 (2017)

28. Netherlands Normalization Institute, (NEN 8100), Wind Comfort and Wind Danger in the Built Environment (In Dutch), Dutch standard (2006)

29. Celen Ayse Arkon & Ünver Özkol: Effect of urban geometry on pedestrian-level wind Velocity. Archit. Sci. Rev. **57**(1), 4–19 (2014)

30. Ng, E., Yuan, C., Chen, L., et al.: Improving the wind environment in high-density cities by understanding urban morphology and surface roughness: A study in Hong Kong. Landsc. Urban Plan. **101**, 59–74 (2011)

31. Oke, T.R.: Canyon geometry and the nocturnal urban heat island: comparison of scale model and field observations. J. Climatol **1**, 237–254 (1981)

32. Lin, P., Gou, Z., Lau, S., et al.: The impact of urban design descriptors on outdoor thermal environment: a literature review. Energies **10**, 2151 (2017)
33. Lyu, T., Buccolieri, R., Gao, Z.: A numerical study on the correlation between sky view factor and summer microclimate of local climate zones. Atmosphere **10**, 438 (2019)

Measurement of Spatial Openness of Indoor Space Using 3D Isovists Methods and Fibonacci Lattices

Yu Cao[ID], Hao Zheng[ID], and Shixing Liu[(✉)][ID]

School of Design, Shanghai Jiao Tong University, 800 Dongchuan RD. Minhang District, Shanghai, China

{cao_yu,liushixing}@sjtu.edu.cn, zhhao@design.upenn.edu

Abstract. The measurement of spatial openness in three dimensions has gained much popularity for recent years because it is helpful both in architecture theory research and actual design practices. Traditional tools have been developed to solve related problems but they have various limitations. In this paper, based on the 3D isovists theory, we propose a new index called spatial openness degree index (SODI). First, the paper introduces a feasible discrete sampling and computing method for the index. Second, the index is applied in several simple spatial configurations and one actual architecture case to verify its effectiveness. The new index has good adaptability to complex spatial environment, and it proves closely related to architectural factors such as openings. The study has the potential to make contributions to the development of quantitative spatial assessment system and the establishment of rational design criteria in practical design process.

Keywords: Spatial openness · 3D isovists · Fibonacci lattices

1 Introduction

Spatial openness is an important feature for an indoor spatial environment. Further studies on spatial openness have significance in both theory and practice. On one hand, it provides architects a deeper understanding of the relationship between architectural elements and human's actual experience. On the other hand, it helps solve practical design problems, for example, finding the most spacious solution under certain given conditions.

The concept of spatial openness can be explained in different perspectives. Some researchers regard it as human's subjective feedbacks of the surrounding spatial elements. Subjects are asked to give a rating after experiencing a certain spatial environment and those data are then collected to draw the conclusions [1–3]. Other researchers treat it as a purely objective property of the space, paying more attention to the spatial information such as scale, shape or spatial configuration [4–6].

Various methods have been used to extract metrics to represent the spatial openness. Yoshinobu Ashihara proposed the D/H index, the ratio of the distance between observation points and buildings (D) and the height of the buildings (H), to describe urban

© Springer Nature Singapore Pte Ltd. 2022
D. Gerber et al. (Eds.): CAAD Futures 2021, CCIS 1465, pp. 419–435, 2022.
https://doi.org/10.1007/978-981-19-1280-1_26

spaces [7]. Benedikt proposed the concept of isovists, which is defined as the set of all visible points from a single given observation point [4, 8]. The isovists analysis can be applied in the two-dimensional space and illustrated as a polygon on the floor plan. The size of the polygon is used to represent the spaciousness at that observation point [9]. Turner and Penn et al. further developed the theory and proposed methods like Visibility Graph Analysis and Agent Analysis to solve more sophisticated problems [10–12].

For last decades, related analysis in three dimensions has gained much popularity. By extending traditional two-dimensional method to the third dimension, researchers have provided several different feasible algorithms [13–16]. Other approaches and concepts have been proposed to accomplish similar purposes, for example, the Spatial Openness Index (SOI) proposed by Fisher-Gewirtzman et al. [5, 17, 18] and the Viewsphere Index proposed by Yang et al. [19]. Some other efforts have been made to adjust the calculation model, making it closer to human perception [20, 21].

2   Research Significance and Objectives

Tools mentioned above can quantitatively describe certain characteristics of a spatial environment. However, the limitation can be found in these traditional methods.

Table 1. The mathematical expressions and explanations of indices for spatial openness.

Index	Expression	Explanation
Opening rate	$OR = \frac{A_O}{A}$	A_O denotes the area of openings, A is the total surface area of the spatial environment. The higher the value, the opener the space
Sky view factor	$SVF = \frac{A_{sky}}{A_{sph}}$, or $\frac{\sum_{i=0}^{N} Bool(r_i)}{N}$	A_{sky} denotes the visible sky area while A_{sph} is the total surface area of the view sphere. r_i refers to different sight directions. $Bool(r_i) = 1$ if the sky is visible in this direction while $Bool(r_i) = 0$ otherwise. N is the total number of sight directions. The greater the value, the more exposed the space
Average viewing distance	$AVD = \frac{\sum_{i=0}^{N} R(r_i)}{N}$	r_i refers to different sight directions while $R(r_i)$ is the visible distance in this direction. N is the total number of sight directions. The greater the value, the opener the space
Spatial openness index	$SOI = \sum_{i=1}^{n} V(r_i)$	r_i refers to different sight directions while $V(r_i)$ is the visible volume in this direction. The greater the value, the opener the space

Four different tools are taken as examples and their mathematical expressions and meanings are listed in Table 1. The opening rate is easy to calculate but is not effective enough when applied in complex environment where the area of openings is hard to defined. The sky view factor is not suitable for describing indoor environment since it mainly focuses on the sky rather than the obstacles. The average viewing distance and the spatial openness index have advantages in applying in complex environment but they require a massive amount of calculation.

Also, the relation between the indices and the specific architectural elements in the spatial environment has not been fully discussed yet in related work. A further research will be especially helpful in actual architecture design process.

Therefore, the main objective of this research is to present a universal and reliable computational method for measuring spatial openness in three dimensions. This paper mainly focuses on the indoor environment. The index is intentionally designed to be fully geometrical and morphological, so the human physiological or psychological factors are not included in this research.

The secondary objective of the research is to analytically describe the relation between the index and the corresponding architectural elements, thus providing a scientific tool for designers, urban planner and related policy makers.

3 Research Method

3.1 Three Dimensional Isovists Method

This research uses the volume of visible space to represent the openness of the space. A discrete method is applied to generate the three-dimensional isovists and to calculate the volume. First, sight directions from the observation point are intentionally sampled to reduce the amount of calculation. Second, points are projected radially along the sampled directions until hitting the obstacles (Fig. 1). The distances between the observation point and the generated visible points, known as radials, are then collected and used to calculate the visible volume from the observation point.

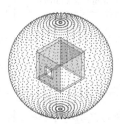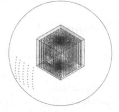

Fig. 1. The three-dimensional isovists sampling of a given space from a given observation point. The test space model and the maximum radius (left), the sampled direction (middle) and the isovists results (right).

Generally speaking, different observation points result in different values of visible volume even in the same spatial environment, which should be taken into consideration when describing the overall property of an entire environment. Under this circumstance, the observation points should be set all over the space (Fig. 2). The average value of visible volumes from all observation points are calculated to represent the global openness property of the entire environment.

Fig. 2. The diagram of multiple observation points set in the test room. The subdivision of the room (left), the observation points generated (middle) and the isovists results (right).

3.2 Sampling Method of Sight Directions

A proper sampling method helps improve the calculation efficiency, and meanwhile ensures the validity of the research. The traditional sampling method uses the latitude–longitude lattice, generated by meridians and parallels on a sphere. However, this lattice shows high anisotropy, and therefore the data need weighting for further analysis. To avoid the complicated calculation, an alternative method that can generate a set of evenly-distributed sight directions is required.

The Fibonacci lattice effectively meets the requirement [22, 23]. As demonstrated in Table 2, this method can generate an odd number (2k+1) of points. For the lattice on a unit sphere, the vertical distance between every two successive points constantly equals to $2/(2k + 1)$ and the horizontal angular spacing constantly equals to $2\pi/\varphi$. The golden ratio φ can avoid forming a periodic pattern due to its mathematical property, making the lattice points distribute evenly on the sphere as a result.

Table 2. The comparison between two different lattice methods. The extreme points are intentionally offset from the pole.

	The latitude–longitude lattice	The Fibonacci lattice
Diagram		
The spherical coordinates of each point	$\begin{cases} latitude_{i,j} = \dfrac{2i}{2k+1} \cdot \dfrac{\pi}{2} \\ longitude_{i,j} = \dfrac{2j}{2K+1} \cdot \pi \\ i = \text{-}k, \text{-}k\text{+}1, \dots, k\text{-}1, k \\ j = \text{-}k, \text{-}k\text{+}1, \dots, k\text{-}1, k \end{cases}$	$\begin{cases} latitude_i = arcsin(\dfrac{2i}{2N+1}) \\ longitude_i = \dfrac{i}{\varphi} \cdot 2\pi \\ i = \text{-}k, \text{-}k\text{+}1, \dots, k\text{-}1, k \\ \varphi = (\sqrt{5}+1)/2 \end{cases}$
The number of points	$(2k + 1)^2$	$2k + 1$

3.3 Volume Approximate Calculation Method

To simplify the calculation, the visible volume is partitioned into numbers of small poly-hedrons. Each visible point generated from the observation point can form a small poly-hedron correspondingly. The volume of each small polyhedron approximately equals to $\frac{1}{3} \cdot \Omega \cdot r^3$, where Ω is the solid angle and r is the radial length (Fig. 3).

Fig. 3. A diagram of visible volume of the $3 \times 3 \times 3$ m room with a skylight window of 1 square meter on the center of the ceiling (left). The observation point is located at the center of the room and the number of sight directions in the diagram is 1089 (middle). The volume can be approximately calculated as the sum of the volumes of small pyramid-like polyhedrons (right).

For a latitude–longitude lattice of M points, the solid angle of each small polyhedron equals to the area of spherical rectangle on a unit sphere whose center locates on the corresponding lattice point, longitude spans $2\pi/\sqrt{M}$ radians and latitude spans $\pi/\sqrt{M}$ radians. The expression turns into Eq. 1 after several mathematical calculations.

$$\Omega_i = \frac{4\pi}{\sqrt{M}} \cdot \sin\left(\frac{\pi}{2\sqrt{M}}\right) \cdot \cos(latitude_i). \tag{1}$$

While for a Fibonacci lattice of M point, due to the even distribution, the expression is much simpler (Eq. 2).

$$\Omega_i = \frac{4\pi}{M}. \tag{2}$$

For either lattice method, if a total of M sight directions is sampled for one observation point, the visible volume of the j(th) observation point can approximately be expressed as Eq. 3:

$$V_j = \sum_{i=1}^{M} \left(\frac{1}{3} \cdot \Omega_i \cdot r_i^3\right). \tag{3}$$

The visible volume is an absolute amount, which is not intuitive enough for a comparison analysis. To solve this issue, a relative index is required, and the theoretical maximum visible volume of the observation point needs to be calculated. In most cases, the maximum value is achieved when all the obstacles are removed (Fig. 4).

The $MAX\left(V_j\right)$ can be calculated with Eq. 3, or alternatively by a geometrical approach to get the precise result. If the sight is only blocked by the ground, the maximum visible

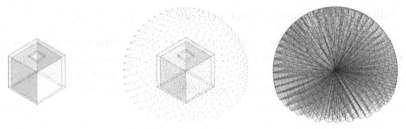

Fig. 4. A diagram of maximum visible volume in the same case as in Fig. 3. The maximum value is reached when all the obstacles except the ground are ignored. And therefore, the shape of the corresponding visible volume turns into a spherical cap.

volume will form a spherical cap. According to related geometry formulas, the maximum value can be expressed as Eq. 4.

$$MAX\left(V_j\right) = \frac{1}{3} \cdot \pi \cdot \left(2r_{max}^3 + 3r_{max}^2 z_j - z_j^3\right).\tag{4}$$

In Eq. 4, r_{max} is the maximum viewing range, and z_j is the z coordinate of the j(th) observation point. The r_{max} needs to be large enough to include the longest sight length. One feasible option is to set r_{max} as the diagonal length of the bounding box of the entire environment.

The relative form of visible volume of the j(th) observation point can be defined as the ratio of the actual visible volume V_j to the corresponding maximum value $MAX\left(V_j\right)$, which goes as Eq. 5. A higher ratio refers to a broader view, and when the value reaches 1, the environment becomes as open as a vast plain. The ratio is defined as the spatial openness degree index (SODI) of the given observation point in this paper.

$$SODI_j = \frac{V_j}{MAX\left(V_j\right)} \times 100\%.\tag{5}$$

To describe the property of the entire spatial environment, if a total of N observation points is sampled in the analyzed space, the spatial openness degree index (SODI) of the entire spatial environment is then defined by Eq. 6.

$$SODI = \frac{\frac{1}{N} \cdot \sum_{j=1}^{N} V_j}{\frac{1}{N} \cdot \sum_{j=1}^{N} MAX\left(V_j\right)} \times 100\% = \frac{\sum_{j=1}^{N} V_j}{\sum_{j=1}^{N} MAX\left(V_j\right)} \times 100\%.\tag{6}$$

The SODI value ranges from 0 to 1. A greater value indicates that averagely speaking, the environment makes a better use of the potential visual space. Therefore, the environment seems more open.

3.4 Parameter Settings: The Number of Sight Directions

Both M (the number of sight directions) and N (the number of observation points) influence the calculation results, and therefore they should be carefully decided.

First, to determine the value of M, two simple rooms are designed for the test. One is a room without any opening at the size of 3 × 3 × 3 m while the other is the same room but with a skylight window of 1 square meter on the center of the ceiling. The observation point is set in the exact center of the room. Different numbers of sampling directions are applied and the visible volume of the observation point is then calculated to check the accuracy of the algorithm. Theoretically, the volume should be exactly 27 m³ in the first case and about 41.54 m³ in the second. The calculation results are shown in Fig. 5 and Fig. 6.

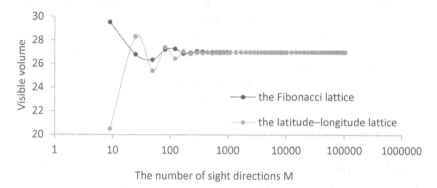

Fig. 5. The calculated results of visible volume of the 3 × 3 × 3 m room with no openings.

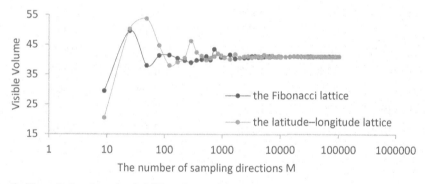

Fig. 6. The calculated result of visible volume of the 3 × 3 × 3 m room with a skylight window of 1 square meter on the center of the ceiling.

Consequently, both lattice methods can provide a correct result as long as the number of sampling sight directions is large enough. The Fibonacci lattice decays faster to a stable value than the latitude-longitude lattice, and thus it has advantages when the value of M is small. Moreover, the value fluctuates more in the second case because of the existence of the opening.

The previous test also shows that the algorithm leads to some slight errors. To find out the relation between the error and the value of M, a number of spherical caps spaces are designed and the Monte Carlo simulation method is applied.

Seven lattice configurations of each lattice type are analyzed ($M \approx 10^2$, $10^{7/3}$, $10^{8/3}$, 10^3, $10^{10/3}$, $10^{11/3}$, 10^4). 500 different cap-shape rooms with different opening rates (from 0% to 50%) and different orientations are randomly generated for each configuration. In total, 7000 different rooms are measured in this test and the errors are calculated by Eq. 7.

$$Error = \frac{|V_j - V_{precise}|}{V_{precise}} \times 100\%. \tag{7}$$

In the equation, V_j refers to the calculated result with the discrete method and $V_{precise}$ is the precise value obtained by geometry approach. The analysis results are shown in Table 3. It points out that the average errors for both types gradually reduce to about 1.2% and the maximum errors reduce to about 2.4%. This systematic error is inevitable but it makes little invalidation to the conclusions since it is within the acceptable range for architecture research. Consequently, the number of sight directions is suggested to be above 10^3 to obtain a relative accurate result.

Table 3. The error analysis in the case of randomly generated spherical cap rooms.

The number of sampling directions M	The average value of errors		The average value of top 5% errors	
	The Fibonacci lattice	The latitude–longitude lattice	The Fibonacci lattice	The latitude–longitude lattice
121 ($\approx 10^2$)	4.57%	5.26%	19.32%	22.63%
225 ($\approx 10^{7/3}$)	3.07%	3.51%	12.67%	16.39%
441 ($\approx 10^{8/3}$)	2.16%	2.26%	8.46%	9.89%
961 ($\approx 10^3$)	1.40%	1.57%	4.62%	6.55%
2209 ($\approx 10^{10/3}$)	1.22%	1.21%	3.49%	3.90%
4761 ($\approx 10^{11/3}$)	1.17%	1.19%	2.75%	3.19%
9801 ($\approx 10^4$)	1.19%	1.15%	2.44%	2.40%

3.5   Parameter Settings: The Number of Observation Points

Next test is to determine the value of N. For this task, different densities of observation points are set in the previous $3 \times 3 \times 3$ m room with a skylight window of 1 square meter on the center of the ceiling. The observation points are set by a three-dimensional orthogonal grid instead of randomly distributed. Various observation point configurations ($N = 1^3$, 2^3, 3^3, 4^3, etc.) are applied and the SODI of the room is then calculated. Only the Fibonacci lattice method is used in this test and the number of sight directions is set as 1089. The calculated results are demonstrated in Fig. 7.

The SODI result gradually gets stable as the density of observation points goes higher. The number of observation points is suggested to be above 8 points per cubic meter to

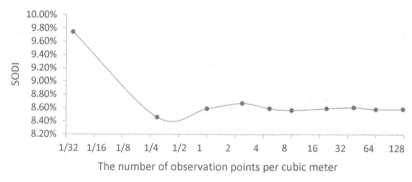

Fig. 7. The calculated result of SODI of the $3 \times 3 \times 3$ m room with a skylight window of 1 square meter on the center of the ceiling and with different number of observation points.

obtain a relative accurate result. However, if the environment analyzed is large enough and the accuracy requirement is not that strict, the sampling density of observation points can be lower to reduce the calculation amount.

4   Case Study

Before choosing any real buildings, some spatial prototypes are designed as study cases to verify the validity of the index. The cases are modelled in Rhinoceros, a 3D computer graphics and computer-aided design application software developed by Robert McNeel & Associates. And to improve the computational efficiency and accuracy, a Grasshopper program is made to perform the data generating and processing.

4.1   The Size of the Opening

In the first set of cases, opening of different sizes are set on the center of a wall or the center of the ceiling, and the room is $3 \times 3 \times 3$ m. The sizes of the openings are set

Table 4. The SODI of rooms with an opening with different sizes.

Indices	The result values				
	1m×1m	1.5m×1.5m	2m×2m	2.5m×2.5m	3m×3m
SODI (opening on the ceiling)	8.56%	11.77%	15.75%	20.94%	25.62%
SODI (opening on the wall)	8.35%	11.09%	14.43%	18.59%	22.00%

respectively as 1 m × 1 m, 1.5 m × 1.5 m, 2 m × 2 m, 2.5 m × 2.5 m and 3 m × 3 m. The Fibonacci lattice is applied in this test. M is set as 1089 and N is set as 216. The results are demonstrated in Table 4.

It can be seen that for the cases with openings at the same location, the SODI shows an increasing trend as expected. A larger opening generates a higher SODI value and vice versa. As for the cases with openings of the same size, the result from a skylight window will be higher than that from a window on the wall. And with the opening gets larger, the gap between two scenarios goes more prominent.

4.2   The Location of the Opening

The previous case study shows that the location of the opening has influence on the value of SODI. In next set of cases, an opening of 1 m^2 is set in the room of 3 × 3 × 3 m, and the location of the opening varies from one case to another. The Fibonacci lattice is applied in this test. M is set as 1089 and N is set as 216. The calculated values are listed in Table 5.

Generally speaking, the opening on the ceiling generates a higher SODI than that on the wall. The opening near the surface center generates a higher SODI than that around the room corner, and the value decays gradually faster as the opening moves farther away from the surface center. Moreover, due to the ground plane, the result from a low window on is smaller than that from a high window.

In addition, it can also be found in the table that symmetric spatial configurations will result in similar SODI values. It can be further deduced that within the range of permitting error, the horizontal Euclidean transformation including moving, rotating and mirroring will not influence the SODI value, while the vertical transformation will influence the value prominently.

4.3   The Configuration of the Opening

In this set of cases, half of the walls is removed from the room of 3 × 3 × 3 m, while the configuration of the openings varies from one case to another. Some cases form broader walls while others form narrow bars. The Fibonacci lattice is applied in this test. M is set as 1089 and N is set as 216. The results are shown below in Table 6.

As the table shows, the more openings are set in the room, the lower SODI value it generates, and different spatial configurations with same number of openings lead to pretty close results. Consequently, a concentrated and unified opening generates a higher SODI result than several scattered and dispersed openings.

The openings in this test only change in the horizontal direction and their height keep constant as the height of the room. If the distribution of openings varies in the vertical direction as well, the conclusions are supposed to be more complicated.

4.4   The Size and Shape of the Room

Any complex space can be derived from a single space with the outer boundary surface alone, no openings or any other obstacles. The SODI of the boundary-only room can be

Table 5. The SODI of rooms with a same-sized window but at different locations.

	The result values				
Skylight windows	8.25%	8.50%	8.56%	8.50%	8.26%
	8.19%	8.44%	8.49%	8.44%	8.20%
	7.99%	8.20%	8.25%	8.19%	7.99%
High windows	7.98%	8.18%	8.23%	8.18%	7.97%
	8.12%	8.35%	8.41%	8.34%	8.11%
Eye-level windows	8.07%	8.29%	8.35%	8.29%	8.06%
	7.85%	8.04%	8.08%	8.03%	7.84%
Low windows	7.38%	7.48%	7.51%	7.48%	7.37%

Table 6. The SODI of rooms with different opening configurations with the same opening area.

	The result values				
Layout diagram					
The number of openings	2	2	3	3	4
SODI	37.41%	37.40%	36.44%	36.44%	35.44%
Layout diagram					
The number of openings	4	4	8	12	32
SODI	35.43%	35.43%	29.96%	27.88%	20.21%

regarded as a baseline value in a comparison analysis. The value grows higher as more openings are set on the room surface, while gets lower as more obstacles are put inside the room. The outer boundary merely contains the information of size and shape, which also has certain influence on the SODI value and their relation has not been discussed yet. In the following set of cases, different sizes of cuboid room are studied as examples.

For a cuboid room having width $= W$, depth $= D$ and height $= H$, the numerator in Eq. 6 equals to the volume of the room (Eq. 8).

$$\frac{1}{N} \cdot \sum_{j=1}^{N} V_j = WDH. \tag{8}$$

And the denominator in Eq. 6 can be written in the form of integral. By inserting the Eq. 4, it can be further expressed as Eq. 9:

$$\frac{1}{N} \cdot \sum_{j=1}^{N} MAX(V_j) = \frac{1}{H} \cdot \int_0^H MAX(V_j) = \frac{1}{3}\pi \left(2r_{max}^3 + \frac{3}{2}r_{max}^2 H - \frac{1}{4}H^3 \right). \tag{9}$$

Therefore, the expression of SODI can then be obtained by Eq. 10:

$$SODI = \frac{\frac{1}{N} \cdot \sum_{j=1}^{N} V_j}{\frac{1}{N} \cdot \sum_{j=1}^{N} MAX(V_j)} \times 100\% = \frac{12WDH}{\pi \cdot (8r_{max}^3 + 6r_{max}^2 H - H^3)} \times 100\%. \tag{10}$$

By making $W = \alpha \cdot H$ and $D = \beta \cdot H$, the r_{max} defined in this paper then equals to $\sqrt{1 + a^2 + b^2} \cdot H$. Therefore, Eq. 9 becomes Eq. 11:

$$SODI = \frac{12}{\pi} \cdot \frac{\alpha\beta}{8\sqrt{1 + \alpha^2 + \beta^2}^3 + 6\sqrt{1 + \alpha^2 + \beta^2}^2 - 1} \times 100\%. \tag{11}$$

According to Eq. 11, a plot of the SODI value mapped in the $\alpha - \beta$ plane can be generated and shown as Fig. 8.

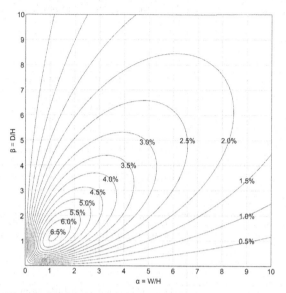

Fig. 8. Two-dimensional plot in the α–β plane of the SODI value of cuboid rooms. The corresponding SODI values are labelled on the level curves.

Several conclusions can be drawn from the plot. First of all, the SODI value reaches its maximum (about 6.60%) when $\alpha = \beta = 1.15$, forming a slightly flattened cuboid shape.

The plot is symmetrical along the line $\alpha = \beta$, which means two shapes (α, β) and (β, α) will always share the same SODI result. It is obvious since one cuboid can be obtained by rotating 90° horizontally and scaling uniformly from the other.

Moreover, each level curve in the plot represents a set of shapes sharing the same SODI value. This indicates the fact that even if there is a limitation on the SODI value, architects still have a wide range of choice for the shape of the room.

4.5   A Complicated Case: The Barcelona Pavilion

In this section, the index is applied to an actual building models, the Barcelona Pavilion, to demonstrate its effectiveness. The Barcelona Pavilion was designed by Mies, which is famous for its flowing space.

The analysis target only focuses on the main part of the architecture. The Fibonacci lattice is applied in this test and the number of sight directions is set as 1089. One thousand randomly distributed observation points are sampled in the model to generate the SODI. All the material properties are ignored in the analysis and the glass windows are removed from the model. Apart from the original pavilion designed by Mies, some variants are generated for a comparison analysis to test the effectiveness of the index. The results are listed respectively in Table 7.

Table 7. The SODI of the Barcelona Pavilion and its variants.

	Diagram	Explanation	SODI
Origin		The virtual model of the pavilion is used for the analysis.	19.75%
Variant #1		Only the plinth of the architecture is remained. The visible volumes are supposed to reach the maximums in this situation.	100.00%
Variant #2		This variant turns to pavilion into a simple cuboid room with no openings or inner obstacles. This situation can be regarded as a baseline in the comparison analysis.	2.04%
Variant #3		The walls are removed from the pavilion, making the architecture more exposed to the surrounding environment.	37.66%
Variant #4		The roof is extended a little from the pavilion, covering the small pond in the architecture and thus making the architecture less exposed to the sky.	17.67%

As the table shown, the SODI functions effectively in the complex environment as well. The index of the pavilion reaches 19.75%, nearly ten times larger than the baseline value. Consequently, the index can correctly describe the spatial openness property of the pavilion.

The table also shows that index grows about 18% by removing the walls in the pavilion, and it drops about 2% by extending the roof of the pavilion. The SODI can directly reflect the adjustments made in the architecture, so it can be useful during the design progress.

5 Discussion

5.1 The Relation Between SODI and Human Perception

The spatial openness is a topic that has a strong connection with anthropological, social and psychological factors. Therefore, it is worth discussing the relationship between the newly-proposed index and the human subjective perception.

The SODI is a purely objective index since it only takes geometrical and morphological factors into considerations. This provides the certainty for the index, which means the SODI will be fixed as long as the spatial environment and related basic parameters are settled and determined. The certainty and reproducibility make the index suitable for group decision-making and systematic space assessment where the consensus is hard to achieve.

In most cases, the SODI matches human perception correctly. For example, a room with a lager window certainly feels more open and spacious, and it results in a higher SODI value indeed, as discussed in Sect. 4.1. However, the index sometimes seems counterintuitive.

One reason for the difference lies in the selection of observation points. For example, as discussed in Sect. 4.1 and 4.2, the SODI from a skylight window is higher than that from a window on the wall when their sizes of opening are the same. However, some empirical experiments have demonstrated that a room with the skylight appears more enclosed because the opening is far from the eye level and hard to be noticed [3]. The difference appears because SODI takes all the observation points into calculation, including those beyond human's reach. If observation points are restricted at the level of our eyes, the conclusion will be closer to our intuitive expectations.

Another reason lies in the relative form of the index. As shown in Eq. 11, the SODI is scale independence since the length factors are eliminated in both numerator and denominator. Some would argue that the spatial opening is closely related to the architecture scale. A larger scale of space should be more open by intuition. The problem can be solved by using both the absolute index (V_j) and the relative index (SODI) in the research so that the scale factor is also taken into consideration.

5.2 The Characteristics of SODI

The SODI has several characteristics. First, the index has good universality and adaptability. Once the space is given, the index of any complex spatial environment can be then calculated, which is especially helpful when analyzing free-form space or flowing space whose boundary is hard to define.

Another characteristic is that the index is correctly related to the spatial configuration of the space, including the location and the number of opening. This results in an advantage over other traditional measurements like opening rate in describing indoor spatial environment.

Furthermore, SODI has extreme values, ranging from 0 to 1. For a given three-dimensional region, SODI = 0 means that the space is completely solid and nothing can be seen, while SODI = 1 means that there is no obstacle or all obstacles are transparent. Any other value of SODI can be understood as an intermediate situation in between. The fixed range of the index is helpful for comparative research between different cases.

Besides, the index and the algorithm can be easily programmed into parametric software tools to quickly generate the result, which provides the convenience for the combination with other digital design tools in architecture field.

6   Outlook

In this paper, we have defined the primary algorithms to build further works on. In the future, the index can be applied to practical applications, individually or in combination with other metrics. For example, it can be used during the optimization work in the architecture design process. By setting the index as the fitness input in parametric design tools such as Grasshopper, optimal solutions can be then generated to meet certain demands. The index can also potentially work as one of the criteria for decision-makers and urban designers to control the quality of space.

In this paper, we mainly focus on the openings of the space, which is the most significant factor that increases the spatial openness in buildings. However, there are other factors that play important roles, such as inner walls and furniture. These can be taken into consideration in our future works to figure out how the index will change when more elements are factored in. Besides, more actual architecture cases will be put into test to further demonstrate the usefulness of the index and its relation with human empirical perception.

References

1. Sadalla, E.K., Oxley, D.: The perception of room size: The rectangularity illusion. Environ. Behav. **16**, 394–405 (1984). https://doi.org/10.1177/0013916584163005
2. Stamps, A.: Effects of Area, Height, Elongation, and Color on Perceived Spaciousness. Environ. Behav. **43**, 252–273 (2011). https://doi.org/10.1177/0013916509354696
3. Sanatani, R.P.: An Empirical Inquiry into the Perceptual Qualities of Spatial Enclosures in Head Mounted Display Driven VR Systems Quantifying the 'Intangibles' of Space. In: Blucher Design Proceedings, pp. 125–132. Editora Blucher, São Paulo (2019). https://doi.org/10.5151/proceedings-ecaadesigradi2019_427
4. Benedikt, M.L.: To take hold of space: isovists and isovist fields. Environ. Plan. B Plan. Des. **6**, 47–65 (1979). https://doi.org/10.1068/b060047
5. Fisher-Gewirtzman, D., Wagner, I.A.: Spatial openness as a practical metric for evaluating built-up environments. Environ. Plan. B Plan. Des. **30**, 37–49 (2003). https://doi.org/10.1068/b12861
6. Wu, X., Oldfield, P., Heath, T.: Spatial openness and student activities in an atrium: A parametric evaluation of a social informal learning environment. Build. Environ. **182**, 107141 (2020). https://doi.org/10.1016/j.buildenv.2020.107141
7. Ashihara, Y.: The aesthetic townscape. The MIT Press, Cambridge (1983)
8. Davis, L.S., Benedikt, M.L.: Computational models of space: Isovists and isovist fields. Comput. Graph. image Process. **11**, 49–72 (1979). https://doi.org/10.1016/0146-664X(79)90076-5
9. Benedikt, M., Burnham, C.A.: Perceiving Architectural Space: From Optic Arrays to Isovists. In: Persistence and change: Proceedings of the first international conference on event perception, pp. 103–115. Psychology Press (1985). https://doi.org/10.4324/9780203781449
10. Turner, A., Doxa, M., O'Sullivan, D., Penn, A.: From Isovists to Visibility Graphs: A Methodology for the Analysis of Architectural Space. Environ. Plan. B Plan. Des. **28**, 103–121 (2001). https://doi.org/10.1068/b2684
11. Turner, A., Penn, A.: Making isovists syntactic: isovist integration analysis. In: 2nd International Symposium on Space Syntax, Brasilia, pp. 103–121. BRASILIA (1999)
12. Penn, A., Turner, A.: Space syntax based agent simulation. In: 1st International Conference on Pedestrian and Evacuation Dynamics, pp. 99–114. Springer-Verlag, Duisburg (2002)
13. Derix, C., Gamlesæter, Å., Carranza, P.M.: 3d Isovists and Spatial Sensations: Two methods and a case study. In: Movement and Orientation in Built Environments: Evaluating Design Rationale and User Cognition, pp. 67–72. Bremen (2008)

14. Suleiman, W., Joliveau, T., Favier, E.: A new algorithm for 3D isovists. In: Advances in Spatial Data Handling. pp. 157–173. Springer (2013). https://doi.org/10.1007/978-3-642-32316-4_11

15. Díaz-Vilariño, L., González-deSantos, L., Verbree, E., Michailidou, G., Zlatanova, S.: From point clouds to 3D isovists in indoor environments. Int. Arch. Photogramm. Remote Sens. Spat. Inf. Sci. **42**, 149–154 (2018). https://doi.org/10.5194/isprs-archives-XLII-4-149-2018

16. Krukar, J., Manivannan, C., Bhatt, M., Schultz, C.: Embodied 3D isovists: A method to model the visual perception of space. Environ. Plan. B Urban Anal. City Sci. **48**(8), 2307–2325 (2020). https://doi.org/10.1177/2399808320974533

17. Fisher-Gewirtzman, D., Burt, M., Tzamir, Y.: A 3-D visual method for comparative evaluation of dense built-up environments. Environ. Plan. B Plan. Des. **30**, 575–587 (2003). https://doi.org/10.1068/b2941

18. Fisher-Gewirtzman, D., Wagner, I.A.: The spatial openness index: An automated model for three-dimensional visual analysis of urban environments. J. Archit. Plann. Res. **23**, 77–89 (2006)

19. Yang, P.P.J., Putra, S.Y., Li, W.: Viewsphere: a GIS-based 3D visibility analysis for urban design evaluation. Environ. Plan. B Plan. Des. **34**, 971–992 (2007). https://doi.org/10.1068/b32142

20. Bittermann, M., Ciftcioglu, Ö.: Systematic measurement of perceptual design qualities. In: Proceedings of the ECCS 2005 Satellite Workshop: Embracing Complexity in Design, pp. 15–22. The Open University (2005)

21. Bittermann, M.S., Ciftcioglu, O.: Visual perception model for architectural design. J. Des. Res. **7**, 35–60 (2008). https://doi.org/10.1504/JDR.2008.018776

22. González, Á.: Measurement of Areas on a Sphere Using Fibonacci and Latitude–Longitude Lattices. Math. Geosci. **42**, 49–64 (2010). https://doi.org/10.1007/s11004-009-9257-x

23. Saff, E.B., Kuijlaars, A.B.J.: Distributing many points on a sphere. Math. Intell. **19**, 5–11 (1997). https://doi.org/10.1007/bf03024331

An Info-Biological Theory Approach to Computer-Aided Architectural Design: UnSSEUS

Provides Ng$^{(\boxtimes)}$ ⓘ, Baha Odaibat ⓘ, David Doria ⓘ, and Alberto Fernandez ⓘ

Bartlett School of Architecture, UCL, London WC1H 0QB, UK
rationalenergyarchitects@gmail.com

Abstract. In Computer-Aided Architectural Design (CAAD), generative and predictive methods are two major practices. This paper discusses how an information theory approach to bio-design reconfigures our means of integrating the two practices, focusing on enhancing solar energy capacities in cities. PART I analyses how information is defined with respect to Shannon and Wiener's work in information and cybernetics theories and considers how they provoke changes in fields of biology, including the Information Theory of Individuality and the Free Energy Principle. From these theoretical findings, PART II tabulates 3 design principles - negentropic, preemptive, and network, and proposes 'Unattended and Smart Solar Energy Urban System' (UnSSEUS). Part III exemplifies the argument with CAAD experiments - form-finding for Carbon Nanotube backed solar cells to formulate sub-systems that can aggregate on a macroscale - UnSSEUS. From which, a combinatorial design production pipeline was developed that synthesizes between rule-based and machine learning systems, with Work-in-Progress examples of solar design optimized by AI. Finally, it discusses the possibilities of personalization and prototyping.

Keywords: Information theory · Bio-design · Computer aided architectural design · Unattended and smart solar energy urban system

1 Introduction

The earliest record of the use of the word "information" was in the 1300s; it is derived from Latin through French by combining the word 'inform', meaning to give a form to the mind, with the ending "ation", which denotes a noun of action (Logan 2012). Even though the word 'information' is a noun, in fact, it is a process - a construct that arises in the process of trying to reason about our complex environment (Ng et al. 2020). An increasing amount of biological research tries to understand complex systems using information theory, which enables them to be captured computationally and transforms the discipline from a biology-of-things to a biology-of-processes. Notably, Krakauer et al.'s (2020) Information Theory of Individuality (ITI) and Friston's (2019) Free Energy Principles (FEP) consider how biological systems employ generative and predictive

© Springer Nature Singapore Pte Ltd. 2022
D. Gerber et al. (Eds.): CAAD Futures 2021, CCIS 1465, pp. 436–455, 2022.
https://doi.org/10.1007/978-981-19-1280-1_27

methods via information feedback to perform bio-intelligence that sustain individuality within a complex environment.

In the context of CAAD, such info-biological principles are helping to rethink design approaches in the face of our energy crisis, which pose an urgency for enhancing renewable resources. For cities, increasing the proximity to renewable energy helps to reduce energy absorption by the power grid and dissipation during transmission. Nonetheless, renewable energy sources that can adapt to urban environments are limited; there are also significant challenges in building local socio-economic capacities, including infrastructural provisions. Among the six types of renewable energy, solar power remains one of the most abundant and readily available for most urban areas worldwide (Leung and Hui 2004).

The advancements in molecular engineering are opening new avenues to which light-harvesting materials can be used in densely populated areas; in particular, Carbon Nanotube (CNT) backed solar cells may integrate with building design to cover large areas in cities (Rowell et al. 2006). While the science of carbon-based material is still in its infancy, the CAAD community should prepare itself for the design challenges to come. This would demand the design of both hard and soft infrastructures, across scales from nano to macro, to ensure sustainability and efficiency. For instance, form-finding can help minimise the physical pathways that electrons circulate to reduce energy dissipation (Braun et al. 2013). This can be measured and evaluated using information feedback but would require network designs to help ensure the speed capacity of synchronising such information between mechanical and social components - a feedback loop between energy supply and demand from a network of light-harvesting devices and users.

Along these lines, biological systems provide prospective models for us to search for sustainable designs that are efficient within the Earth's energy chains, which are open-systems that exchange information to maintain their equilibrium, from DNA to natural languages. This research aims at studying such information exchanges, translating info-biological principles into CAAD. How can bio-intelligence inspire us in designing sustainable and efficient energy systems using information feedback?

2   PART I - Information Theory Approach to Biology

2.1   Information and Cybernetics Theories

The notion of information as a mathematically defined quantity that can be stored, transferred, or communicated to inanimate objects did not arise until the 20th century (Logan 2012). The OED (2020) cited two sources: The first regards the transmission of signals to television, which gave us a sense of the socio-economic context to which Claude Shannon's and Weaver (1948) pioneering work - information theory - was born. The second regards feedback processes between human and machine interaction show us a part of history that encompasses Norbert Wiener (1948) and cybernetics theories.

The year 1948 saw the birth of two publications that changed the way we think of information: Shannon's 'A Mathematical Theory of Communication' and Wiener's 'Cybernetics: Or Control and Communication in the Animal and the Machine'. The former dealt with the nature of information and its relationship to the tools that operate it; the latter dealt with how information can become the tool. While both considered how information becomes valuable at the point of interaction, the former saw value as compression and bandwidth - how much can be expressed within a transaction - as opposed to semantics or scarcity; the latter saw value as the ability to preempt within a statistical structure (Kaiser 2020).

There are three main ideas in Shannon's and Weaver (1948) paper. The first one was dedicated to the compression and information entropy problem, which measures the number of surprises or non-redundancy in some signals. The second was on the bandwidth problem, which has physical constraints (e.g. how do we represent the amount of noise added in a created channel, perhaps, from heat produced by electrons circulating?). This offered a means to represent physical infrastructure as a mathematical model. The third was dedicated to the probabilistic structure of some signals (e.g. the structural redundancy in natural languages) and gave rise to the concept of 'bits' - binary digits. Together, they are concerned with finding the limits to communication - the entropy limit - the minimum number of bits needed to retrieve a piece of information, where the semantics to the language are irrelevant. 'Can we find a measure of how much "choice" is involved in the selection of the event or of how uncertain we are of the outcome?'; in other words, information is measured relative to the number of choices within a stochastic process, and Shannon and Weaver (1948) referred specifically to Markov processes.

The Markov chain is a random process in which the future is independent of the past, given the present (Kirchhoff et al. 2018). The time relationships are summarized by the notion of a state, which evolves over time according to some probability distribution. Thus, knowing the state of a system at any point in time not only provides a good description of the system at that time, but it does seem to capture the critical information that we need to answer questions about the future evolution of the system (Kirchhoff et al. 2018). Wiener (1968) stated, 'the more probable an event, the less information it gives' - the prediction of less likely events yields more information. This is different to Shannon and Weaver (1948), who thought that the communication of more likely events gives more information. Although Wiener and Shannon studied the transmission of signals using Markov processes, Wiener's (1948) work on prediction in time series - Fourier transform (FT) and Brownian motion - studied signals not from the frequency of the time domain. This established their shared understanding of information in terms of probability distribution within a statistical structure, but their differing views on how information can be quantified and its relationship to entropy. Wiener (1968) poetically put: 'this amount of information is a quantity which differs from entropy merely by its algebraic sign and a possible numerical factor'.

Shannon and Weaver (1948) correlated information with entropy positively, whereas Wiener (1948) defined information as negative entropy. This gave rise to Wiener's (1948) cybernetics, which seeks to model self-organization through information feedback, and enables cyberneticians to describe socio-biological systems from within, from autopoiesis to financial markets (Krippendorff 1984). Within a world that always decays into chaos (third law of thermodynamics), Wiener held that information is negentropic: our environment contains no information, but the information is always the construct of living beings who export entropy - 'organization maintained by extracting "order" from the environment' (Schrodinger 1944).

2.2   Information Theory of Individuality and Free Energy Principle

Schrodinger (1944) first introduced the notion of negentropy in his prescient book 'What Is Life?'. Inspiring a self-organizational understanding to biological systems, which give emergence to complexity through recursive interactions between simpler sub-systems via information feedback across various system scales: 'at the micro-scale (e.g., dendritic self-organization and morphogenesis), across intermediate scales (e.g., cultural ensembles), and at the macro-scale (e.g., natural selection)' (Ramstead et al. 2018). The ITI and FEP are two emerging fields that approach socio-biological studies using information theory. The former can be seen as first-order cybernetics - "the study of observed systems"; while the latter can be seen as second-order cybernetics - "the study of observing systems" - considering the observer as part of the system (von Foerster 1992). In this respect, time information gives a statistical meaning to entropy.

In the ITI, Krakauer et al. (2020) argues that "individuals are aggregates that preserve a measure of temporal integrity, i.e., "propagate" information from their past into their futures", where we may begin to describe the performance and qualities of an individual by their variation "in the degree of environmental dependence and inherited information" - a measure of entropy. Krakauer et al. (2020) hypothesized three types of individuality: organismal, colonial, and environment-driven, each differs in their level of dependence on the information that is transmitted from external states; thus, the ITI 'is analogous to figure-ground separation ... or computer vision. The background of an image carries as much if not more information than the object'. Such an entropy approach can be used to understand nonlinear power scaling. West and Brown (2005) stated that biological systems are observed to have an economy of scale embedded - the larger the organism, the less energy it needs to metabolize for survival (~¼ less), and the pace of life (e.g. heart rate) is systematically slower with increased size. This idea proposes an entropy limit that can be used to guide the development of urbanism. Theoretically, the bigger the city, the less energy it should consume per capita - a 50% savings should be achieved every time an urban system doubles, and growth should be "sigmoidal reaching a stable size at maturity"; unfortunately, we are consuming superlinearly with an average exponent 1.15 (West and Brown 2005).

The FEP takes the study of observed systems one step further and emphasizes the system's communication to itself - 'explaining the observer to himself' (von Foerster 1992). It argues that individuals have access to information on their individuation. The "consideration of how an individual maintains the boundary that delimits itself" - Markov blanket - is the key to studying self-organization (Ramstead et al. 2018). Using the Markov blanket, Friston (2019) hypothesized how an individual delimits oneself while having access to information on one's environment; thereby, generates predictions and minimizes errors in one's predictions by recursively influencing one's environment - "by acting in ways that maintain the integrity of those expectations over time, the organism defines itself as an individual apart from its surroundings" - a form of preemption based on a statistical boundary. Friston (2019) stated that such processes are analogical to a neural network, where optimization is achieved through maximizing marginal likelihood in some parameter variables. This provides a means to identify intelligence out of information flow. Thus, biological systems tend to minimize entropy (the average level of disorder or surprises in the information) through active inference, a socio-biological behaviour that "uses an internal generative model to predict incoming sensory data" - a negentropic process (Friston 2012).

3   PART II - UnSSEUS: CAAD + NPN

3.1   NPN - Negentropic, Preemptive, and Network

From these info-biological theories, three core principles are formulated (Table 1):

Table 1. NPN principles

Negentropy	Negentropic design thinks about the conservation of energy by creating order, where the order is not the cartesian grid but the statistical boundary of a system and how energy is transformed and transferred from one to another and is manifested and quantifiable transmission of information. In CAAD, the negentropic design considers the use of both entropic and negentropic strategies. The former assists information communication and generation amongst a network of nodes, whereas the latter assists recursive control and prediction; together, they formulate a preemptive feedback system
Preemptive	By defining the statistical boundaries of interacting systems over time (e.g. using the Markov blanket), an individual can predict and influence its immediate environment via iterative feedback (i.e. active inference), where information ensures an individual's first right to act to minimise entropy - preemptive design. Preemptive mechanisms have been widely studied in economics and game theories using perfect, complete, and incomplete information and have the potential to be applied in CAAD via preemptive computing and predictive analysis to minimise energy dissipation and tackle scalability issues with a network of communicative nodes
Network	In the context of complex systems, negentropy and preemption operate on network designs that enable us to hypothesize optimal information communication using topologies, which can be translated or compared as a datum to physical constructs, from architecture to infrastructure. For instance, swarm topologies have been rapidly studied by CAAD communities as they can be used to deduce some of the most densely packed and highly symmetrical topologies and have immense potential in the design of solar energy systems within urban environments. Within any large-scale, self-organizational information system that gives emergence to complexity, network designs also have to consider how knowledge-based information can be directed to nodes in need. Preemptive strategies such as personalization can be used to tackle information overload

3.2   UnSSEUS - Unattended and Smart

This research exemplifies NPN with a design experiment: Unattended and Smart Solar Energy Urban System (UnSSEUS). "Unattended" means that a system can be left unattended to minimise energy dissipation due to excessive control and communication processes to external systems because it has self-sufficient (i.e. perpetual) networks and predictive capacities. Within a solar energy system, the immediate advantage is a reduction in energy absorption by the power grid, which is usually obtained from carbon fossil, thus, becomes environmentally friendly (Miozzo et al. 2014). "Smart" means that a system has sensing, actuating, and control components that enable it to be preemptive to steer towards its preferred outcome before undesirable conditions are realised (Rodrigues et al. 2020). Thus, an unattended and smart system can be attributed to its NPN operations in minimising information and energy entropy (negentropy) (i.e. reducing surprises in system information and energy that are unavailable to do valuable work), predicting and acting before certain solar or climatic conditions are realised (preemptive), and efficiency in control and communication (network).

UnSESUS is an energy system and an information system with three main goals: leverage between aesthetic, climatic and microclimatic, and structural concerns, minimise energy dissipation through active inference and preemptive designs and distribute infrastructural power, not every building has the same solar capacity. In UnSESUS, negentropy is performed through form-finding to search for optimal physical pathways of energy transmission. Negentropy is equally essential in the information that represents energy data to minimise uncertainty and maximise efficiency between a network of subsystems (i.e. building components and buildings) using active inference (i.e. machine learning and Markov models). This facilitates a solar energy system that is not reactive but preemptive to climatic and microclimatic conditions and will need the help of media platforms that direct and derive information exchanges between a network of individuals (Fig. 1).

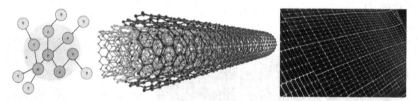

Fig. 1. Markov blanket, CNT, and solar cell (Ramstead 2020; AZoNano 2013; ISEN 2014).

3.3 CAAD Challenges

This research identified three CAAD challenges in UnSSEUS: topological structure, nonlinear power scaling, and growth capacity. They can be understood across three scales: micro-scale (i.e. building components and their formal organizations), across intermediate scales (i.e. building components ensembles into building colonies), and at the macro-scale (i.e. building and infrastructural populations adapt, change, and grow). Each scale varies by its dependency on information from its external environment. NPN offers a prospective model for tackling these challenges.

At the micro-scale, this research focuses on a topological approach to solar busbar design and panel arrangements to help boost module efficiency. Braun et al. (2013), Tan et al. (2018), and their teams had respectively proved that a multi Busbar design could help to reduce "the total series resistance of the interconnected solar cell" while saving costs in "metal consumption for front side [Ag] metallization", and cells arranged in a golden ratio spiral produces ~10% more voltage, ~14% more current, and ~26% more power. Nonetheless, in both research, the prototypes produced are aesthetically not so pleasing; also, structural concerns towards instalment on existing buildings and shading problems that might be caused when installed on building facades or amongst greenery are not being taken into consideration. CAAD form-finding techniques can be applied to generate a spectrum of designs that attend to these limits to identify optimal topological structures by considering negentropic limitations.

At the intermediate scale, this research focuses on fitting the exponential power scaling of cities with quarter-power scaling of organisms as an efficiency target, attributed to both the topology of building components and the infrastructural provisions that help make populations self-organize. The transition from micro to intermediate scale and intermediate to macro-scale should adapt to each other so there would be no blockage caused in energy and information transmission. For example, Hong Kong's recent 'Feed-in Tariff' program promotes that rural village houses should install solar panels on their rooftop and sell the power generated back to electricity companies. Nonetheless, the infrastructural provision did not catch up, causing difficulties transmitting power to electricity stations (SCMP 2018). Thus, quartering the power scaling depends on reducing energy consumption and the efficiency in energy distribution, which would require an NPN approach to identify an optimal in information and energy exchange, predict demand and supply and distribute across the network accordingly. PART III will focus on discussing the CAAD experiment conducted at the micro and intermediate scale (Fig. 2).

Fig. 2. Braun et al.'s (2013) multi busbar design and Tan et al.'s (2018) panel arrangement using the golden ratio have both proved to have increased light-harvesting efficiency.

4   PART III - Form-Finding for CNT Solar Cell

This design experiment aims to formulate subsystems of local feedback groups that form an open system, which can aggregate into larger complex urban systems. Their local topological structure will guide such subsystems; thus, form-finding becomes a crucial exercise. Form-finding is the search for 'structurally appropriate geometries that satisfy equilibrium conditions, under the constraints of a prescribed stress-state of the surface, user-defined boundary conditions and external forces' (Bhooshan and Sayed 2011). Form-finding for aggregatable components also gives considerations to adaptive performance, demanding an understanding of information exchanges between subsystems. For instance, algae are carbon-based light harvesters that are photosynthetic efficient through biochemical reactions, branching properties, and dynamic interactions with their environments (Qiang and Richmond 1996). All of which is achieved via inherited and inferential information - from DNA to their localised shade-avoidance responses - a form of bio-intelligence (Friston 2012).

The developments of carbon-based materials provide prospects in mimicking sustainable biological energy systems and increase the urgency for form-finding exercises specific to their qualities. This research is particularly interested in CNT for low voltage stability, lightweight qualities, and flexible and economical construction (e.g. thin-film techniques). Form-finding may help in developing topologies that can be carved, and 3D printed into different forms of solar cell canopies. This ongoing design experiment aims at providing a holistic approach in generative to production processes and is performed across three subscales of UnSSEUS: busbar/interconnection, panel, and installation arrangements.

4.1 Busbar/Interconnection Designs in Solar Cells

A busbar is a strip of metal used to conduct electricity within solar cells. Form-finding for busbar patterns may help minimise fabrication costs of front-side metallisation and enhance light-harvesting efficiency by increasing current and minimising series resistance. UnSSEUS's approach is to locally reduce bar/wire and cable lengths using the golden ratio, a circular subdivision of surfaces connecting and arranging smaller parts together at the global scale.

While a rectangular arrangement to solar cells is practical, rectilinear figures are rarely found in nature. In most cases, biological organisations are often limited by their volume while needing a larger surface area to exchange energy. Spherical or circular organisations reduce the distance necessary to travel from the center to the edge of an object. The surface-to-surface boundaries are crucial to objects that form by growing (Swanson et al. 1991). In the chemical vapour deposition (CVD) method, CNT is fabricated by growing (Kumar and Ando 2010). Thus, Euclidean geometry may not be the best fit for describing natural shapes or designing carbon-based topologies. On the other hand, a circular approach might be more efficient in maximising the surface area to building volume ratio to enhance solar cell coverage in some cases. This form-finding exercise does not stop at spherical/circular organisations, but it provides a prospective starting point for the above reasons (Fig. 3).

The golden ratio can help us think about how to subdivide a circle efficiently and generate grid points that are self-symmetrical and irreducible, to which sunflower seeds are very proficient. Sunflowers have seeds of varying sizes that divide their territory by drawing from the center outwards to achieve an optimal subdivision. The distribution of this circle helps maintain structural integrity while maintaining equilibrium in solar energy gain between units. Such a distribution is a tessellation using the golden ratio,

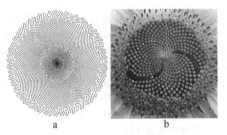

a   b

Fig. 3. (a) Tessellation topology found using an oscillating golden ratio to maintain modular integrity and proportion between units of a solar cell. (b) Sunflower tessellation (Lisa 2015)

which is the science of subdividing sections without overlaps or gaps to maximise the use of any surfaces. The challenge is that circular arrangements can become confined and lose agility. Suppose we imagine travelling across a circular city. In that case, it gets thinner as one proceeds to the center, which makes it inefficient in circulation. Hence, the key is to maintain modular integrity and proportion between units through oscillating the tessellation.

The experiment began by subdividing a circle with its UV vectors, oscillating clockwise and counterclockwise, where the counts of units and distance are Fibonacci-based in any direction. This generates an opposite to a perfect close circle, which is a pattern with no end; this is not confined. Surfacing technique based on golden ratio efficiency has proven itself in many organic systems, while rectilinear approaches may benefit specific environments. Thus, the golden ratio is being hypothesised as a high truth in UnSSEUS, and form-finding based on the Fibonacci number may act as a datum to chaotic structures in an ever-changing natural environment where there is not only one solar or wind vector but many. In this interpretation, the symmetrical and identical curves can be scalable and adaptable to different conditions with recalibration of CAAD, under the principle of not comprising the purity of the Fibonacci topology while working with it.

Although this design experiment was conducted based on the assumption of busbar design for circular solar cells, the same topology may be applied to multi-wire interconnections for busbar-free solar cells, like CNT thin films.

4.2 Solar Panel Arrangements

Solar cell modules arranged as solar panels using interconnecting technologies are often set with a spreading approach to maximise exposure to the sun. One problem we identified is that this promotes a competitive model between the solar panels and their adjacent environment, creating a large shading footprint over its surroundings. For instance, having solar panels covering large areas of high land might affect local vegetation. Also, large flat surfaces are not always available in mountain regions; the same applies to urban environments (Fig. 4a). This design experiment aims to maintain light-harvesting performance without comprising microclimatic qualities and adaptable to all kinds of conditions.

Fig. 4. (a) Solar farm in direct competition with vegetation (Hubbuch 2020). (b) Diverse coral topologies that adapt to each other to win over all kinds of solar vectors (GBR 2020).

The objective is to compose a library of possibilities to generate various panel arrangement designs that echo with different contexts. Instead of developing one design for every problem, the library comprises one family of designs that can be fabricated in similar methods using the same core concept. The experiment began by referencing biodiversity and its variations. In a coral reef, each coral shows a different topological solution for maximizing surface area to harvest as much nutrition. As the surrounding conditions can be taken as largely uniform with each individual being different, the local topologies work together to distribute energy through diversity. Most corals in this reef (Fig. 4b) have been observed to have a backbone with fractal spreading in both vertical and horizontal dimensions. Some of them contract into its backbone when sensing stimulus and give variation to shading strategies according to time factors; others develop static spherical canopies; together, they adapt to each other to win over localized problems of solar vectors. Research has shown that corals grow with shading responses, enabled by sensory components to backscattered infrared lights from one another (Porter et al. 1984). The same logic can be applied to building facades to leverage external climatic and internal microclimatic conditions while each building reflects excessive solar rays.

We tend to go out of the city to plant solar farms, but this contributes to a loss of solar energy in cities from direct and indirect sources (e.g. reflections from buildings and reflective surfaces, corners unused, large open spaces at public areas like universities, etc.). The US has found a plan to embed light-harvesting capacities in historical buildings. Still, a large percentage did not install enough solar panels because they change the appearance of these historical sites (Kandt et al. 2011). The challenge is to make solar energy attractive for diverse building component topologies (e.g. handrails on balconies, roof tiles, etc.). Instead of transporting solar energy from afar locations, panel arrangements with more organic designs can help embed devices in urban structure and relieve energy loss and transport congestion (e.g. charging your phone directly from light harvested on a plaza bench).

Here, we present two designs in progress that attend to the golden ratio principle. The golden ratio arrangement ensures the slightest self-shading at a specific time frame, maximum solar victor projection with the most negligible footprint, and optimized cable connection. Both designs attempted to aggregate repeating solar canopy units inspired by corals into local feedback groups. The unit is formed by extruding the oscillating

Fig. 5. Top, side, and axo view of the proposed golden ratio canopy solar unit that emerges from an integration between sunflower and coral topologies (Woodyatt 2020).

golden ratio topology into a single surface, creating a canopy structure in 3D. At the top view (Fig. 5), we can see that the canopy almost disappears as such a canopy unit is not covering all of its understory areas to minimize its shading footprint. Thus, such units can be installed on landscapes and vegetation while letting grass and other plants survive underneath. This canopy can be fabricated by bending thin films, where one side is coated with CNT, and the other is coated with a reflective material. Such a vertical extrusion between two different profiles enables light to hit at different angles, reflects and cascades down the structure.

In design 1, local canopy units were aligned to the global golden ratio topology with a line constant that performs tessellation in maximising surface area without overlapping. The units all differ in size and density towards the center, achieving a sunflower structure. The golden ratio also acts as a wireframe to electricity, where the units collect light and transmit electricity towards the center to minimise the path of current. For areas with low sun angles, such units can be configured for installation around building facades or on the street side to provide aesthetical structures, without sealing wind ventilation or having an ugly backside, where the rear can absorb reflective light from the facade or light-coloured pavements. Nonetheless, in areas where the sun angle has a significant difference during the day, the production of the canopy should enable customization (Fig. 6).

We have realised that using only one layer of alignment in maximising footprint might be a negative strategy in terms of adaptation. We performed another experiment that aimed at increasing verticality, keeping the projection area at maximum without reducing the total footprint. Thus, in design 2, we stacked the units into 7-layer towers, and arranged them according to the golden ratio formation. We find 7 layers to be the most optimal based on entropy in energy due to the tracking of the sun, where the stacked canopies are working as trees that cascade light down to the understory, and help absorb sun rays hitting on different surfaces at different angles. Using the vertical direction and the golden ratio to create this global distribution helps to minimise shadow dropping

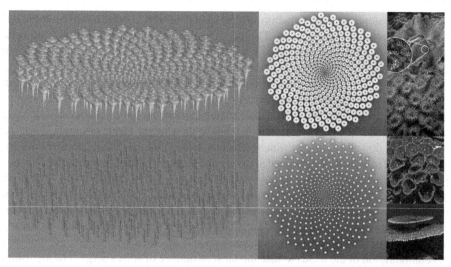

Fig. 6. Design 1 tessellation with every unit differs in size and density channeling electricity towards the center. Design 2 reduced shading footprint, maintaining surface area by stacking.

from the elements on themselves. In contrast, any other pattern will have specific times of the day where all components are shadowing each other. At the same time, the footprint is highly reduced by applying an organic distribution and creating the opportunity where these modules can be installed in areas like paddy fields without developing competition for light between the machine and the plants. Each crop maintains a distance from the other, allowing these solar modules to be installed in between.

4.3 Solar Installation Designs

In the search for a global structure that can adapt to new buildings and existing buildings, there are three main challenges: 1.-Structural concerns regarding weight and costs, where a self-standing structure can be fabricated at lost prices, is needed. 2.-Microclimatic problems, where installing facades does not produce a large footprint and instalment in public areas can provide a decent shading strategy for people to gather and use the solar power generated. 3.-Aesthetic concerns, where the form can act as an art object to attract users.

This ongoing experiment aims at finding a global structure that can be paired with the golden ratio explorations. The local geometry of the golden ratio reduces the relevance of solar vectors through adaptability to ensure variety in configurations, compactness, high exposure to light, and minimal footprint. At the same time, global geometry should provide flexibility to geometrical and functional contexts for integration with diverse and often chaotic environments. We proposed a 'solar swarm' structure based on the dynamics of the Hadley cell - the Earth's tropical atmospheric circulation that is a consequence of the Sun's illumination and the laws of thermodynamics - which 'transports energy polewards, thus reducing the resulting equator-to-pole temperature gradient' for more equal distributions in energy transport (UCI 2016). In the mathematical field of dynamical systems, strange attractors are the spatial collective iterations of multiple initial conditions of a chaotic system. The fact of being chaotic suggests an indeterministic quality, although that is true locally, yet it's stable globally, with a typology that could be rationalized.

The solar swarm is an attempt to rationalise the global geometry of a chaotic system by ensuring the surface developability. A vector approximation is performed to resolve issues in fabrication, and the surfaces are made of an orthogonal unified subdivision due to the solar panel tessellation technology, hence the straight bands principle. With the help of CNT thin-films, it's possible to bend the bands into the global geometry needed actively. The solar cells are uniformed in size and shape, arranged to form straight bands regularly constructed, with a relatively low fabrication cost compared to crystalline solar cells using spin-coating or inkjet printing methods. The bands are actively bent through local junctions that collectively freeze the band shape in the correct double-curved global geometry, resulting in a self-standing structure. The form is achieved using an iterative computing algorithm that reacts to specific solar parameters, generating a 3D configuration of the bands to take the most advantage of solar energy. The algorithm reacts with the global and local geometry to the solar vectors in 2 conditions (direct and reflected).

The possibility of backscattering is considered by layering the back surface of the band with a reflective material that reflects the rays to the CNT cells at the side of another band. The global geometry can absorb tension created by the heat expansion of the two different materials. In turn, the swarm is encapsulating the potential projections of solar conditions that could morph and reintegrate in multiple dynamic environments. As such, efficiency is enhanced by the arrangements and orientations of solar cells, the routes to which electricity circulates, and an equal energy distribution from patterning and reflection.

As the "solar swarm" can be achieved physically through bending, the installation can be easily reused in different forms in different conditions, hosting possibilities of temporary structures to engage in a circular economy. The Swarm installation can be designed and developed as self-standing installations on building facades, art objects in lounges with skylights, furniture on open landscapes at universities, where connecting

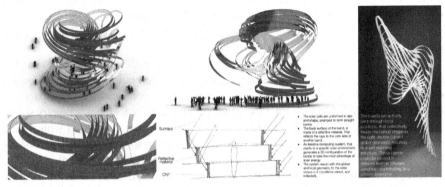

Fig. 7. Solar Swarm - a self-standing structure proposed based on Hadley cell dynamics.

and charging ports can be integrated to offload solar energy harvested spot to minimise energy dissipation during transmission. This forms a collective landscape of energy circulation in a complex environment that counts on existing models in nature to channel chaotic systems (Fig. 7).

5   Work-in-Progress

For a solar energy system to be intelligent and unattended, form-finding alone is not sufficient and demands parallel sensing, actuating, and control components, formulating an information feedback system with Negentropic, preemptive and network capacities.

The ability in smart systems to self-regulate or even self-learn is critical. The immediate advantage is reducing energy consumption by the power grid through active inference. Preemptive computing, where each local feedback group relinquishes control of its own accord, and only relevant information needs to be synchronized at the global scale, helps relieve network load and excessive computing power. Feedback loops facilitate self-correction during the next iteration. The golden ratio, solar swarm, and solar parameters can act as placeholder variables.

5.1   Sudoku Distribution and Personalisation

In applying our research to the urban scale, personalization processes for energy sharing are needed to coordinate between buildings and their users. Particularly, this research looked into predicting and preempting supply and demand to optimize energy logistic operations. In this context, a prediction algorithm was investigated to assist in the negotiation between different actors - a network of users and solar devices, minimizing information entropy and addressing the data overload problem.

Our strategy is based on Sudoku, a combinatorial game predicated on a 9 by 9 grid that nine different items can populate. We used the same grid to discretize and encode solar units into an architectural system, each computationally associated with a number from 1 to 9. These were: north-facing facade on the lower floor, south-facing facade on the lower floor, east-facing facade on the lower floor, west-facing facade on the lower floor, north-facing facade on the higher floor, south-facing facade on the higher floor, east-facing facade on the higher floor, west-facing facade on the higher floor, and roof. As each of these items would be harvesting different levels of solar energy at other times of the day, the computational game-playing strategy can then be used to solve the organization of energy distribution by finding optimal solutions for the games created and modified through informational feedback. Thereby, the system can consider time and energy factors to define different organizational solutions for different games and states. This can go beyond a single building and can be applied to a broad spectrum of inter-building configuration problems (Fig. 8).

5.2   Combinatorial Design Pipelines with Machine Intelligence

By developing a critical understanding of what computational intelligences are modelling behind, from natural to artificial processes, the proposed combinatorial pipeline helps

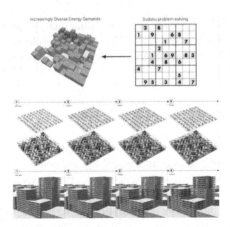

Fig. 8. Solar supply and demand data cascade up connected buildings using sudoku.

the AI learn within a solar environment, a chaotic complex system that translates the designer intuition into algorithmic generation human-machine interaction.

AI-informed workflows are being explored where form-finding is not simply taken as an optimal resultant shape but information that acts as protocols or sets of instructions in component aggregation and distribution, energy harvesting and circulation. Since all of these algorithmic tools are weak AI that focuses on narrow tasks, the understanding of which intelligence is better at what tasks and which algorithm can be accelerated with machine learning is critical. This research proposes a means of how various intelligence can come to a consensus, integrating rule-based and machine learning systems, accelerating the sudoku distribution logic with AI (Fig. 9).

Artificial intelligence is the scientific field that "builds intelligent entities" through virtual agents, which is often associated with the automation of intellectual labour; yet, they are not necessarily designed by mimicking what humans would do in the same scenario, and their development is done around the abilities to reason over complex cases artificially. A poorly constructed dataset or environment can lead to the reinforcement of prejudice. However, this research chooses to take advantage of biases in machine learning to formulate human-machine interaction.

Natural patterns from form-finding are perfectly simulated with Cellular Automata (CA) as a rule-based system, with results based on interactions of simple growing rules in a strict relationship with an environment and the rational use of energy and matter process. This aggregation logic promotes options over optimal solutions, working with aggregated geometries, adding features that can define a CA to achieve variational outputs. The CA informs the final result speculatively by iterating possibilities in an evolutionary way, starting from seeds with geometrical outputs as aggregation data, tested and optimised with solar radiation inputs, operating design in a hybrid model.

The aggregative output can be stored in a series of lists to be used as input for the next steps - a feedback loop that helps machine learning to iterate the computation processes without the necessity to rerun complex rule-based operations. Data produced from computationally intensive and intuition-based processes create the training material that assists the machine in learning hidden patterns and predicts results relatively

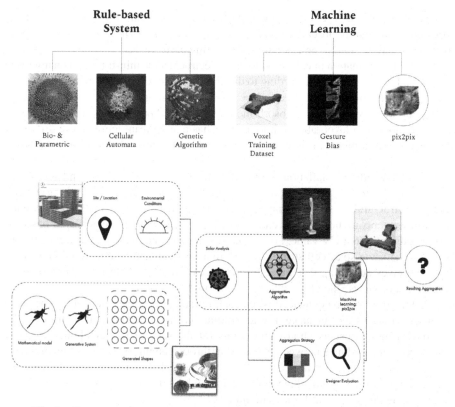

Fig. 9. The proposed pipeline integrates rule-based and machine learning systems.

instantaneously. Given a context as input, the trained model can define a 3-dimensional aggregation strategy that accumulates the most solar incidences on a target structure.

6   A Next Step: Physical Prototyping and Smart Materials

Physical prototyping helps collect real-world data, which can be compared with a simulated lab condition to understand deviations (i.e. uniform lighting in a controlled environment). The simulation can act as a datum to the optimal, compared with design variations under different physical boundary conditions. Automation may be used to attend to both climatic (e.g. incoming solar data) and microclimatic (e.g. indoor light intensity) conditions while maintaining low energy consumption. By 'low', it means that solar power generated by the system should be able to sustain the automation needed - an unattended system.

This research took inspiration, again, from corals in how they react to physical conditions. A study by Bollati et al. (2020) has shown that coral bleaching is, in fact, an optical feedback loop. After healthy corals lose the symbiont pigment and turn white, it causes internal blue light fluxes to increase, further stressing the corals, which induces upregulation of pigments in bleached corals - they make themselves colourful again to relieve stress. While a moment should be taken to reflect on how human impact has stressed corals out all around the world, it is worthy to note that such photo-adaptive responses may be used to enhance light-harvesting performance.

One way automation may be used to have electrochromic thin-films that change their colour and/or transparency according to different solar conditions. From an engineering perspective, the most simple and effective way of harnessing solar power using buildings is to paint every wall black to absorb the most energy. Apart from aesthetic concerns, this may not be so favourable in hot areas or has a significant temperature difference between day and night. In these circumstances, electrochromic thin-film installations may be desirable.

For self-standing installations on building facades, both solar and wind vectors are essential to consider. For corals, some develop soft tentacles that can dance along currents to relieve structural stress; others have extended tentacles that retract in the face of stimuli. Such movements do not always require extra electricity but can be achieved based on simple physics about UV value. For instance, when the materials used to create the two sides of the same solar cell strip have different heat capacity, the material can shrink by themselves according to temperature changes. This may help to prevent clogging ventilation or views when light intensity is below harvestable.

Automation may be enhanced when such retracting becomes an adaptive behavior between buildings. For instance, if one building has harvested enough solar power for the day, it may reject its photovoltaic units for the sake of its neighbours. Suppose structures can coordinate between one another through an information system or having photosensors to actuate locally. In that case, they may collaborate in changing the shapes and position of their photovoltaic units relative to how the sun tracks across the sky - a strategy often used in solar farms with motors.

Another strategy used by many carbon-based organisms is temporality; animals like corals and sloths move very slowly to save energy. Such a strategy is also very suited to solar energy systems as the current might not always sustain very sudden movements. SlothBot is a recent project that pertains to similar thinking. It is deployed by GeorgiaTech (2020) using a robot as slow as a sloth that uses solar power to move around and collect data on animals and plants.

7   Conclusion

This paper discussed the use of info-biological theories with the aim of translating existing models in nature to channel emergent systems in CAAD. More specifically, it considered how the Information Theory of Individuality (ITI) and Free Energy Principle (FEP) define individuality through information transfer, which helps to analyse relationships between individual entities and their environments as an emergence. This paper argued three main points of design implications that emphasised the circularity of energy and information in a system - negentropic, preemptive, and network strategies (NPN). These were then applied at three scales: micro-scale (i.e. building components and their formal organisations), across intermediate scale (i.e. components ensembles into colonies at the building scale), and at macro-scale (i.e. building and infrastructural populations adapt to grow). Each scale varies by its dependency on information from external environments.

The paper exemplified its argument with a series of CAAD experiments at the micro-intermediate-scale - form-finding for Carbon Nanotube (CNT) backed solar cells. It aimed at using complex geometries modelled from nature to enhance efficiency in the modules, leveraging structural, climatic, and aesthetic concerns. It tried to identify optimal surface patterning and coverage through thinking about the boundary of a building system according to internal/external climatic demands. Since the form was found by defining how solar cells can be uniformly modular while adaptable to a broad range of nonlinear global geometries, the parametric model can be used as placeholder variables in Markov models to predict external/internal climatic needs.

The paper discussed work-in-progress at the macro-scale, and described a CAAD method explored using sudoku gameplay. Such a rule-based approach translates energy harvested by individual solar modules to a game. It may help to achieve personalisation that preempts the supply and demand of solar energy to facilitate agility for rigid physical buildings. Individuality is being also proposed to be taken at the scale of building ensembles, where if the game of one building is flipped according to data input, the decision cascades down to trigger a new configuration - an ecology of buildings that adapts themselves into a self-sufficient energy system - an Unattended and Smart Solar Energy Urban System (UnSSEUS).

The sudoku distribution logic is then being adopted for the training of AI to accelerate computational processes. From this, a combinatorial design pipeline that synthesises rule-based and machine learning systems are proposed. Combinatorial means that it focuses not on final designs, but many critical variations of the same pipeline, with the capacity to adapt to diverse solar environments, seeking to compensate for our modest renewable energy capacities in different cities and optimising solar incidence accumulation on targeted photovoltaic elements through iterative methods, taking form not merely as shapes but instructions for component aggregation and distribution.

This research was merely a starting point that narrowed down to specific problems in architectural design. NPN strategies can be translated to a broad spectrum of info-biological projects to help us critically revise our practices in generative and predictive algorithms and raise further questions in the relationship between design and our environments.

References

AZoNano, Multi-Walled Carbon Nanotubes: Production, Analysis, and Application. https://www.azonano.com/article.aspx?ArticleID=3469. Accessed 20 Sep 2020

Bhooshan, S., El Sayed, M.: Use of sub-division surfaces in architectural form-finding and procedural modelling. In: Proceedings 2011 Symposium on Simulation for Architecture and Urban Design, Boston, Massachusetts, pp. 60–67 (2011)

Bollati, E., D'Angelo, C., Alderdice, R., Pratchett, M., Ziegler, M., Wiedenmann, J.: Optical feedback loop involving dinoflagellate symbiont and scleractinian host drives colorful coral bleaching. Curr. Biol. 30(13), 2433–2445 (2020)

Braun, S., Hahn, G., Nissler, R., Pönisch, C., Habermann, D.: The multi-busbar design: an overview. Energy Procedia 43, 86–92 (2013)

Friston, K.: A free energy principle for biological systems. Entropy 14(11), 2100–2121 (2012). MDPI, Basel Switzerland

Friston, K.: A free energy principle for a particular physics. arXiv preprint arXiv:1906.10184. Cornell Tech, Ithaca NY (2019)

GBR. Great Barrier Reef - Coral Nurture Program - Tourism & Science Together. https://www.coralnurtureprogram.org/. Accessed 15 Jan 2021

GeorgiaTech. Georgia Institute of Technology, from https://news.gatech.edu/2020/06/16/slothbot-garden-demonstrates-hyper-efficient-conservation-robot. Accessed 15 Jan 2021

Hubbuch, C.: Wisconsin State Journal: Construction at halfway mark for Wisconsin's first large-scale solar farm. https://madison.com/wsj/business/construction-at-halfway-mark-for-wisconsins-first-large-scale-solar-farm/article_a3153d9a-e93e-5807-a975-036f241702ff.html. Accessed 15 Jan 2021

ISEN. Breakthrough for Carbon Nanotube Solar Cells - Institute for Sustainability and Energy at Northwestern. https://isen.northwestern.edu/breakthrough-for-carbon-nanotube-solar-cells. Accessed 20 Sep 2020

Kaiser, D.: "Information" for Wiener, for Shannon, and for us. In: Possible Minds: Twenty-Five Ways of Looking At AI, pp. 151–159. Penguin Books, London (2020)

Kandt, A., Hotchkiss, E., Walker, A., Buddenborg, J., Lindberg, J.: Implementing solar PV projects on historic buildings and in historic districts, No. NREL/TP-7A40-51297. National Renewable Energy Lab. (NREL), Golden, CO (2011)

Kirchhoff, M., Parr, T., Palacios, E., Friston, K., Kiverstein, J.: The Markov blankets of life: autonomy, active inference and the free energy principle. J. R. Soc. Interface 15(138), 20170792 (2018). London England

Krakauer, D., Bertschinger, N., Olbrich, E., Flack, J.C., Ay, N.: The information theory of individuality. Theory Biosci. 139, 1–15 (2020). Cornell Tech, Ithaca NY

Krippendorff, K.: An epistemological foundation for communication. J. Commun. 34(3), 21–36 (1984). University of Pennsylvania, Philadelphia Pennsylvania

Kumar, M., Ando, Y.: Chemical vapor deposition of carbon nanotubes: a review on growth mechanism and mass production. J. Nanosci. Nanotechnol. 10(6), 3739–3758 (2010)

Leung, K.M., Hui, J.W.: Renewable energy development in Hong Kong. In: 2004 IEEE International Conference on Electric Utility Deregulation, Restructuring and Power Technologies. Proceedings, vol. 1, pp. 398–404. IEEE, Piscataway New Jersey (2004)

Lisa: Fibonacci in a sunflower. https://thesmarthappyproject.com/fibonacci-in-a-sunflower/. Accessed 16 Jan 2021

Logan, R.K.: What is information?: why is it relativistic and what is its relationship to materiality, meaning and organization. Information 3(1), 68–91 (2012)

Miozzo, M., Zordan, D., Dini, P., Rossi, M.: SolarStat: modeling photovoltaic sources through stochastic Markov processes. In: 2014 IEEE International Energy Conference (ENERGYCON), pp. 688–695. IEEE, Piscataway New Jersey (2014)

Ng, P., Odaibat, B., Doria, D.: An Information Theory Application to Bio-design in Architecture. 1st CCBDT. Dalian Polytechnic University, Dalian China (2020)

OED: The Oxford English Dictionary. Clarendon Press, Oxford (2020)

Porter, J., Muscatine, L., Dubinsky, Z., Falkowski, P.: Primary production and photoadaptation in light- and shade-adapted colonies of the symbiotic coral, Stylophora Pistillata. Proc. R. Soc. London Ser. B Biol. Sci. 222(1227), 161–180 (1984)

Qiang, H., Richmond, A.: Productivity and photosynthetic efficiency of *Spirulina platensis* as affected by light intensity, algal density and rate of mixing in a flat plate photobioreactor. J. Appl. Phycol. 8(2), 139–145 (1996). https://doi.org/10.1007/BF02186317

Ramstead, M.J.D., Badcock, P.B., Friston, K.J.: Answering Schrödinger's question: a free-energy formulation. Phys. Life Rev. 24, 1–16 (2018)

Rodrigues, J.M., Cardoso, P.J., Monteiro, J., Ramos, C.M.: Smart Systems Design, Applications, and Challenges. Global. Engineering Science Reference, Hershey (2020)

Rowell, M.W., et al.: Organic solar cells with carbon nanotube network electrodes. Appl. Phys. Lett. **88**(23), 233506 (2006)

Schrödinger, E.: What Is Life? Cambridge University Press, Cambridge (1944)

SCMP. Electricity firms get 1,100 renewable energy applications. https://www.scmp.com/news/hong-kong/health-environment/article/2179224/renewable-energy-applications-skyrocket-attaching. Accessed 15 Jan 2021

Shannon, C.E., Weaver, W.: The Mathematical Theory of Communication. University of Illinois Press, Urbana, Champaign (1948)

Swanson, J.A., Lee, M., Knapp, P.E.: Cellular dimensions affecting the nucleocytoplasmic volume ratio. J. Cell Biol. **115**(4), 941–948 (1991)

Tan, D., Benguar, A.N., Casiano, P., Valdehueza, T.: Golden ratio applied in the orientation of solar cells in a golden spiral solar panel. Int. J. Dev. Res. **08**, 20416–20420 (2018)

UCI. University of California, Irvine - How does the Hadley Cell help spread energy around in the global climate system? https://sites.uci.edu/. Accessed 25 Nov 2020

von Foerster, H.: Ethics and second-order cybernetics. Cybern. Human Knowing **1**(1), 40–46 (1992)

West, G.B., Brown, J.H.: The origin of allometric scaling laws in biology from genomes to ecosystems: towards a quantitative unifying theory of biological structure and organization. J. Exp. Biol. **208**(9), 1575–1592 (2005)

Wiener, N.: Cybernetics. Wiley, New York (1948)

Wiener, N.: The Human Use of Human Beings. Sphere Books, London (1968)

Woodyatt, A.: The Great Barrier Reef has lost half its corals within 3 decades. https://edition.cnn.com/travel/article/great-barrier-reef-coral-loss-intl-scli-climate-scn/index.html. Accessed 16 Jan 2021

Risk Assessment for Performance Driven Building Design with BIM-Based Parametric Methods

Fatemeh Shahsavari$^{(\boxtimes)}$ ⓘ, Jeffrey D. Hart ⓘ, and Wei Yan ⓘ

Texas A&M University, College Station, TX 77843, USA
{Fatemeh.shahsavari,j-hart,wyan}@tamu.edu

Abstract. A growing demand for handling uncertainties and risks in performance-driven building design decision-making has challenged conventional design methods. Thus, researchers in this field lean towards viable alternatives to using deterministic design methods, e.g., probabilistic methods. This research addresses the challenges associated with conventional methods of performance-driven building design, i.e., ignoring the existing uncertainties and lacking a systematic framework to incorporate risk assessment in building performance analysis. This research introduces a framework (BIMProbE) to integrate BIM-based parametric tools with building probabilistic performance analysis to facilitate uncertainty analysis and risk assessment in performance-based building design decision-making. A hypothetical building design scenario is used to demonstrate the application of the proposed framework. The results show that the probabilistic method leads to a different performance ranking order than the deterministic method. Also, the probabilistic method allows evaluating design options based on different risk attitudes.

Keywords: Performance driven building design · Uncertainty analysis · Risk assessment · Building information modeling · Parametric design

1 Introduction

Architectural design decision-making begins with identifying design problems and objectives. It sets boundaries for designers' potential problem-solving methods. Performance-based building design focuses on methods and strategies that integrate and optimize different aspects of building performance. Using the computer power allows designers to explore a broader range of solutions, efficiently. Building performance simulation (BPS) tools have been extensively used by the architects and engineers to simulate building performance [1].

The BPS tools usually produce building performance predictions based on a set of input data including building physical characteristics, interior conditions, weather data, and mechanical specifications [2, 3], which commonly come with outstanding uncertainties [4]. The Building Energy Software Tools Directory provides an ongoing list of 200 simulation tools [5], but most of these tools are not designed to deal with

© Springer Nature Singapore Pte Ltd. 2022
D. Gerber et al. (Eds.): CAAD Futures 2021, CCIS 1465, pp. 456–472, 2022.
https://doi.org/10.1007/978-981-19-1280-1_28

uncertainties. The majority of the BPS tools only collect deterministic input data and run a single deterministic simulation to evaluate the building performance [6, 7].

The various sources of uncertainties associated with building performance and its potential impact on the human health and environmental crisis demands for new methods of design thinking to deal with uncertainties. The advancement of computer technologies allows the application of data-driven methodologies in the field of building performance analysis to achieve resiliency and sustainability.

This research introduces a framework to incorporate probabilistic models into the building design decision making process, demonstrated with a design test case. The goal of this research is to tackle data uncertainties and potential risks in architectural design decision-making with a focus on building energy performance. For that purpose, the results obtained from the proposed probabilistic framework are compared with those from a conventional deterministic method. Also, three design decision making criteria including expected value, maximax, and maximin are applied to discuss the simulation results based on different attitudes towards risk.

2 Literature Review

Deterministic methods in performance-driven building design decision-making fail to address existing uncertainties in design decision making. Uncertainty analysis techniques such as Monte Carlo coupled with risk assessment methods introduce a potential solution to tackle uncertainties and make robust design decisions [8].

2.1 Monte Carlo and Building Performance Simulation

As Saltelli et al. [9] stated, variance-based methods, e.g., Monte Carlo, have shown more effectiveness and reliability when working with uncertainties. The Monte Carlo simulations use random variables and input probability density functions to address the stochastic status of the problem. Let a mathematical modeling $Y = f(x)$ define correlations between a vector of input variables $X = \{X_1, X_2, \ldots, X_k\}$ and an output Y, where f is a deterministic integrable function which translates from a $k - D$ space into a $1 - D$ one, i.e., $\mathbb{R}^K \rightarrow \mathbb{R}$. The model produces a single scalar output Y when all input variables are deterministic scalars. However, if some inputs are uncertain or undecided, the output Y will also associate with some uncertainties. An input variable X_i, is defined by a mean value μ_i, a variance σ_i, and a probability distribution, such as Normal, Uniform, Poisson, etc. In the Monte Carlo methods, a set of samples from possible values of each input variable are generated. These input values are inserted into the simulation model to generate the probability distribution of the output Y. Processing the output range Y delivers a mean value and a frequency distribution of the output.

The challenge of mapping between simulation tools and probabilistic techniques in the process of Monte Carlo simulations has received a lot of attention in the literature. Lee et al. [10] introduced an uncertainty analysis toolkit explicitly for building performance analysis referred to as the Georgia Tech Uncertainty and Risk Analysis Workbench (GURA_W). The identification and modification of input variables are possible using GURA_W, and the uncertainty quantification repository available in this tool allows the energy modelers to access the uncertainty distributions of previous parameters being modeled. Hopfe et al. [6] studied uncertainty analysis in an office building energy consumption and thermal comfort assessment through connecting MATLAB with a building performance simulation tool, called VA114. They declared that including uncertainties in building performance analysis could support the process of building design decision making. Macdonald and Strachan [11] integrated uncertainty analysis with building energy simulation using Esp-r software. They studied the uncertainties in thermal properties of building materials and building operation schedules and concluded that uncertainty analysis facilitates risk assessment and improves building design decision making. de Wit and Augenbroe [12] studied the impact of variations in building material thermal properties, along with model simplifications, on building thermal behavior, using two simulation tools, ESP-r and BFEP, integrated with Monte Carlo uncertainty analysis technique. They applied expected value and expected utility decision making criteria to discuss how designers can use the extra information obtained from uncertainty analysis. Asadi et al. [13] used Python programming to automate the integration of Monte Carlo simulations into energy analysis. They developed a regression model as a pre-diagnostic tool for energy performance assessment of office buildings to identify the influence of each design input variable.

2.2 BIM and Parametric Tools Integrated with Uncertainty Analysis and Risk Assessment

BIM technologies may be useful in performance analysis and risk management since they facilitate transferring data from BIM authoring tools to other analysis tools, also enable designers to automate iterations in design and analysis processes [14, 15]. Kim et al. [16] applied the Monte Carlo uncertainty analysis in building energy analysis through the integration of BIM and MATLAB platforms. A set of software applications including Revit Architecture 2010, ECOTECT 2010, and EnergyPlus 6.0 were used for modeling and simulation. For uncertainty analysis, the MATLAB Graphical User Interface (GUI) platform was applied to develop a self-activating Monte Carlo simulation program. Rezaee et al. [17] provided a CAD-based inverse uncertainty analysis tool to estimate the unknown input variables and improve design decision-makers confidence in the early stage of building design. They created two energy models, one in EnergyPlus and another one in a spreadsheet-based energy analysis tool to run the energy calculations.

Although BIM and parametric design tools allow the iteration of building performance simulations, the integration of Monte Carlo into the platform of these applications have not been broadly studied. For the performance-based building design, further studies are required to provide clear guidance on the mapping between BIM authoring tools, parametric analysis tools, and probabilistic techniques such as Monte Carlo to provide probabilistic outcomes for building energy analysis [14]. This research intends to apply variance-based methods such as Monte Carlo in the performance driven design decision making process using BIM and parametric tools.

2.3 Design Decision Making and Risk Assessment

The international ISO standard (ISO 31000:2009) defines "risk" as the effect of "uncertainty" on objectives, that could be positive consequences as well as negative impacts [18]. Different decision-making criteria including expected value, maximax, and maximin for risk assessment and design decision making under uncertainties are applied in this research. In this approach, two major terms are introduced:

- Key performance indicators (KPI) [19], and
- Key risk indicators (KRI) [20].

While KPI is a metric to measure the performance or objective, KRI measures deviations from the target and depicts the threat of breaching a specific threshold [21]. KPI and KRI are useful metrics to measure the objective aspect of risk (likelihood), also facilitate decision making based on the risk attitude of decision makers and evaluate the performance subjectively.

A building energy model is too complex to estimate the variance of the output by just reviewing the variance of the input variables. In building energy projects, KPI could be defined as building annual energy consumption or energy saving, and KRI may show the possibility of occurrence of discrepancy between expected energy performance during the design stage and real energy performance after project completion [22]. Thus, performing a probabilistic analysis and uncertainty propagation for identifying the probability distribution of the output could be helpful in the process of risk assessment and design decision making.

Expected Value Criterion

The Expected Value (EV) of a variable is the return expected or the average benefit gained from that variable. This statistical measure, the sum of all possible gains, each multiplied by their probability of occurrence, demonstrates the cost-benefit analysis of a design option, considering the input uncertainties [23].

De Wit and Augenbroe [12] applied the criterion of EV to evaluate competing design alternatives. Their example describes a situation where a designer needs to make a decision on whether or not to use a cooling system in their building design. The designer would decide to use a cooling system only if the indoor temperature excess of the building without a cooling system is more than 150 h. Based on their probabilistic results, they concluded that the most likely value of the outcome or EV would be well below 150 h and the designer could make their decision comfortably not to include the cooling system in their design.

Maximin and Maximax Criteria

Design options are not always assessed objectively, as suggested by the EV theory. It is important to know how likely different design outcomes are to occur for a design option, but decision-makers' preferences affect their decision in selecting the optimal design option as well [23]. The best choice for one design decision-maker might not be the best for another one with different preferences, thus it is not always the best decision to select the design option with the maximum EV. There may be a situation which demands risk-averse decision-making and selecting a design option which can go the least wrong. This decision-making strategy is known as maximin and suggests selecting the design option maximizes the minimum payoff achievable [24, 25]. On the other hand, taking a risk-seeking approach towards risk might lead to selecting a design option with the most optimistic possible outcome. This decision-making strategy is termed maximax and searched for a design option to maximize the maximum payoff available [24, 26].

3   Methodology

This study presents a new framework to implement probabilistic methods in the field of performance-driven building design decision-making, using Building Information Modeling (BIM) and parametric tools. Figure 1 presents the workflow including the key steps of this process, data being transferred, variables, and the software:

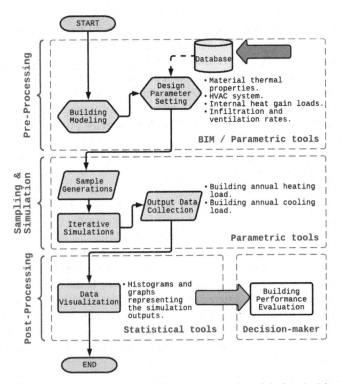

Fig. 1. BIMProbE workflow for probabilistic performance-based design decision-making.

This framework considers the uncertainties in building energy simulations including material properties, internal heat gains, and infiltration and ventilation rates (more details about this framework are presented in [27]). This framework targets the Revit-based building design process and has three main steps described below:

3.1 Pre-processing

The building geometry is modeled in a BIM authoring tool such as Revit. An external application named BIMProbE is developed to set the probability distributions for thermal properties of building materials based on the real-world material properties learned from literature studies and retrieved from an external database (Microsoft Excel). The mapping between the BIM model and the Excel-based database is performed through BIM API and using object-oriented programming.

Building elements consist of exterior walls, interior walls, roofs, floors, and windows. This program will find and collect opaque components including walls, roofs, and floors. The material IDs associated with each material is identified to collect the thermal properties. The thermal properties of building materials including thickness, thermal conductivity, and R-value can be collected from the BIM model. For instance, the thickness can be accessed as the width of each layer of the building components. As an example, a Structurally Insulated Panel (SIP) wall type consists of six layers of Plasterboard, Wood, Timber Insulated Panel – OSB, Timber Insulated Panel – Insulation, Timber Insulated Panel – OSB, and Sand/Cement Screed. The width of each layer of an SIP wall can be identified as the thickness of that specific material. If the thermal properties of a material are missing in the Revit model, the program will create a thermal asset for that material and will set the thermal properties according to the corresponding values in the Excel database.

BIMProbE add-in will automatically create the required shared parameters to add the probability distributions to the building material thermal properties and bind them to each building material in the project. This application searches for the material names in the Excel database to assign the correct value to each parameter. This process begins with an attempt to open the external Excel database. Once the user starts this addin, a window will pop up asking the user to select an external Excel file. If the user selects an Excel file, the program will automatically start reading it cell by cell. The program will check the sanity of the data first and if there is no error found, it will continue with the rest of this process. The probability distributions (including the mean and standard deviation) for each material type are set according to the corresponding values in the Excel database. At the end of this process, BIMProbE allows writing the mean and standard deviation values of the thermal properties to a new Excel spreadsheet to be used for later steps of sampling and simulation.

Other design input parameters required for building energy simulations including Heating, Ventilation, and Air Conditioning (HVAC) system specifications, internal heat gain loads, infiltration and ventilation rates, and operational schedules are added using parametric tools such as Grasshopper, which is the visual programming environment for Rhinoceros.

3.2 Sampling and Simulation

The information prepared in the pre-processing step is used for sample generation and iterative simulations. Using Latin Hypercube Sampling method, N samples for each design input variable are generated and N energy simulations are run for each design option (building forms) in Grasshopper. The simulation outputs (building annual heating and cooling loads) are recorded in Grasshopper and written to Excel using TT Toolbox plugin for Grasshopper [28]. For each round of the energy simulation, a corresponding value from the sample pool of each design variable (x_{uncer_i}) is selected and inserted into the simulation model, and as a result, N output values are obtained. The set of outputs whose elements correspond to the samples are $Y = \{y_1, y_2, y_3, \ldots, y_m\}$, $m = N$, where Y denotes the space of output values, which are the results of building thermal load simulations.

In the deterministic approach, all the design variables, regardless of being deterministic or uncertain, are assigned to their associated mean (μ_i) as a fixed value ($x_i = \mu_i$). Fixing all the input variables at their mean values, the energy simulation is run once and a single output $y_{det} = f(\mu_1, \mu_2, \mu_3, \ldots, \mu_n)$ is obtained, where ($\mu_1, \mu_2, \mu_3, \ldots, \mu_n$) are the means of the design variables.

3.3 Post-processing and Design Decision-Making

The post-processing phase consists of data analysis and graphical presentation of the simulation results. This phase is conducted using a statistical software known as JMP. The data are collected in Grasshopper and exported to Excel for post-processing. The JMP [29] add-in for Excel provides interactive graphics and tables that enable the user to identify relationships visually and examine patterns. The histogram demonstrations, normality plots and box plots are used for risk assessment and data visualization.

The output results are represented by the values of deterministic and probabilistic outputs using two metrics of KPI and KRIs. In addition, design options are ranked according to the criteria of expected value, maximax, and maximin. The information provided in this process could help design decision-makers with building performance evaluation including uncertainties and comparing the probabilistic results with the deterministic results.

4 Test Case

This test case studies the energy performance of a hypothetical mid-size office building with four design options. For this test case, the building annual thermal load (the sum of heating and cooling loads) in kWh/m^2 is defined as the target. The effects of some of the existing uncertainties including the thermal properties of building materials, internal heat gain loads, ventilation rates, infiltration rate, and occupant behavior are studied.

4.1 Test Case Model

This test case presents a hypothetical 5-story office building, 54 m by 93 m (Total gross area of 25,110 m^2) in Chicago, Illinois (climate zone 5A). Figure 2 shows the 3D view of the building base model in Revit, and the building visualization in Rhino using Ladybug tools in Grasshopper, from left to right. The building geometry is created in Revit and the materials are defined in the same model. The building information including the materials' thermal properties is exported from Revit to Excel. This information is used

 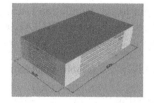

Fig. 2. Geometric representation of the test case.

in Rhino/Grasshopper to generate samples and run multiple energy simulations using the parametric capabilities of Grasshopper.

The building geometry, layout, orientation, operation schedules, internal heat gain sources, and HVAC type are assumed to be identical throughout this experiment. On the other hand, exterior wall construction, floor construction, roof construction, Window-to-Wall Ratio (WWR), glazing properties, and temperature set points vary among the proposed design options (see Table 1).

Table 1. Design variable assumptions.

Input variable	Design Option 1	Design Option 2	Design Option 3	Design Option 4
RSI of Exterior wall construction [m²K/W]	3.7	2.76	3.35	3.35
RSI of Floor construction [m²K/W]	4.59	4.59	4.59	3.35
RSI of Roof construction [m²K/W]	5.88	4.17	5.88	5.88
Window to Wall Ratio (WWR)%	70/65/60/20 (N/S/E/W)	40 overall	50/70/40/45 (N/S/E/W)	50/70/40/45 (N/S/E/W)
Glazing U-value [W/m²K]	1.7	3.12	1.2	1.2
Glazing SHGC	0.20	0.42	0.20	0.15

The materials in this test case are specified with some attributes such as RSI value and U-value. These terms are commonly used in the building industry to describe the thermal resistance and thermal conductivity of building materials. The mean values of these variables are defined based on the ASHRAE 90.1-2010 requirements for this test case's climate zone.

The internal heat gain loads, ventilation rates, and infiltration rate, along with the HVAC system description and building operation schedules are used to convert the building mass into thermal zones. The realizations of the mean values for the internal heat gain loads, ventilation rates, and infiltration rate for all four design options are specified based on the ASHRAE 90.1-2010 requirements for climate zone 5A (Table 2).

Table 2. Description of the system design assumptions.

Variable [Unit]	μ
Equipment loads per area [W/m^2]	10.765
Infiltration (air flow) rate per area [m^3/s-m^2]	0.0003
Lighting density per area [W/m^2]	10.55
Number of people per area [ppl/m^2]	0.07
Ventilation per area [m^3/s-m^2]	0.0006
Ventilation per person [m^3/s-m^2]	0.005

The standard deviations of all uncertain variables are assumed to be equal to 10% of the corresponding mean values, due to a lack of information. This assumption is based on the previous research done by [30]. de Wit [30] has estimated this percentage to be up to 10% in his report. For the future research, it is encouraged to conduct experiments or use expert knowledge to provide an appropriate standard deviation for each input variable with uncertainties.

The other system design inputs are assumed to be fixed and described as shown in Table 3.

Table 3. HVAC system specifications

	Design Option 1	Design Option 2	Design Option 3	Design Option 4
HVAC system type	VAV with reheat	VAV with reheat	VAV with reheat	VAV with reheat
Temperature set points for cooling/heating	23 °C/2 °C	23 °C/21 °C	23 °C/21 °C	23 °C/21 °C
Supply air temperature for cooling/heating	12.78 °C/32.2 °C	12.2 °C/32.2 °C	12.78 °C/32.2 °C	12.2 °C/32.2 °C
Chilled water temperature	7.2 °C	6.7 °C	7.2 °C	6.7 °C
Hot water temperature	60 °C	82.2 °C	60 °C	82.2 °C
Maximum heating supply air temperature	40 °C	40 °C	40 °C	40 °C

(*continued*)

Table 3. (*continued*)

	Design Option 1	Design Option 2	Design Option 3	Design Option 4
Minimum cooling supply air temperature	14 °C	14 °C	14 °C	14 °C
Maximum heating supply air humidity ratio	0.008 kg-H$_2$O/kg-air	0.008 kg-H$_2$O/kg-air	0.008 kg-H$_2$O/kg-air	0.008 kg-H$_2$O/kg-air
Minimum cooling supply air humidity ratio	0.0085 kg-H$_2$O/kg-air	0.0085 kg-H$_2$O/kg-air	0.0085 kg-H$_2$O/kg-air	0.0085 kg-H$_2$O/kg-air
Recirculated air per area	0 m^3/s-m^2	0 m^3/s-m^2	0 m^3/s-m^2	0 m^3/s-m^2

The occupancy, lighting, and equipment schedules are matched with the office schedules in the ASHRAE 90.1-2010 [31] for all four design options.

4.2 BIMProbE and Probabilistic Energy Analysis

The building element modeling environment in Revit is used to model the base model, assign the associated materials, and add the probability distributions to the thermal properties of building materials. BIMProbE add-in for Revit allows creating required shared

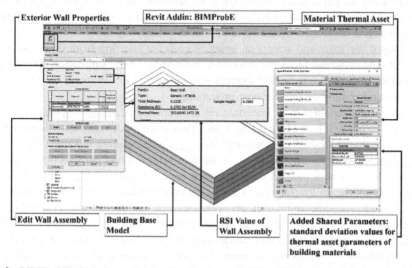

Fig. 3. BIMProbE add-in for getting and setting the probability distributions of thermal properties for building materials.

parameters to add probability distributions of the thermal properties of building materials and export this information to Excel for further analysis. Figure 3 demonstrates the Revit model and the workflow to set the probability distributions for thermal properties of building materials.

The visual programming tool in Rhino, Grasshopper, is used to get the base model from Revit and develop the four different design options. The building mass was converted to thermal zones in each case with defining the adjacency types, WWRs, internal heat gain loads, and HVAC system specifications. The building mass is split to floors and thermal zones using the existing components in Ladybug tools. The WWR for each façade is parametrically set in Grasshopper.

In this test case, the uncertain variables for which the variations take place due to unpredictable changes during construction, climate change, age, and maintenance are sampled using normal probability distribution. On the other hand, bases on the findings of [32] the input variables related to occupant behavior or presence can be best described with Poisson distributions.

The mean and standard deviation values of the thermal properties of building materials are set according to the Excel data inventory (created using BIMProbE).

The probabilistic simulations are conducted in Grasshopper using the statistical tools programed in CPython and simulation applications available in Ladybug tools (with OpenStudio simulation engine). The CPython component in Grasshopper is used to import the statistical tools such as Numpy and Scipy into Grasshopper (Abdel Rahman, 2018) to generate input samples with normal and Poisson distributions.

The internal heat gain sources and the HVAC settings are set in Grasshopper based on the findings of the previous research. The occupancy, lighting, and equipment operation schedules are set, and the thermal zones are exported to IDF files and run through OpenStudio in Grasshopper. The energy simulations are programmed to start automatically and run using the generated input samples. To automate the random value selection for each input variable and run OpenStudio for 500 (number of samples) times, a number slider, which is controlled by the Fly component in Ladybug tools, is connected to the list of input variables and selects an index of each list (starting from 0 and ending at 499) automatically and feeds the associated input value to the simulation. The simulation output is the building annual thermal load calculated in kWh/m^2. The outputs are collected and stored in Excel for post-processing.

The design variables with uncertainties are denoted as $X_{uncer} = \{x_{uncer_1}, x_{uncer_2}, x_{uncer_3}, \ldots, x_{uncer_k}\}$, k = 11. In this test case, the eleven input variables with uncertainties include the RSI values of exterior walls (x_{uncer_1}), the glazing U-value (x_{uncer_2}), the RSI values of floor construction (x_{uncer_3}), the RSI values of roof constructions (x_{uncer_4}), the three internal heat gain loads (equipment (x_{uncer_5}), lighting (x_{uncer_6}), and people loads (x_{uncer_7}), the infiltration rate (x_{uncer_8}), the two ventilation rate factors (ventilation per person (x_{uncer_9}) and ventilation per area ($x_{uncer_{10}}$), and the infiltration schedule that is strongly correlated with the possibility of opening or closing windows by occupants ($x_{uncer_{11}}$) for which the mean and standard deviation values are obtained from the literature studies [33, 34]. In this test case, the probability distributions of physical input variables $\{x_{uncer_1}, x_{uncer_2}, x_{uncer_3}, x_{uncer_4}\}$, also ($x_{uncer_5}$), ($x_{uncer_9}$), and ($x_{uncer_{10}}$) are set as normal, while the internal heat gain loads $\{x_{uncer_5}, x_{uncer_6}, x_{uncer_7}\}$, also the infiltration schedule ($x_{uncer_{11}}$) are assumed to be highly dependent on the occupant presence and sampled using the Poisson distribution.

5 Results

In this research, the main design objective is improving building energy performance, thus KPI and KRI are defined to rank the predicted energy performance of different design options. KPI in this study is building annual thermal (heating and cooling) loads, and the lowest possible value would be desired. KRIs include the mean, standard deviation, and variance of building annual thermal loads. The mean value shows the average of the samples, while standard deviation and variance are presented as additional measures of risk. The probabilistic framework works with quantifying the uncertainties in design inputs and allows predicting the probability distribution of the simulation outcome and the risks threatening building energy performance for each design option.

Using a processor Intel® Core™ i7-4770 CPU at 3.40 GHz speed, 500 energy simulations were run for each design option, each simulation taking 11 s. Figure 4 illustrates deterministic and probabilistic results of annual thermal loads (kWh/m^2) for the four design options.

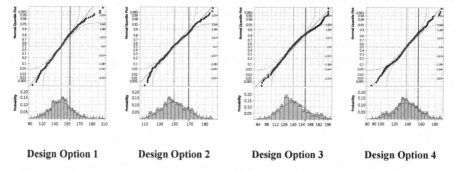

Design Option 1   **Design Option 2**   **Design Option 3**   **Design Option 4**

Fig. 4. Simulation results in terms of building annual thermal load (Color figure online).

Figure 4 shows the distribution of results with histograms (bottom), and normality plots (top). In each top graph shown in Fig. 4, the black points representing the data points are distributed around a diagonal red line. The more the datapoints are falling along the diagonal line, the closer is the distribution to the normal distribution. The horizontal green dashed lines depict the median point in the data set for each design option and the red dashed lines show the confidence limits. The confidence interval is set to 0.95 for all the data sets. The histograms (bottom graphs) show the relative frequency of the results. The vertical green lines show the mean values, and the vertical blue lines represent the deterministic results associated with each design option.

Looking at the plots, there is a clear indication of lack of fit to normal distribution in most design options. Especially, the main difference from normality is evident in the tails rather than in the middle. Furthermore, the normality of the distributions is assessed using the Shapiro Wilk W test (goodness of fit test). In this test, the null hypothesis (H0) is that the data are forming a normal distribution. A small p-value rejects the null hypothesis, meaning there is enough evidence that the data are drawn from a non-normal population. The test results are listed in Table 4.

Table 4. Shapiro Wilk W test results.

	Design Option 1	Design Option 2	Design Option 3	Design Option 4
W	0.985816	0.991713	0.976497	0.995059
Prob < W	<.0001	0.0069	<.0001	0.1110

Note: H0 = The data is from the Normal distribution. Small p-values reject H0.

The null hypothesis for this test is that the data are normally distributed. The Prob < W value listed in the output is the p-value. If the chosen alpha level is 0.05 and the p-value is less than 0.05, then the null hypothesis that the data are normally distributed is rejected. If the p-value is greater than 0.05, then the null hypothesis is not rejected [29]. The results show that the p-values in the first three design options are less than the predefined significance level (0.05). Thus, we can reject the null hypothesis and conclude that the data are not from populations with normal distributions in those design options. The reason could be drawing some input variables from Poisson distribution, also the nonlinear nature of equations in the building energy simulations. On the other hand, in Design Option 4 we cannot reject the null hypothesis, since the p-value is larger than 0.05.

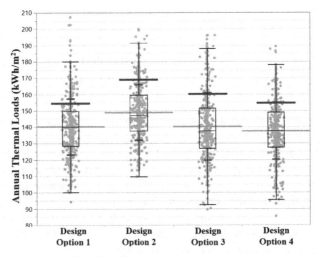

Fig. 5. Box Plot of building annual thermal load vs. design type.

Figure 5 shows the boxplots to further discuss the probability distributions of the results for each design option. The data points, quantiles, mean values, standard deviations, and deterministic result for each design option are superimposed on the quintile box plot.

The gray points illustrate the data points and the boxplots (shown in black lines) depict the quantiles, dividing the range of the data into four continuous intervals with equal probabilities (25%). The red lines on each boxplot show the standard deviation of the results. The green lines show the mean value for each design option, compared to the blue lines that show the deterministic results. The summary of the results for each design option is described as follows:

1. Design Option 1 shows a range of expected annual thermal load from 94.19 kWh/m² to 207.17 kWh/m² with a mean value of 140.16 kWh/m², a standard deviation of 17.19 kWh/m², and a variance of 295.50 (kWh/m²)². The deterministic result shown by the KPI value predicts the building annual thermal load to be equal to 154.42 kWh/m², which is in the last quartile (located in the fourth 25% of the data).
2. Design Option 2 shows a range of thermal load from 109.85 kWh/m² to 199.93 kWh/m² with a mean value of 148.88 kWh/m², a standard deviation of 16.96 kWh/m², and a variance of 287.64 (kWh/m²)². The deterministic result shown by the KPI value is equal to 168.19 kWh/m², which is in the last quartile (located in the fourth 25% of the data).
3. Design Option 3 shows a range of thermal load from 89.48 kWh/m² to 196 kWh/m², with a mean value of 140.29 kWh/m², a standard deviation of 20.52 kWh/m², and a variance of 421.07 (kWh/m²)². The deterministic result indicated by the KPI value shows the value of 159.71 kWh/m², which is in the last quartile (located in the fourth 25% of the data).
4. Design Option 4 shows a range of thermal load from 85.45 kWh/m² to 189.22 kWh/m² with a mean value of 137.58 kWh/m², a standard deviation of 17.4 kWh/m², and a variance of 302.76 (kWh/m²)². The deterministic result shown by the KPI value is equal to 154.76 kWh/m², which is in the last quartile (located in the fourth 25% of the data).

Building annual thermal load is identified with quantiles in Fig. 6. Using this data, the building thermal loads could be compared under different decision-making criteria.

According to Fig. 6, the minimum and maximum thermal loads are lower in the case of Design Option 4 (85.45 kWh/m² and 189.22 kWh/m², respectively) compared to the other design options.

The effect of deterministic and probabilistic results on the ranking of the design options based on different decision-making criteria are summarized as follows:

1. Deterministic: based on the results shown as blue lines in Fig. 4, Design Option 1 has the best performance, followed by Design Options 4, 3, and 2.
2. Probabilistic:

   2.1 Expected value criterion: Design Option 4 has the best performance, followed by Design Options 1, 3, and 2.

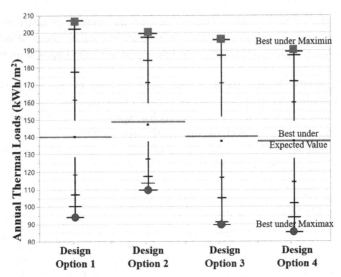

Fig. 6. Design decision making suggestions under expected value, maximax, and maximin criteria for test case 3, phase 1.

2.2 Maximax criterion: Design Option 4 has the best performance, followed by Design Options 3, 1, and 2.

2.3 Maximin criterion: Design Option 4 has the best performance, followed by Design Options 3, 2, and 1.

Based on the deterministic results (predicted KPIs), it can be concluded that Design Option 1 has the best energy performance, followed by Design Options 4, 3, and 2. However, the probabilistic results provide a more comprehensive outcome with KRIs. Design Option 4 has the lowest mean value of thermal load and is the best design option based on the expected value decision making criterion. Based on the maximax and maximin criteria Design Option 4 is confirmed to have the best performance, while the ranking of other design options is different. The major finding here is that including the uncertainties of input variables in the simulations can lead to probability distributions of the output and may change the performance ranking of design options according to the risk attitudes of design decision makers.

6 Conclusion

This research proposes a new framework to implement probabilistic methods in the field of building thermal energy analysis. The proposed framework introduces BIMProbE, a new tool (Revit add-in) to create probability distributions for thermal properties of building materials. This framework integrates a building design process with Monte Carlo uncertainty analysis using parametric tools (Grasshopper and its add-ins). This study demonstrated the application of the framework with a test case, considering several sources of uncertainties in building energy analysis with two different probability distributions: normal and Poisson as examples. The test case showed that it takes only about two hours to prepare the data and run the additional simulations to provide probabilistic results and predictions about the possible range of building thermal energy consumption.

The result of this study shows that probabilistic building performance analysis versus deterministic models provides important information about the consequences of design decisions under different possible conditions and may change the performance ranking order of design options. The major finding of this research is that compared with the existing deterministic method for architectural design, using probabilistic methods can result in significantly different design decisions to be made. In addition, using different decision-making criteria including expected value, maximax, and maximin may suggest different design options to select based on different attitudes towards risk. This framework can be widely applied to other design problems and domains to enhance the process of design decision-making. For the future research, the authors intend to work on maximax and maximin criteria based on, for example, first and 99th percentiles instead of the most extreme values in the output range. In this way, inference can be applied to find statistically significant differences between design options.

Acknowledgement. The authors would like to thank Perkins+Will, who provided the gift funding for this research.

References

1. Shaghaghian, Z., Yan, W.: Application of deep learning in generating desired design options: experiments using synthetic training dataset. In: Proceedings of the 2020 Building Performance Analysis Conference and Simbuild, pp. 535–544 (2020)
2. Eisenhower, B., O'Neill, Z., Fonoberov, V.A., Mezić, I.: Uncertainty and sensitivity decomposition of building energy models. J. Build. Perform. Simul. **5**(3), 171–184 (2012)
3. Tian, W.: A review of sensitivity analysis methods in building energy analysis. Renew. Sustain. Energy Rev. **20**, 411–419 (2013)
4. Ding, Y., Shen, Y., Wang, J., Shi, X.: Uncertainty sources and calculation approaches for building energy simulation models. Energy Procedia **78**, 2566–2571 (2015)
5. IBPSA-USA. Best Directory | Building Energy Software Tools (2021)
6. Hopfe, C., Hensen, J.: Uncertainty analysis in building performance simulation for design support. Energy Build. **43**(10), 2798–2805 (2011)
7. Tian, W., et al.: A review of uncertainty analysis in building energy assessment. Renew. Sustain. Energy Rev. **93**, 285–301 (2018)
8. Hopfe, C., Augenbroe, G., Hensen, J.: Multi-criteria decision making under uncertainty in building performance assessment. Build. Environ. **69**, 81–90 (2013)
9. Saltelli, A., Annoni, P., Azzini, I., Campolongo, F., Ratto, M., Tarantola, S.: Variance based sensitivity analysis of model output. Design and estimator for the total sensitivity index. Comput. Phys. Commun. **181**(2), 259–270 (2010)
10. Lee, B., Sun, Y., Augenbroe, G., Paredis, C.J.: Towards better prediction of building performance : a workbench to analyze uncertainty in building simulation. In: 13th International Building Performance Simulation Association Conference, Chambéry, France (2013)
11. MacDonald, I., Strachan, P.: Practical application of uncertainty analysis. Energy Build. **33**(3), 219–227 (2001)
12. de Wit, S., Augenbroe, G.: Analysis of uncertainty in building design evaluations and its implications. Energy Build. **34**(9), 951–958 (2002)

13. Asadi, S., Amiri, S., Mottahedi, M.: On the development of multi-linear regression analysis to assess energy consumption in the early stages of building design. Energy Build. **85**, 246–255 (2014)
14. Volk, R., Stengel, J., Schultmann, F.: Building information modeling (BIM) for existing buildings—literature review and future needs. Autom. Constr. **38**, 109–127 (2014)
15. Du, J., Zou, Z., Shi, Y., Zhao, D.: Zero latency: real-time synchronization of BIM data in virtual reality for collaborative decision-making. Autom. Constr. **85**, 51–64 (2018)
16. Kim, Y., Oh, S., Park, C., Kim, I., Kim, D.: Self-activating uncertainty analysis for BIM-based building energy performance simulations. In: 12th Conference of International Building Performance Simulation Association (2011)
17. Rezaee, R., Brown, J., Augenbroe, G., Kim, J.: A new approach to the integration of energy assessment tools in CAD for early stage of design decision-making considering uncertainty. In: eWork and eBusiness in Architecture, Engineering and Construction, ECPPM, pp. 367–373 (2015)
18. ISO: 31010: Risk management–risk assessment techniques. Event (London). Geneva 552 (2009)
19. Smart A. Creelman, J.: Risk-Based Performance Management: Integrating Strategy and Risk Management. Springer, Heidelberg (2013). https://doi.org/10.1057/9781137367303
20. Institute of operational risk_Sound practice guidance: key risk indicators (2010)
21. Pruvost, H., Scherer, R.: Analysis of risk in building life cycle coupling BIM-based energy simulation and semantic modeling. Procedia Eng. **196**, 1106–1113 (2017)
22. Sun, S., Kensek, K., Noble, D., Schiler, M.: A method of probabilistic risk assessment for energy performance and cost using building energy simulation. Energy Build. **110**, 1–12 (2016)
23. Arrow K., Lind, R.: Uncertainty and the evaluation of public investment decisions. In: Uncertainty in Economics, pp. 403–421 (1978)
24. Bae, N.: Influence of uncertainty in user behaviors on the simulation-based building energy optimization process and robust decision-making, Doctoral dissertation (2016)
25. Fargier, H., Guillaume, R.: Sequential decision making under ordinal uncertainty: a qualitative alternative to the Hurwicz criterion. Int. J. Approx. Reason. **116**, 1–18 (2020)
26. Fargier, H., Guillaume, R.: Sequential decision making under ordinal uncertainty: a qualitative alternative to the Hurwicz criterion. Int. J. Approximate Reason. **116**, 1–18 (2020)
27. Shahsavari F., Yan, W.: Integration of probabilistic methods and parametric tools for performance-based building design decision-making. In: Conference of International Building Performance Simulation Association (2021)
28. TT Toolbox 2017. https://www.food4rhino.com/app/tt-toolbox. Accessed 31 Jan 2021
29. JMP. https://www.jmp.com/support/notes/35/406.html
30. de Wit, S.: Uncertainty predictions of thermal comfort in buildings (2001)
31. ASHRAE. https://www.ashrae.org/
32. Zhou, X., Yan, D., Hong, T., Ren, X.: Data analysis and stochastic modeling of lighting energy use in large office buildings in China. Energy Build. **86**, 275–287 (2015)
33. Lomas, K., Eppel, H.: Sensitivity analysis techniques for building thermal simulation programs. Energy Build. **19**(1), 21–44 (1992)
34. Hopfe, C.: Uncertainty and sensitivity analysis in building performance simulation for decision support and design optimization, Doctoral dissertation (2009)

Outdoor Comfort Analysis in a University Campus During the Warm Season and Parametric Design of Mitigation Strategies for Resilient Urban Environments

Francesco De Luca[(✉)] [iD]

Department of Civil Engineering and Architecture, Tallinn University of Technology, Ehitajate tee 5, 19086 Tallinn, Estonia
`francesco.deluca@taltech.ee`

Abstract. Cities are one of the major contributors of climate change. The built environment urgently needs to significantly reduce its impact on resource depletion and its CO_2 emissions. At the same time, urban environments must adapt to guarantee livability and safety in increasingly frequent severe conditions. To aid this process, assessment methods and indexes have been developed to help designers and researchers investigate optimal solutions for outdoor thermal comfort. Temperature increase during summer is a growing concern also in northern European cities such as Tallinn, Estonia. This paper presents a study on the comfort conditions of the outdoor areas of the TalTech campus in Tallinn during summer and investigates the cooling potential of vegetated surfaces and trees in the local microclimate. A parametric design workflow was developed that integrates building and climate modeling, environmental and building simulations and outdoor comfort assessment through the metrics of Universal Thermal Climate Index and Outdoor Thermal Comfort Autonomy. The results show that heat stress can be experienced on the outdoor areas of the campus. The quantity and the optimal location of vegetated surfaces and trees to provide comfort were determined through the developed algorithm. The methods and the generated vegetation patterns are presented and discussed.

Keywords: Climate change · Resilient urban environments · Outdoor comfort · Mitigation strategies · Environmental simulations · Parametric design

1 Introduction

Climate change is one of the biggest contemporary challenges that our society is facing. Buildings and the urban environment are one of the biggest causes of climate change. On a global level, they consume 36% of the energy produced and contribute directly and indirectly for almost 40% of the carbon dioxide emissions [1]. The Intergovernmental Panel on Climate Change (IPCC) has defined possible climate future scenarios, the Representative Concentration Pathways (RCP), related to predicted amounts of Greenhouse Gas (GHG) emissions [2]. RCP 8.5 is the scenario with the highest prediction of GHG

© Springer Nature Singapore Pte Ltd. 2022
D. Gerber et al. (Eds.): CAAD Futures 2021, CCIS 1465, pp. 473–493, 2022.
https://doi.org/10.1007/978-981-19-1280-1_29

emissions (Fig. 1). Without effective climate change reduction initiatives, consistent population growth and the slow adoption of new technology solutions will lead to an increase of energy demand and related GHG emissions. On these bases, IPCC scientists have produced climate models related to northern European countries which predict a minimum increase of temperatures during the warm season of 2 °C by 2100. This aligns with a constant trend in air temperature increase during the warm season in Europe in the last decades [3].

The increase of air temperatures on a global scale due to climate change is more severe in cities due to the Urban Heat Island (UHI) effect [4]. The materials of paved areas, streets and buildings that capture longwave solar radiation and reach high surface temperatures, the difficulty of air to flow freely in dense urban fabric configurations and around clustered buildings without the possibility to dissipate the stagnant heat, the high intensity cooling of commercial buildings during the warm season that releases heat in their proximities, the anthropogenic activities such as car traffic and buildings operation, and the scarcity of green areas and trees, all contribute to increase the temperatures in cites compared to the surrounding country areas and to produce warmer summers and heatwaves (Fig. 1). The result is an increased sense of discomfort in outdoor spaces in cites, and in some cases illnesses and respiratory difficulties.

Climate studies for the northern European city of Tallinn, Estonia (Lat. 59.43° N – Lon. 24.75°E), where this research was conducted, show a consistent increase of air temperatures that significantly harms outdoor thermal comfort. According to the Köppen-Geiger climate classification, Tallinn presents the typical northern European climate characterized by cold winters and warm summers. Despite being comfortable for most of the warm season, in Tallinn air temperatures increased more than 2 °C on average during the last decades, which is more than the average in cities in the north of Europe

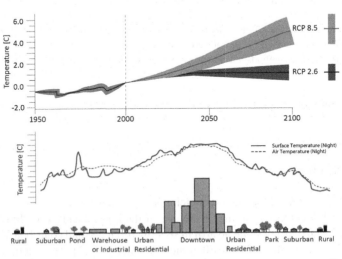

Fig. 1. Predicted global average surface temperature increase with and without climate change reduction initiatives (top – source: IPCC). Diagram of the urban temperature increase due to the UHI effect (bottom – source: U.S. Environmental Protection Agency).

[5]. Further, climate change studies have predicted the alarming doubling of the number of hot days on an annual basis in Tallinn before 2050 [6].

Outdoor thermal comfort is significantly influenced by the local urban microclimate. The main microclimatic factors that determine the comfort of pedestrians are air temperature, solar radiation, wind and surface temperature of the surrounding environment [7]. Shortwave solar radiation, the main factor for human outdoor comfort, can be blocked, reflected and absorbed by buildings. Longwave solar radiation is responsible for significant thermal exchanges in urban environments [8]. Wind can be accelerated by differences in height and by the short distance between buildings, or blocked by dense urban fabrics and articulated buildings. The first effect increases, whereas the latter decreases outdoor comfort during the warm season. Thus, the building materials and form, the materials of outdoor paved surfaces and the presence of vegetated surfaces and trees are critical factors to take into account to improve the outdoor thermal comfort during the warm season [9].

Architects and planners are urged on the one hand to design buildings characterized by low energy consumption and indoor comfort through passive design strategies to reduce resource depletion and the impact on the GHG emissions thus on the climate, and on the other hand to design taking into account climate change scenarios by adopting mitigation strategies to create resilient urban environments and to improve the outdoor thermal comfort and livability of pedestrian areas.

2 Background

Previous studies show the potential of an environmentally conscious urban design to improve indoor comfort by limiting excessive air temperatures, and to reduce energy consumption of residential buildings during the warm season by taking into account building heights, distances and orientations at the north latitudes and climate of Estonia [10, 11]. Existing research conducted for the same region underlines the significant influence of building envelope design, i.e., orientation and size of windows, and presence of shading devices and balconies for the fulfillment of indoor visual and thermal comfort and cooling energy consumption requirements of the local building code [12, 13].

Further, recent studies developed in the city of Tallinn integrated multi-domain and multi-performance analysis to propose building cluster configurations to improve the indoor and outdoor comfort. The pattern layout, the distances and the orientations of apartment buildings located in a residential area were analyzed to guarantee the solar access of the building interiors as required by the Estonian daylight standard and the pedestrian wind comfort according to established criteria [14]. Geometric variations of a commercial buildings cluster located in various central areas of the city were investigated by different studies to determine optimal configurations to improve the indoor and the outdoor thermal comfort during winter and summer [15, 16]. Results show that the distances between tall buildings and their orientation strongly influence daylight availability in the interiors and pedestrian discomfort due to lower perceived temperatures during the cold season. Building cluster patterns and the relation with the different urban environments are also responsible for significant variations in the indoor overheating and consequent cooling energy consumption needed to maintain comfort, and in the outdoor thermal comfort in the commercial district during summer in the Nordic climate of Tallinn. The mentioned studies underline the importance and show how to take into account the climatic factors and the microclimatic modifications to design resilient buildings and urban environments and cities with low climate impact.

While the impact of the buildings massing, the distances and the orientations, and the morphology and density of urban environments was investigated in order to reduce energy consumption and to increase indoor/outdoor comfort in Tallinn, the analysis of the potential of vegetation to modify the urban microclimate in order to increase the outdoor thermal comfort during the warm season represents a research gap.

Trees have significant potential to increase outdoor thermal comfort during the warm season mainly through the shading provided by the crown and also through the air temperature reduction by transpirative cooling. The cooling effect of trees can be localized or can have a wider effect depending on the tree size and quantity, and on different environmental factors. Studies showed that the shadowing effect of trees can reduce the perceived temperature from 36 °C to 26 °C UTCI and that the transpiration of the leaves can further reduce the temperature of more than 1 °C [17]. If planted in adequate quantities and in critical locations, trees can help to reduce the UHI effect [18]. Though they yield smaller impacts, vegetated surfaces also can mitigate the urban microclimate in warm conditions through solar radiation absorption and evapotranspiration [19].

A study conducted during summertime in the hot climate of Bilbao, Spain to analyze how the local microclimate conditions in the urban canyons are affected by the building distances and orientations, the street materials, and the size and patterns of trees determined that large crown trees arranged linearly reduce the perceived temperature by up to 15.3 °C of the Physiological Equivalent Temperature thermal comfort index [20]. A study conducted through measurements and simulations investigating the cooling effects of different vegetation scenarios in a residential quarter in the city of Beijing, China during a day of mid-August between 9 a.m. and 6 p.m. found that the use of the vegetated areas and trees can reduce the sensible heat by 2% to more than 14% depending on the hour of the day and the vegetation arrangement [21]. A study conducted to quantify the positive effect of trees and green cover on the outdoor thermal comfort in two cities with different climate, Lecce, Italy, and Lahti, Finland, compared the current vegetation scenario with a new scenario with higher density vegetation and found that during summer the increased area covered by trees and green features improved the comfort of humans by up to two points of the Predicted Mean Vote thermal sensation scale, thus from the hot level to the slightly warm sensation level [22].

This paper presents initial work to fill the mentioned research gap concerning the potential mitigation of uncomfortable microclimatic conditions through the use of trees and green cover in the city of Tallinn during the warm season. The investigation, conducted on the campus of Tallinn University of Technology (TalTech), had three main objectives: 1) to quantify the level of discomfort in the outdoor areas during the warmest summer days; 2) to determine the density and location of the trees and the vegetated surfaces to guarantee outdoor thermal comfort levels; and 3) to provide the university with information for the planned renovation of the areas as part of the redevelopment plan toward a sustainable and resilient campus. The paper addresses researchers, practitioners and administrators to help them to increase the quality of the built environment and to reduce climate change.

3   Materials and Methods

In order to fulfill the objectives of the research a parametric design workflow was developed that integrates: a) three-dimensional data of the urban environment of the TalTech campus buildings and of the outdoor areas used for the analysis; b) trees and vegetated surfaces properties; c) climatic and microclimatic data; d) wind and surface temperature simulations; and e) outdoor thermal comfort analysis (Fig. 2).

The urban context, the campus buildings, the trees and the ground were modeled in Rhinoceros [23] and instantiated into the parametric model realized in Grasshopper [24]. The weather file climatic data and the microclimatic data generated through Urban Weather Generator (UWG) [25] were used for the wind speed and the surface temperature simulations performed through the grasshopper plug-ins Eddy [26] and Honeybee of Ladybug Tools [27], respectively. The first used the Computational Fluid Dynamics (CFD) software OpenFOAM [28] and the latter used the energy simulation software EnergyPlus [29] to finally assess outdoor comfort using the Universal Thermal Climate Index (UTCI) [30] and the Outdoor Thermal Comfort Autonomy (OTCA) [31] metrics.

The parametric workflow takes advantage of an automated process, realized through the algorithm developed by the author, for the placement of the vegetated surfaces and the trees in correspondence to the more critical locations inside the analyzed areas.

The analysis period used for the research is the week from 3rd to 9th of August. This was selected because it represents the warmest week of the year in Tallinn according to the statistical weather data and thus represents a worst-case scenario. It must be noted that although summer holidays take place during August, the university campus is used for summer schools and international student events involving outdoor activities.

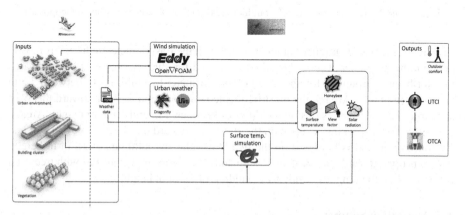

Fig. 2. The parametric design workflow developed for the research.

3.1 Analysis Areas

The TalTech campus consists of connected buildings surrounded by paved, parking and green areas (Fig. 3). As part of a redevelopment plan to improve the sustainability and resilience of the campus, the parking areas will be relocated outside of the campus, and the paved and green areas will be refurbished with new walkable and vegetated surfaces and trees. The analyzed areas were selected because they are currently the most used and in the future plan are the most suitable for outdoor activities. In addition, they present a diversity of orientations and openness (distance and height of surrounding buildings) which makes them useful for analyzing different outdoor conditions.

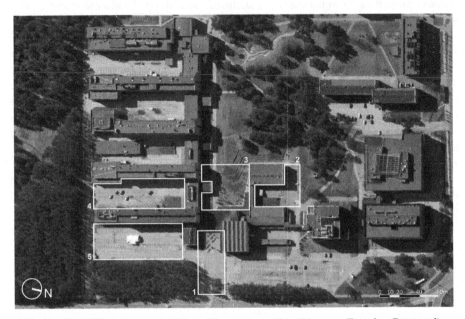

Fig. 3. The TalTech campus and the outdoor areas analyzed (source: Estonian Geoportal).

Area 1, the paved entrance square, is located on the East of the campus and is mostly open on the perimeter. The L-shaped Area 2, used for student events, is open towards the North and surrounded for about half of the perimeter by buildings of between 7.5 m and 10 m in height. Area 3, the most frequented outdoor area, is open toward the West and surrounded on three sides by buildings of between 4 m and 10 m in height. Area 4, currently used as parking, is open on the short side toward the South, faces a large and vegetated area, and is bordered on the two long sides by buildings of 18 m in height. Area 5 is open towards the East and South and faces large areas covered with trees and is bordered on the long West side by a building of 18 m in height.

3.2 Vegetation Settings

The city of Tallinn is populated by a large variety of urban tree species [32]. For the study, the lime tree (*Tilia cordata*) (Fig. 4) was selected because it is common in the area and because its crown presents a significant shading capacity, allowing only 15% of sunlight transmittance [33], due to a high density of medium sized leaves. This value was used to model the trees for the surface temperature and the outdoor comfort simulations. The trees were modeled as polysurfaces using a height of 8 m and a radius of 2.85 m (Fig. 4). Only the shading effect of the trees was taken into account in the study. The cooling effect of the trees by evapotranspiration was not considered because this complex process can not be modeled by the simulation software used. The grass surfaces were modeled using patches of 6 m in size in correspondence to each tree. The use of the patches and the properties of the grass are presented in the following sections.

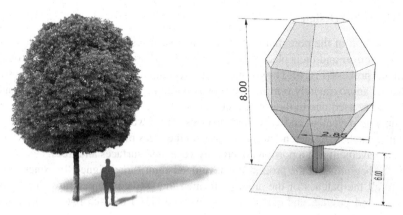

Fig. 4. Figure depicting a lime tree (left). 3D model of the tree and ground patch (right).

3.3 Urban Weather

The TalTech university campus is located in the residential quarter of Mustamäe. The medium density quarter is populated by concrete housing blocks with heights from 6 to 10 floors realized during the Soviet era, by industrial facilities and commercial buildings and by a network of large driveways. Further, the apartment buildings are commonly surrounded by large parking areas. The quarter also contains green areas with trees between the buildings but not along the roads.

In order to accurately assess outdoor thermal comfort, the possible UHI effect was taken into account by morphing the available weather data collected at the weather station outside the city in urban weather data. To perform this operation, UWG was used in Grasshopper through the plug-in Dragonfly of Ladybug Tools.

UWG takes into account the building characteristics such as typologies, heights, footprint coverage and façade-to-site ratio, and relevant anthropogenic activities such as traffic and building operation and generates the morphed weather data of air temperature and relative humidity. Figure 5 shows a significant increase of air temperature of up to 1.4 °C of average dry bulb temperature during the analysis period.

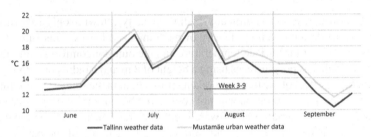

Fig. 5. Weekly average urban dry bulb temperature increase.

3.4 Algorithm Design

The evaluation of the potential of the vegetated surfaces and trees to provide outdoor thermal comfort and their placement in the analyzed areas was performed through an automated process (Fig. 6). The ground of each rectangular area was subdivided in square patches of approximately 6 m in size. For Area 2 and Area 4 some larger rectangular patches were used due to their non-rectangular shape. The patches were used to simulate concrete pavement and vegetated surfaces. Trees were located in the center of the vegetated patches. At the end of the process all the areas presented a portion covered by concrete pavement and a portion covered by vegetated surfaces and trees.

At the beginning of the process the developed algorithm assigned the concrete pavement to all the patches of the area, performed the surface temperature simulation of buildings and ground, and calculated the average UTCI of each patch, where higher values indicated higher discomfort, considering climate and microclimate data, wind velocities and surface temperatures. In two different processes, one using only the vegetated patches and the other using the vegetated patches and the trees, the algorithm incrementally substituted one concrete patch starting from the one with higher UTCI with a vegetated one, and after having performed again the simulations including the vegetated surfaces and the trees, calculated the OTCA from the updated UTCI values. The automated process stopped when a sufficient quantity of vegetated patches and trees was placed in order to achieve the OTCA threshold (Fig. 6). The algorithm then determined the quantity and the optimal layout of the vegetation to provide outdoor thermal comfort. The UTCI and OTCA are discussed in Sects. 3.6 and 3.7.

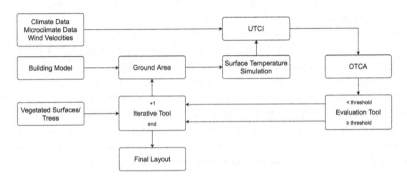

Fig. 6. Flowchart of the automated process for the placement of vegetated patches and trees.

3.5 Wind Simulations

Wind velocities are one of the main factors influencing outdoor thermal comfort. In order to obtain accurate results, in this research wind velocities were simulated using CFD following the methods of the Grasshopper CFD plug-in Eddy and simulation best practices [34]. The simulations were conducted using a cylindrical CFD domain (wind tunnel) with a radius of 1 km. The size of the domain was $15 \cdot H_{max}$ in both flow directions (H_{max} = maximum height of the modeled buildings). Such domain dimension was required upstream for flow establishment and downstream for flow redevelopment. The size of the domain also allowed obtaining a blocking ratio (the ratio between the projection of the modeled buildings and the width of the domain) of 3%, which is a value recommended not to be exceeded. The height of the CFD domain was $5 \cdot H_{max}$, appropriate to avoid wind acceleration above the buildings. Buildings were modeled and included in the domain up to a distance of 500 m from the analysis areas. The existing trees in the vicinity of the analysis areas were also modeled and included in the domain.

The size of the domain mesh cells was set to 20 m, and progressively refined up to 1 m for all the buildings and the encompassing ground surface for a total of more than 7.2 M cells. The wind simulations were conducted from 16 cardinal directions using a wind velocity of 5 m/s at the domain inlet and at the reference height of 10 m and using a logarithmic wind profile. Such an amount of wind directions was necessary to obtain accurate simulations of hourly wind velocities. However, the cylindrical domain allowed for the saving of computational time because only one domain mesh was generated, a process that requires considerable time [26]. Due to the uniformity of the areas surrounding the TalTech campus the terrain surface roughness parameter used for all the simulation directions was $Z_0 = 1$, which corresponds to an environment characterized by buildings with similar heights and green areas with trees.

To obtain the wind velocities, every area was divided in an analysis grid of square cells of approximately 3 m in size, each one corresponding to ¼ of a ground patch, located at 1.5 m from the ground, the human body reference height. The result of the wind simulations were 16 wind velocities probed at the points located in the center of the grid cells related to the fixed wind velocity of 5 m/s used at the domain inlet (Fig. 7). Finally, wind factors were used to calculate the wind velocities at the points of the analysis grid for each hour of the analysis period. The use of the wind factors is presented in Sect. 3.7.

3.6 Surface Temperature Simulations

In order to accurately assess outdoor thermal comfort during the warm season, the surface temperatures of the environment were taken into account. Surface temperatures were simulated considering the campus buildings' use and thermal settings, and the building envelope and the ground properties presented in Table 1. Thermal zones were modeled for offices, classrooms, corridors and stairs, canteens, and technical rooms (Fig. 8). The thermal settings of the offices and the classrooms are those used for the TalTech buildings. For the other zones the EnergyPlus parameters were used. The building envelope properties are those of the campus buildings. The vegetated surfaces were modeled

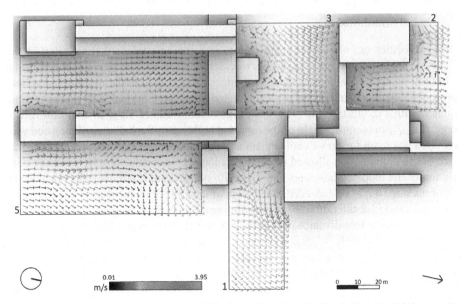

Fig. 7. Plots of the wind patterns for the CFD simulation from the South direction (right arrow).

Table 1. Simulation parameters. U_t = Total thermal transmittance, EW/IW = external/internal walls, W = window, F = floor, GF = ground floor, R = roof, VT = visible transmittance, A = albedo, H = height of plants, LAI = leaf area index.

Thermal zone settings		Building envelope properties					
People density	0.06 p/m^2		EW	W	IW/F	GF	R
Equipment density	8 W/m^2	U_t (W/m^{2}K)	0.13	0.55	3.6	0.19	0.1
Lighting density	7 W/m^2	Windows	VT 70%		SHGC 0.4		
Infiltration (x10^{-5})	5.6 m^3/sm^2	**Ext. Ground Properties**			A	H	LAI
Mech. ventilation	0.003 m^3/sm^2	Concrete pavement			0.3	–	–
Schedule Off./Class	8–18/8–20	Vegetated surface			0.25	0.1 m	5

using EnergyPlus vegetation material definition. The surface temperature simulations were performed using the generated urban weather data.

As discussed in Sect. 3.4, the algorithm automated the simulations using an incrementally larger quantity of vegetated square patches of approximately 6 m in size, iteratively substituting the concrete ones. The trees were located in the center of the patches. The ground grid used for the surface temperature simulations was approximately 3 m in size, the same size as the wind simulations, meaning that each ground patch was divided in four analysis cells. The simulation sensor points were located in the center of the analysis grid cells surrounding the tree trunk. The thickness used for the soil beneath the paved and vegetated surfaces was 1.5 m.

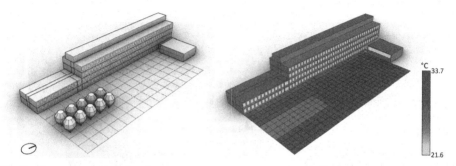

Fig. 8. Thermal zones, ground and trees used in the surface temperature simulations of Area 5 (left). Average surface temperatures between 8 a.m. and 8 p.m. during the analysis period (right).

3.7   Outdoor Thermal Comfort Analysis

The analysis of the outdoor thermal comfort and the assessment of the potential of the vegetated surfaces and trees to provide comfort was performed using the UTCI [30] thermal index and the OTCA [31] temporal and spatial method. The assessments were performed using analysis grids of the same size as the wind and surface temperature simulations, considering the center points of the cells at 1.5 m from the ground as human body locations. The hourly analysis period (occupied time) considered was from 8 a.m. to 8 p.m. as these are the activity hours of the university campus.

The UTCI is an efficient model for the prediction of the temperature perceived by a person at a specific time and location, in both building interiors and outdoor spaces, for different climatic conditions and periods of the year. The UTCI is based on the climatic and microclimatic factors of air temperature, solar radiation, wind and humidity and on a physiological and clothing model. The UTCI index identifies different cold and heat stress levels and one no thermal stress level between 9 and 26 °C UTCI Equivalent Temperature. For the study, the threshold of 26 °C UTCI was considered.

A specific section of the parametric design workflow developed for the study calculated the UTCI taking into account all the factors required for an accurate assessment through the plug-in Honeybee of Ladybug Tools. The workflow considered the solar radiation form the weather data, the dry bulb temperature and the relative humidity from the generated urban weather data, the simulated wind velocities and the surface temperatures of the building envelopes and the ground and the view factors for the Mean Radiant Temperature (MRT) calculation. The MRT was calculated for each point of the analysis grid by computing the longwave radiation from the surrounding buildings and the ground surface temperatures weighted by the view angle, i.e., by the visible surface area from each point. Further, the MRT was adjusted in consideration of the shortwave solar radiation using the SolarCal model of the ASHRAE-55 standard [35].

The wind velocities for every hour and sensor point of the analysis areas were calculated using the results of the CFD simulations through wind factors [35]. The 16 simulated wind velocities, one for each wind direction, were divided by the fixed wind velocity used for the simulations, generating a dimensionless matrix of wind factors. Finally, among the 16 wind factors, one was selected with the wind direction closest to that of the wind data of the Tallinn weather file for each analyzed hour and multiplied

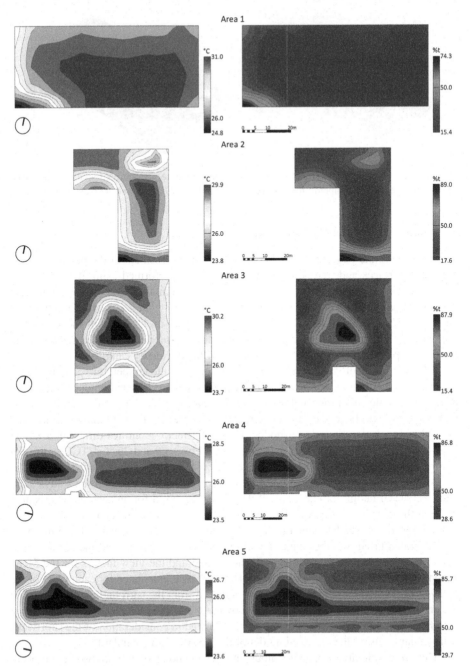

Fig. 9. Analysis period av. UTCI (left) and OTCA (right) contour maps for Area 1 calculated using concrete pavement, Area 2, Area 3, Area 4 and Area 5 calculated using 25%, 50%, 75% and 100%, respectively of the vegetated surfaces and trees necessary to fulfil the OTCA criteria.

by the measured wind velocity of the wind data. The results were hourly UTCI values for each sensor point showing the potential comfort or discomfort in the analysis area. Figure 9 presents average (av.) UTCI comfort contour maps for the analysis period.

For this study, the OTCA method was used to assess the comfort level of the analyzed areas and the ratio of the vegetated surfaces and the quantity of the trees necessary to provide comfort during the analysis period on the basis of the hourly UTCI assessment. The OTCA is defined as the percentage of occupied time during which a location meets specific thermal comfort criteria and defines 50% as the threshold to consider the location comfortable. Thus, at each point of the analysis grid the percentage of time the UTCI was below or above 26 °C was calculated in the parametric workflow and OTCA comfort maps were generated (Fig. 9). The method to assess the comfort of the entire analyzed areas using the OTCA is presented in the next section.

4   Results

This section presents the levels of outdoor thermal comfort of the analyzed areas in the different conditions, i.e., when the entire area was covered with concrete pavement and when the vegetated surfaces and the trees were used. The scope was to quantify the level of discomfort of the actual paved areas during the analysis period and to determine the percentage of each area that should be covered with vegetated surfaces and trees in order to guarantee comfort and their optimal location to improve the resilience of the TalTech campus outdoor areas. It must be noted that although only Area 1 is currently paved with concrete, Area 2, 4 and 5 are paved with aged asphalt, which presents a similar albedo as concrete pavement, and although Area 3 is currently mostly covered with grass, all the areas will be refurbished, thus concrete pavement covering an entire area was used in the study as a worst-case scenario.

In order to assess the level of thermal comfort of the outdoor areas, the Spatial OTCA (sOTCA) method was used. This is defined as the ratio of an outdoor space that meets the comfort criteria for at least 50% of the occupied time. The threshold to consider an area comfortable depends on the activities, e.g., for dining terraces a threshold of 80% is suggested due to the difficulty of rearranging the layout of tables according to variable microclimatic conditions, whereas for generic outdoor activities for which people can change location inside the area 50% is considered acceptable [31]. Thus, the automated algorithm determined the area to be covered either only by the vegetated surface or also by the trees, i.e., the vegetation to ground ratio to achieve 50% of the area in state of comfort for 50% of the occupied time according to the UTCI index.

The results of the research are presented using the vegetation to ground ratio and the different variations of area av. UTCI (of the analysis period averages) and sOTCA in the analyzed areas (Fig. 10). Additionally, the influence of the wind and solar radiation affected by the area orientation and openness are also presented.

4.1   Vegetation to Ground Ratio

The vegetation necessary to guarantee comfort is presented as the vegetation to ground ratio (VGR), i.e., the ratio of the vegetated surfaces and trees cover (m^2) to the area size (m^2). In all the areas the sole use of soil, grass and plants was not sufficient to provide comfort even when covering the whole area. The sOTCA increase and the UTCI decrease were modest in all areas (Fig. 10), because although during summer the dry soil absorbs less solar radiation than concrete and consequently re-emits less heat, the grass has a higher albedo, then it reflects more shortwave radiation. This increases the MRT, which is one of the factors influencing outdoor thermal comfort during summer.

Half of Area 1 had to be covered by the vegetated surfaces and the trees to achieve the required level of comfort. The algorithm replaced 27 out of a total of 55 concrete ground patches with the vegetated ones. The initial sOTCA, when concrete paved the entire area, was 1.8% and the av. UTCI was 30.2 °C. After 27 iterations the sOTCA was 52.2%, the av. UTCI was 26.5 °C for a VGR of 49.1% (Fig. 10). The use of the sole grass surfaces for the entire area allowed the av. UTCI to decrease to 29.5 °C and the sOTCA to increase to a mere 2.3%. It must be noted that the av. UTCI threshold of 26 °C for the analyzed area was not an objective of the study because it assesses the comfort of a single location. Nevertheless, for all the areas an average value of approximately 26 °C UTCI was obtained.

In Area 2 the VGR to guarantee thermal comfort was 43.8%, corresponding to 17 vegetated patches with trees out of 36. The discrepancy between the VGR and the patches ratio (17/36) was due to the necessity of using several larger rectangular patches to subdivide the L-shaped area (see Sect. 3.4), whereas most of the 17 are the square patches. The sOTCA and the av. UTCI changed from 14.6% to 53.9% and from 28.6 °C to 26.4 °C, respectively. As for Area 1, using only vegetated surfaces facilitated a decrease of av. UTCI of only 0.2 °C and an increase sOTCA of only 1.1% (Fig. 10).

In Area 3 the automated algorithm stopped after 30 iterations out of 68. The resulting VGR necessary to guarantee the outdoor comfort was 44.1%, which allowed an increase of sOTCA from 6.6% to 51.1% and a decrease of av. UTCI from 29.6% to 26.2% when using vegetated surfaces and trees. When all the area was covered only by the vegetated surfaces the sOTCA increased to a mere 8.1% and the av. UTCI decreased only to 29.0 °C (Fig. 10).

The VGR necessary to guarantee comfort in Area 4 was 24.1% using 22 patches with the vegetated surface and the tree out of a total of 85 ground patches. The reason of the discrepancy between the VGR and the patches ratio (22/85) is the same as for Area 2, though in Area 4 a smaller quantity of larger rectangular patches was used. The 22 vegetated surfaces and the trees were able to guarantee a sOTCA of 52.4%, from the initial 15.2%, and an av. UTCI of 26.4 °C, from the initial 28.0 °C. Using only vegetated surfaces the sOTCA did not change, though slight increases were recorded for intermediate iterations, and the av. UTCI decreased by a mere 0.4 °C (Fig. 10).

In Area 5 the quantity of vegetated surfaces and trees to guarantee comfort was 38 out of a total of 90, for a VGR of 42.2%. The sOTCA increased from 0.3% and the av. UTCI decreased from 28.8 °C recorded in the initial condition of concrete pavement for the entire area to 51.7% and to 26.1 °C, respectively. As for the other areas, when only vegetated surfaces were used the sOTCA increased to 4.7% and the av. UTCI decreased minimally to 28.3 °C (Fig. 10). Figure 11 shows the minimum layout of the vegetated surfaces and trees location to obtain thermal comfort in the analyzed areas.

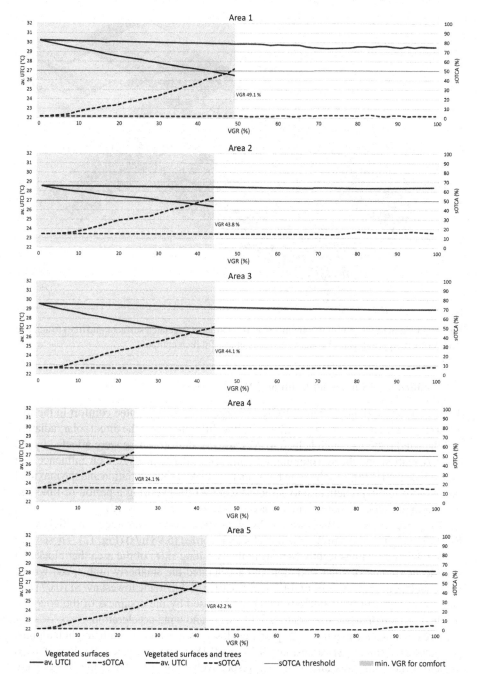

Fig. 10. Minimum VGR to achieve outdoor comfort (sOTCA ≥ 50%) using vegetated surfaces and trees, and variations of av. UTCI and sOTCA at the increase of only the vegetated surfaces cover and of the vegetated surfaces and trees cover (the latter corresponds to VGR).

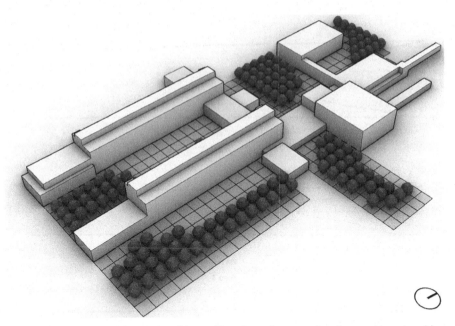

Fig. 11. Layout of the minimum quantities and locations of vegetated surfaces and trees to achieve outdoor thermal comfort in the analyzed areas of the TalTech campus.

4.2 Influence of Solar Radiation and Wind Patterns

The results showed that different VGR were necessary to guarantee comfort in the five areas (Fig. 12). The factors most influencing the difference were the direct solar radiation and the wind patterns which helped to decrease discomfort. These were affected by the area orientation and the surrounding buildings. In order to analyze their influence, the Shade Index (SI) [36], i.e., the ratio between the blocked solar radiation and the maximal radiation of the hypothetical unobstructed scenario (0–1) during a period of time and the maximum wind velocities (m/s) for each area were used (Fig. 12).

Evidence shows that the lowest VGR of 24.1% in Area 4 was due to the highest av. SI (0.33) and the second highest wind velocity recorded (5.98 m/s) (Fig. 12). These were caused by the buildings tightly bordering the two long sides of the area that blocked a significant amount of solar radiation and accelerated the southerly high-speed winds. Conversely, the highest VGR of 49.1% in Area 1 was due to the lowest av. SI (0.07) and an average value of wind velocity (4.41 m/s) caused by the openness of the area that resulted in a small direct solar radiation blockage and wind acceleration.

Areas 2 and 3 located in spaces partly enclosed by low and medium height buildings presented medium av. SI values, i.e., 0.15 and 0.18, respectively, and the lowest wind velocities, i.e., 3.88 m/s and 2.94 m/s, respectively (Fig. 12). Thus, the VGRs were among the highest, i.e., 43.8% and 44.1%, respectively. Area 5, despite being bordered on the long west side by a close building and being the most open to the southerly winds that caused the second highest av. SI (0.27) and the highest wind velocity (7.57 m/s), presented a high VGR, i.e., 42.2%. Comparing this result with that of Area 4 which has a similar shape and the same orientation, it is possible to underline the largest influence of direct solar radiation compared to wind on outdoor thermal comfort.

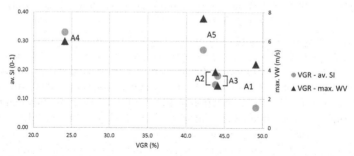

Fig. 12. Relations of the VGR (%), with the average Shade Index (av. SI 0–1), and with the maximum wind velocities (max. WV m/s) for the analyzed areas. (A1–5 = Area 1–5).

5   Conclusions

The presented study analyzed the outdoor thermal comfort conditions during summer of the pedestrian areas of the TalTech university campus located in Tallinn, Estonia with three main objectives: 1) to quantify the severity of possible thermal stress; 2) to investigate the potential of vegetated surfaces and trees to provide comfort quantifying the required vegetation and optimal layout for the analyzed areas; and 3) to provide the university with support data to use for the redevelopment plan to increase the sustainability, the efficiency and the resilience of the campus. The novelty of the study resides in the scarcity of research investigating thermal discomfort in urban environments during the warm season due to climate change and temperature increase in Tallinn, and in the lack of analysis of the mitigation potential of green areas and trees.

A parametric design workflow was developed to integrate the university buildings and the urban environment three-dimensional models, the climatic and microclimatic data, the wind and surface temperature simulations, and the outdoor comfort assessment through the metrics of Universal Thermal Climate Index and Outdoor Thermal Comfort Autonomy. An automated algorithm, developed for the study, located the vegetated surfaces and the trees in the appropriate quantity and location in the analyzed areas through an iterative process of simulation and analysis. The algorithm considerably shortened the total assessment time for each analyzed area reducing the large number of iterations, each of which is computationally intensive. This allowed the investigation of a larger number of outdoor areas than initially planned.

The results show that during the warmest period of summer the outdoor areas of the TalTech campus in the existing situation present severe pedestrian discomfort conditions. The average UTCI among the analyzed areas was between 28.0 °C and 30.2 °C with maximum values between 38.6 °C and 39.5 °C, which correspond to the moderate, strong and very strong heat stress levels of the UTCI scale. Additionally, the area with the most comfortable condition, presented only 15% of the space in state of comfort according to the Spatial OTCA assessment, while the minimum required is 50%. Providing the areas with vegetated surfaces and trees significantly improved the comfort conditions. Among all areas, using soil and grass instead of the concrete pavement and planting trees in approximately 24% to 49% of the walkable surface facilitates comfortable outdoor spaces as required by the Spatial OTCA.

The study presents two main limitations. The wind simulations were performed for the current situation, thus including only the existing trees surrounding the campus, and not also when the trees were placed in the areas by the automated algorithm. The reason lies in the computational expensive CFD simulations, which took approximately 72 h (4.5 h for each of the 16 wind directions). Ideally, the CFD simulations should have been performed for each tree added, making the workflow impracticable. The second limitation is the lack of consideration of the cooling potential of the trees through the evapotranspiration effect. The bias produced by the first limitation could have been an underestimation of the perceived temperatures because the trees slow the airflow increasing discomfort. Conversely, not taking into account the trees' evapotranspiration could have produced an overestimation of thermal discomfort. Thus, the author argues that the two possible biases balanced each other producing realistic results.

However, future work will be conducted with software capable of taking into account the evapotranspiration of trees and the algorithm will be developed further to include the trees in the wind simulations, optimizing computation time and accuracy. Additionally, the study will be conducted investigating the shading and cooling potential of different tree sizes and species and utilizing additional urban areas in the city of Tallinn to obtain a larger dataset of possible solutions to improve urban resilience.

The outcomes of the study underline on the one hand the efficacy of the mitigation strategies through vegetation to improve urban resiliency in the increasingly uncomfortable summer conditions of Tallinn, and on the other hand the potential of the integration of computational design and environmental simulations to develop methods and solutions for the adaptation of the urban environment to adverse conditions and to allow the designers and researchers to contribute to mitigating climate change. The study will be used to raise the awareness of the Tallinn city planning department of the necessity of adopting mitigation strategies and the need to develop a resilient urban environment.

Acknowledgements. This research was supported by the ZEBE grant 2014–2020.4.01.15–0016 funded by the European Regional Development Fund and by the H2020 grant Finest Twins 856602 funded by the European Commission.

References

1. Global Alliance for Buildings and Construction: 2019 global status report for buildings and construction: Towards a zero-emission, efficient and resilient buildings and construction sector. IEA, UN Environment Programme (2019). https://globalabc.org/resources/publications/2019-global-status-report-buildings-and-construction. Accessed 06 June 2021
2. Intergovernmental Panel on Climate Change (IPCC): Annex I: Atlas of Global and Regional Climate Projections Supplementary Material RCP8.5, in Climate Change 2013: The Physical Science Basis. Working Group I, Bern (2013). https://www.ipcc.ch/site/assets/uploads/2018/07/WGI_AR5.Annex_I_RCP8.5.1.20.14.pdf. Accessed 06 June 2021

3. The European Data Journalism Network (EDJNet). https://www.europeandatajournalism. eu/eng/News/Data-news/Europe-is-getting-warmer-and-it-s-not-looking-like-it-s-going-to-cool-down-anytime-soon. Accessed 06 June 2021
4. Santamouris, M.: Heat-island Effect. In: Santamouris, M. (ed.) Energy and Climate in the Urban Built Environment. Routledge, Abingdon (2011)
5. Eensaar, A.: Temporal and spatial variability of air temperatures in Estonia during 1756–2014. J. Climatology **2016**, 9426791 (2016). Doi: https://doi.org/10.1155/2016/9426791
6. Remmelgas, L.: The Resilience of Tallinn urban landscapes to a changing climate: land surface parameters and their impact on urban heat island effect. Estonian University of Life Sciences, Tartu (2020). http://hdl.handle.net/10492/5801. Accessed 06 June 2021
7. Park, S., Tuller, S.E., Jo, M.: Application of Universal Thermal Climate Index (UTCI) for microclimatic analysis in urban thermal environments. Landsc. Urban Plan. **125**, 146–155 (2014). https://doi.org/10.1016/j.landurbplan.2014.02.014
8. Santamouris, M.: Cooling the cities – a review of reflective and green roof mitigation technologies to fight heat island and improve comfort in urban environments. Sol. Energy **103**, 682–703 (2014). https://doi.org/10.1016/j.solener.2012.07.003
9. Roaf, S., Crichton, D., Nicol, F.: Adapting Buildings and Cities for Climate Change. A 21st Century Survival Guide. 2nd edn. Architectural Press, Oxford (2009)
10. Voll, H., Thalfeldt, M., De Luca, F., Kurnitski, J., Olesk, T.: Urban planning principles of nearly zero-energy residential buildings in Estonia. Manage. Environ. Quality Int. J. **27**(6), 634–648 (2016). https://doi.org/10.1108/MEQ-05-2015-0101
11. Voll, H., De Luca, F., Pavlovas, V.: Analysis of the insolation criteria for nearly-zero energy buildings in Estonia. Sci. Technol. Built Environ. **22**(7), 939–950 (2016). https://doi.org/10.1080/23744731.2016.1195657
12. De Luca, F., Voll, H., Thalfeldt, M.: Horizontal or vertical? Windows' layout selection for shading devices optimization. Manage. Environ. Qual. Int. J. **27**(6), 623–633 (2016). https://doi.org/10.1108/MEQ-05-2015-0102
13. De Luca, F., Dogan, T., Kurnitski, J.: Methodology for determining fenestration ranges for daylight and energy efficiency in Estonia. In: Simulation Series, vol. 50(7), pp. 47–54. 9th Annual Symposium on Simulation for Architecture and Urban Design, SimAUD 2018, Delft, The Netherlands, 5–7 June 2018. SCS, San Diego (CA), USA (2018). https://dl.acm.org/doi/ https://dl.acm.org/doi/10.5555/3289750.3289757
14. De Luca, F.: Sun and wind: Integrated environmental performance analysis for building and pedestrian comfort. In: Simulation Series, vol. 51, Issue 8, pp. 3–10. 10th Annual Symposium on Simulation for Architecture and Urban Design, SimAUD 2019, Atlanta (GE), USA, 7–9 April 2019. SCS, San Diego (CA), USA (2019). https://dl.acm.org/doi/10.5555/3390098.339 0099
15. De Luca, F.: Environmental performance-driven urban design: parametric design method for the integration of daylight and urban comfort analysis in cold climates. In: Lee, J.-H. (ed.) CAAD Futures 2019. CCIS, vol. 1028, pp. 15–31. Springer, Singapore (2019). https://doi.org/10.1007/978-981-13-8410-3_2
16. De Luca, F., Naboni, E., Lobaccaro, G.: Tall buildings cluster form rationalization in a Nordic climate by factoring in indoor-outdoor comfort and energy. Energy Build. **238**, 110831 (2021). https://doi.org/10.1016/j.enbuild.2021.110831
17. Manickathan, L., Defraeye, T., Allegrini, J., Derome, D., Carmeliet, J.: Parametric study of the influence of environmental factors and tree properties on the transpirative cooling effect of trees. Agric. For. Meteorol. **248**, 259–274 (2018). https://doi.org/10.1016/j.agrformet.2017. 10.014
18. Bowler, D.E., Buyung-Ali, L., Knight, T.M., Pullin, A.S.: Urban greening to cool towns and cities: a systematic review of the empirical evidence. Landsc. Urban Plan. **97**, 147–155 (2010). https://doi.org/10.1016/j.landurbplan.2010.05.006

19. Alexandri, E., Jones, P.: Temperature decreases in an urban canyon due to green walls and green roofs in diverse climates. Build. Environ. **43**, 480–493 (2008). https://doi.org/10.1016/j.buildenv.2006.10.055

20. Lobaccaro, G., Acero, G., Sanchez Martinez, A. Padro, A., Laburu, T., Fernandez, G.: Effects of orientations, aspect ratios, pavement materials and vegetation elements on thermal stress inside typical urban canyons. Int. J. Environ. Res. Public Health **16**, 3574 (2019). https://doi.org/10.3390/ijerph16193574

21. Wu, Z., Chen, L.: Optimizing the spatial arrangement of trees in residential neighborhoods for better cooling effects: Integrating modeling with in-situ measurements. Landsc. Urban Plan. **167**, 463–472 (2017). https://doi.org/10.1016/j.landurbplan.2017.07.015

22. Gatto, E., et al.: Impact of urban vegetation on outdoor thermal comfort: comparison between a mediterranean City (Lecce, Italy) and a Northern European City (Lahti, Finland). Forests **11**, 228 (2020). Doi:https://doi.org/10.3390/f11020228

23. Rhinoceros Homepage. https://www.rhino3d.com/. Accessed 06 June 2021

24. Grasshopper Homepage. https://www.grasshopper3d.com/. Accessed 06 June 2021

25. Bueno, B., Norford, L., Hidalgo, J., Pigeon, G.: The urban weather generator. J. Build. Perf. Sim. **6**(4), 269–281 (2013). https://doi.org/10.1080/19401493.2012.718797

26. Kastner, P., Dogan, T.: A cylindrical meshing methodology for annual urban computational fluid dynamics simulations. J. Build. Perform. Simul. **13**(1), 59–68 (2019). https://doi.org/10.1080/19401493.2019.1692906

27. Sadeghipour Roudsari, M., Pak, M.: Ladybug: a parametric environmental plugin for grasshopper to help designers create an environmentally-conscious design. In: Wurtz, E. (ed.) Proceedings of BS2013: 13th Conference of IBPSA, pp. 3128–3135. IBPSA, Chambéry (2013). http://www.ibpsa.org/proceedings/BS2013/p_2499.pdf. Accessed 06 June 2021

28. OpenFOAM Homepage. https://www.openfoam.com/. Accessed 06 June 2021

29. EnergyPlus Homepage. https://energyplus.net/. Accessed 06 June 2021

30. Bröde, P., Jendritzky, G., Fiala, D., Havenith, G.: The universal thermal climate index UTCI in operational use. In: Proceedings of Conference Adapting to Change: New Thinking on Comfort, pp. 1–6. NCEUB, Windsor (2010). http://utci.org/isb/documents/windsor_vers05.pdf. Accessed 06 June 2021

31. Nazarian, N., Acero, J.A., Norford, L.: Outdoor thermal comfort autonomy: performance metrics for climate conscious urban design. Build. Environ. **155**, 145–160 (2019). https://doi.org/10.1016/j.buildenv.2019.03.028

32. Sander, H., Elliku, J., Läänelaid, A., Reisner, V., Reisner, Ü., Rohtla, M., Šestakov, M.: Urban trees of Tallinn, Estonia. Proc. Estonian Acad. Sci. Biol. Ecology **52**(4), 437–452 (2003). https://www.researchgate.net/publication/271965708_Urban_trees_of_Tallinn_Estonia. Accessed 06 June 2021

33. Balakrishnan, P., Jakubiec, J.A.: Measuring light through trees for daylight simulations: a photographic and photometric method. In: Proceedings of Building Simulation and Optimization, pp. 1–8. IBPSA-England, Newcastle (2016). http://www.ibpsa.org/proceedings/BSO2016/p1152.pdf. Accessed 06 June 2021

34. Franke, J., Hellsten, A., Schlünzen, H., Carissimo, B.: Best Practice Guideline for the CFD Simulation of Flows in the Urban Environment. COST, Brussels (2007). https://www.researchgate.net/publication/257762102_Best_Practice_Guideline_for_the_CFD_Simulation_of_Flows_in_the_Urban_Environment_COST_Action_732_Quality_Assurance_and_Improvement_of_Microscale_Meteorological_Models. Accessed 06 June 2021

35. ASHRAE: Standard 55-2013 – Thermal Environmental Conditions for Human Occupancy. ASHRAE, Atlanta (GE), USA (2013). https://www.techstreet.com/ashrae/standards/ashrae-55-2013?gateway_code=ashrae&product_id=1868610. Accessed 06 June 2021

36. Kastner, P., Dogan, T.: Predicting space usage by multi-objective assessment of outdoor thermal comfort around a university campus. In: Chronis, A., Wurzer, G., Lorenz, W.E., Herr, C.M., Pont, U., Cupkova, D., Wainer, G. (eds.) Proceedings of the Symposium on Simulation for Architecture and Urban Design, SimAUD 2020, pp. 91–97. SCS, Vienna (online) (2020). http://simaud.org/2020/preprints/89.pdf. Accessed 06 June 2021

37. Aleksandrowicz, O., Zur, S., Lebendiger, Y., Lerman, Y.: Shade maps for prioritizing municipal microclimatic action in hot climates: learning from Tel Aviv-Yafo. Sustainable Cities Soc. **53**, 101931 (2020). https://doi.org/10.1016/j.scs.2019.101931

AVM Pavilion: A Bio-inspired Integrative Design Project

Kenneth Tracy[1](✉) ⓘ, Mahdi Jandaghimeibodi[2](✉) ⓘ, Sara Aleem[3](✉),
Rahil Gupta[4](✉), and Ying Yi Tan[1](✉) ⓘ

[1] Singapore University of Technology and Design, Singapore 487372, Singapore
`kenneth_tracy@sutd.edu.sg`, `yingyi_tan@alumni.sutd.edu.sg`
[2] MJM Computation, Tehran, Iran
`jm_mahdi@yahoo.com`
[3] IBI Group, Toronto, Canada
`sara.aleem92@gmail.com`
[4] Tara Metal Group, Sharjah, UAE
`rahil_gupta@hotmail.com`

Abstract. The AVM pavilion is a lightweight shading structure inspired by the plant Dipcadi Serotinum and translated into a hybrid, double-layered monocoque shell structure using digital design tools. In this paper, a design-to-fabrication framework is presented to parametrically design, analyze, and manufacture the structure. It also documents the different stages of the project, such as heuristic prototyping, computational design, fabrication, and assembly. The resultant pavilion is a catenary shell discretized into petal-like, porous panels with bioinspired coloration that infill and brace an internal beam network. These components were designed as a kit-of-parts to facilitate manufacture and on-site assembly

Keywords: Bio-inspired design · Computational design · Metal fabrication

1 Introduction

With the rapid development of computational design tools in recent years, there has been an increased drive in realizing freeform structures that have been optimized for structural, environmental, and aesthetical criteria [4, 7]. Some of these designs have taken inspiration from biology, which has led to the development of customized workflows in translating geometries derived from nature into architectural forms [2, 5, 9].

In recent decades, design-build studios have investigated experimental methodologies to derive geometric forms and structural principles from nature or natural processes, convert them into digital models that have been built as full-scaled pavilions. Some examples include the 2011 ICD/ITKE's pavilion that took inspiration from the plate skeletons of sea urchins [8] and Studio One's pavilion originating from a banana leaf stalk [14]. These studios first involved extensive trial-and-error prototyping to understand and test the fabrication feasibility of these bio-inspired geometries and structural principles on a smaller scale. Informed by these physical prototypes, these forms were later converted

D. Gerber et al. (Eds.): CAAD Futures 2021, CCIS 1465, pp. 494–512, 2022.
https://doi.org/10.1007/978-981-19-1280-1_30

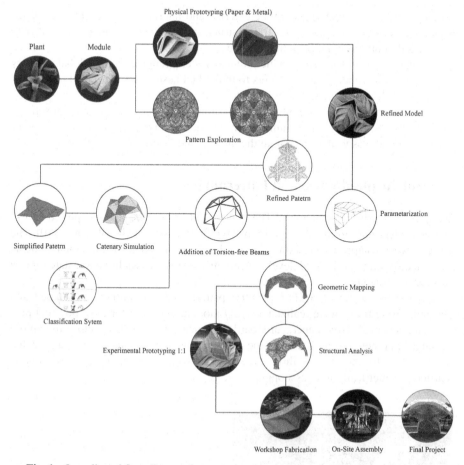

Plant   Module   Physical Prototyping (Paper & Metal)   Refined Model

Pattern Exploration

Refined Patetrn   Parametarization

Simplified Patetrn   Catenary Simulation   Addition of Torsion-free Beams

Classification Sytem   Geometric Mapping

Experimental Prototyping 1:1   Structural Analysis

Workshop Fabrication   On-Site Assembly   Final Project

Fig. 1. Overall workflow diagram from prototyping, computational design to fabrication

into digital models and machining instructions to physically realize their student-driven installations on a one-to-one scale.

The AVM Pavilion[1] positions itself in this domain as a product of a design-build studio. It implemented an experimental workflow to translate the form of a native flower, Dipcadi Serotinum, into a fabricate-able full-scale 30 m^2 pavilion (Fig. 1). The process relied on both heuristic prototyping and parametric design tools to derive the form, from the global massing to the individual panel units, while considering the real-life material constraints of working with folded sheet metal.

This led to the design, analysis, and fabrication of a hybrid monocoque shell and steel frame structure [12]. It consisted of a thin continuous steel frame stiffened by 44

---

[1] In this case, the AVM pavilion loosely alludes to the 18th-century Linnaean classification system of "Animal", "Vegetable" and "Mineral". The name was given on hindsight to the project's completion and will be explained in the conclusion.

discrete panel units which make up the external and internal envelope surfaces. Thus, the integration of both stacked and frame structural typologies stabilize each other to create a self-supporting lightweight structure.

The pavilion was designed as a hierarchical system, ranging from a global catenary massing that was subdivided into facets to individual nested modular panel units. This also allowed the constituent components to be decomposed into a kit-of-parts which facilitated fabrication and assembly of the pavilion. Moreover, an exterior color scheme derived from the Dipcadi Serotinum flower, was added on the panels to create a color bleed effect that was visible from both the interior and exterior of the structure.

2   Plant-Inspired Form and Patterning

Taking reference from Moussavi's "The Function of Form", the studio explored "novel forms and identities through the repetition and differentiation of virtual forms" [10]. The native flower of Dipcadi Serotinum was used as a starting point from which students were encouraged to derive and develop geometric forms that could be translated into a 3D shading structure.

The process began as a 2D geometric pattern that was initially tessellated and repeated. These forms were realized as a 3D modular unit in a series of physical prototype models made from paper, cardboard, and metal (Fig. 4). This helped to explore the range of potential forms that could be built to create the modular panel structure that could be used for the final pavilion. These modules were later modeled in Rhino and adjusted parametrically in Grasshopper.

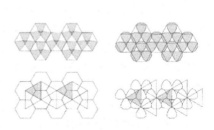

Fig. 2. The Dipcadi Serotinum and 2D geometric patterning

The preliminary massing form was envisioned as a pinwheel formation which was referenced from the global form of the Dipcadi Serotinum flower (Fig. 2). This consisted of three petal legs and its center lines form the edges of the triangular primary frame. These also served as guiding lines to create a faceted triangular mesh which was further subdivided and converted into planar faces (Fig. 3). Modular panels were nested within each triangular face and attached to peripheral framing panels. This created a double-layered envelope skin that was affixed to an internal beam network to provide shear rigidity to the individual modules.

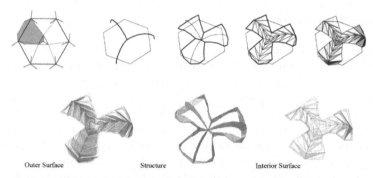

Outer Surface   Structure   Interior Surface

Fig. 3. Development of the pinwheel formation and constituent components

Fig. 4. Folded paper and metal prototype models

3 Computational Design

The modeling of this project utilized digital computational tools to create the overall geometry and its structure, inclusive of its massing, modular panel geometry, and the internal beam network. Grasshopper 3d and its plug-in, the physics engine Kangaroo, were used to form-find the global shape of the pavilion. The modeling sequence first started with design exploration of various massing forms using particle spring simulations. This led to a morphological classification system based on different geometric features. Subsequently, triangular frames were derived from the intended massing mesh, which served as tentative positions for the modular panels. The final positions of the frames were calibrated to optimize structural performance by minimizing stresses in the beams. Finally, the modular panel unit was parameterized to accommodate the intended form.

3.1 Massing Generation: Global Form Finding

The mesh pattern was converted into a particle-spring system [1, 6] and subjected to a reverse gravitational force to generate a stable catenary shell. This digital form-finding approach allowed for the rapid exploration and creation of multiple structurally feasible design solutions that can be adopted as the massing form. Input parameters used in this study included altering the initial mesh face patterns, mesh boundary shape, position of anchor points, different spring target lengths, etc.

Various grid shells were made from a variety of regular and irregular mesh face patterns. Voronoi and other irregular patterns were disregarded because it was possible for each individual mesh face to have a variable number of edges. This discrepancy was inconsistent with the hexagonal tiling pattern developed for the modular aggregation of the unit. Triangular mesh faces were instead chosen because they could be configured within and subdivide hexagonal faces. These also automatically ensure planarity of the mesh faces which reduces the geometric complexity of the panel units and eases fabrication.

The resultant 3d models served as data set for a morphological classification system that helped recognize and group the output forms based on different geometric features. These were sorted into a generational timeline which displays the procedural refinement of the massing form (Fig. 5). The forms were categorized based on mesh face pattern, curvature, symmetry, and boundary layout.

The final form was selected based on several geometric criteria, namely the size of openings, mesh face geometry, and fabrication criteria, such as laser-cut bed restrictions, site dimensions, and budget.

The geometric criteria were based on meeting qualitative aesthetical appearance expectations that the studio had for the pavilion. First, the pavilion had to have adequately tall and wide openings for occupants to enter. These opening sizes also related to views, allowing the pavilion's interior to be visible from most angles.

Additionally, the mesh face geometry contributed to the appearance of the massing models as modular units were to be fitted within the mesh faces. The triangular faces that were either equilateral or close to being equilateral were closer to the intended geometry of the module as compared to faces with disproportionate edge lengths.

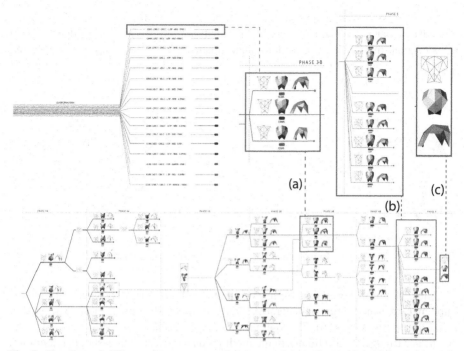

Fig. 5. Timeline of different forms generated using particle-springs: (a) initial selection; (b) refinement of the form using similar input parameters; (c) final selected form

On the other hand, the fabrication criteria were hard quantifiable requirements. This involved making the triangular mesh faces within allowable limits of the laser cutter bed of 2.54 × 1.27 m to reduce the need for partitioning components for fabrication. Site dimensions of shading an area of 30 m^2 and ensuring that the project used material that was within budget were also critical selection criteria.

3.2 Structural Optimization: Minimizing Stresses

The structural challenge of realizing this type of freeform envelope shape comes mainly from the positioning of its internal beams and connecting nodes. In this case, the objective was to rely on thin sheets of metal to create a structurally stable self-supporting frame while maintaining a low self-weight. Thus, it was necessary to rationalize the mesh structures to minimize internal stresses as a result of its geometry.

To increase stiffness, the design of the frame was decided as a two-layer envelope with an internal beam network that connects both inner and outer surfaces. Each modular panel unit was designed to be adequately stiff on its own to reduce shear deformation, which was documented to be more susceptible in single-faced panel units. Moreover, the average width of the panels was more than 1.0 m which further made it imperative to include a second layer for reinforcement.

Thus, this configuration required both the mesh faces and edges to be parallel for both layers to minimize the occurrence of torsion. However, triangular face meshes "do not possess offsets at constant face-face or edge-edge distance" [13]. This means that a triangular mesh that is offset outwards will not have its faces nor edges parallel to its original mesh. This causes non-planar quadratic beam layouts when connecting edges of the inner and the outer mesh. This non-planarity translates to warp at the end cross-sections of the individual beam and induces torsional shear stress to the system.

Beams With Torsion   Torsion Free Beams   Mapped Surfaces

Fig. 6. Implementation of torsion-free beams simulation

Thus, to minimize torsion from the longitudinal axis of the beams, a script was implemented using Kangaroo to iteratively alter the outer edges of the mesh to be parallel to the inner edges (Fig. 6). The optimized network is then converted into structural beams and triangular base frames of the modules on the mesh. A Grasshopper definition references the central plane of the beams to generate the side ribs profiles of each module. The new profiles of main beams are created by combining the edge layout of side ribs which are positioned alongside the longitudinal axis of the main beams (Fig. 7).

■■■■ Sub-beams
■ Main Beams

Fig. 7. New profile of the main beams

3.3 Structural Analysis

The beams and frames were converted into meshes for the Finite Element Analysis model and analyzed to determine the structural performance of the pavilion (Fig. 8). The analysis also determined the thickness of the metal sheets for each type of component.

This analysis was run on the beams only and later was run with the beams and panels. The results demonstrated that it was necessary to integrate the curved internal beams with the outer and inner layers of thin surface panels as both component types supported each other interdependently. This ensured that the overall structure was sufficiently stiff to be both self-supporting and resist wind loads.

From the analysis, the material was assigned as steel sheets and the thicknesses were varied and the model was analyzed to ensure that the structure was within allowable deflection limits and within a factor of safety. The final specifications were 5 mm thickness for the main beams, 3 mm thickness for the sub-beams, and 1 mm thickness for the surface panels.

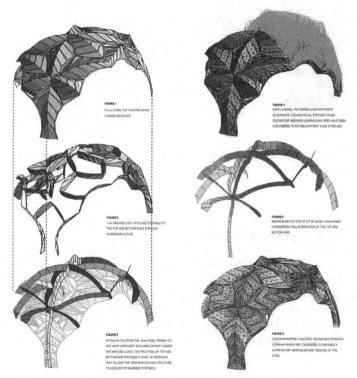

Fig. 8. Finite Element Analysis model reveals that the combination of frame and panels allow the overall structure to be adequately stiff under dead weight and wind loads

3.4 Hierarchical Subdivision System

The next step was to fit the smaller petal-like panel units within the modular panels. These needed to fit within the boundary constraints of its designated module while maintaining planarity to facilitate fabrication.

Thus, a hierarchical subdivision system was conceived to simplify and partition the computational design processes into smaller and more manageable segments. The pavilion's complex geometry was expressed through four levels of a hierarchical subdivision system, nested inside of their parent network, adding geometrical resolution and features within their respective level. These are, from the highest order to lowest: (a) the global massing form that was divided into triangular modules; (b) the modules were divided into three folded units; (c) the folded units were divided into 5 discrete faces and; (d) the faces were further divided into flat polygonal curves to create petal-like surfaces (Fig. 9).

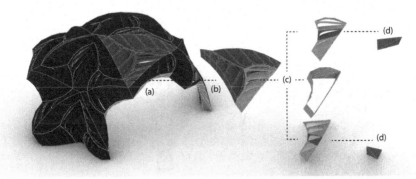

Fig. 9. Hierarchical system: (a) Triangular modules; (b) 3 folded units; (c) discrete faces; (d) polygonal curves.

3.5 Geometric Mapping of Modules

The base module panels were individually modeled by converting a base panel module model into a parametric version by decomposing it into its key components and recomposing the panel based on the geometrical subdivision of the massing surface using an adjustable network of curves and points (Fig. 10). The Anemone plugin was used to map the parametric definitions on all other triangular frames of the massing shape individually, which ensured that the smallest panel units were nested within the input facets.

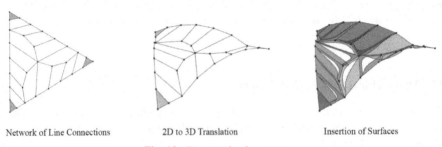

Network of Line Connections   2D to 3D Translation   Insertion of Surfaces

Fig. 10. Parametrization process

Graph mappers were further utilized to manage the numerous parameters and partition them within isolated systems. Calibrate and constraining the range of these parameters could be done easily in a visual manner by adjusting a curve borne from a quadratic equation. This allowed control over the edge profiles of the unit to create desired incremental changes to the geometry of the modules. This change in edge profile curvature further led to the variations of depth and porosity within each panel.

4 Fabrication and Assembly

To build the 1:1 scale pavilion, the knowledge gained from the initial prototyping stage allowed the studio to check for assembly tolerances and develop an assembly sequence. This was further verified as a full-scale prototype module unit to develop the detailing and connections (Fig. 12). The pavilion was segmented into a kit-of-parts to facilitate manufacturing and on-site assembly. Additionally, the modular panels are pasted with vinyl stickers to emulate the color scheme of the pavilion's starting point – the Dipcadi Serotinum flower.

4.1 Learning from Prototyping

The initial prototyping stage involved 50 small-scale folding tests to determine the tolerances and assembly sequence and predict the potential pitfalls of this installation. The later models were made from metal sheets and allowed the students to iterate and re-model the components to minimize possible assembly errors for the full-scale model.

The prototyping phase considered the folding strategy documented in [11], which creates full planar contact between beams and panels to removes kink angles between them. The studio addressed this issue by adding trapezoidal tabs to the edges of the parts that fold inward and positioned face-to-face with the connecting ribs which would then be riveted to hold them in place. This was tested across 12 full-scale prototypes.

Moreover, multiple laser-cut folding patterns were tested to account for discrepancies between digital and physical models caused by folding deformation. These tests led to different distinctive notch patterns which were assigned based on the degree of the bend. This feedback from the physical prototyping also helped with refining and assigning an edge offset value for the laser cut layouts.

4.2 Kit-of-Parts

The two main components, frame beams and panel modules, each was made into a kit-of-parts of smaller manageable parts. This hastened the manufacturing and assembly processes and offered the possibility of disassembling and resituating the pavilion in another location if necessary.

Both main components each had a series of their own group of parts: The 'frame beams' group had a central metal disc as a top connector and metal base plates and anchor bolts. This group also consisted of sub-frames that form the triangular faces within which the panel modules would be affixed to. The panel modules group consisted

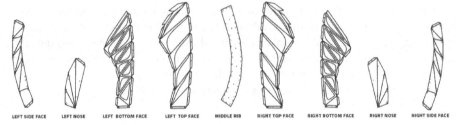

| LEFT SIDE FACE | LEFT NOSE | LEFT BOTTOM FACE | LEFT TOP FACE | MIDDLE RIB | RIGHT TOP FACE | RIGHT BOTTOM FACE | RIGHT NOSE | RIGHT SIDE FACE |

Fig. 11. Unrolled 2-dimensional line drawings of parts

of 44 half-individual modules of which 22 units belonged to the left and right halves and were numbered starting from L1 to L22 and R1 to R22 correspondingly.

The 3D panel modules were unrolled into a 2D strip that was converted into a vector file. Side ribs, tabs, and rivet holes were parametrically added in each strip and sent for laser cutting to obtain the shapes out of sheet metal (Fig. 11). Notch patterns were also added to enable the flat cut-out steel sheet to fold up into its intended 3-dimensional form. These strips were positioned to align the tabs of its adjacent face which can then be riveted together.

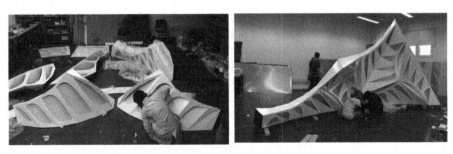

Fig. 12. Fabrication of the full-scale prototype

4.3   Coating and Color Strategies

All components went through a metal surface pretreatment process and were later coated. This pretreatment involved the surfaces and edges being rinsed by a solution of zinc phosphate and later moved to a dry-off oven to remove all the moisture. Since the shade pavilion was to be permanently installed out in the open, the parts were thermoset powder coated which improved their weather resistance compared to a conventional paint finish. The parts were then sprayed with a coat of zinc oxide, pre-heated at a softening temperature of about 80 °C, to prevent the surface from corroding and spread to the final coat, after which the metal parts were sprayed on with the final coat and cured in the oven at 200 °C. The primary color chosen for the final coat was part of the collection RAL Classic, specifically RAL 9010, which is Pure White, to provide a good contrast to the colors of the vinyl stickers pasted on the panel surfaces.

The studio decided to incorporate a color scheme derived from the Dipcadi Serotinum plant (Fig. 13). The color, applied through vinyl stickers, went through 40 iterations of different saturations and hues to create a color bleed effect. This color bleed phenomenon creates a dynamic interplay between the exterior and interior of the pavilion as it interacts with the shadows and light cast on the pavilion's interior. This allows the colors to be visible both from outside and within the pavilion and changes throughout the day based on the angle of the sun (Fig. 14).

Fig. 13. Color study and its implementation on the tiling pattern (Color figure online)

Fig. 14. Coloration effect on interior and exterior of the pavilion (Photo Credit: Michael Hughes) (Color figure online)

5   Site Preparation and Assembly

Construction documents were created for the project starting from foundation drawings leading to a calculated sequence of assembly drawings with all the necessary bracing details predetermining the points of stress where shoring and steel props would be required, which significantly helped reduce any accidents, human error, and time for the site installation.

Fig. 15. Site preparation

5.1 Foundation

Calculations and drawings done by the structural engineers determined the design, layers, and thickness of the ground-bearing slab for the foundation of the shade structure. The designated area was marked and excavated 30 cm below the finish floor level and the perimeter of the excavated part was fitted with formwork to prevent the soil from falling in. The area within the formwork was then leveled and covered with a layer of insulation sheet above which were positioned two layers of welded steel reinforcement mesh (6 mm rods) placed at 10 cm and 20 cm below finish floor level to avoid shrinkage cracking at the surface of the concrete. The concrete was then poured into the formwork and leveled again before the concrete was set in place. For the finished floor, the surface of the concrete was showered with a mixture of handpicked pieces of sandstone, limestone, and gravel before the concrete was set to finally cool, to give it a rough finish rendering our concept of a flower blossoming from the ground.

5.2 Sequence of Assembly

The pavilion was assembled from the base of the legs and installing components upwards until the pinwheel legs met at the top (Fig. 16). This first involved a pair of identical base modules being installed on either side, which holds the front while 4 load-bearing modules, and the back of the structure. Before they were anchored down to the ground, the beams were conjoined in between the loadbearing modules to accurately define the structural lattice. A circular disc was inserted at the top intersection of the beams to stabilize them with a good degree of installation tolerance. This disc was held in place with appropriate shoring.

Next, the triangular sub-frames were installed within which a 3-unit grouped component would fit by reverse bolting the conjoining holes of the parallel surfaces together. Finally, the remaining few units were added individually with the help of metal props that held up the grouped units.

Set up of Main Beams and Propping   Addition of Modules   Assembled Pavilion

Fig. 16. Sequence of assembly

Fig. 17. Assembly site

6 Conclusion

The resultant structure is an amalgamation of design, aesthetic, and contextual character. It serves as a demonstrator of how a pavilion, with its geometry abstracted from a native plant, could be explored formally using scaled models and translated into a structurally stable and fabricate-able model using parametric design tools. This hierarchical system further eased the design of multiple individual panel components while enabling them to be decomposed into a kit-of-parts to facilitate manufacture and assembly. Thus, the result is a lightweight monocoque and frame structure that could be installed on-site by students of the studio.

In hindsight, the pavilion was termed AVM due to subtle connections to "Animal", "Vegetable", and "Mineral" characteristics: "Animal" from its hierarchical patterning of its envelope skin; "Vegetable" from its starting inspiration of a plant; "Mineral" from its materiality. The strangeness of this pavilion allows it to be interpreted individually by its occupants and its effect is not only derived from but also in its representation of the plant and an allusion to the context it is situated in.

6.1 Further Research

Despite the use of an automated workflow, a good deal of manual labor was still required in all three stages of design, computation, and fabrication of the AVM pavilion. There are several directions for further research to improve the pavilion's design and construction workflow.

Transparent ⟶ Opaque

Fig. 18. The transition from transparent to opaque pattern

In terms of computation, recent advances in the field of artificial intelligence could potentially further enhance this workflow by generating fabricate-able geometric forms from image sampling. Instead of manual 2D to 3D translation analyses, inverse graphic generative adversarial networks could be trained to translate images of plants into 3D models. To satisfy metal sheet fabrication constraints, a custom neural network could be developed and trained to map the generated 3d panel models into the subdivided massing. Furthermore, the data set, which was created from the design exploration of the massing models, could be used to investigate the creation of a supervised classification system to predict the massing class of the newly generated models using graph neural networks [3].

The modular computational design strategy of the whole pavilion and partitioned parameters of the individual units have created a parametric setup that can be further tested with dimension-distributed optimization algorithms. Subsequently, the result can be compared with current popular multi-objective optimization workflows.

One potential direction that the studio initially attempted was to create a wall composed of tessellated varied modular units akin to an abstraction of floral petals curling and unfolding. This generated a screen of variable porosity (Fig. 18) where the amount of permissible light can be controlled to create an adjustable sun-shading or ventilation façade system.

This wall could be made kinetic in response to environmental conditions by adding rotary actuators to the metal sheets. The actuators can be used to dynamically adjust the amount of light and airflow passing through the structure by altering the angles of the metal sheets based on real-time changes in the surrounding. These could be linked to environmental sensors that are controlled and calibrated by machine-learning algorithms.

Fig. 19. Interior and exterior photos of the AVM pavilion (Photo Credit: Juan Roldan)

Fig. 19. continued

Acknowledgment. The following students of the AVM pavilion design-build studio are acknowledged for their contribution to the project: Anaisabel Olvera Alacio, Arwa Alnasser, Donna Ashraf, Fay Elmutwalli, Forough Abadian, Mai Elhossiny, Mahdi Jandaghimeibodi, Nada Abushaqra, Nasser Alzayani, Noor Abu Shammalah, Petra Aji, Rahil Gupta, Rita Isshak, Sally Mekhaeil, Sara Aleem, Yara Gamal. The following professionals from Buro Happold engineering are acknowledged for performing structural analyses of the pavilion: Christopher Wodzicki (Director), Richard Cerfontyne (Associate). We would like to thank Juan Roldan (Associate Professor, AUS) for taking photographs of the pavilion. We would like to thank Ammar Kalo (Director of CAAD Labs, AUS), Habib Bitar (Student Volunteer), and Idrees Kanchwala (Student Volunteer) for helping us with the fabrication of the pavilion. We would like to thank Michael Hughes (Professor and former Department Head of Architecture, AUS) for his guidance and support of this project. We would like to thank the College of Architecture, Art and Design (CAAD) of the American University of Sharjah for funding this research project.

References

1. Adriaenssens, S., Ney, L., Bodarwe, E., Williams, C.: Finding the form of an irregular meshed steel and glass shell based on construction constraints. J. Archit. Eng. (2012). https://doi.org/10.1061/(asce)ae.1943-5568.0000074

2. Ahmar, S.E., Fioravanti, A., Hanafi, M.: A methodology for computational architectural design based on biological principles. In: Proceedings of eCAADe 2013, pp. 539–548 (2013)
3. Ahmed, E., et al.: A survey on deep learning advances on different 3D data representations (2018)
4. Bhooshan, S.: Parametric design thinking: a case-study of practice-embedded architectural research. Des. Stud. **52**, 115–143 (2017). https://doi.org/10.1016/j.destud.2017.05.003
5. Dörstelmann, M., Parascho, S., Prado, M., Menges, A., Knippers, J.: Integrative computational design methodologies for modular architectural fiber composite morphologies. In: ACADIA 2014 - Design Agency: Proceedings of the 34th Annual Conference of the Association for Computer Aided Design in Architecture, pp. 219–228 (2014)
6. Kilian, A., Ochsendorf, J.: Particle-spring systems for structural form finding. J. Int. Assoc. Shell Spat. Struct. (147), 77–84 (2005)
7. Kolarevic, B.: Post-digital architecture : towards integrative design. In: First International Conference Crit. Digital What Matter(s)? (2008)
8. La Magna, R., Waimer, F., Knippers, J.: Nature-inspired generation scheme for shell structures. In: Proceedings of the International Symposium of the IASS-APCS Symposium, Seoul, South Korea (2012)
9. Moussavi, F., Lopez, D.: Muqarnas domes. In: Lopez, D., Ambrose, G., Fortunato, B., Ludwig, R., Schricker, A. (eds.) The Function of Form, pp. 326–341. ACTAR, Harvard Graduate School of Design (2009)
10. Louth, H., Reeves, D., Bhooshan, S., Schumacher, P., Koren, B.: A prefabricated dining pavilion. In: Menges, A., et al. (eds.) Fabricate 2017: Rethinking Design and Construction, pp. 58–67. UCL Press, London (2019). https://doi.org/10.2307/j.ctt1n7qkg7.12
11. Poirriez, C., Wortmann, T., Hudson, R., Bouzida, Y.: From complex shape to simple construction: fast track design of "the future of us" gridshell in Singapore. In: IASS Annual Symposium 2016 Spatial Structures 21st Century (2016)
12. Pasternak, H., Krausche, T.: The porsche pavilion in the autostadt Wolfsburg. Germany. In: Procedia Engineering (2013). https://doi.org/10.1016/j.proeng.2013.04.010
13. Pottmann, H., Liu, Y., Wallner, J., Bobenko, A., Wang, W.: Geometry of multi-layer freeform structures for architecture. ACM Trans. Graph. (2007). https://doi.org/10.1145/1276377.1276458
14. Schleicher, S., Kontominas, G., Makker, T., Tatli, I., Yavaribajestani, Y.: Studio one: a new teaching model for exploring bio-inspired design and fabrication. Biomimetics **4**, 34 (2019)

Architectural Automations and Augmentations: Spatial Computing

Learning Geometric Transformations for Parametric Design: An Augmented Reality (AR)-Powered Approach

Zohreh Shaghaghian[1] , Heather Burte[2] , Dezhen Song[3] , and Wei Yan[1](✉)

[1] Department of Architecture, Texas A&M University, College Station, TX 77843, USA
{zohreh-sh,wyan}@tamu.edu
[2] Department of Psychological and Brain Sciences, Texas A&M University, College Station, TX, USA
heather.burte@tamu.edu
[3] Department of Computer Science and Engineering, Texas A&M University, College Station, TX, USA
dzsong@cs.tamu.edu

Abstract. Despite the remarkable development of parametric modeling methods for architectural design, a significant problem still exists, which is the lack of knowledge and skill regarding the professional implementation of parametric design in architectural modeling. Considering the numerous advantages of digital/parametric modeling in rapid prototyping and simulation most instructors encourage students to use digital modeling even from the early stages of design; however, an appropriate context to learn the basics of digital design thinking is rarely provided in architectural pedagogy. This paper presents an educational tool, specifically an Augmented Reality (AR) intervention, to help students understand the fundamental concepts of parametric modeling before diving into complex parametric modeling platforms. The goal of the AR intervention is to illustrate geometric transformation and the associated math functions so that students learn the mathematical logic behind the algorithmic thinking of parametric modeling. We have developed BRICKxAR/T, an educational AR prototype, that intends to help students learn geometric transformations in an immersive spatial AR environment. A LEGO set is used within the AR intervention as a physical manipulative to support physical interaction and improve spatial skill through body gesture.

Keywords: Geometric transformation · Mathematics · Education · Augmented Reality · Parametric modeling

1 Introduction

The superficial knowledge that architectural students have in "digital design thinking" can cause them to adopt excessive trial-and-error approaches when using parametric modeling software, which leads to them not taking full advantages of what the software has to offer [1]. Most students are exposed to parametric modeling software without having the fundamental knowledge of computational design concepts including variables (as

© Springer Nature Singapore Pte Ltd. 2022
D. Gerber et al. (Eds.): CAAD Futures 2021, CCIS 1465, pp. 515–527, 2022.
https://doi.org/10.1007/978-981-19-1280-1_31

the properties of the geometry), parameters (as the members of the function family), and functions (as the math operations). Understanding geometric transformation as one of the essential components in parametric modeling and learning the associated math functions would allow students to improve their knowledge of the mathematical logic behind digital design modeling and utilize parametric modeling software more efficiently and professionally [2]. However, studies have confirmed the difficulty of learning geometric transformations and the related math concepts through common traditional methods [3, 4]. Despite the close relationship between spatial thinking and math problem solving, much of mathematics is still taught in a number sense - a collection of mathematical rules and procedures - in many educational systems, and geometry courses often focus on shape attributes rather than spatial reasoning [5]. Spatial methods can be useful for solving mathematical problems when diagrams, drawings, graphs and conceptualization are applicable [5]. Based on APOS (Action-Process-Objects-Schema) theory [6], students with only an action conception in math knowledge are more likely to use a trial-and-error technique to find function outputs instead of conducting an analysis and in-advance prediction to solve a problem. Such superficial techniques could cause architectural students to face difficulty in properly analyzing and understanding the logical algorithmic process behind sophisticated parametric geometry modeling. Modeling software may not provide a competent context to educate the fundamentals, specifically due to the extraneous cognitive load that GUIs (Graphical User Interfaces) impose on learners [7]. On the other hand, physical model interaction has shown significant impact on spatial visualization and reduction of extraneous cognitive load [8, 9]. Hence, due to the close relationship between spatial reasoning and mathematics, physical interaction could affect learning math concepts positively.

Augmented Reality (AR) as a mediator tool with the ability to superimpose abstract information over the physical environment provides a spatial intervention that could support embodied learning and virtual augmentation. We have developed an AR educational prototype for teaching the fundamentals of parametric modeling including geometric transformations and the associated math functions using graphical elements (e.g., arrows, tags, highlighting). The prototype, named as BRICKxAR/T is developed on top of the prior work - an AR instruction tool for LEGO assembly [10]. We have developed BRICKxAR/T as a Rotation/Translation/Scale (RTS) puzzle game for teaching geometric transformations and the associated math functions within the constructionist learning environment. A tangible LEGO model is employed as a physical manipulative to support physical interactions.

2 Background

2.1 Physical Interaction in Education

In various areas of STEAM (Science, Technology, Engineering, Arts/Architecture, and Mathematics), the effect of physical models on cognitive learning and spatial abilities has been explored. The studies suggest that physical activities increase the creativity of students in design ideation [8, 11], reduce the extraneous cognitive load in the creative design process [12], promote students' spatial skill in understanding scale relations between geometries [13], and improve interaction and communication in a collaborative working environment [14]. Physical interaction supports embodied spatial awareness and helps students in mental visualization skills [15]. Psychological studies argue that physical interaction facilitates the epistemic behavior of students, enables them to form embodied abstract metaphors to internalize the data, and helps them strengthen memory retrieval [16]. Multiple research projects have studied the advantage that physical models have over textbooks or computer-based 3D models [8, 9], revealing the impact of physical model/interaction on spatial cognition and understanding of complex spatial relations. The results of a study exploring the impact of Tangible User Interface (TUI) versus Graphical User Interface (GUI) found that TUI provides more epistemic actions for designers, promotes spatial cognition, and stimulates design creativity, while GUI restricts designers to only following the design briefs and causes less exploration and discovery between design and solution spaces [8].

2.2 Digital and Parametric Modeling in Architectural Education

Digital modeling is an essential tool for effective design generation and design exploration. Digital modeling brings a new paradigm of design thinking named "digital design thinking." Parametric design and its ability to generate multiple design alternatives through anchoring design parameters is one of the most innovative achievements of digital modeling. However, during the digital design thinking process, students need to internalize an understanding of parametric design elements (variables, parameters, and math functions) and algorithmic analytical thinking in order to successfully take the full advantage of a parametric design process. Although the current visual programming platforms in parametric modeling may not require in-depth knowledge of computer language syntax, they urge for a proper understanding of input data (design variables), data structure, function components (parameters and equations), and output data for an efficient process and valid results. The importance of spatial transformations and their mathematics - linear algebra including vectors and matrices - for design professionals who wish to utilize and develop efficient computational techniques has been emphasized clearly in "Essential Mathematics for Computational Design" by Rajaa Issa, 2010 [2], a textbook used widely in courses of parametric modeling and in learning of the modeling tools Rhino/Grasshopper.

2.3 AR in Education

AR applications have been studied in many educational fields, such as physics, mathematics, chemistry, and biomedical sciences, to enhance learning, especially in the situations where students cannot feasibly achieve real-world experiences [17, 18]. In these experiments AR is used as an instrument to embody the interpretation of artifacts or abstractions in the physical world to ease the learning process. The intrinsic capability of AR to automatically align the perspective view with the user's relative position provides a context for viewing the digital model from any arbitrary, natural perspective without mentally translating 2D to 3D [19]. Hence, the automatic natural perspective alignment could reduce the extraneous cognitive load that may impose through interpreting 3D models from 2D desktops. A couple of studies have investigated the applications of AR in learning geometry and related mathematics and demonstrated positive impact of AR intervention in geometry perception [20–22]. The results of the study done by Dünser et al. (2006) showed insignificant, yet positive, impacts from AR intervention on learning geometry and math [20]. Part of the reason for the insignificant impact might be due to user's interaction with the digital model, which was realized through pencils and panels instead of direct interaction. GeoGebra AR [22] is another AR educational tool for learning geometry and algebra based on the widely used GeoGebra application. GeoGebra AR has a limitation of anchoring virtual geometric models to only physical surfaces instead of arbitrary 3D physical objects (in contrast with BRICKxAR/T that enables physical-virtual model registration, tracking, and interplay). Studies show that AR can enhance teaching quality and contribute to architectural design [23, 24], improve the understanding of building science through immersive visualization of building information [25–27], advance the understanding of design sustainability and performance-based design [28–30], and encourage collaboration and engagement among Architecture/Engineering/Construction (AEC) students [26, 31]. The overall results from studies in literature confirm that students' learning effectiveness improves when the related information is situated spatially and temporally close to the real-world experiment.

Most studies in literature have investigated AR's impact in the architectural industry, design ideation, collaboration, and communication. To the best of authors' knowledge, no research has studied the educational impact of an AR intervention in improving the learning of geometric transformations and allied math concepts in order to enhance architecture students' understanding of parametric and digital thinking.

3 Methodology

In this paper, we present an educational AR application to augment abstract, mathematical information found in geometric transformations into a physical environment. The prototype is developed to support three levels of 'motion, mapping and function' in order to create a "network of learning path" for progressive learning of transformation conceptions [32]:

- Motions: AR supports physical actions (using LEGO model as a physical manipulative) to enable embodied learning through a spatial learning experiment.
- Mappings: AR supports visualizing the transformation mapping process through demonstration of the transformation image (physical model) and pre-images (virtual models) with input and output variables (points, lines and surfaces), and transformation parameters (rotation axis, distance, and angle) through displaying of arcs, lines, and notions.
- Functions: AR supports augmentation of mathematical functions of transformation matrices and their multiplications through spatiotemporal alignment of information and physical actions to enhance students understanding of geometric transformations as mathematical functions.

Figure 1 describes the framework of the developed prototype. The AR App is developed for iOS device using the Unity game engine, Apple's ARKit, C# programming, and Xcode. The last section of the framework, i.e. Deploying, is in the process and will be added in a future study.

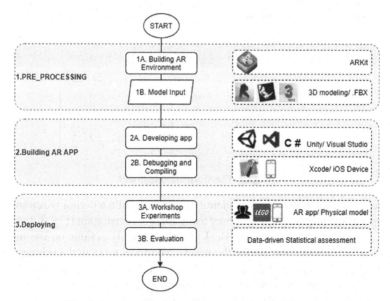

Fig. 1. Proposed workflow for the AR app development

3.1 Prototype

We have developed a primary prototype for an RTS (Rotation, Translation, and Scale) puzzle game in an AR environment. The prototype is presented in Research Symposium 2020, College of Architecture, Texas A&M, and is publicly visible at YouTube [33]. In this prototype, the three levels of geometric transformations, i.e. "motion, mapping, and function," are realized by visualizing pre-images and images of transformations in the AR environment and displaying the associated math functions that are dynamically aligned with transformations conducted in the integrated physical and AR environment. Graphical information such as dimension lines, rotation arcs, coordinate systems, and notations are displayed in the AR environment in real-time to assist users in visualizing math concepts and tracking the transformations matched with the users' perspective in the physical environment. Two rows of matrices displayed in AR demonstrate the math functions associated with the geometric transformations conducted in the physical and AR environment. The matrices are the common 4×4 transformation matrices that could represent all types of linear transformations such as rotation, translation, scale, reflection, and shear [5], as well as perspective transformation. Figure 2 shows the transformation matrices used in this prototype, including translation, rotation, and scale. The blue highlighted region of the 4×4 transformation matrix (column 1–3 and row 1–3) would be filled by any of the 3×3 matrices of rotation/scale, while the translation vector will fill the orange region (column 4 and row 1–3). The result of the combination of transformations will be conducted through matrix multiplication accordingly.

In BRICKxAR/T, the first row of matrices shows a transformation matrix and associated mathematical functions (i.e. linear algebraic operation of matrix multiplication) applied on the physical model in the physical environment in real-time. The second row corresponds to the transformations of the digital model in the AR environment.

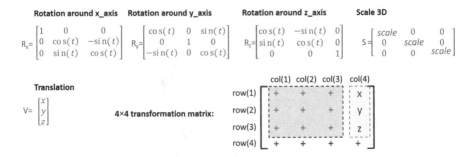

Fig. 2. Transformation matrices

In the beginning, both matrices are identity matrices, while a digital wireframe model superimposes the physical LEGO model using marker registration (Fig. 3. Left). When the user starts to play with the physical model and apply actions on the model, for example, move or rotate the model, the wireframe model stays in the primary position representing the pre-image of the transformation. The tracking graphics (i.e., dimension line, rotation arc, and notations) appear to represent the mapping relation between the pre-image (wireframe model) and the image (physical model). At the same time, the user can observe the mathematical functions of the applied transformations where numbers match the notations displayed on dimension line and rotation arc (Fig. 3. Right) for developing an intuitive understanding of the transformations. This approach generates a meaningful context for abstract mathematical functions aligned with the studies that reveal the benefits of contextualization in presenting mathematics content [34, 35].

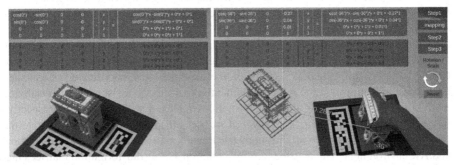

Fig. 3. Left: The wireframe model superimposed on the LEGO model using marker registration. Right: Transforming (translating/rotating) the physical model freely with hand.

Later, students can apply transformations on the virtual model for matching the transformations of the physical model through interacting with the members of the function family using sliders: for example, touching x, y, and z of the translation vector or tapping the transformation logo to apply rotation and scale through modifying the values of the selected parameters. Figure 4 shows the parameters corresponding to the translation operation. The colors of the translation vector parameters match with the associated coordinate system to visually help students in matching the math information to the action.

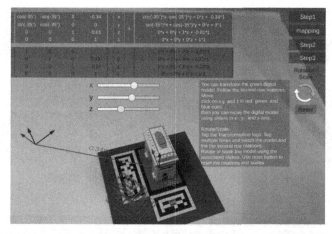

Fig. 4. Interacting with parameters of the matrix function to transform the virtual model.

To apply other transformations, i.e., rotation and scale, students can tap on the transformation logo displayed on the screen. Each time the logo is tapped, the associated transformation matrix and the corresponding slider to change the selected parameter (i.e., rotation angle or scale factor) appear on the screen. For rotation, the corresponding rotation axis and plane are graphically highlighted as well as the color of the logo that get matched with the corresponding rotation axis. These graphical representations intend to catch students' visual attention. Figure 5 shows the graphical information and associated matrices for rotation around x- and y-axis, and Fig. 6 shows the same information for rotation around z-axis and 3D scale. The three coordinate systems in the figures belong to the physical model (image), the original position of the model (pre-image) and the digital model respectively. The last two coordinates are overlapped in Figs. 5, 6 and 7.

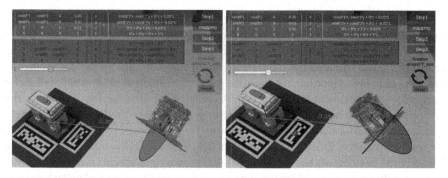

Fig. 5. Left: Rotation around x-axis; Right: Rotation around y-axis

In addition, the spatial mapping concept can be visualized using a set of representative points on the pre-image and their corresponding mapped points on the image (Fig. 7).

In this phase, the students can play with multiple parameters such as translation axes (x, y, and z), rotation axes (x, y, and z), and the scale factor, and trace the associated math

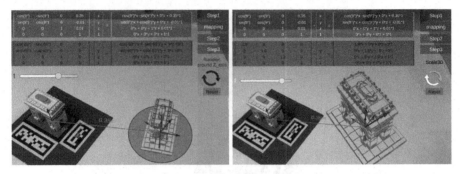

Fig. 6. Left: Rotation around z-axis; Right: 3D scale

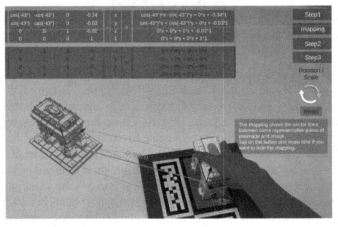

Fig. 7. Visualize mapping operation between pre-image and image using mapped points on the models or in the space.

matrices corresponding to the wireframe model, i.e., the second row of matrices in green color. The task assigned to this part is to practice the AR registration (physical and virtual model alignment) method that is normally done automatically by the AR technology using the camera and motion sensors on the AR device. Students need to modify function parameters to implement the correct transformations so that the two models (virtual and physical) align again. In order to exactly match the two models, students need to compose the transformation matrix of the virtual model (second row) so it matches with the transformation matrix of the physical model (first row) in the transformed position. The original position of the LEGO model (the primary pre-image) is displayed with a world coordinate system in this prototype where the model is registered in the beginning. Figure 8 shows from a third-person view a student playing with BRICKxAR/T using an iPad with AR enabled.

Fig. 8. Student playing with BRICKxAR/T through an iPad

Figure 9 shows screen shots of the App where the student is practicing AR registration.

Fig. 9. Practicing AR registration through playing with function parameters

4   Discussion

The AR-enabled RTS prototypes intend to support architecture students learning parametric modeling, specifically geometric transformations and allied, essential mathematical concepts [3]. During the RTS tasks, spatial relations help students intuitively understand the math concepts behind geometric transformations. Some examples of the relations between spatial reasoning and math in the RTS tasks are described as follows:

- The spatial rotation in different directions (clockwise or counterclockwise) is associated with and displayed through the - and + signs in math. The arc graphically visualizes the rotation quantity with the corresponding degree notation.
- The measuring line graphically demonstrates the distance between the images of the transformations (pre-image and image). The distance notation also numerically matches with the physical distance. For example, when the distance is doubled, the measuring line and the distance notation demonstrate a doubled value.

- The transformation matrices numerically match with the graphical information of the transformation dynamically displayed through lines and arcs. The linear algebraic equations corresponding to matrix multiplication also numerically match with the graphical information of the corresponding transformations.

We expect that providing a spatial context for mathematical information associated with geometric transformations will positively help students in spatial cognition associated with geometric transformations and mathematics. We hope that such spatial contextualization help students understand math subjects such as matrices as an integrated procedure with spatial motions rather than solely numbers and equations. This positive effect may be most pronounced in students with medium and low spatial skills because this intervention may help them in constructing visual imagery of transformation matrices that could lead to understanding the math concepts more effectively in an intuitive way. Ultimately, we anticipate that this intervention may boost students' understanding of parametric modeling methods and use of the software tools accordingly.

5   Conclusion and Future Work

In this paper, we have presented an AR educational prototype, named BRICKxAR/T, to improve learning fundamentals of parametric modeling. The goal of the BRICKxAR/T is to assist students in learning geometric transformations and their associated math functions as one of the essential components of parametric modeling. In this paper, we broke down the geometric transformation process into three levels of 'motion, mapping, and function' to help students understand the mathematical functions behind transformation actions. The AR educational prototype is designed to help students learn about digital modeling components named as variables, parameters, and functions through a transformation puzzle game. We believe that learning the fundamentals of parametric modeling and understanding the mathematical reasoning behind "digital design thinking" in an intuitive way may positively impact students in the professional use of parametric modeling methods for architectural design. We are going to evaluate the impact of BRICKxAR/T on students' spatial and math skill learning in future studies upon approved IRB (number: IRB2020-1213M). We are going to hold workshops and conduct user studies to compare experimental and control groups, using AR and non-AR learning methods, respectively. We expect that contextualizing math functions in a spatial and physical scenario and visualizing the related information and notations may help students improve their understanding of geometric transformations as mappings and functions rather than motions. Also, we believe that BRICKxAR/T may reduce the mental load in learning geometric transformations and increase students' motivation in learning the targeted subject. In future studies, we are going to conduct multiple tests and surveys, including Purdue Visualization of Rotations Test [36], math tests on transformation matrices, NASA_TLX survey [37], and motivation surveys [38] to assess our claims.

Acknowledgements. This material is based upon work supported by the National Science Foundation under Grant No. 2119549 and Texas A&M University's PTTG and Innovation [X] grants.

References

1. Dino, İ.G.: Creative design exploration by parametric generative systems in architecture. 1, 207–224 (2012). https://doi.org/10.4305/METU.JFA.2012.1.12
2. Rajaa Issa: Essential Mathematics for Computational Design - Third Edition (2010)
3. Hollebrands, K.F.: High school student's understanding of geometric transformation in the context of a technological environment. J. Math. Behav. **22**, 55–72 (2003). https://doi.org/10.1016/S0732-3123(03)00004-X
4. Gülkılıka, H., Uğurlub, H.H., Yürükc, N.: Examining students' mathematical understanding of geometric transformations using the pirie-kieren model. Kuram ve Uygulamada Egit. Bilim. **15**, 1531–1548 (2015). https://doi.org/10.12738/estp.2015.6.0056
5. Wheatley, G.H.: Spatial sense and mathematics learning. Natl. Counc. Teach. Math. **37**, 10–11 (1990)
6. Dubinsky, E., Harel, G.: The nature of the process conception of function (1992)
7. Tang, A., Owen, C., Biocca, F., Mou, W.: Comparative effectiveness of augmented reality in object assembly. In: Conf. Hum. Factors Comput. Syst. - Proc., 73–80 (2003). https://doi.org/10.1145/642625.642626
8. Kim, M.J., Maher, M.: Lou: The impact of tangible user interfaces on designers' spatial cognition. Hum. Comput. Interact. **23**, 101–137 (2008). https://doi.org/10.1080/0737020802016415
9. Preece, D., Williams, S.B., Lam, R., Weller, R.: "Let's Get Physical": advantages of a physical model over 3D computer models and textbooks in learning imaging anatomy. Anat. Sci. Educ. **6**, 216–224 (2013). https://doi.org/10.1002/ase.1345
10. Yan, W.: Augmented reality instructions for construction toys enabled by accurate model registration and realistic object/hand occlusions. Virtual Reality (2021). https://doi.org/10.1007/s10055-021-00582-7
11. Viswanathan, V.K., Linsey, J.S.: Physical models and design thinking: a study of functionality, novelty and variety of ideas. J. Mech. Des. Trans. ASME. **134**, 1–13 (2012). https://doi.org/10.1115/1.4007148
12. Chandrasekera, T., Yoon, S.Y.: The effect of tangible user interfaces on cognitive load in the creative design process. In: Proc. 2015 IEEE Int. Symp. Mix. Augment. Real. - Media, Art, Soc. Sci. Humanit. Des. ISMAR-MASH'D 2015, pp. 6–8 (2015). https://doi.org/10.1109/ISMAR-MASHD.2015.18
13. Sun, L., Fukuda, T., Tokuhara, T., Yabuki, N.: Differences in spatial understanding between physical and virtual models. Front. Archit. Res. **3**, 28–35 (2014). https://doi.org/10.1016/j.foar.2013.11.005
14. Seichter, H.: Augmented reality and tangible interfaces in collaborative urban design. Comput. Archit. Des. Futur. **2007**, 3–16 (2007). https://doi.org/10.1007/978-1-4020-6528-6_1
15. Ha, O., Fang, N.: Development of Interactive 3D Tangible Models as Teaching Aids to Improve Students ' Spatial Ability in STEM Education, pp. 8–10 (2013)
16. Bujak, K.R., Radu, I., Catrambone, R., MacIntyre, B., Zheng, R., Golubski, G.: A psychological perspective on augmented reality in the mathematics classroom. Comput. Educ. **68**, 536–544 (2013). https://doi.org/10.1016/j.compedu.2013.02.017
17. Kaufmann, H., Schmalstieg, D.: Mathematics and geometry education with collaborative augmented reality. Comput. Graph. **27**, 339–345 (2003). https://doi.org/10.1016/S0097-8493(03)00028-1
18. Taçgin, Z., Uluçay, N., Özüağ, E.: Designing and Developing an Augmented Reality Application: A Sample Of Chemistry Education. Turkiye Kim. Dern. Derg. Kisim C Kim. Egit. 1, 147–164 (2016)

19. Shaghaghian, Z., Yan, W., Song, D.: Towards Learning Geometric Transformations through Play: an AR-powered approach. In: Proc. 2021 5th Int. ICVARS Conf. Virtual Augment. Simulations (2021). https://doi.org/10.1145/3463914.3463915

20. Dünser, A., Steinbügl, K., Kaufmann, H., Glück, J.: Virtual and augmented reality as spatial ability training tools. ACM Int. Conf. Proceeding Ser. **158**, 125–132 (2006). https://doi.org/10.1145/1152760.1152776

21. Martin, J., Saorín, J., Contero, M., Alcaniz, M., Perez-lopez, D.C., Mario, O.: Design and validation of an augmented book for spatial abilities development in engineering students. Comput. Graph. **34**, 77–91 (2010). https://doi.org/10.1016/j.cag.2009.11.003

22. Khalil, M., Farooq, R.A., Çakiroglu, E., Khalil, U., Khan, D.M.: The development of mathematical achievement in analytic geometry of grade-12 students through GeoGebra activities. Eurasia J. Math. Sci. Technol. Educ. **14**, 1453–1463 (2018). https://doi.org/10.29333/ejmste/83681

23. Wang, X.: Experiential mixed reality learning environments for design education. 272–277 (2007)

24. Keshavarzi, M., Parikh, A., Zhai, X., Caldas, L.: SceneGen : Generative Contextual Scene Augmentation using Scene Graph Priors. arXiv Prepr. arXiv2009.12395 (2020). https://doi.org/10.1145/nnnnnnn.nnnnnnn

25. Messadi, T., Newman, W.E., Fredrick, D., Costello, C., Cole, K.: Augmented Reality as Cyber-Innovation in STEM Education. J. Adv. Educ. Res. **4**, 40–51 (2019). https://doi.org/10.22606/jaer.2019.42002

26. Vassigh, S., et al.: Teaching building sciences in immersive environments: a prototype design, implementation, and assessment. Int. J. Constr. Educ. Res. **00**, 1–17 (2018). https://doi.org/10.1080/15578771.2018.1525445

27. Ashour, Z., Yan, W.: BIM-Powered Augmented Reality for Advancing Human-Building Interaction. In: Proceedings of the 38th eCAADe Conference, TU Berlin, Berlin, Germany, vol. 1, pp. 169–178 (2020)

28. Ayer, S.K., Messner, J.I., Anumba, C.J.: Augmented reality gaming in sustainable design education. J. Archit. Eng. **22**, 1–8 (2016). https://doi.org/10.1061/(ASCE)AE.1943-5568.0000195

29. Birt, J., Manyuru, P., Nelson, J.: Using virtual and augmented reality to study architectural lighting. ASCILITE 2017 - Conf. Proc. - 34th Int. Conf. Innov. Pract. Res. Use Educ. Technol. Tert. Educ. 17–21 (2019)

30. Zhao, S., Zhang, L., DeAngelis, E.: Using augmented reality and mixed reality to interpret design choices of high-performance buildings. Proc. 2019 Eur. Conf. Comput. Constr. 1, 435–441 (2019). https://doi.org/10.35490/ec3.2019.142

31. Figen Gül, L., Halıcı, S.M.: Collaborative design with mobile augmented reality. Complex. Simplicity - Proc. 34th eCAADe Conf. 1, pp. 493–500 (2015)

32. Fife, J.H., James, K., Bauer, M.: A learning progression for geometric transformations. ETS Res. Rep. Ser. **2019**, 1–16 (2019). https://doi.org/10.1002/ets2.12236

33. BRICKxAR/T, an educational AR prototype to Learn Geometric Transformation through play, demo video. https://www.youtube.com/watch?v=igKIvhOspEw&t=81s&ab_channel=BRICKxAR

34. Casey, B., Kersh, J.E., Young, J.M.: Storytelling sagas: an effective medium for teaching early childhood mathematics. Early Child. Res. Q. **19**, 167–172 (2004). https://doi.org/10.1016/j.ecresq.2004.01.011

35. Casey, B.M., Andrews, N., Schindler, H., Kersh, J.E., Samper, A., Copley, J.: The development of spatial skills through interventions involving block building activities. Cogn. Instr. **26**, 269–309 (2008). https://doi.org/10.1080/07370000802177177

36. Bonder, G.M., Guay, R.B.: The purdue visualization of rotations test. Chem. Educ. **2**, 1–17 (1997). https://doi.org/10.1007/s00897970138a
37. Hart, S.: Human Performance Research Group. https://humansystems.arc.nasa.gov/groups/TLX/
38. Pintrich, P.R., Groot, E.V.D.: Motivational and Self-Regulated Learning Components of Classroom Academic Performance. **82**, 33–40 (1990)

Designing with Your Eyes

Active and Two-Way Employment of Our Visual Sense

Cameron Wells, Marc Aurel Schnabel[(✉)] ⓘ, Andre Brown ⓘ, and Tane Moleta ⓘ

Victoria University of Wellington, Wellington 6140, New Zealand
MarcAurel.Schnabel@vuw.ac.nz

Abstract. Our eyes consume the visual field around us, and the brain processes that information as sight and visual content. By incorporating eye-tracking into the design process, we can in some way reverse that process and begin to rethink the role our eye-brain interaction plays instead of as a method of active external input rather than internal processing. Our research developed a prototype application, the Eye-Tracking Voxel Environment Sculptor (EVES), that incorporates eye-tracking as the designed actuator, extending our visual sense as an active design generator. The eye-tracking data garnered from the designer when interacting in EVES is directly utilized as an input within a modelling virtual environment to manipulate and sculpt voxels. We have tested how eye-tracking offers novel possibilities as a Human-Computer Interface within the Virtual Reality Aided Design (VRAD) realm and presents such methodology's potentials.

Keywords: Human-Computer Interface (HCI) · Eye-tracking · Virtual reality · Modelling · Sketching

1 Introduction

Using active eye-tracking within architectural space design, we employ Virtual Reality (VR) unconventionally in this body of research. Carreiro and Pinto [1] noted that through VR, the visual representation of architecture could facilitate a more robust understanding of (virtual) spaces. However, by combining VR and eye-tracking, we propose a novel method to extend the Human-Computer Interface (HCI) of Virtual Reality Aided Design (VRAD), extending the research of Donath and Regenbrecht [2].

Up to date, research has focused on eye-tracking only as a passive or investigative tool. Studies that objectively analyze how we perceive various types of information and how we build assessments of our surroundings [3] discuss the conventional implementation of eye-tracking. However, no investigations have been made exploring the possibility of using eye-tracking technology as an active generation technology within a 3D space. Our research presents the development of a novel prototype design tool, the Eye-Tracking Voxel Environment Sculptor (EVES), which explores how our eyes can be active, enabling generators and receptors in the design process.

© Springer Nature Singapore Pte Ltd. 2022
D. Gerber et al. (Eds.): CAAD Futures 2021, CCIS 1465, pp. 528–537, 2022.
https://doi.org/10.1007/978-981-19-1280-1_32

1.1 Passive and Active Eye-Tracking

To begin with, we have categorized eye-tracking into two different implementations, Passive and Active. The Passive implementation happens 'within' the design process approach. It is therefore used solely as a method that does not manipulate the designed environment. Instead, it can simply produce overlaid layers of information that informs the designer or investigator about visual attention. On the other hand, we propose an Active implementation, utilized 'as' part of the design process, where eye-tracking is used to manipulate the experienced environment directly. While both use similar technologies, the applications are distinct from each other. This paper and the development of EVES explores the potential variety and limits of eye-tracking within the design process.

1.2 Gaze Heatmaps

Our investigation into the implementations of eye-tracking within VR environments began initially with the generation of 'gaze-heatmaps' within Virtual Environments (VE). By analyzing where, when, and what a person is looking at within an environment, a designer can make design decisions and amendments retrospectively (Fig. 1).

Fig. 1. Gaze-heatmap using voxel generations for spatial analysis

When eye-tracking is incorporated in the design process as a passive method, a designer can track users' visual experience within any given space. By knowing what experiencers of space users are visually drawn to, a designer can reliably validate and evaluate their behaviours within the architectural realm [4]. This passive method enables designers to confirm how their schemes are interpreted and thus facilitate a connection between users and designers through VR and Mixed Reality (XR) interfaces. The data gained from the eye-tracking (as well as head- and movement-tracking) highlights visually active areas to allow the designers to revise or redesign if applicable.

1.3 EVES

We realized the potential contribution eye-tracking could make to the design process as a real-time active design generation instrument from these voxel-based heatmaps.

When eye-tracking is incorporated in the design process as an active method, we can understand how to use our eyes and extend their VE capabilities. The development of EVES investigated active eye-tracking in two ways: how the eye can influence a virtual experience via the control and navigation of the Virtual Reality User Interface (VRUI) and how vision (as sightlines) can shape and mould the space a user inhabits.

EVES is a VR design instrument that utilizes eye-tracking as the input for a voxel-based modelling environment. The voxels allow for ease of sculpting and simplify the use of EVES as much as possible to allow users to learn how to use it quickly. Implementation of eye-tracking has been utilized within EVES as extensively as possible to understand its capabilities. In addition to the modelling input being controlled by the users' gaze, so is tool/brush selection and User Interface (UI) navigation. In general, bar some other features such as slight head and hand movements, EVES is controlled primarily via the users' gaze data within the immersive VE.

1.4   Precedents

The development of EVES can be broken down into two parts. The active use of eye-tracking and the essential application as a VR Design Tool. As such, these categories form the outline for which EVES takes a reference. The first part focuses more so on artists and their works that have experimented with Active Eye-Tracking before. In contrast, the VR Design Tools section explores existing instruments that share Human-Computer Interface (HCI) traits and other operational similarities to EVES.

Active Eye-Tracking. In "Active Vision: Controlling Sound with Eye Movement", Polli [5] explores the use of active eye-tracking within music and notes on her ability to" control her eye in an exact way to create specific sounds". She exemplifies humans' ability to utilize their eyes in a non-consummatory manner (despite the differences between auditory and visual art) by shifting the traditional method of melodic input from mouth and hand to that of the gaze. Farahi [6] explores the usage of this idea of active vision through her project Caress of the Gaze, whereby eye-tracking is implemented to manipulate a garment worn by a person whenever an observer looks at it.

In the realm of drawing, artists Graham Fink and Sarah Ezekiel have both utilized eye-tracking as a direct method of input onto a digital canvas. While Fink explores the medium as an opportunity for 'purer' artistic expression directly from the subconscious to the canvas [7], Ezekiel's exploration comes from necessity. Ezekiel is diagnosed with ALS (otherwise known as Motor Neuron Disease), severely inhibiting her ability to use her arms and body [8]. While these technologies allow for increased interaction between humans and computers, they also open gateways for disadvantaged people with disabilities. In "Active Vision: Controlling Sound with Eye Movement", Polli [5] explores the use of active eye-tracking within music and notes on her ability to "control her eye in an exact way to create specific sounds". She exemplifies humans' ability to utilize their eyes in a non-consummatory manner (despite the differences between auditory and visual art) by shifting the traditional method of melodic input from mouth and hand to that of the gaze. Farahi [6] explores the usage of this idea of active vision through her project Caress of the Gaze, whereby eye-tracking is implemented to manipulate a garment worn by a person whenever an observer looks at it.

In the realm of drawing, artists Graham Fink and Sarah Ezekiel have both utilized eye-tracking as a direct method of input onto a digital canvas. While Fink explores the medium as an opportunity for 'purer' artistic expression directly from the subconscious to the canvas [7], Ezekiel's exploration comes from necessity. Ezekiel is diagnosed with ALS (otherwise known as Motor Neuron Disease), severely inhibiting her ability to use her arms and body [8]. While these technologies allow for increased interaction between humans and computers, they also open gateways for those disadvantaged with disabilities [9].

Virtual Reality Aided Design (VRAD) Tools. The concepts behind VRAD and their supporting technologies greatly influence the fields of architecture, design, and construction [8]. Early VR programs like 'HoloSketch' and 'DDDoolz' have explored design tools' usability through many different facets, such as HCI and modelling input. More recent examples like those of 'SculptVR', 'Tilt Brush' and 'Gravity Sketch' also explore various means of modelling and designing within VEs.

'HoloSketch', while not utilizing a Head-Mounted Display (HMD) like more modern examples of VR, explores the ability to interact and design with virtual 3D objects in front of a user. Through a 'wand', these forms are visualized on a screen [10]. DDDoolz [11], while similarly limited in comparison to modern VR technology, also investigated the applicability of VR within early design stages through the use of voxel-like massing. Concerning more modern implementations of VR design tools, SculptVR takes a similar approach to modelling as DDDoolz with its integration of voxel and marching cubes. Other recent VR programs have different techniques, such as Tilt Brush's planar stroke-based modelling and Gravity Sketch's NURBS-based modelling [12 The concepts behind VRAD and their supporting technologies greatly influence the fields of architecture, design, and construction [10]. Early VR programs like 'HoloSketch' and 'DDDoolz' have explored design tools' usability through many different facets, such as HCI and modelling input. More recent examples like 'SculptVR', 'Tilt Brush' and 'Gravity Sketch' also explore various means of modelling and designing within VEs.

'HoloSketch', while not utilizing a Head-Mounted Display (HMD) like more modern examples of VR, explores the ability to interact and design with virtual 3D objects in front of a user. Through a 'wand', these forms are visualized on a screen [11]. DDDoolz [12], while similarly limited in comparison to modern VR technology, also investigated the applicability of VR within early design stages through the use of voxel-like massing. Concerning more modern implementations of VR design tools, SculptVR takes a similar approach to modelling as DDDoolz with its integration of voxel and marching cubes. Other recent VR programs have different techniques, such as Tilt Brush's planar stroke-based- and Gravity Sketch's NURBS-based modelling [13].

1.5 Development Environment

Through EVES development, two leading eye-tracking hardware were used, the *Tobii Eye-Tracker 4C* and *HTC's Vive Pro Eye*. The Tobii Eye-Tracker 4C is a monitor-mounted bar eye-tracker that gathers the users' gaze data as they look at the screen. HTC's Vive Pro Eye, on the other hand, is an entirely in-built eye-tracker within the VR HMD. EVES was developed within the *Unity3D* game engine. The eye-tracking was implemented via HTC Vive and Tobii Software Development Kits (SDK's) for Unity3D.

2   Eye-Tracking 'Within' the Design Process (Passive)

While not intrinsically related to EVES development, it is still salutary to explore eye-tracking within VRAD. Our research's intermediary program employing eye-tracking to generate voxel heatmaps consists of two phases: i) the Recording Phase and ii) the Analysis Phase.

The Recording Phase involves a user or client experiencing a space or environment within VR. During this time, the program gathers and stores their visual experience within an Octree data structure. Almost everything is recorded, from where, when, what and when, to how long a user looks at objects within the environment. The Analysis Phase then utilizes this data recorded within the Recording Phase to generate a 3D voxel heatmap of the users' visual experience (Fig. 2). In this phase, the user or designer can freely move around the environment to analyze the generated heatmap. Information about where and how long a user has looked at particular places is visualized as the heatmap. In addition, information such as what is being looked at (along with other specifics) is displayed by hovering over individual voxels.

Fig. 2. Within the "Analysis Phase" - displaying voxels and direct gaze rays.

3   Eye-Tracking 'As' the Design Process (Active)

3.1   Voxel Sculpting

At its core, EVES is essentially a voxel modelling tool controlled by the eye within VR. Voxels offer the opportunity for a user to sculpt as they might clay with their hands. This method allows for rough means of input without the user's overwhelming need to make precise modifications to their models. Further, despite their simplicity, voxels are capable of creating complex wholes due to their modular nature. Voxels can be generated and erased as well as modified via colour (Fig. 3).

Other sculpting methods within the digital realm were also considered, such as marching cubes. Although marching cubes essentially work in the same way as voxels and boast smoother forms, the process tends to require a certain amount of precision from the user. This precision is needed when sculpting results in small meshes. It would be problematic to manipulate and often make modelling difficult and tedious in our instantiation (Fig. 4).

Fig. 3. Voxels in EVES.

Fig. 4. Marching Cubes in EVES with small tedious meshes.

3.2   Methods of Input

Active Eye-Tracking and HCI. EVES utilizes eye-tracking similarly to how both Fink and Ezekiel draw with their eyes [8]. However, rather than drawing on a 2-dimensional canvas, we extend the application to a 3D VR environment. Within EVES, a ray is projected into the VR environment. This ray is informed by the gaze data from the eye-tracking hardware. It is then utilized as the input method to sculpt voxels and navigate in-program menus. Wherever the user looks within the environment, a guide-cursor follows to indicate where they will sculpt. The guide-cursor takes the form of whatever brush currently is selected, and by pressing the trigger on the controller, voxels are sculpted and manipulated.

As previously mentioned, in addition to gaze data being utilized for direct sculpting input, it is also the primary UI navigation and tool/brush selection method. In EVES, the controller is used as an anchor point for the UI. When looking at the controller, a user can focus on any icon in the menus to select it. Indicators display whether the icon is being focused on or if it has been chosen. Interaction with the controller and the user's hand is limited as much as possible to allow for simple operation. With the hand-controlled interaction bound to the thumb, the user controls simple values such as brush size and distance. With this operation method, the user does not have to continually look back to the UI, reducing interruptions to the primary modelling input.

Brushes and Tools. Brushes and tools in EVES are simply the differing ways a user's gaze generates and manipulates voxels. While brushes act as the base shapes that voxels are generated in (such as spheres and boxes (Fig. 5 top)), tools alter how brushes act (such as single, line, and erasure (Fig. 5 bottom)). Both brushes and tools allow for various uses that can be used in different ways to generate forms and environments through simple inputs. Like the 'Hollow' variants and 'MorphBox', other brushes grant the ability to generate more complex spatial forms with relative ease. MorphBox specifically allows for multiple uses in and of itself. For example, walls, roofs, floors, windows, columns, and more can be generated via a two-point input system whereby the user defines a box's extent by selecting two opposing corners.

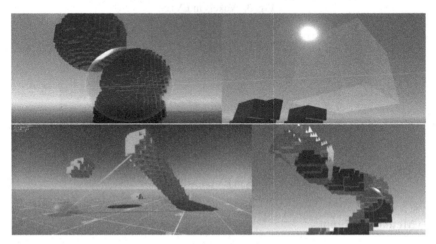

Fig. 5. Sphere brush (top-left) and box brush (top-right), line tool (bottom-left) and curved line tool/paint tool (bottom-right).

4   Results

The development of EVES prompted further understanding of the potential for eye-tracking as a Human-Computer Interface within the realm of digital architecture and design. As explained, Eye-tracking's usability can play various roles within or as the design process. Where Passive incorporation covers a more grounded approach, Active incorporation explores the more radical and artistic. By investigating the possibilities that eye-tracking can offer in these broader senses, we can understand how it may affect and benefit design.

4.1   The Passive ('Within' the Design) Approach

We analyzed points of visual interest, visual experience, and behaviours. In particular, one space that we tested was an intersection of two roads in an urban context – with plenty of distractions and potential hazards. We found that typically visual attractions tend to be moving objects, colours and shapes that stood out from the rest of the scene, such as cars, traffic lights, trees, and advertisements. They all tended to take up more visual interest than that any architectural feature in particular. Although the scene favoured these aspects regarding its specific setting, the passive analysis of eye-tracking highlighted these visual experiences. Here, eye-tracking is helpful, yet it does not add anything to the process of designing the scene in the first place.

4.2 The Active ('As' the Design) Approach

When eye-tracking is incorporated in the design process as an active method, we can understand how to use our VR/XR eyes and extend their VE capabilities. The development of EVES investigated active eye-tracking in two ways: how the eye can influence a virtual experience via the control and navigation of the VRUI and how vision can shape and mould the space a user inhabits.

Controlling or navigating anything with the eyes, whether it be the VRUI or the act of sculpting voxels, tends to take a bit of practice. It is because our bodies often run on performative or procedural movements; in other words, 'muscle memory' [14]. Arguably, the act of looking and using our eyes would fall under this idea of effortless, spontaneous skill performance. However, shifting the eye's usage to manual control can offer a slightly steeper learning curve. For example, when inputting the two points to draw a line, one would often look towards the following action before completing the previous. This action would then cause the line to be generated so that we did not intend. In contrast, the VR UI operation tended to be the easiest obstacle to overcome, as it relied primarily on the user's existing muscle memory. Despite this, we found that once learning how to control our eyes, the operation of EVES became more instinctive and streamlined.

Despite the eventual ability to control EVES, there remained a relative warping of perceived space and environments when modelling. These distortions would often manifest through the shape and scale of modelled architectures. In "Vision and Touch", Rock and Harris discuss vision's influence over the sense of touch. This dominance over touch is present even if the eye is fed distorted images. Rock and Harris [15] note an observation by James J. Gibson whereby a subject runs their hand along a straight rod while looking through a prism. Despite the rod's linear form, the participant was said to have felt it as curved. In this instance, the subject's vision overruled the touch of their hand, in contrast to reality. Law et al. [16] discuss a scalar distortion when designing VR spaces whereby subjects are often more precise yet less accurate. Often favouring the space to be much smaller (albeit more consistently) than it was when compared to other traditional sketching methods.

In respect to EVES, vision and touch are inherently blended. Vision is the touch. Therefore, when considering the visual inconsistencies of designing within VR, such as those discussed by Law et al., it is not surprising that the users' perceived reality is modelled into a true reality. EVES takes in the user's perceived shapes and scales and models them accordingly, not necessarily in a way we may consider accurate. Consequently, this creates the architectures and spaces that our eyes inhabit, not our bodies. The three dimensions of reality are translated into our minds as a 2D plane and reinterpreted back into the third dimension via EVES.

Through EVES's development, we have made the possibility for a user to utilize their eyes within the virtual realm to 'consume' and extend its capabilities to 'generate'

- a novel way to use one's senses. As a result, the HCI becomes more than just a person interacting with a computer and connecting and integrating with it [17]. The integration with VR technologies means that we can still explore new ways people and designers can interact with the digital world [18].

4.3 Further Action

As EVES is currently a prototype, further development is undertaken to facilitate the operation of VR (and similar technologies) by users who are less able at controlling these mediums (such as those with ALS or other affections). Despite the level at which eye-tracking has been implemented within EVES, some hand interaction within its operation is used (i.e. set up, switch on/off, etc.). In the next steps of EVES, these items are developed and implemented.

5 Conclusion

EVES is a novel way to interact actively within VEs using eye-tracking technology. We have developed eye-tracking as the instrument that both generates and interprets the designed product. Using both passive and active means of implementation, eye-tracking treats gaze as an extension of how design can be developed, perceived and analyzed. Applying gaze as both the means of navigation in the VRUI and gaze as a sculpting hand is integral to the scope of the proposed active eye-tracking environment. Allowing a user's eye to sculpt and manipulate form and spaces directly amplifies the capabilities of what our eyes can do beyond that of what is possible within the conventional VE context. Our design method also reveals potential for the future of architectural design, whereby we begin to directly utilize our senses bi-directionally to design the spaces that we inhabit. In the context of our research, in particular, our eyes are 'touching the untouchable' [19] and become actors in the design generation. Our research goes beyond architectural designing within VE. It extends any experience in immersive VE, ranging from gaming to learning and from social to human-computer interactions.

References

1. Carreiro, M., Pinto, P.L.: The evolution of representation in architecture. In: Future Traditions – 1st eCAADe Regional International Workshop, pp. 27–38. eCAADe, Porto (2013)
2. Donath, D., Regenbrecht, H.: VRAD (virtual reality aided design) in the early phases of the architectural design process. In: Tan, M., Teh, T. (eds.) Sixth CAADFutures, pp. 313–322. The National University of Singapore, Singapore (1995)
3. Lisinska-Kusnierz, M., Krupa, M.: Suitability of eye tracking in assessing the visual perception of architecture - a case study concerning selected projects located in cologne. Buildings 10(2), 20 (2020)
4. Wang, B., Moleta, T., Schnabel, M.A.: The new mirror – reflecting on inhabitant behaviour in VR and VR VISUALIZATIONS. In: Intelligent & Informed (24th CAADRIA Conference), pp. 535–544. CAADRIA, Hong Kong (2019)

5. Polli, A.: Active vision: controlling sound with eye movements. Leonardo **32**(5), 405–411 (1999)

6. Farahi, B.: Caress of the gaze: a gaze actuated 3D printed body architecture. In: ACADIA: Posthuman Frontiers: 36th ACADIA, pp. 352–361. ACADIA, Ann Arbor (2016)

7. Roet, L.: VICE: Graham Fink is Drawing with His Eyes. https://www.vice.com/en/article/bmyvjv/graham-fink-is-drawing-with-his-eyes, Accessed 10 Nov 2020

8. Page, T.: CNN: Sarah Ezekiel: Artist paints with her eyes. https://edition.cnn.com/style/article/sarah-ezekiel-artist-als-eye-gaze-spc-intl/index.html, Accessed 12 Nov 2020

9. Cooper, Z.: Unseen - digital interactions for low vision spatial engagement. Master's Thesis, Victoria University of Wellington (2020)

10. Schnabel, M.A.: Framing mixed realities. In: Wang, X., Schnabel, M.A. (eds.) Mixed Reality in Architecture, Design, and Construction, pp. 3–12. Springer, Heidelberg (2009). https://doi.org/10.1007/978-1-4020-9088-2_1

11. Deering, M.F.: HoloSketch: A Virtual reality sketching/animation tool. ACM Trans. Comput.-Human Interact. **2**(3), 220–238 (1995)

12. Achten, H.H., de Vries, B., Jessurun, J.: DDDoolz: a virtual reality sketch tool for early design. In: 5th CAADRIA Conference, pp. 451–460. National University of Singapore (2000)

13. Arnowitz, E., Morse, C., Greenburg, D.P.: vSpline: physical design and the perception of scale in virtual reality. In: ACADIA Disciplines & Distribution (37th ACADIA Conference), 110–117. Massachusetts Institute of Technology, Cambridge (2017)

14. Shusterman, R.: Muscle memory and the somaesthetic pathologies of everyday life. Human Mov. **12**(1), 4–15 (2011)

15. Rock, I., Harris, C.S.: Vision and touch. Sci. Am. **216**(5), 96–107 (1967)

16. Law, T., Lam, D., Endang, J.: How rad(-ical) is VRAD (virtual reality aided design)? In: IV2020 – 24th International Conference Information Visualisation, pp. 731–736. Victoria University, Australia & Technische Universitat, Austria (2020)

17. Schnabel, M.A., Chen, R.: Design interaction via multi-touch. In: Luo, Y. (ed.) CDVE 2011. LNCS, vol. 6874, pp. 14–21. Springer, Heidelberg (2011). https://doi.org/10.1007/978-3-642-23734-8_3

18. Rogers, J., Schnabel, M.A., Moleta, T.: Digital design ecology to generate a speculative virtual environment reimagining new relativity laws. In: Lee, J.-H. (ed.) Computer-Aided Architectural Design on "Hello, Cultures", Communications in Computer and Information Science (CCIS), vol. 1028, pp. 120–133. Springer, Singapore (2019). https://doi.org/10.1007/978-981-13-8410-3_9

19. Schnabel, M.A., Wang, W., Seichter, H., Kvan, T.: Touching the untouchables: virtual-, augmented- and reality. In: 13th International Conference on Computer-Aided Architectural Design Research in Asia (CAADRIA), Chiang Mai, pp. 293–299 (2008)

Synthesizing Point Cloud Data Set for Historical Dome Systems

Mustafa Cem Güneş[1]([⊠]) [ID], Alican Mertan[2] [ID], Yusuf H. Sahin[2] [ID], Gozde Unal[2] [ID], and Mine Özkar[3] [ID]

[1] Architectural Design Computing Graduate Program, Istanbul Technical University, 34437 Istanbul, Turkey
gunesmus@itu.edu.tr

[2] Faculty of Computer and Informatics Engineering, Computer Engineering, Istanbul Technical University, 34469 Istanbul, Turkey
{mertana,sahinyu,gozde.unal}@itu.edu.tr

[3] Department of Architecture, Istanbul Technical University, 34437 Istanbul, Turkey
ozkar@itu.edu.tr

Abstract. This paper offers a workflow for generating synthetic point cloud data sets to be used in deep learning algorithms in tasks of modeling historical architectural elements. Documentation of cultural heritage is a time-consuming process that requires high precision. Computational and semi-automatic tools enhance conventional methods to shorten the duration of the documentation phase and increase the accuracy of the output. Photogrammetry and laser scanning are how geometrical data is acquired and delivered as a point cloud with position, color, and optionally normal vector information. Segmenting architectural elements based on our interpretations of this data is possible using deep neural networks but is limited when, despite the millions of points from one building, the data is insufficient in terms of variance and quantity. To overcome this limitation, we propose a semi-automatic synthetic data set generation using parametric definitions of historic architectural elements. We create a synthetic dataset, namely the Historical Dome Dataset (HDD), consisting of nearly 1000 dome systems with four semantic classes. We quantitatively and qualitatively analyze the usefulness of the HDD by training a number of modern neural networks on it. Our method of synthesizing point clouds can quickly be adapted into similar cultural heritage projects to prepare relevant data to accurately train deep neural networks and process the collected cultural heritage data.

Keywords: Cultural heritage · Synthetic data set · Point clouds · Deep learning · Training data

1 Introduction

This paper presents a method to synthesize point cloud data specifically for dome systems in architectural heritage. The broader aim is to be able to deliver sufficient data in both

© Springer Nature Singapore Pte Ltd. 2022
D. Gerber et al. (Eds.): CAAD Futures 2021, CCIS 1465, pp. 538–554, 2022.
https://doi.org/10.1007/978-981-19-1280-1_33

quantity and variety to train deep neural networks used for the automatic segmentation of collected point cloud data in cultural heritage sites.

In heritage studies, reliable documentation is essential in preserving the existing state of architectural elements. Nonetheless, it requires an extensive workforce and time since it is a detailed task to complete. Digital photogrammetry and laser scanning, now commonly used for efficiency, deliver point cloud data which may be enhanced with building information through some automated processes. Heritage-BIM (HBIM), building information modeling for historic buildings, handles architectural and constructional information beyond surface geometry. The breakdown of scanning historic buildings and representing them as BIM involves data preparation, segmentation, classification, clustering, and reconstruction (Bassier et al. 2020). Our work targets the preparation of data for deep learning tools to automate segmentation and classification in the scan-to-HBIM process (Fig. 1).

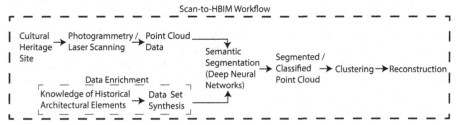

Fig. 1. Typical Scan-to-HBIM workflow. The presented work focuses on the data enrichment step in the process.

Existing work on segmentation of architectural elements in 3D can be classified into two groups based on the input data type. In the first group are works that use images of the scene to segment the point cloud of the scene semantically. Riveiro et al. (2016) tackle the problem of segmenting masonry blocks in a point cloud acquired by LiDAR technology. They create 2D images based on the intensity information obtained by LiDAR systems, use geometrically-constrained marker controlled watershed segmentation to detect masonry blocks on the created images, and back-project the results to the original 3D point cloud. However, this solution is problem-specific with focus on block detection and hard to generalize for segmenting different architectural elements. Stathopoulou and Remondino (2019) apply Convolutional Neural Networks (CNNs) to semantically label pixels of historic building façade images used during the photogrammetry process. These labels can be transferred to the point cloud to create a semantically segmented point cloud. They also show that the semantic labels can be used to constrain the search for correspondence. Collecting and labeling their own dataset, they deliberately exclude the unusable classes such as "sky" to speed up the photogrammetry process and increase its performance. Grilli and Remondino (2019) obtain orthoimages or UV maps of the 3D objects. Obtained images get segmented, classified, and back-projected into the original geometry. Due to the lack of a dataset that serves their purpose, they manually annotate small parts of the images. While segmentation algorithms are well established for 2D inputs, using 2D data is only an indirect solution which adds extra steps in the workflow.

2D data also lacks the depth and normal vector information that 3D data contains and potentially allows for more informed predictions.

In the second group are works that use the 3D data as input. Brodu and Lague (2012) tackle the problem of segmenting point clouds of natural environments obtained by terrestrial laser scanning (TLS). Their main insight is that the dimensionality of local neighborhood of a point in different scales can be a very representative and discriminative feature for semantic segmentation tasks. By applying Principal Component Analysis (PCA) to a local neighborhood of a point and examining its eigenvalues, they decide whether the local neighborhood is 1D, 2D, or 3D. This procedure can be repeated for multiple scales and the most discriminative features determined on a training set can be used to classify the point semantically. Farella (2016) uses this method to remove vegetation from the point cloud of World War 1 military structures at Monte Celva automatically. Hackel et al. (2016) propose a downsampling strategy that allows faster computation of the nearest neighbors of a point. They voxelize the point cloud with cubes of different sizes and replace the points inside a cube with its centroid to downsample the original point cloud. This procedure can be paralleled and results in a more homogeneous point cloud. Since it is faster to calculate neighbors in the downsampled point cloud, they propose to calculate a number of hand-crafted features at different scales to capture the characteristics of the point cloud better. Subsequently, they proceed to use the random forest classifier to classify each point given its features. These studies all use standard machine learning algorithms on top of hand-crafted features. Nevertheless, modern machine learning approaches, namely deep learning, have recently outperformed standard machine learning methods, and are capable of learning high-level (e.g. semantically complex) features from simpler features or even from raw data (Zhang et al. 2019).

As a state-of-the-art method, semantic segmentation of scanned point clouds using deep neural networks provides fast and accurate solutions. The first studies focusing on classifying the 3D objects preferred to voxelize the objects and use the voxelized occupancy map as an input to a deep belief (Wu et al. 2015) or convolutional (Maturana and Scherer 2015) network. PointNet (Qi et al. 2017a) can be counted as the first attempt to use point clouds as raw input to a neural network. As might be expected, the main difficulty of using the points directly instead of view renders or voxelized 3D maps comes from the set representation since all permutations on a point set describes the same entry. Hence in PointNet, all points are processed in multi-layer perceptrons in a parallel way to obtain point features. PointNet++ (Qi et al. 2017b) focuses on the fact that the original PointNet loses local features since all points are treated independently until the maximum selection step. Thus, they use sampling and grouping of the points. In DGCNN (Wang et al. 2019), the authors handled the point cloud as a graph and used the k-nearest neighbor approach to define the connections. Then, they define a new operation called edge convolution to make a convolution operation centered on the selected point according to its nearest neighbors. In ODFNet (Sahin et al. 2020), orientation distribution for all points are calculated. In the literature cited above, deep learning methods show promising results for semantic segmentation of point clouds in multiple benchmarks. Naturally, deep learning based semantic segmentation methods for point clouds are frequently used to document architectural elements in the cultural heritage domain as well.

To the best of our knowledge, one of the essential works that applies deep learning techniques to segment architectural elements in cultural heritage domain is the work of Malinverni et al. (2019). They use PointNet++ to semantically segment the point cloud of architectural buildings acquired by a real survey and annotated manually. Pierdicca et al. (2020) experiment with a dynamic graph convolutional neural network (DGCNN) using the ArCH data set. They observe that architectural elements such as walls, roofs, and vaults have enough points in the data set to train the model. Nevertheless, there is only a small number of points for doors, windows, and stairs in the data set. The unbalanced distribution of the acquired data impacts the quality of the learning model negatively. Indeed, deep learning models require a large amount of different data to recognize architectural elements successfully (Grilli et al. 2017). As each cultural heritage has a limited number of architectural elements, deep learning models struggle with limited samples instead of generalizing the characteristics of the parts.

Using synthetic data to address the limited data problem is a widely used approach in computer vision. Several methods are developed to create synthetic data for further use in various domains (Müller et al. 2018; Hamarneh and Jassi 2010; Yang et al. 2020; Jerman et al. 2016; Cetin et al. 2012). Similarly, there are some cultural heritage projects that focus on synthetic data set generation. For instance, Morbidoni et al. (2020) propose a synthetic dataset of historical buildings and elements to facilitate deep learning-based research in the cultural heritage domain, which requires a vast amount of annotated data. They use a standard BIM-based approach to obtain models from parametric libraries. They use CloudCompare (Girardeau-Montaut 2020) to sample points from 3D models of historical buildings and annotate the resulting point clouds. Additionally, they combine these models from the online archives and point clouds of real-world surveys. Finally, they modify DGCNN using a radius-based approach and name the new method as RadDGCNN. While their work uses existing 3D models, inherently limited in terms of quantity, we focus on synthesizing 3D models parametrically without relying on existing 3D models, which allows us to create virtually unlimited amount of data. In another example, Pierdicca et al. (2019) use Python scripts to create synthetic column geometries in the Blender platform. Nevertheless, the solution is problem-specific and lacks simplicity. In contrast to their work, where they only focus on the generation of a simple architectural element without much variation, we focus on creating an entire dome system and controlling its variation parametrically.

This paper focuses on the dome systems in 13th–15th century Anatolian architecture for the purposes of defining a scope based on select and limited features to parameterize. A dome system consists of a spherical dome, transition elements, arches and walls, all of which are primitive geometries or their basic combination. Semantic segmentation, as a state-of-the-art method in a scan-to-HBIM process in documenting dome systems, faces a problem of recognition due to the similarity of these geometries and the insufficient sampling from actual sites. We propose to synthesize architectural elements for an alternative sampling. In our study with domes, we focus on the elements of the dome system and model each element type with flexible variables to produce a diverse range of outputs. A semi-automatic exportation process quickly delivers hundreds of labelled mesh models. Additionally, we can convert these models to point clouds as dense as the machine learning models require. The dataset and the algorithm used to generate is available online[1]. With sufficient and various data, it is expected that the quality of the semantic segmentation improves in scan-to-HBIM processes. In order to measure the effectiveness of HDD, we trained neural networks on HDD for semantic segmentation and tested it on both synthetic and real-world data. We were able to show that networks trained on HDD are capable of segmenting real-world data without the need for any fine-tuning.

2   Methodology

Computational design tools offer considerable potentials to represent and generate shapes using the defined relations between the architectural elements. Visual programming tools such as Rhinoceros-Grasshopper and Revit-Dynamo are handy for defining shapes parametrically, particularly when design follows mathematical and algorithmic rules. Studies on historic Renaissance domes in Campania with a similar method (Capone and Lanzara 2019) show that historically, using geometric relations and combinations of primitive shapes have helped create holistic spaces in aesthetics and structurally resilient buildings in medieval. Expressing shapes as parametric geometries discloses reliable and fast solutions to holistic representations. In our work, we incorporate multiple combinations of the varied parameters of plan dimensions of the dome system, types of the supporting structural elements, i.e. a wall or an arch, types and dimensions of the arches, types of the transition elements, and whether a drum exists or not based on the architectural history literature on the said period and geography. Our parametric model and algorithm have delivered approximately 1000 dome sets so far and can extend to above 10000 examples by using all the combinations with varying size parameters.

2.1   Level of Detail

Representations of objects may have several levels of detail (LOD). Different scales carry out different types of information on the represented architectural elements in AEC models. While LOD 500 (for as-built models) contains all the local geometric properties, colors, and nongeometrical details, LOD 100 (for concept models) expresses only generalized box representations using height, volume, position, direction without colors, and details (BIMForum 2019). In our model, we adhere to the LOD 200 standard with approximate geometrical properties combined with class information.

2.2   Dome Systems in Anatolian Seljuk Architecture

Anatolian Seljuk period and early Ottoman period buildings are well documented within restoration and documentation projects, and the properties of their domes and transition element are known (Okçuoğlu 1995). There are studies specifically on the transition elements of mosques and prayer rooms in Bursa in the 14th and 15th centuries (Özcan

---

[1] https://github.com/HistoricDomeDataset/HistoricDomeDataset.

2008), transition elements in Anatolian Seljuk town prayer rooms in 13th century Konya (Turan 2018), and transition element systems in early Ottoman architecture (Şimşek 2010). A study of the restoration problems of Anatolian Seljuk town prayer rooms in Konya provides authentic documents (Baş 2008).

This project examines the domes of the aforementioned period and geography in three parts: domes, transition elements, structural elements. This triplet takes a square plan shape, transforms it into a polygon that converges to a circle to carry a dome (Şimşek 2010). With a few exceptions, these domes have a half-sphere shape. Transition elements can be classified into four main types: planar triangle, triangular band, pendentive, squinch (Turan 2018). These may or may not support a drum to ascend the dome. The structural elements that carry the dome can be either a wall or an arch. Several types of arches such as semi-circular, equilateral, drop single centered tangent, double centered tangent are available. Our data set classifies all the parts of the dome geometries into four classes, namely the dome, the transition element, the wall, and the arch, and includes all of their aforementioned subclasses.

2.3   The Parametric Definition of the Elements of the Dome System

In our study, we model a dome system that includes a dome, transition elements, and either walls or arches. We create the architectural elements and their variant types separately and define the parameters that designate their unique properties. Below is a list and explanation of these properties.

The Dome
The height and radius of a dome are dependent on the cutting plane of a sphere (Fig. 2, Table 1). Flatness ratio (f) determines the radius and center of the sphere.

Table 1. Parameters, properties and values of dome samples.

	r_{base} (cm)	f	r_{sphere} (cm)
Full Dome	$w_{wall}/2$	$1=f$	r_{base}
Flat Dome	$w_{wall}/2$	$0<f<1$	$\dfrac{r_{base}(1+f^2)}{2f}$

The Transition Element
In Anatolian architecture, building elements that translate the geometry of the plan to the dome are characteristic. This paper focuses on all variants of the transition elements mentioned in the above Sect. 2.2. In dome systems, the drum often takes on the role to place a dome circle on the support structures below. Drums come in different heights. A drum may have a circular or polygonal plan with multiple edges. Other transition elements are simple squinch, complex squinch, corner triangle (planar triangle), Turkish triangle (planar triangle), pendentive, planar triangular band, simple prismatic triangular band, and complex prismatic triangular band. We model all as variants of a general transition element. Shapes in each variant are dependent on the height of the transition part, the edge number of the drum polygon, and the width of the supporting wall (Fig. 3).

The Supporting Wall
Each supporting wall can be a plain wall, an arch, or a combination of the two, and has

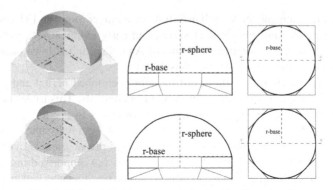

Fig. 2. Full and flat domes created using the flatness ratio.

width and height properties (Fig. 4). These features lead to numerous combinations in the data set.

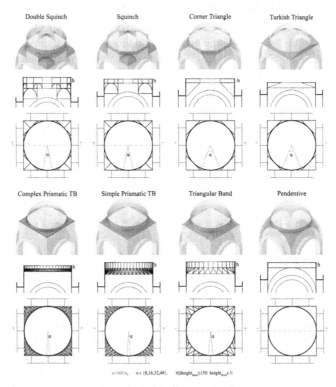

Fig. 3. Transition elements use drum height and polygon number parameters.

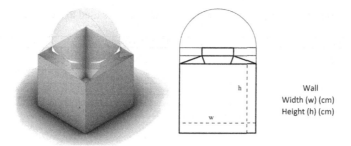

Fig. 4. Walls have width and height parameters.

The Supporting Arch

Lastly, an arch may be semi-circular, drop, flat, single-centered tangent, or double-centered tangent. Configurable properties include the length between the wall edge and the arc columns, depth, thickness, radius, and height (Fig. 5, Table 2).

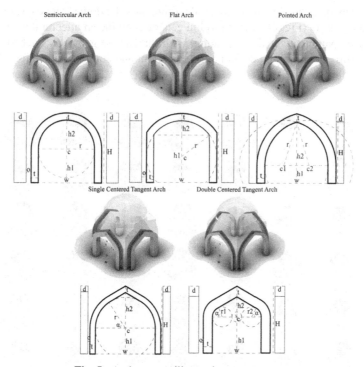

Fig. 5. Arch types utilize various parameters.

Table 2. Descriptions of parameters and properties used to create arches.

Parameter	Properties	Parameter	Properties
d = depth of the arch (cm)	$50 \leq d \leq 150, d \in \mathbb{R}$	W = inner width (cm)	$W = W_{wall} - 2(o + t),\ W_{wall} \in \mathbb{R}$
o = offset from the wall (cm)	$0 \leq o, o \in \mathbb{R}$	r = radius of circle	$r \in \mathbb{R}$
t = thickness of the arch (cm)	$20 \leq t \leq 70,\ t \in \mathbb{R}$	c = center of an arch (cm)	$c_{x,y,z} \in \mathbb{R}$
h_1 = height of the linear part (cm)	$50 \leq h_1,\ h_1 \in \mathbb{R}$	α = angle	$0 < \alpha < 90, \alpha \in \mathbb{R}$
h_2 = height of the curved part (cm)	$0 \leq h_2,\ h_2 \in \mathbb{R}$	f = flatness ratio	$f \in \mathbb{R}$
H = total height of arch (cm)	$H = h_1 + h_2 + t,\ H \in \mathbb{R}$		

2.4 Combining Parts of the Dome System

Once we separately create each architectural element, we appropriately merge them. Firstly, we use the transition elements to cut the wall. "Split breps" function separates the part to be kept and thrown away on each wall. Our algorithm selects the central part of the wall end leaves the rest. Then, for the wall, we have a second control state. If a wall has an arch on it, the code splits the geometry into parts. As in the previous step, the algorithm excludes unnecessary surfaces. Secondly, in the case that a dome rests on a drum, transition elements merge with the drum, and the dome extends as high as the height of the drum (Fig. 6). Our algorithm generates all the architectural element types for the given parameters in the beginning. With shared base geometry, each element fits in place. Still, we may exclude an unpicked variant and combine elements only of the type(s) we choose.

Fig. 6. Samples of combined mesh dome geometries. Used colors represent different architectural elements. Geometries do not carry color information.

2.5 User-Friendly Semi-automatic Mesh Geometry Exportation

In a standard workflow, one needs to define and bake the layers of each class in Grasshopper in order to select them in Rhinoceros. In Rhinoceros, all the mesh geometries are selected and exported as obj files containing labelled meshes. However, this process is manual and time-consuming. Instead, we can automate the labelling, baking, naming and mesh exportation steps by using a C# module. Bypassing manual works and expressing almost everything visually creates a user-friendly and practical tool to synthesize required models quickly. It is also possible to support our export method with a recursive function that calls each possible parameter to create excessive options. We focus on examining each exported model. Hence our exportation function works each time we manipulate the parameters manually and generates models rapidly.

2.6 Point Cloud Sampling

Lastly, since we focus on enriching our data acquired from photogrammetric or laser scanning surveys, we need to convert our mesh models into points clouds. Even though Grasshopper includes functions to sample point clouds such as "populate geometry", the software does not give the expected result rapidly when the desired number of points are excessively high. To accomplish the task robustly, we import the created mesh geometries into CloudCompare software. Using the "sample on a mesh" tool, we get randomly distributed point clouds with a certain amount of density (Fig. 7). This operation also creates surface normal vector of the points. After this step, our models are ready to be exported in a format that the deep learning model requires.

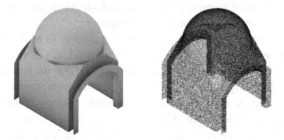

Fig. 7. Transformation of a model from mesh to point cloud in CloudCompare.

3   Evaluation

To evaluate the effectiveness and usability of our synthetic data generation procedure, we generate different sets using the sampled point clouds, where each point of each instance contains the location and semantic class information, namely dome, transition element, wall, and arch. Then, we train three different neural network architectures, which have shown accurate results for segmentation tasks: PointNet, DGCNN and ODFNet. We train each of these networks to estimate the semantic classes of points of the objects by feeding point location information as an input.

The rest of the section describes the dataset variants created from the 3D models, experimental results on the generated dataset variants, and the quantitative and qualitative analysis of deep learning models when trained on our proposed synthetic datasets and tested on real-world data.

3.1 Dataset Variants

We create two different variants by sampling different numbers of points from each object model to enrich the generated data. The variants help to create different difficulty levels, which are important for analyzing and comparing different deep learning methods for the task at hand. Particularly, we sample 2048 and 10000 points from each object model and create two sets of variants. While 2048 points per object model are pretty close to the number of points per object in standard point cloud segmentation datasets such as ModelNet40 (Wu et al. 2015) and ShapeNet (Yi et al. 2016); we also create a variant with 10000 points per object model since the historical buildings are quite bigger than objects in the standard datasets and more points are needed to preserve the details. We name the variants with 2048 and 10000 points as HDD_2048 and HDD_10000, respectively.

3.2 Experimental Results on HDD

We split both of the dataset variants into train, validation, and test sets based on the type of transition elements used in the objects. Instances with Turkish Triangles are used for validation, instances with Complex Prismatic Turkish Bands are used for testing, and the rest of the instances are used for training purposes.

Table 3 shows the results on the test splits of datasets. To evaluate the results, two main criteria are used. We can briefly define accuracy (acc) as the ratio of correctly segmented points and mean intersection over union (mIoU) as the intersection over union values averaged according to different classes. All of the experimented network models can successfully segment the generated synthetic data. Notably, ODFNet significantly outperforms other network models. ODFNet's emphasis on orientation of neighboring points helps the network model to capture the structure of point clouds. We conjecture that this feature becomes more prominent as the number of points increases.

3.3 Evaluation on Real-World Data

To test our method with real data, we refer to a photogrammetry-based point cloud for the dome system in Konya Ince Minareli Madrasah. We sample two main point clouds containing 2048 and 10000 points to feed the mentioned networks trained on the HDD_2048 and HDD_10000, respectively. Figure 8 illustrates the ground truth labels of the main point cloud and the subsampled data. In the ground truth point clouds, green represents the dome class, blue represents the transition element class, red represents the wall class, and yellow represents the arch class. This color scheme is consistent throughout all figures below.

Table 3. The quantitative results of experimented network models on the two HDD variants. The best results are shown in boldface.

	HDD_2048		HDD_10000	
	mIoU (%)	acc (%)	mIoU (%)	acc (%)
PointNet	85.96	93.64	57.13	74.26
DGCNN	**87.77**	**94.77**	87.15	94.45
ODFNet	86.74	94.27	**94.14**	**97.43**

Fig. 8. Evaluation on Konya Ince Minareli Madrasah, (Left) Ground truth of the main point cloud, (Center) cloud with 10000 points subsampled from the main cloud, (Right) ground truth of the subsampled data. (Color figure online)

The visual and quantitative results of our methods are given in Table 4. According to the results, ODFNet has the best mIoU score for the 2048 points dataset and DGCNN obtained the best mIoU score for 10000 points. From our findings, we can conclude that our networks trained with the synthetic data can mostly segment a real dome from the Seljuk era.

Figure 9 shows the confusion matrices of the trained network models on real-world data. Each row of the confusion matrices shows the prediction distribution for the given class. Both networks trained on HDD_2048 and HDD_10000 are capable of segmenting dome and arch classes precisely. It can be seen that points belong to the transition element class are the most challenging to segment and commonly mistaken for wall and dome classes. Since the transition elements connect the wall and dome, and they share similar geometrical features, especially at the boundaries of the architectural elements, it is very challenging for models to correctly identify the classes without additional information such as color.

We test one of the deep learning models (DGCNN trained on HDD_10000) on a part of Konya Sahip Ata Mosque and Complex built in the Anatolian Seljuk era. We use a section of a dome system of the building as an arbitrary example considering the blind arch which, combines a wall and an arch and atypical transition elements it has. Figure 10 shows the original point cloud, annotated ground truth and the prediction of our model, which achieves 81.27% accuracy. The result is not optimal due to problems of recognizing an arch that is embedded in a wall and has gradient thickness in contrast

Table 4. The quantitative and qualitative results on real-world data. The best results are shown in boldface. Better viewed digitally.

Training Set

Networks	HDD_2048	HDD_10000
PointNet	mIoU: 29.27 acc: 49.37 	mIoU: 50.96 acc: 64.89
DGCNN	mIoU: 58.38 acc: 77.69 	mIoU: **52.45** acc: **78.15**
ODFNet	mIoU: **67.81** acc: **80.91** 	mIoU: 49.45 acc: 70.38

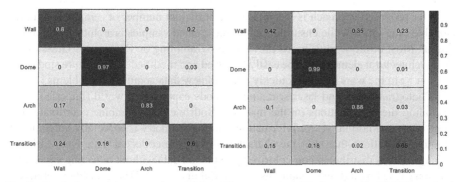

Fig. 9. Confusion matrices on real-world data. The one on the left belongs to the ODFNet, trained on HDD_2048; the one on the right belongs to the DGCNN, trained on HDD_10000.

to examples in the training sets. In addition to that, experimented point cloud consists of a single wall while the training instances consist of full dome systems. Still, we believe that the model trained on our proposed synthetic dataset learns robust features. It can recognize wall and transition elements, even when the dome is not present. And it can segment the arch, even though it is a blind arch.

Fig. 10. Evaluation on Konya Sahip Ata, Mosque and Complex, (Left) Ground truth of the main point cloud, (Center) cloud with 10000 points subsampled from the main cloud, (Right) ground truth of the subsampled data. (Color figure online)

4   Conclusion

This work offers a method to create synthetic point cloud data of historic dome systems by controlling the parameters of the architectural elements. Deep learning methods are not easily applicable to the documentation and preservation process of cultural heritages since it is costly, in terms of time and effort, to acquire enough labelled data to train deep learning models for this task. Data augmentation methods such as rotation, scale, flip, giving noise to data could be used to overcome this problem. Nonetheless, even if it helps to generalize the results with minimal data, this method lacks in identifying the different variations of architectural elements, and over-augmentation can lead to the problem of missing the architectural relations between the parts. Similarly, generative adversarial networks can produce non-existing relevant samples. However, initial data that lacks in different size and scale, as well as varying detail could be problematic, and it can be challenging to ensure the coherency of the generated architectural elements.

Our proposed approach is to synthesize a sufficient number of samples within reasonable time limits while ensuring each of the created samples is architecturally meaningful. Most of the work is based on visual programming on Rhinoceros-Grasshopper, and architectural elements can be easily followed and modified so that future studies can benefit from the workflow and quickly adapt to new data acquisition problems.

Photogrammetry and laser scanning methods capture real-world (imperfect) data, including the deformations on the material. Both methods combine coordinates data with color and normal vector information. Photogrammetry is especially vulnerable to lighting conditions that alter the characteristics of the collected data. Our work uses homogenously distributed non-colorized point cloud data without reflecting any malformation caused by lighting conditions or material distortions. Thus, this data set may be limited for detecting deformed materials or irregular details of architectural elements.

Most of the architectural elements we use are well-defined to create a sufficient number of combinations representative of each element to train the artificial intelligence model accurately. Nevertheless, some detailed elements, such as the muqarnas and the recursive squinches, and elements such as doors, windows and their openings are left out to maintain the complexity of the problem at controllable levels.

Despite the differences between the proposed synthetic HDD and real-world data, our results show that networks trained with HDD are applicable to real-world data, as demonstrated on Konya Ince Minareli Madrasah. Additionally, the model trained on HDD learns generalizable features as demonstrated on the sample from the Konya Sahip Ata Mosque and Complex, which is an example considerably different than the training instances. The proposed workflow of acquiring a synthetic data set for cultural heritage can be adopted into new studies to generate new sets to train deep learning models. Further studies could extend to and benefit from adding discarded elements, blending real and synthetic data, and architecturally plausible texturing.

Acknowledgements. This work is supported by TÜBİTAK (The Scientific and Technological Research Council of Turkey) Project Number: 119K896. A very special thanks to Demircan Taş and Berkay Öztürk for providing photogrammetric data. Lastly, we would like to thank our research group for their valuable discussions.

References

Baş, T.: Anadolu Selçuklu dönemi Konya mahalle mescitlerinin restorasyon sorunları. Selçuk University, Konya, Turkey (2008)

Bassier, M., Yousefzadeh, M., Vergauwen, M.: Comparison of 2D and 3D wall reconstruction algorithms from point cloud data for as-built BIM. J. Inf. Technol. Constr. (ITcon) **25**(11), 173–192 (2020)

Brodu, N., Lague, D.: 3D terrestrial lidar data classification of complex natural scenes using a multi-scale dimensionality criterion: applications in geomorphology. ISPRS J. Photogramm. Remote. Sens. **68**, 121–134 (2012)

BIMForum: Level of Development Specification: For Building Information Models (2019). http://bimforum.org/lod/

Capone, M., Lanzara, E.: Scan-to-BIM vs 3D ideal model HBIM: parametric tools to study domes geometry. Int. Arch. Photogramm. Remote Sens. Spat. Inf. Sci. **42**, 219–226 (2019)

Cetin, S., Demir, A., Yezzi, A., Degertekin, M., Unal, G.: Vessel tractography using an intensity based tensor model with branch detection. IEEE Trans. Med. Imaging **32**(2), 348–363 (2012)

Farella, E.M.: 3D mapping of underground environments with a hand-held laser scanner. In: Proceedings of the SIFET Annual Conference (2016)

Girardeau-Montaut, D.: CloudCompare (2020). https://www.danielgm.net/cc

Grilli, E., Remondino, F.: Classification of 3D digital heritage. Remote Sens. **11**(7), 847 (2019)

Grilli, E., Menna, F., Remondino, F.: A review of point clouds segmentation and classification algorithms. Int. Arch. Photogramm. Remote Sens. Spat. Inf. Sci. **42**, 339 (2017)

Hackel, T., Wegner, J.D., Schindler, K.: Fast semantic segmentation of 3D point clouds with strongly varying density. ISPRS Ann. Photogramm. Remote Sens. Spat. Inf. Sci. **3**, 177–184 (2016)

Hamarneh, G., Jassi, P.: VascuSynth: simulating vascular trees for generating volumetric image data with ground-truth segmentation and tree analysis. Comput. Med. Imaging Graph. **34**(8), 605–616 (2010)

Jerman, T., Pernuš, F., Likar, B., Špiclin, Ž: Enhancement of vascular structures in 3D and 2D angiographic images. IEEE Trans. Med. Imaging **35**(9), 2107–2118 (2016)

Malinverni, E.S., et al.: Deep learning for semantic segmentation of 3D point cloud. Int. Arch. Photogramm. Remote Sens. Spat. Inf. Sci. **vol. XLII-2/W15**, 735–742 (2019)

Maturana, D., Scherer, S.: VoxNet: a 3D convolutional neural network for real-time object recognition. In: 2015 IEEE/RSJ International Conference on Intelligent Robots and Systems (IROS), pp. 922–928. IEEE (2015)

Morbidoni, C., Pierdicca, R., Paolanti, M., Quattrini, R., Mammoli, R.: Learning from synthetic point cloud data for historical buildings semantic segmentation. J. Comput. Cult. Heritage (JOCCH) **13**(4), 1–16 (2020)

Müller, M., Casser, V., Lahoud, J., Smith, N., Ghanem, B.: Sim4CV: a photo-realistic simulator for computer vision applications. Int. J. Comput. Vis. **126**(9), 902–919 (2018). https://doi.org/10.1007/s11263-018-1073-7

Okçuoğlu, T.: Anadolu Selçuklu mescitlerinde kubbeye geçiş alanının değerlendirilmesi. Istanbul University, Istanbul, Turkey (1995)

Özcan, A.: 14.15. Yüzyıl Bursa cami ve mescitlerinde kubbeye geçiş elemanları. Erciyes University, Kayseri, Turkey (2008)

Pierdicca, R., Mameli, M., Malinverni, E.S., Paolanti, M., Frontoni, E.: Automatic generation of point cloud synthetic dataset for historical building representation. In: De Paolis, L., Bourdot, P. (eds.) AVR 2019. LNCS, vol. 11613, pp. 203–219. Springer, Cham (2019). https://doi.org/10.1007/978-3-030-25965-5_16

Pierdicca, R., et al.: Point cloud semantic segmentation using a deep learning framework for cultural heritage. Remote Sens. **12**(6), 1005 (2020)

Qi, C.R., Su, H., Mo, K., Guibas, L.J.: PointNet: deep learning on point sets for 3D classification and segmentation. In: Proceedings of the IEEE Conference on Computer Vision and Pattern Recognition, pp. 652–660 (2017a)

Qi, C.R., Yi, L., Su, H., Guibas, L.J.: PointNet++: deep hierarchical feature learning on point sets in a metric space. In: NIPS (2017b)

Riveiro, B., Lourenço, P.B., Oliveira, D.V., González-Jorge, H., Arias, P.: Automatic morphologic analysis of quasi-periodic masonry walls from LiDAR. Comput.-Aided Civ. Infrastruct. Eng. **31**(4), 305–319 (2016)

Sahin, Y.H., Mertan, A., Unal, G.: ODFNet: using orientation distribution functions to characterize 3D point clouds. arXiv preprint arXiv:2012.04708 (2020)

Şimşek, H.: Erken osmanlı mimarisinde kubbeye geçiş sistemlerinden üçgenler kuşağı. Yüzüncü Yıl University, Van, Turkey (2010)

Stathopoulou, E.K., Remondino, F.: Semantic photogrammetry: boosting image-based 3D reconstruction with semantic labeling. Int. Arch. Photogramm. Remote Sens. Spat. Inf. Sci. **42**(2), W9 (2019)

Turan, Ş.N.: Türk mimarisinde kullanilan kubbeye geçiş elemanları; 13.Yy. Anadolu Selçuklu dönemi Konya mahalle mescitleri örneği. Necmettin Erbakan University, Konya, Turkey (2018)

Wang, Y., Sun, Y., Liu, Z., Sarma, S.E., Bronstein, M.M., Solomon, J.M.: Dynamic graph CNN for learning on point clouds. ACM Trans. Graph. (TOG) **38**(5), 1–12 (2019)

Wu, Z., et al.: 3D ShapeNets: a deep representation for volumetric shapes. In: Proceedings of the IEEE Conference on Computer Vision and Pattern Recognition, pp. 1912–1920 (2015)

Yang, X., Xia, D., Kin, T., Igarashi, T.: Intra: 3D intracranial aneurysm dataset for deep learning. In: Proceedings of the IEEE/CVF Conference on Computer Vision and Pattern Recognition, pp. 2656–2666 (2020)

Yi, L., et al.: A scalable active framework for region annotation in 3D shape collections. ACM Trans. Graph. (ToG) **35**(6), 1–12 (2016)

Zhang, J., Zhao, X., Chen, Z., Lu, Z.: A review of deep learning-based semantic segmentation for point cloud. IEEE Access **7**, 179118–179133 (2019)

Author Index

Printed in the United States
by Baker & Taylor Publisher Services